Selection of Materials and Manufacturing Processes for Engineering Design

MAHMOUD M. FARAG BSc, MMet, PhD

Professor of Engineering
American University in Cairo

Prentice Hall

New York   London   Toronto   Sydney   Tokyo

First published 1989 by
Prentice Hall International (UK) Ltd,
66 Wood Lane End, Hemel Hempstead,
Hertfordshire, HP2 4RG
A division of
Simon & Schuster International Group

Printed and bound in Great Britain at
the University Press, Cambridge

Library of Congress Cataloging-in-Publication Data

Farag, Mahmoud M., 1937–
    Selection of materials and processes for engineering
design/by Mahmoud M. Farag.
        p.   cm.
    Bibliography: p.
    Includes index.
    ISBN 0-13-802208-9
    ISBN 0-13-802216-X (pbk)
    1. Engineering design. 2. Materials. 3. Manufacturing
processes.
    I. Title.
    TA174.F27 1989
    620'.0042 – dc20   89-8518
                                            CIP

British Library Cataloguing in Publication Data

Farag, Mahmoud M., 1937–
    Selection of materials and processes for engineering
    design.
    1. Engineering. Design. Optimization's
    I. Title
    620'.00425

    ISBN 0-13-802208-9
    ISBN 0-13-802216-X (pbk)

---

1 2 3 4 5 93 92 91 90 89

OCR-A ISBN

ISBN 0-13-802208-9
      0-13-802216-X PBK

To Penelope, Sherif and Sophie

Preface

The introduction of a new engineering product or the changing of an old model involves reaching economic decisions, making designs, selecting materials and choosing manufacturing processes. These activities are interdependent and should not be performed in isolation from each other. This is because the materials and processes used in making the product can have a large influence on its design, shape, cost and performance in service. For example, making a part from injection-molded plastics instead of formed sheet metal is expected to involve large changes in design, new production facilities and widely different economic analysis. The further the design process proceeds the more difficult it is to consider alternative materials and manufacturing processes. Thus careful consideration should be given to materials and process selection in the earliest stages of design and decision making.

With the increasing pressure to produce cheaper and more reliable components and with the greater number of new engineering materials and manufacturing processes that are now available, there is a growing need for an integrated approach to economic analysis, design and materials and process selection. The integrated approach will make it easier to achieve the optimum component that will combine the functional requirements with reliability at a competitive cost. However, this task is not easy, especially in the context of today's technical and social climate where a large number of factors, not all of which are necessarily compatible, have to be taken into consideration. These factors are usually so diverse that it is seldom possible for one individual to be thoroughly conversant with all of them. At the same time, however, the engineer cannot afford to overlook any of these factors. Above all, the engineer must know how all these considerations fit together, what interactions are possible and what sort of tradeoffs can be made. The increasing use of computers in the various stages of product development has made the integrated approach easier to attain. Computer aided design, computer aided manufacture, computer aided economic analysis and computerized materials properties data banks are among the tools that are now available to the engineer.

The objective of this book is to provide both the technical and economic backgrounds that will enable the engineer to integrate the various activities involved in product development in order to arrive at the optimum solution for a given application. The book contains 25 chapters which are divided into four parts. Part I discusses the behavior and processing of engineering materials including metals, polymers, ceramics and composite materials; a discussion of the different causes of failure of components in service is also included. Part II introduces the elements of engineering design, reviews the different methods of decision making and discusses the effects of material properties and manufacturing processes on design. A review of the concepts of computer aided design and computer aided manufacture is also included. Part III reviews the economic concepts that are involved in design, materials selection and manufacturing. Part IV reviews the different methods of selection and uses case studies to illustrate the integration of design principles, economic analysis, manufacturing methods and materials selection. In view of the breadth of the subjects covered and in order to keep the length of the text within reasonable limits, only information that has direct relation to the objective of the book is presented. Selected references are given at the end of each chapter to allow the reader to find more detailed information. Whenever possible, examples and case studies are given to illustrate the practical application of the presented material, while questions and problems are given to help in reviewing the material. Appendices which give the properties of selected engineering materials and principles of engineering statistics are also included.

This book is written at the level of senior undergraduate or graduate engineering students; however, practicing engineers will also find the subject matter interesting and useful. Although the text is mainly written in metric units, English units are also given whenever possible. Appendices are also provided to give easy conversion between the two systems of units.

Mahmoud M. Farag

Contents

Chapter 1   The Activities Involved in Developing a Concept into a Finished
Product

1.1	Introduction	1
1.2	Stages of product development	1
1.3	The feasibility study	4
1.4	Developing the design and selecting materials and processes	5
1.5	Project planning and scheduling	7
1.6	Launching the product	7
1.7	The product life cycle	8
1.8	Review questions and problems	10
	Bibliography and further reading	10

PART I   BEHAVIOR AND PROCESSING OF ENGINEERING MATERIALS

Chapter 2   Metallic Materials

2.1	Classification and specification of metallic materials	13
2.2	Strengthening of metallic materials	16
2.3	Carbon steels	21
2.4	Alloy steels	24
2.5	Cast irons	27
2.6	Light nonferrous metals and alloys	29
2.7	Heavy nonferrous metals and alloys	32
2.8	Design and selection considerations for metallic materials	35
2.9	Review questions and problems	36
	Bibliography and further reading	37

Chapter 3   Processing of Metallic Materials

3.1   Classification of manufacturing processes   38
3.2   Dimensional accuracy and surface finish   40
3.3   Manufacturing by casting   47
3.4   Powder metallurgy processes   52
3.5   Bulk forming processes   52
3.6   Sheet metal forming processes   55
3.7   Fastening and joining processes   59
3.8   Machining processes   65
3.9   Surface treatment   69
3.10 Review questions and problems   72
      Bibliography and further reading   72

Chapter 4   Polymeric Materials and their Processing

4.1   Classification of polymers   74
4.2   Parameters affecting the behavior of plastics   75
4.3   General characteristics of plastics   77
4.4   Thermoplastics   83
4.5   Thermosetting plastics   86
4.6   Elastomers   87
4.7   Adhesives   88
4.8   Processing of plastics by molding   90
4.9   Extrusion of plastics   93
4.10 Thermoforming   94
4.11 Casting of plastics   96
4.12 Fastening and joining of plastic parts   97
4.13 Finishing of plastic parts   98
4.14 Design and selection considerations for polymers   100
4.15 Review questions and problems   101
      Bibliography and further reading   101

Chapter 5   Ceramic Materials and their Processing

5.1   Classification of ceramic materials   103
5.2   General characteristics of ceramics   103
5.3   Refractory ceramics   106
5.4   Whitewares   107
5.5   Clay products   108
5.6   Glass   108
5.7   Processing of ceramic products   110
5.8   Forming of glass products   113
5.9   Design considerations for ceramic products   114

5.10 Review questions and problems   115
      Bibliography and further reading   116

Chapter 6   Composite Materials and their Processing

6.1   Introduction   117
6.2   Dispersion-strengthened composites   118
6.3   Particulate-strengthened composites   120
6.4   Mechanics of fiber reinforcement   121
6.5   Materials for fiber reinforcement   125
6.6   Laminated composites   127
6.7   Manufacturing of components made of composite materials   130
6.8   Designing with composites   134
6.9   Selection and use of composite materials   136
6.10 Review questions and problems   140
      Bibliography and further reading   140

Chapter 7   Failure of Components in Service

7.1   Causes of failure of engineering components   141
7.2   Types of mechanical failure   142
7.3   Fracture toughness and fracture mechanics   143
7.4   Ductile and brittle fractures   149
7.5   Fatigue fracture   153
7.6   Wear failures   159
7.7   Corrosion failures   160
7.8   Stress corrosion and corrosion fatigue   163
7.9   Elevated temperature failures   164
7.10 Failure analysis   167
7.11 Review questions and problems   170
      Bibliography and further reading   170

Chapter 8   Functional Requirements of Engineering Materials

8.1   Introduction   172
8.2   Selection of materials for static strength   172
8.3   Selection of materials for stiffness   174
8.4   Selection of materials for fatigue resistance   179
8.5   Selection of materials for toughness   181
8.6   Selection of materials for corrosion resistance   185
8.7   Selection of materials for temperature resistance   191
8.8   Selection of materials for wear resistance   194
8.9   Selection of materials for protective coatings   198
8.10 Review questions and problems   202
      Bibliography and further reading   202

**PART II   DESIGN AND MANUFACTURE OF ENGINEERING
COMPONENTS**

Chapter 9   Elements of Engineering Design

9.1   Introduction   207
9.2   Factors influencing design   207
9.3   Major phases of design   208
9.4   Design codes and standards   211
9.5   Probabilistic design   213
9.6   Factor of safety and derating methods   214
9.7   Modeling and simulation in design   217
9.8   General considerations in mechanical design   219
9.9   Review questions and problems   219
      Bibliography and further reading   220

Chapter 10   Decision Making

10.1   Introduction   221
10.2   Decision matrix   222
10.3   Decision trees   224
10.4   Planning and scheduling models   225
10.5   Optimization methods   229
10.6   Optimization by differential calculus   231
10.7   Search methods of optimization   231
10.8   Linear programming   234
10.9   Sensitivity analysis   236
10.10 Geometric programming   237
10.11 Review questions and problems   238
       Bibliography and further reading   239

Chapter 11   Effect of Material Properties on Design

11.1   Factors affecting the behavior of materials in components   240
11.2   Statistical variation of material properties   240
11.3   Stress concentration   242
11.4   Designing for static strength   246
11.5   Designing with high-strength low-toughness materials   250
11.6   Designing against fatigue   255
11.7   Designing under high-temperature conditions   263
11.8   Review questions and problems   268
       Bibliography and further reading   268

Chapter 12   Effect of Manufacturing Processes on Design

12.1	Introduction	269
12.2	Design considerations for cast components	269
12.3	Design considerations for molded plastic components	271
12.4	Design considerations for forged components	274
12.5	Design of powder metallurgy parts	275
12.6	Design of sheet metal parts	278
12.7	Designs involving joining processes	279
12.8	Designs involving heat treatment	285
12.9	Designs involving machining processes	287
12.10	Designing for corrosive environments	289
12.11	Designs involving automated assembly	292
12.12	Review questions and problems	293
	Bibliography and further reading	293

Chapter 13   Reliability of Engineering Components

13.1	Introduction	294
13.2	Assessment of reliability	296
13.3	Service life	301
13.4	Hazard analysis	302
13.5	Fault tree analysis	302
13.6	The role of design in achieving reliability	305
13.7	The role of materials and manufacturing in achieving reliability	309
13.8	The role of the user in achieving reliability	311
13.9	Maintenance and condition monitoring	311
13.10	Product liability	312
13.11	Review questions and problems	314
	Bibliography and further reading	315

Chapter 14   Computer Aided Design

14.1	Introduction	316
14.2	The components of CAD systems	317
14.3	Geometric modeling	319
14.4	Automated drafting	321
14.5	Finite element analysis	322
14.6	Design review and evaluation	325
14.7	Creating the design and manufacturing data base	325
14.8	Applications of CAD in industry	326
14.9	Benefits of CAD	327
14.10	Review questions and problems	328
	Bibliography and further reading	328

Chapter 15   Elements of the Production Function

15.1   Introduction   329
15.2   Types of manufacturing system   331
15.3   Production planning and control   332
15.4   The manufacturing shop   335
15.5   Modeling of production systems   336
15.6   Improving productivity in manufacturing   337
15.7   Review questions and problems   339
        Bibliography and further reading   340

Chapter 16   Computer Aided Manufacture

16.1   Fundamentals and applications of CAM   341
16.2   Numerical control   341
16.3   Computer aided process planning   343
16.4   Computer aided quality control   345
16.5   Group technology   346
16.6   Industrial robots   351
16.7   Flexible manufacturing systems   353
16.8   The automated factory   355
16.9   Review questions and problems   356
        Bibliography and further reading   356

PART III   ECONOMIC CONSIDERATIONS

Chapter 17   Concepts of Economic Analysis

17.1   Types of costs in manufacturing   361
17.2   Break-even analysis   362
17.3   Time value of money   365
17.4   Comparing alternatives on cost basis   367
17.5   Depreciation and tax considerations   370
17.6   Benefit-cost analysis   373
17.7   Cost-effectiveness analysis   375
17.8   Minimum cost analysis   377
17.9   Value analysis   380
17.10 Review questions and problems   381
        Bibliography and further reading   381

Chapter 18   Economics of Manufacturing Processes

18.1   Introduction   382
18.2   Methods of cost estimation in the process industries   382
18.3   Methods of cost estimation in manufacturing industries   384

18.4   Manufacturing time   385
18.5   Manufacturing costs   388
18.6   Economic justification of jigs and fixtures   390
18.7   Economics of metal cutting   391
18.8   Standard costs   394
18.9   Learning curve   395
18.10 Selling price of a product   396
18.11 Life cycle costing   399
18.12 Review questions and problems   399
       Bibliography and further reading   400

Chapter 19   Economics of Materials

19.1   Introduction   401
19.2   Elements of the cost of materials   402
19.3   Factors affecting material prices   404
19.4   Comparison of materials on cost basis   407
19.5   Value analysis of material properties   407
19.6   Economics of material utilization   410
19.7   Competition in the materials field   412
19.8   Review questions and problems   413
       Bibliography and further reading   413

**PART IV   INTEGRATION OF DESIGN AND ECONOMIC ANALYSIS
WITH MATERIALS AND PROCESS SELECTION**

Chapter 20   The Selection Process

20.1   Introduction   417
20.2   The nature of the selection process   418
20.3   Analysis of the material performance requirements   419
20.4   Development and evaluation of alternative solutions   421
20.5   Cost per unit property method   422
20.6   Weighted properties method   424
20.7   Incremental return method   429
20.8   Limits on properties method   430
20.9   Computer aided materials and process selection   433
20.10 Materials data bases   435
20.11 Sources of information on materials properties   437
20.12 Review questions and problems   438
       Bibliography and further reading   443

Chapter 21   Design and Selection of Materials for a Turnbuckle

21.1   Introduction   445
21.2   Design of the turnbuckle   446

21.3   Candidate materials and manufacturing processes   450
21.4   Sample calculations   451
21.5   Selection of optimum materials   453
       Bibliography and further reading   456

Chapter 22   Design and Selection of Structural Parts of a Cargo Trailer

22.1   Introduction   457
22.2   Design of the cargo trailer   457
22.3   Candidate materials and manufacturing processes   459
22.4   Evaluation of candidate materials   461
22.5   Selection of optimum materials   463
       Bibliography and further reading   466

Chapter 23   Design and Materials Selection for Lubricated Journal Bearings

23.1   Introduction   467
23.2   Design of the journal bearing   468
23.3   Analysis of bearing material requirements   470
23.4   Classification of bearing materials   473
23.5   Selection of the optimum bearing alloy   474
       Bibliography and further reading   475

Chapter 24   Analysis of the Requirements and Selection of Materials for
             Tennis Rackets

24.1   Introduction   477
24.2   Analysis of the functional requirements of the tennis racket   477
24.3   Design considerations   478
24.4   Analysis of the racket material requirements   481
24.5   Classification of racket materials   482
24.6   Evaluation of racket materials   484
       Bibliography and further reading   485

Chapter 25   Design and Selection of Materials for a Surgical Implant

25.1   Introduction   486
25.2   Design considerations   486
25.3   Analysis of implant material requirements   489
25.4   Classification of materials for the prosthesis pin   491
25.5   Evaluation of candidate materials   492
       Bibliography and further reading   493

PART V   APPENDICES

Appendix A   Properties and composition of selected engineering materials   497
Appendix B   Conversion of units and hardness values   513
Appendix C   Principles of engineering statistics   516

Index   528

Chapter 1

The Activities Involved in Developing a Concept into a Finished Product

1.1 INTRODUCTION

The introduction of a new product or changing an old model involves carrying out a feasibility study, making designs, reaching economic decisions, selecting optimum materials, choosing appropriate manufacturing processes, planning and scheduling of various activities, developing the market, selling the product and arranging for after-sales service. These diverse activities are interdependent and should not be performed in isolation from each other. This is because it is not sufficient that the design of the product should satisfy the technical requirements, it must also be possible to manufacture it with the available facilities and to sell it at a competitive price.

The main objective of this chapter is to outline the spectrum of activities that are involved in developing a new product, starting from the conception of the idea and ending with the marketable product. This chapter will also help in showing how the different topics that are discussed in this book fit into the total picture of the industrial enterprise.

1.2 STAGES OF PRODUCT DEVELOPMENT

An industrial product is normally expected to satisfy a certain demand and to give satisfaction to the user. A product usually starts as a concept which, if feasible, develops into a design, then into a finished product. While each engineering product has its own individual character and its own sequence of development events, there is a general pattern for the various stages that accompany the introduction of a new product, as shown in Fig. 1.1. To illustrate how the various stages could apply in practice, let us take a hypothetical case of a motor car company considering the introduction of an inexpensive fuel-efficient two-passenger (two-seater) model. This is based on the statistics that on about 80 percent of all trips American cars carry no more than two people and that in a little more than 50 percent of all trips the driver is alone. Such a car will be predominantly driven in city traffic, where the average vehicle speed is about 55 km/h (30 mph). Based on this concept and function, a feasibility study could be started. As a first phase of the project, assume it is decided to select a design concept which is based on the present traditional internal combustion engine technology. The company expects, however, that

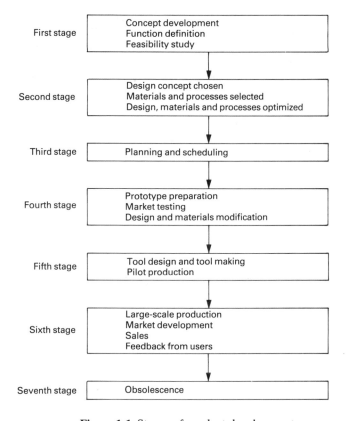

First stage — Concept development / Function definition / Feasibility study

Second stage — Design concept chosen / Materials and processes selected / Design, materials and processes optimized

Third stage — Planning and scheduling

Fourth stage — Prototype preparation / Market testing / Design and materials modification

Fifth stage — Tool design and tool making / Pilot production

Sixth stage — Large-scale production / Market development / Sales / Feedback from users

Seventh stage — Obsolescence

Figure 1.1 Stages of product development.

in view of the growing pressure to reduce pollution in large cities, a battery/solar cell-driven car could be in demand in the future. This concept is adopted as a second phase.

Having defined the overall concept and product function, the second step is the feasibility study where social, economic and legal issues related to the nature and functions of the new model are analyzed and questions related to the market and competition are posed. Important design features as well as the main manufacturing processes and materials requirements should be broadly outlined at this stage. More details about the feasibility study will be given in Section 1.3.

With such a new product (which has no precedent, since at present the only two-passenger models available are the relatively expensive and energy-inefficient sports cars) the first stage involves a large amount of creative work and innovation. In many cases of product development, however, an existing product is modified to suit other applications, to take advantage of new processes and materials or to improve its service performance. In the latter cases, the innovative part of the first stage may not constitute an important phase in product development, although economic analysis would still be required.

As a result of the feasibility study and the comparisons between the various design concepts, a final design concept is selected (Fig. 1.1, second stage). For the hypothetical case of the two-passenger car, let us assume that for the first phase of the project it was decided to select a design concept which would not require a major break with traditional automotive technology. The design limits that were imposed on the design for the first phase are shown in Table 1.1. The relevant figures for a four-passenger car which is produced by the same company are included for comparison. Based on the imposed limits, a workable design will be developed and used as a basis for a more accurate estimation of the development costs. Optimization techniques are then performed to refine the design and to select the optimum material and processing route. Other related departments within the company, e.g. purchasing, quality control, industrial engineering, production and marketing, should be consulted to determine the optimum material procurement, manufacturing methods and sales of the product. The design of engineering components will be discussed in detail in Part II of this book while the selection process will be discussed in Part IV.

The third stage of product development is planning and scheduling in preparation for production. Planning consists of identifying the key activities and ordering them in the

Table 1.1 Comparison of design parameters for the proposed two-passenger car and an average four-passenger model

Design parameter	Range for 4-passenger car[a]	2-passenger car
Fuel consumption	10.4–25 km/l (25–60 mile/gal)	35 km/l (84 mile/gal)
Mass	900–1600 kg (2000–3600 lb)	500 kg (1100 lb)
Acceleration time from 0–90 km/h (0–50 mile/h)	10–15 s	15 s
Speed maintained on 5% gradient	100 km/h (55 mile/h)	90 km/h (50 mile/h)
Cost of the car	$5000–12 000	$4000
Safety requirement in US	30 mile/h crash test	same
Engine emission	EPA test limits	same

[a] Based on figures given by Gray and Hippel (see bibliography).

sequence in which they should be performed; scheduling consists of putting the plan on a calendar timetable, as will be discussed in Section 1.5.

The fourth stage involves preparation of prototypes, preliminary production and market tests. In the case of our two-passenger car, the prototype is used for measuring aerodynamic drag forces, crash tests and consumer reaction. As a result of this development work, some design or materials modifications may have to be made. The fifth stage consists of tool design and tool making as well as pilot production. Even at this stage, some design and materials modifications may have to be made in order to suit large-scale production.

The sixth stage is commercial or large-scale production of the product, which is carried out concurrently with market development. The activities involved in launching the product will be discussed in Section 1.6. Feedback from users to evaluate the reliability of the product and its effectiveness in performing the intended function is useful in determining future modifications or developments. The availability of spare parts and maintenance facilities are also important factors that could influence the final level of product use. The factors affecting the reliability of engineering components will be discussed in Chapter 13.

The seventh and final stage is reached when the appearance of new models or technological advances render the product obsolete. This causes the sales volume and production to decrease to uneconomical levels and the product is normally discontinued, thus ending its life cycle, as discussed in more detail in Section 1.7.

1.3   THE FEASIBILITY STUDY

The primary purpose of the feasibility study is to seek information and to evaluate all possible alternative answers to the questions which are normally asked when launching a new product or any other engineering project. It is normally started after identifying the project goals and the need for the product. The feasibility study should address not only the technical and economic aspects of the project but also its social and legal aspects. This is because the increasing influence of technology on everyday life and the increasing awareness of modern society have led to the development of social forces that have an important impact on the practice of engineering and the rate of innovation. As a result of these forces more time is now spent in planning and predicting the future effects of engineering projects on environment and consumer safety. Engineering products have to comply with an increasing number of regulations which are intended to protect the public and this has had a major impact on the economic payoff for new technologically oriented ventures. As an example, consider the case of the two-passenger car discussed in Section 1.2. If the car is intended for US market, it would not be enough for the design to meet the limits on fuel consumption, weight, power and cost. It will also have to meet the 30-mph crash-test standards and the Environmental Protection Agency (EPA) test limits on engine emission. Product liability considerations will be discussed in Chapter 13.

Market research and product survey are important steps in the development of a new product. First, it is necessary to identify competing products and the mechanism of their operation. This is an important parameter in industries where technology is moving rapidly and new processes or materials could greatly affect the marketability of the product under consideration. For each competing product, it is necessary to identify:

1. The technical advantages and disadvantages.
2. The range of applications.
3. Patent or license coverage.
4. The reasons for any modifications which have been carried out recently.
5. Market share by value and volume.

In addition to competition analysis, market research should provide the following information, many items of which are relevant to the case of the two-passenger car considered earlier:

1. Consumer needs in the area covered by the product under consideration.
2. Characteristics of prospective users (income, age, sex, etc.).
3. What improvements should be made over existing products?
4. Past and anticipated market growth rate.
5. Market share anticipated for the product under consideration.
6. Product price that will secure the intended volume of sales.
7. The number of companies entering and leaving the market over the past few years, with reasons for those movements.

8. What will be the effect of new technology on the product?
9. How long will it take the competition to produce a competitive product?
10. What is the optimum packaging, distribution and marketing method?

The above information is essential for determining the rate of production, plant capacity, and financial and economic evaluation of the proposed product.

The technical aspects of the feasibility study should start with a statement and suggested solutions to the problem. As it must be physically possible to produce the product, the anticipated design, materials or production difficulties should be evaluated. If all suggested solutions cannot meet the intended functional requirements, because appropriate technology, materials or fabrication methods are not available for example, then the project should not proceed unless these problems are overcome by appropriate research and development work. For example in the earlier case of the two-passenger car, it may not be possible for a vehicle weighing 500 kg to meet the 30-mph crash-test with the available materials. The technical aspects of the study may eventually result in a prototype test which adequately demonstrates the technical feasibility of the product.

If the technical requirements appear possible, then the economics of the various available alternative solutions can be investigated. The initial investment that will be needed, the operating expenses that will occur and the income that will probably result should be estimated for each solution. In each of these estimates there will be risks. In order to estimate the economic feasibility, a realistic appraisal of the return on investment should be made and the prospective earnings compared with the risks. The total investment needed includes the capital required for acquiring additional production facilities which may be needed for the proposed product. The research and development expenses, cost of buying technology and preproduction expenses are usually considered as part of the total investment needed. The sources and cost of financing are then estimated. The rate of interest and schedule of payment are important parameters in the economic evaluation. An important part of the economics of a new product is the production or manufacturing costs which include direct and indirect costs as will be discussed in detail in Part III.

The final stage of the feasibility study is to establish whether the alternative solutions, that were proven technically and economically feasible, are acceptable not only to the consumer of the product but also to the society in general. If other members of the community object to the product, whether for functional or safety reasons or merely because of social customs or habit, then it may not be successful. This part of the study requires an understanding of the structure and the needs of society and any changes that may occur during the intended life time of the product.

After the information has been collected and evaluated, and after the undesirable solutions have been discarded, the engineer may still have several acceptable alternatives to compare. The selection in this case may need previous experience, but there are techniques which have been developed to improve judgment. Some of these techniques will be discussed in Chapter 10.

1.4 DEVELOPING THE DESIGN AND SELECTING MATERIALS AND PROCESSES

Discussion in the previous section has shown that at the end of the feasibility study stage the alternative design concepts are narrowed to the few which fulfil all the requirements. The task of the engineer at this stage is to select the optimum design concept in order to develop it in detail. This could be difficult, since the choice must be based on such diverse factors as physical characteristics of size and weight, expected life and reliability under service conditions, energy needs, maintenance requirements and operating costs, availability and cost of materials, availability and cost of manufacturing processes, quantity of production and expected delivery date. Since it is difficult to make such comparisons subjectively without introducing personal bias, it is useful to adopt a suitable objective methodology. Various decision-making techniques and quantitative methods of selection will be discussed in Chapters 10 and 20.

After selecting the optimum design concept, it is then divided into subassemblies and those are further divided into parts in preparation for making a detailed design. Examples of subassemblies in a motor car include the engine, steering and brake system, body, electrical system, etc. As the design progresses from feasibility study to detail design, the tasks to be accomplished become more narrowly defined. As discussed in Section 1.2, the initial stages are concerned with legal, social and economic issues in addition to the technological questions. In the detail design stage, however, the engineer is concerned with the static and dynamic forces and their effect on the performance of the part under the expected service conditions. A detailed account of the design of mechanical components and the different parameters that may influence the design is given in Part II.

In many cases the different performance requirements that have to be met by a given part present conflicting limitations on the material properties. For example, the material that meets the strength requirements may be difficult to manufacture using the available facilities or the material that resists the corrosive environment may be too expensive. To resolve such problems, compromises or trade-offs have to be made. To arrive at the best compromise, the relationships between different materials properties should be understood. Such relationships are discussed in Part I for the different types of engineering materials.

In addition to the complications caused by the conflicting requirements on the material properties, the designer is faced with the fact that materials like sheet, bar or pipe are commercially available in certain sizes or gages and that stock subassemblies like pumps and electric motors are only available in certain power or capacity ranges. In general, commonly available items should be specified unless the designer is prepared to pay the extra cost of a special mill run with the off-standard dimensions or specifications. The design and materials selection for a subassembly which contains several parts can be even more complicated. Frequently trial and error, iteration and compromises have to be made until a well-matched combination of components is found. It is not sufficient that each individual part is well designed, the assembled components should function together to achieve the design goals. The issue of successfully matching a group of components should also be addressed when redesigning a part in an existing subassembly. If the material of the new part is too different from the surrounding materials, for example, problems resulting from load redistribution or galvanic corrosion could arise.

Design reviews represent an important part of each phase of the design process. They provide an opportunity to identify and correct problems before they can seriously affect successful completion of the design. The design review teams normally include representatives from the manufacturing, quality control, safety, financial and marketing areas, which ensures that the design is satisfactory not only from the performance point of view but also from the manufacturing, economic, reliability and marketing points of view.

1.5   PROJECT PLANNING AND SCHEDULING

Engineering projects normally have a completion date which is part of a contract with penalties for not finishing on time. In order to avoid delays and in view of the complexity of many of the engineering projects, planning and scheduling should play an important role in project development. The first step in planning is to identify the activities that need to be controlled. The usual way to do that is to start with the entire system and identify the major tasks. These major tasks are then divided into sections, and these in turn are subdivided until all the activities are covered. Generally the division of tasks proceeds in a hierarchical order from the system to the subsystem to the simple activities. Normally, scheduling and planning of activities should be done together, which means establishing the timing and interdependence of the various activities during the planning stage.

Several analytical techniques have been developed to facilitate the planning and scheduling of the large number of activities that are usually involved in industrial projects. Using network planning models make it possible to locate those activities that are critical and must be done on time and those activities that have schedule slack. The critical path method, CPM, and the program evaluation and review technique, PERT, are widely used network planning models and will be discussed in Chapter 10.

1.6   LAUNCHING THE PRODUCT

Launching the product covers the activities of manufacturing the product, marketing and after-sales services. This stage is best organized on the basis of the planning and scheduling schemes which are drawn to meet the delivery times, as discussed in the previous section. The sequence of manufacturing processes is first established for each part of the product and recorded on a process sheet. The form and condition of the material as well as the tooling and production machines that will be used are also recorded on the process sheet. These activities are discussed in more detail in Chapters 15 and 16.

Using established standard times and labor costs for each operation, the information in the process sheets is used to estimate the processing time and cost for each part. More detailed discussion of cost estimation in manufacturing is given in Chapter 18. High cost may indicate the need to change the material, the processing methods or even the design of the part. The information in the process sheets is also used to estimate and order the necessary stock materials, to design special tools, jigs and fixtures, to specify the production machines and assembly lines and to plan work schedules and inventory controls. In addition to specifying the stock materials to be used in production, it is necessary at this stage to decide on the work to be subcontracted to outside manufacturers and to select the components and subassemblies that have to be bought from outside vendors. The make-or-buy decisions are discussed further in Chapter 15.

Before large-scale production is started, a pilot batch is usually made to test the tooling and familiarize the production personnel with the new product and also to identify outstanding problems which could affect the efficiency of production.

Quality control represents an important activity in manufacturing and could vary from 100 percent inspection of produced parts to statistical sample inspection, depending on the application and the number of parts produced. In some applications it may be necessary to test subassemblies and assemblies in order to make sure that the product performs its function according to specifications. Both the designer and the manufacturer should keep abreast of the latest legislation, as it is now a criminal offense if designers, manufacturers or suppliers fail to ensure that the product is safe and without risk to health when properly used.

Packaging is meant to protect the finished product during its shipping to the consumer. Secondary functions of packaging include advertising and sales appeal which are important aspects of marketing the product, especially in the case of consumer goods. Marketing of engineering products, however, may require specialized sales brochures supported by technical demonstration and performance test data. Clear installation, operation and maintenance instructions will make it easier for the user to achieve the optimum performance of the product.

The marketing personnel should be involved in the various stages of product development to allow them to develop the publicity material that will help in selling the product. In addition to publicity material, installation and maintenance instructions need to be prepared and distributed with the product. Most products require either regular or emergency service during their useful life. The accuracy and speed of delivering the needed service and the availability of spare parts could affect the company's reputation and the sales volume of the product. Feedback from the user is an important factor to be considered in making an improved version of a product and in developing a new product.

1.7   THE PRODUCT LIFE CYCLE

The product life is defined as the length of time between its appearance on the market for the first time and the time when the company decides to stop its production. The life of a product may vary from a few months to several years depending on its nature, competition, economic and social climates as well as political decisions. Examples of products which normally have a relatively short life cycle are found in the clothing and toy industries where the main determining parameters are fashion and consumer desires, or in the electronic and computer industries where technological development is relatively fast and competition pressures are high. In contrast, machines for power generation and production as well as similar heavy equipment have a relatively long life cycle, as this is determined mostly by the relatively slow advances in their well-established technology.

The life cycle of a successful product can generally be represented by a curve similar to that shown in Fig. 1.2. The sales are low in the introduction stage but gradually increase as the product gains the acceptance of an increasing number of consumers. As the product reaches maturity, the production rates and sales volume reach the design values. While the initiation and growth stages should be as short as possible, the maturity stage should be prolonged as it is during this stage that production rates are most efficient, the investment used in the product development is recovered and most profits are made. As shown in Fig. 1.2, there is a time lag between sales and profit. Heavier investments in product development will take longer to recover, which means a longer time lag. In order to prolong the maturity stage efforts should be made to develop new markets by adopting new marketing strategies. Introducing an improved version of the product can also prolong its life cycle. This can be achieved by introducing design modifications or employing new materials or manufacturing processes which could improve efficiency or extend the use of the product to new applications and environments. In the case of the two-passenger car discussed in Section 1.2, improving fuel economy by reducing the weight, reducing aerodynamic drag forces or improving the efficiency of the engine can extend the maturity stage.

Eventually social change, appearance of other competitive products or technological advances make the product less competitive, causing the sales to decrease and the product to reach the decline stage. When the sales volume causes the production to reach

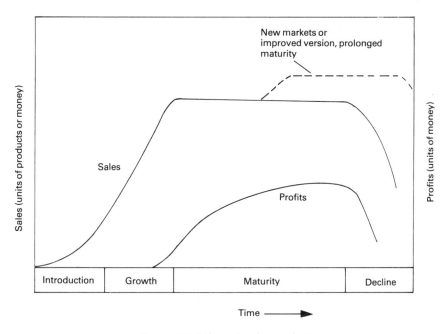

Figure 1.2 Life cycle of a product.

uneconomical rates the product is normally discontinued, thus ending its life cycle. In our two-passenger car example, this stage could correspond to the case where increasing pollution in large cities and the mounting social pressure would force local authorities to ban fuel-burning cars from the inner city. This would be the time when the battery/solar cell-driven car should be ready for production.

The growth curve of a given product and recognition of the factors that cause changes in sales and production rates are helpful in deciding whether to rejuvenate it or to concentrate on a totally different product. Timing in making this decision is important because possible sales and profits could be lost if the product is discontinued too soon, while losses may result due to decreased sales if the decision is delayed.

The above discussion indicates that it is not sufficient to design and manufacture a successful product. Engineers should continue to improve their products and to make use of new technologies in order to keep their products competitive.

1.8   REVIEW QUESTIONS AND PROBLEMS

1.1   In the example of the two-passenger car given in Section 1.2, draw a conceptual design of the battery/solar cell-driven. car. What are the major technological problems that have to be solved before this type of car becomes a mass-produced product?

1.2   A large dairy producer near a large city is considering different expansion schemes for its fresh milk department. Among the various alternatives are: (a) door-step delivery; (b) delivery to supermarkets in the city. Compare the feasibility of each alternative scheme.

1.3   The above dairy producer is also considering the different materials for packaging its product. Among the alternative materials are glass bottles, plastic containers and carton packages. Compare the advantages and limitations of each material. Which material will be most suitable for door-step delivery?

1.4   A group of new engineering graduates are thinking of starting a small business in manufacturing of educational toys. What are the steps that they need to take before they can start production?

1.5   What is the sequence of events involved in building, decorating and furnishing a house with a small garden?

BIBLIOGRAPHY AND FURTHER READING

Beakley, G. C., Evans, D. L. and Keats, J. B., *Engineering: An Introduction to a Creative Profession*, 5th ed., Macmillan, 1986.

Brichta, A. M. and Sharp, P. E. M., *From Project to Production*, Pergamon Press, New York, 1970.

Dieter, G. E., *Engineering Design*, McGraw-Hill, New York, 1983.

Edel, D. H. (ed.), *Introduction to Creative Design*, Prentice Hall, Inc., New Jersey, 1967.

Gray, C. L. and Hippel, F., 'The fuel economy of light vehicles', *Scientific American*, vol. 244, May 1981.

Hubka, V., *Principles of Engineering Design*, trans. W. E. Eder, Butterworth Scientific, London, 1982.

Khane, A. R., *Manual for the Preparation of Industrial Feasibility Studies*, United Nations Publications, New York, 1978.

Polak, P., *A Background to Engineering Design*, The Macmillan Press, London, 1976.

Ray, M. S., *Elements of Engineering Design*, Prentice Hall, New Jersey, 1985.

Wood, T. T., *Introduction to Engineering Design*, McGraw-Hill, New York, 1966.

PART I

BEHAVIOR AND PROCESSING OF ENGINEERING MATERIALS

A civilization is both developed and limited by the materials at
its disposal.
Sir George Paget Thomson

Materials have evolved with mankind since the dawn of civilization. Many materials are found in nature or are grown, but the first material to be made by combining different substances was glass in ancient Egypt. With the progress of time, stone followed by bronze and then iron marked the main materials that shaped the early ages of civilization. The number of materials in the service of man has increased slowly over the years and even by the early twentieth century, timber, concrete, steel and copper and its alloys were still the primary materials used by engineers. This is because the properties of these materials were sufficient to meet the requirements of the traditional engineering structures and machines.

With the increasing demand for higher service temperatures, lighter structures and more reliable components, better materials had to be found. As a result, many ferrous and nonferrous alloys, plastics, ceramics and composite materials were introduced and the modern engineer has a great and diverse range of materials at his disposal. It has been

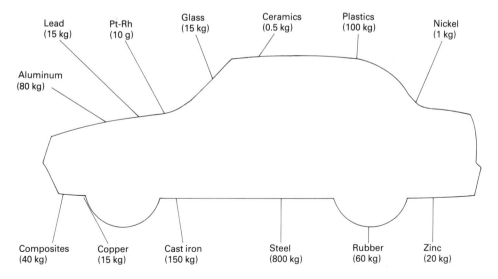

Figure 2.1 Types of materials used in the manufacture of an average motor car. The weights given are approximate average values and could vary considerably from one make or model to another.

reported that there are more than 40 000 currently useful metallic alloys and probably close to that number of nonmetallic engineering materials such as plastics, ceramics, composite materials and semiconductors. One of the distinguishing features of modern industry is the extensive use of this expanding range of materials, many of which are usually combined to make the final product. As an example, consider the case of an average motor car where almost all types of engineering materials are used in its manufacture, as shown in Fig. 2.1. It is estimated that in a wide range of engineering industries the direct cost of materials represents 30 to 70 percent of the value of production. One-third to one-half of the energy consumed by industry goes into the value added to materials in their production and fabrication.

Part I of this book describes the properties and processing of the main engineering materials used by the modern engineer and evaluates their behavior under service conditions.

Chapter 2

Metallic Materials

2.1 CLASSIFICATION AND SPECIFICATION OF METALLIC MATERIALS

Pure metallic elements have a wide range of properties and their alloys are even more versatile and form a large proportion of engineering materials. The major characteristics of metallic materials are their crystallinity, conductivity to heat and electricity and relatively high strength and toughness. Metallic materials can be divided into ferrous and nonferrous alloys. Ferrous alloys occupy the major bulk of world production of engineering materials, as shown in Fig. 2.2. Ferrous alloys have iron as the base metal and range from plain carbon steels, containing more than 98 percent iron, to high alloy steels containing up to about 50 percent of a variety of alloying elements. All other metallic materials fall into the nonferrous category, which can be subdivided into light metals, e.g. aluminum, magnesium and titanium; low melting-point metals, e.g. bismuth, lead and tin; refractory metals, e.g. molybdenum, niobium, tantalum and

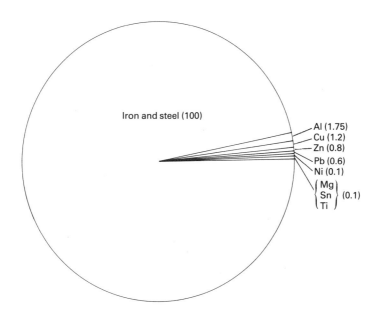

Figure 2.2 Comparison of the relative world production by weight of the main metallic materials.

tungsten; and precious metals, e.g. gold, silver and platinum. Within each group of alloys, classification can be made according to (a) chemical composition, e.g. carbon content and alloy content in steels; (b) finished method, e.g. hot rolled or cold rolled; or (c) product form, e.g. bar, plate, sheet, tubing or structural shape. The method of producing the alloy can be used for further subdivision. For example, carbon steels can be classified into rimmed, semikilled or killed depending on the deoxidation practice.

Metallic materials can also be divided according to the method of production into cast alloys and wrought alloys. Cast alloys form about 20 percent of all industrial metallic materials and are cast directly into shape. Wrought alloys are usually shaped by hot or cold working into semifinished products like plates, sheets, rods, wires or tubes. These semifinished materials provide a starting-point for the fabrication of finished components by further forming or machining processes. A third type, which has gained industrial favor in recent years, is powder metals and alloys. The powders are compacted and sintered to produce ready-to-use components that need very little further machining.

While classification is the systematic arrangement or division of materials into groups on the basis of some common characteristic, designation is the identification of each class by a number, letter, symbol, name or a combination thereof. Designations are normally either based on chemical composition or mechanical properties. Examples of designation systems are given in Tables 2.1 and 2.2 for steels and aluminum alloys respectively. The system used by the American Iron and Steel Institute (AISI) and the Society of Automotive Engineers (SAE) consists of four, or five, digits which designate the composition of the alloy. The alloy system is indicated by the first two digits while the nominal carbon content is given in hundredths of a percent by the last two, or three, digits. The designation system employed by the Aluminum Association (AA) uses four

Table 2.1 Some AISI-SAE standard steel designations and corresponding UNS numbers

No.	Name	Example	UNS	C	Mn	Nominal composition (wt %)					
						P(max.)	S(max.)	Si	Ni	Cr	Other
Carbon steels											
10XX	Plain carbon	1015	G10150	0.15	0.45	0.04	0.05				
11XX	Free machining	1118	G11180	0.18	1.15	0.04	0.13				
15XX	Plain carbon	1524	G15240	0.24	1.35–1.65	0.04	0.05				
Manganese steels											
13XX	Mn	1335	G13350	0.35	1.75	0.035	0.04	0.25			
Molybdenum steels											
40XX	Mo 0.25	4027	G40270	0.27	0.8	0.035	0.04	0.3			0.25 Mo
44XX	Mo 0.52	4419	G44190	0.19	0.55	0.035	0.04	0.22	–	–	0.52 Mo
Chromium-molybdenum steels											
41XX	Cr Mo	4118	G41180	0.18	0.8	0.035	0.04	0.3	–	0.5	0.1 Mo
Nickel-chromium-molybdenum steels											
43XX	Ni Cr Mo	4320	G43200	0.20	0.55	0.035	0.04	0.3	1.8	0.5	0.25 Mo
86XX	Ni Cr Mo	8640	G86400	0.40	0.85	0.035	0.04	0.3	0.55	0.5	0.2 Mo
94BXX	Ni Cr Mo B	94B17	G94171	0.17	0.85	0.035	0.04	0.3	0.45	0.4	0.1 Mo, 0.0005Bmin
Nickel-molybdenum											
46XX	Ni Mo	4620	G46200	0.20	0.6	0.04	0.04	0.3	1.8		0.25 Mo
Chromium steels											
50XX	Cr	5015	G50150	0.15	0.4	0.035	0.04	0.3	–	0.4	
51XX	Cr	5120	G51200	0.20	0.8	0.035	0.04	0.22	–	0.8	
Chromium-vanadium steels											
61XX	Cr V	6118	G61180	0.18	0.6	0.035	0.04	0.3	–	0.6	0.12 V

Table 2.2 Designation system for aluminum and its alloys

No.	Name	Example	UNS	Nominal composition (wt %)
I Wrought alloys				
IXXX	Aluminum 99% and greater	1060	A91060	99.6% Al
2XXX	Two phase Cu alloys	2014	A92014	4.4% Cu, 0.8% Si, 0.8% Mn
3XXX	One phase Mn alloys	3003	A93003	1.2% Mn
4XXX	Two phase Si alloys	4032	A94032	12.5% Si, 1% Mg, 0.9% Cu, 0.9% Ni
5XXX	One phase Mg alloys	5052	A95052	2.5% Mg, 0.25% Cr
6XXX	Two phase Mg-Si alloys	6061	A96061	0.6% Si, 1% Mg, 0.25% Cu, 0.25% Cr
7XXX	Two phase Zn alloys	7075	A97075	5.6% Zn, 2.5% Mg, 1.5% Cu, 0.3% Cr
II Cast alloys				
2XX.X	Al-Cu alloys	108.0	A02080	4% Cu, 3% Si, 1.2% Fe, 1.0% Zn
3XX.X	Al-Si-Cu alloys	333.0	A03330	9% Si, 3.5% Cu, 1.0% Fe, 1.0% Zn
4XX.X	Al-Si alloys	B443.0	A24430	5% Si
5XX.X	Al-Mg alloys	520.0	A05200	10% Mg
7XX.X	Al-Zn alloys	A712.0	A17120	6.5% Zn, 0.7% Mg
8XX.X	Al-Sn alloys	850.0	A08500	6.5% Sn, 1% Cu, 1% Ni

digits to identify wrought aluminum and its alloys and three digits to identify cast alloys, as shown in Table 2.2. Unalloyed aluminum, 99 percent and greater, is designated by the first digit of 1, the second digit indicates a modification of impurity limits and the last two digits indicate purity. For aluminum alloys, the first digit is between 2 and 9 and indicates the group, the second digit indicates a modification of the original alloy and the last two digits indicate the specific alloy.

The Unified Numbering System (UNS) has been developed by the American Society for Testing and Materials (ASTM) and SAE and several other technical societies, trade associations and United States government agencies. The UNS number is a designation of chemical composition and consists of a letter and five numerals. The letter indicates the broad class of the alloy and the numerals define specific alloys within that class. Existing systems, such as the AISI-SAE system for steels and the AA for aluminum, have been incorporated into UNS designations. A sample of the UNS numbering system is included in Tables 2.1 and 2.2.

Another example of designation systems is that used by the British Standard for wrought steels, BS970: Part 1 and employs six digits. The first three digits represent one hundred times the mean manganese content. The fourth digit is given by the letters A, M or H to indicate whether the steel is to be supplied to analysis A or mechanical property M or hardenability H requirements. The fifth and sixth digits represent one hundred times the mean carbon content. As an example BS 970:060A52 describes a steel containing 0.50–0.55 percent carbon, 0.50–0.70 percent manganese supplied according to analysis requirements. Most commercially produced wrought aluminum alloys are described by BS 1470–1475. This system uses one number to describe the alloy composition in addition to prefix and suffix letters to describe the heat treatment, form of the product and temper condition. For example, the first prefix letter indicates whether the alloy is heat treatable (H) or nonheat-treatable (N); the second prefix letter indicates whether the material is in the form of sheet (S), tube (T), forging (F), or extruded (E). The temper condition is described by the suffix letters, e.g. (O) is fully annealed, (H1 to H8) is work-hardened to various degrees, and (T followed by another letter) describes the heat treatment. Thus the number BS 1470:NS3–H4 refers to a nonheat-treatable alloy containing 0.8 to 1.5 percent manganese in the form of half-hard sheet.

The designation systems discussed above are not specifications but are often incorporated into specifications describing products which are made of the designated materials. A standard specification is a published document that describes the characteristics a product must have in order to be suitable for a certain application. Standard specifications will be discussed further in Chapter 9.

In most engineering applications, selection of metallic material is usually based on one or more of the following considerations:

1. Product shape: (a) sheet, strip, or plate; (b) bar, rod, or wire; (c) tubes; (d) forgings; (e) castings.
2. Mechanical properties, as ordinarily revealed by the tensile, fatigue, hardness, creep or impact tests.
3. Physical and chemical properties, such as specific gravity, thermal and electrical conductivities, thermal expansion coefficient and corrosion resistance.
4. Metallurgical considerations, such as anisotropy of properties, hardenability of steels, grain size and consistency of properties, i.e. absence of segregations and inclusions.
5. Processing considerations, such as castability, formability, machinability and weldability.
6. Sales appeal such as color and luster.
7. Cost and availability.

In the following discussion, the main types of metallic materials will be evaluated and compared on the basis of the first five considerations. The factors affecting the cost of materials will be discussed in Chapter 19.

2.2 STRENGTHENING OF METALLIC MATERIALS

Mechanical behavior of metallic materials represents an important reason for their selection in engineering applications where relatively high strength and reasonable toughness are needed. Higher strength means thinner sections and lighter components while higher toughness means less likelihood of sudden failure in service. For many engineering alloys, however, an increase in strength is accompanied by a reduction in toughness and the optimum combination of these properties should be selected to meet the service requirements.

Within the elastic range, the behavior of the material is mainly a function of its modulus of elasticity which is a characteristic of the material and is not markedly affected by alloying or heat treatment. Most steels, for example, have elastic moduli within the relatively narrow range of 190–220 GPa ($27–31 \times 10^6$ lb/in^2) and aluminum and most of its alloys have a modulus of elasticity of about 69 000 MPa (10×10^6 lb/in^2) at room temperature. The modulus of elasticity is an important design parameter as its value affects the deflection of components and structures under load (stiffness). To illustrate this point, consider a tie rod of 1.5 cm^2 area and 3 m long which is subjected to a tensile load of 20 kN. If the tie rod is made of steel, the elongation (e_s) will be:

$$e_s = \frac{20\,000 \star 3000}{150 \star 207\,000} = 1.93 \text{ mm}$$

If the tie rod is made of an aluminum alloy, the elongation (e_a) will be:

$$e_a = \frac{20\,000 \star 3000}{150 \star 69\,000} = 5.8\,\text{mm}$$

which is about three times as much as the elongation in the steel rod. In order to reduce the elongation in the aluminum alloy tie rod to the same value as that of the steel rod, its area will have to be increased to about 4.5 cm^2.

In contrast to modulus of elasticity, other mechanical properties are sensitive to the material composition and its thermomechanical history. This is because beyond the elastic limit most metallic materials deform by slip, which involves dislocation movement. The more difficult it is to move the dislocations in the lattice the more will be the resistance to deformation and the stronger will be the material. Lattice distortions caused by foreign atoms in solid solution, second-phase particles or grain boundaries can act as obstacles to dislocations and are commonly used to increase the strength.

Strengthening by forming a solid solution is used in practice for aluminum and copper alloys. An example is shown in Fig. 2.3 for Al-Mg alloys. The effectiveness of an alloying

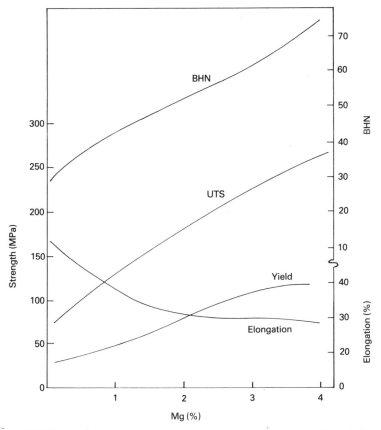

Figure 2.3 Effect of magnesium content on the mechanical properties of aluminum.

Source: Farag, M. M., *Materials and Process Selection in Engineering*, Applied Science Publishers, London, 1979.

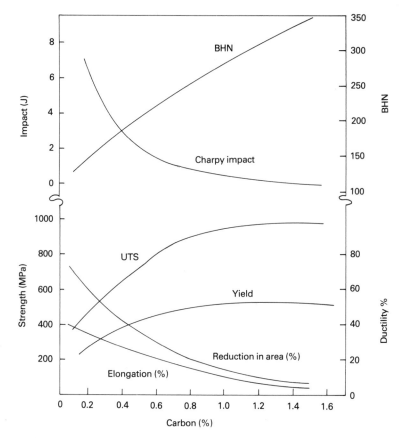

Figure 2.4 Effect of carbon content on the mechanical properties of hot-rolled plain carbon steel.

Source: Farag, M. M., *Materials and Process Selection in Engineering*, Applied Science Publishers, London, 1979.

element in solution hardening is a function of the lattice distortion it causes in the parent metal. Figure 2.3 shows that an increase in strength is accompanied by a decrease in ductility as represented by elongation percent. Alloying elements that form second phases, rather than solid solutions, are also used in practice to increase the strength. In the case of steel the addition of carbon causes iron carbide phase to form, which increases the strength of the ferrite matrix. Increasing the carbon content increases the iron carbide volume fraction, which increases the strength and hardness but decreases the ductility and toughness, as shown in Fig. 2.4. Changing the carbide shape from lamellar to spheroidal reduces the strength but improves the toughness of a given steel.

Precipitation hardening, where a hard phase is finely dispersed in a soft matrix, is another way of strengthening alloys. The precipitation hardening process involves a solution treatment and quenching, to produce a supersaturated solid solution, followed by ageing to precipitate the hard phase. When the precipitates are fine and coherent with matrix lattice, they form effective barriers to dislocations and substantial hardness increases are achieved. Artificially ageing the solution treated alloy by heating accelerates the precipitation and results in higher hardness than naturally ageing at room temperature. Table 2.3 gives the designation system for the different precipitation hardening treatments that are usually used for heat treatable aluminum alloys. The effect

Table 2.3 Temper designations of aluminum and its alloys

Temper designation	Corresponding condition of the material
F	As fabricated, e.g. as extruded
O	Softest temper, e.g. annealed, recrystallized
H 1	Strain-hardening by cold working. A second digit indicates the degree of hardness, e.g. H 12 is quarter-hard, H 14 is half-hard, H 18 is full-hard and H 19 is extra-hard
H 2	Strain-hardening and partial annealing, e.g. H 24 is half-hard
T 2	Annealed casting
T 3	Solution treatment followed by strain-hardening and natural ageing
T 4	Solution treatment followed by natural ageing
T 6	Solution treatment followed by artificial ageing
T 8	Solution treatment followed by strain-hardening and then artificial ageing
T 9	Solution treatment followed by artificial ageing and then strain-hardening

Table 2.4 Effect of precipitation hardening treatment on the mechanical properties of 2104 aluminum alloy

Temper	Tensile strength		Yield strength		Elongation percent	Hardness BHN
	MPa	ksi	MPa	ksi		
O	186	27	96	14	18	45
T4	427	62	290	42	20	105
T6	483	70	413	60	13	135

of precipitation hardening on the mechanical properties of 2014 aluminum alloy is illustrated in Table 2.4.

Strengthening of metallic materials can also be accomplished by replacing softer phases by harder ones through heat treatment, as in the case in steels, where a large variety of microstructures result from variations in the cooling rate from the austenitic region (γ-phase). Table 2.5 and Fig. 2.5 summarize the different heat treatment possibilities and the resulting microstructures and properties. The shape and position of the transformation curves (Ts, Tf) are functions of the chemical composition of steel. The critical cooling rate is defined as the slowest cooling rate that avoids the knee of the Ts curve and can be used as an estimate of the hardenability of a given steel. The lower the critical cooling rate, i.e. the further the knee of the Ts curve from the temperature axis, the higher is the hardenability and the easier it is to achieve the hard martensitic structure in a given steel.

Cold working of wrought alloys is another important method of strengthening engineering alloys. Cold working increases the number of dislocations, causing them to tangle up and become more difficult to move. Figure 2.6 shows the effect of cold working on the mechanical behavior of 3003 aluminum alloy. The cold working designation system for aluminum alloys is given in Table 2.3.

The strength of metallic materials can usually be increased by reducing grain size. The quantitative relation between yield strength, Y, and grain diameter, d, is given by the Hall–Petch relationship:

$$Y = A + Bd^{-1/2} \qquad (2.1)$$

Table 2.5 Transformations and properties of steel

Process	Procedure	Properties	Remarks
Annealing	Slow cool from γ	Soft, ductile and tough	Coarse structure (pearlite)
Normalizing	Air cool from γ	Relatively stronger than annealed steel with lower ductility	Finer structure than annealed steel
Quenching	Rapid cool from γ	Hard and brittle	Martensitic steels must be toughened by a tempering treatment before use
Interrupted quench	Quench followed by slow cool through M_s and M_f	Hard and brittle	This process is called martempering and is less severe and less likely to cause cracking than direct quench to M_s
Ausforming	Deformation just above M_s then slow cool	Hard and brittle	Deformation during the quench interruption adds strain-hardening to martensitic properties
Austempering	Interrupted quench then isothermal transformation	Hard but less brittle than martensite	Bainite is obtained and has similar properties to tempered martensite
Tempering	Reheating of martensite	Hard but less brittle than untreated martensite	The higher the tempering temperature and the more prolonged the tempering time, the less the hardness and the greater the toughness

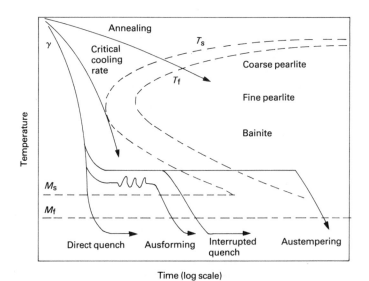

Time (log scale)

Figure 2.5 Temperature—time-transformation diagram for plain carbon steel (0.8 percent C). T_s = transformation starts. T_f = transformation is completed. Ms = martensitic transformation starts. Mf = martensitic transformation is completed. Annealing gives pearlitic structure; austempering gives bainitic structure; and direct quench, ausforming, as well as interrupted quench give martensitic structure.

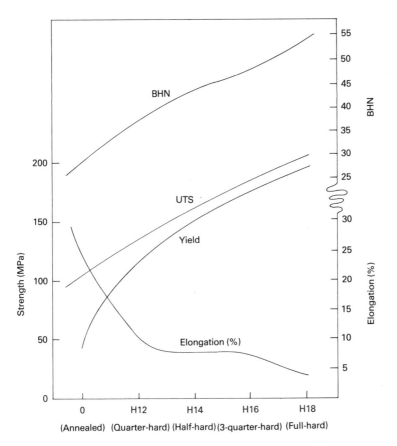

Figure 2.6 Effect of cold working on the mechanical properties of AA 3003 aluminum alloy sheet.

where A and B are constants, usually determined experimentally for a given material.

In ultrafine-grained steels, special treatments produce grain size of about 1 μm and increase the strength by approximately a factor of two. An added advantage of reducing grain size is that it also increases ductility, which distinguishes it from the other strengthening methods where ductility decreases as the strength increases.

Some of the above methods of strengthening can be combined to achieve even higher strength values. An example is T3 temper of aluminum alloys where cold working is combined with natural ageing.

The above discussion shows that specifying the chemical composition of an alloy is not enough to define its mechanical properties; the way it is treated (the temper condition) should be also specified.

2.3   CARBON STEELS

Ferrous materials, of which steels are the major alloys, form about 90 percent of the total usage of metallic materials in the world. The main reasons for this overwhelming preference for ferrous materials are their versatile properties and low cost. Plain carbon steels constitute the major bulk of steels used in industry and are defined as those in which carbon is the alloying element that essentially controls the properties and in which the amount of manganese does not exceed 1.65 percent. These steels can be divided according to composition into:

1. Low-carbon, less than 0.3 percent C
2. Medium-carbon, between 0.3 and 0.6 percent C
3. High-carbon, more than 0.6 percent C.

Specifications usually restrict the maximum allowable amounts of sulfur and phosphorus as they are generally considered deleterious to the mechanical properties of steels. The amount of silicon in carbon steels is affected by the deoxidation practice employed in their manufacture. Killed steels are fully deoxidized by the addition of silicon, aluminum or both, or by vacuum treatment. Silicon deoxidation is the least expensive method and is frequently employed. In this case a range of 0.15 to 0.35 percent Si is specified. Usually no silicon is specified in aluminum-killed or vacuum-deoxidized carbon steels. In some applications, such as the forming of low-carbon steel sheet, the choice of oxidation practice can significantly affect the performance of the steel. In such cases it is often desirable to cite a standard specification where the various ramifications of the different deoxidation methods have been considered. Generally, the ranges and limits of chemical composition depend on the product form and intended use.

Low-carbon steel sheet and strip are used primarily for mass-produced consumer goods, such as automobile bodies and appliances, because they combine ease of fabrication, adequate strength and good finishing characteristics. The steels used for these products are supplied in a wide range of chemical compositions but the most widely used compositions are 0.05 to 0.10 percent carbon, 0.25 to 0.50 percent manganese, 0.035 percent max. phosphorus and 0.04 percent max. sulfur. These sheets are supplied in commercial quality, drawing quality, special killed drawing quality and structural quality. Commercial quality sheet and strip are suitable for moderate forming and are not subject to any mechanical test requirements other than room-temperature bending test. When greater ductility or more uniform properties are required, drawing quality is specified. Drawing quality material is suitable for deep drawing and other applications requiring severe deformation. When deformation is particularly severe or resistance to stretcher strains is required, drawing quality, special killed (DQSK) is specified. Structural quality sheet and strip are specified when certain strength and elongation values are required in addition to bend tests. The range of strength and ductility covered by carbon steels is shown in Fig. 2.7 and a sample of the mechanical properties of some structural quality low-carbon steel sheet and strip is given in Table A.1, Appendix A.

Steel bars cover rounds, squares, hexagons and similar cross-sections while shapes include structural components like angles, channels, I-sections and other shapes that may be designed for specific applications. Hot-rolled steel bars and shapes can be produced according to chemical composition specifications or according to mechanical property requirements in addition to limited composition requirements. Merchant quality bars are commonly produced according to chemical composition and contain 0.50 percent max. carbon, 0.60 percent max. manganese, 0.04 percent max. phosphorus and 0.05 percent max. sulfur. Silicon content, grain size or other requirements that would dictate the method of steel production are not specified for merchant quality steel bars. Special quality bars are produced to specified mechanical properties and are usually employed for structural uses, hot forging, heat treating and cold working. Table A.1 gives a sample

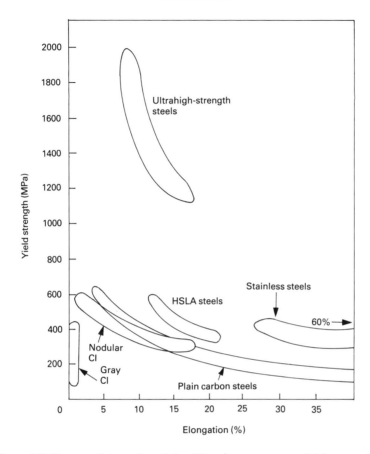

Figure 2.7 Ranges of strength and ductility of some commercial ferrous alloys.

of the mechanical properties of special quality steel bars. Special quality bars can be cold finished either by drawing or machining which gives them attractive combinations of mechanical and dimensional properties for use in mass production of machined and other parts.

Tubular products can be broadly classified as tubes and pipes. Pipes are made according to standard combinations of outside diameter and wall thickness and are grouped into standard pipe, line pipe, oil country tubular goods, water-well pipe and pressure pipe. Tubes are grouped into pressure tubes, structural tubing and mechanical tubing. Pressure pipes are distinguished from pressure tubes in that the latter are suitable for applications where heat is applied externally as in the case of boilers. Tubular products can be made by piercing or extrusion, as in the case of seamless tubes, or by forming and welding of a strip, as in the case of longitudinally or helically welded pipe. A variety of processes and steels of different chemical compositions and mechanical properties can be used in making tubular products which meet the end use requirements described by standard specifications, e.g. ASTM, ASME and API.

Steel castings can be made from many of the steels produced in the wrought form and normally exhibit similar physical and mechanical properties. Steel castings are normally expected to meet specified mechanical properties as well as some restrictions on chemical composition. For example, a maximum limit on carbon and manganese contents may be specified to ensure satisfactory weldability in the low-strength ranges. A sample of specification requirements for steel castings is given in Table A.1.

Free-machining steels have been developed to meet the increasing demand for steels which can be readily cut at high speeds, especially on automatic machine tools. The excellent machinability of these steels is mainly due to the presence of inclusions, such as MnS, which act as discontinuities in the microstructure to form broken chips. This reduces rubbing against the cutting tool and the associated friction and heat. The inclusions may also provide lubrication at the tool–chip interface and reduce the formation of built-up edge on the tool. Free-machining steels are designated by AISI-SAE as 11xx for resulfurized groups, containing 0.08–0.33 percent and 12xx for resulfurized and rephosphorized groups, containing 0.1–0.35 percent S and 0.04–0.12 percent P. It should be noted that the improved machinability of these steels is achieved at the expense of somewhat increased cost and reduced ductility and impact properties.

The strength of plain carbon steels is primarily a function of their carbon content, as shown in Fig. 2.4. Unfortunately, the ductility of these steels decreases as the carbon content is increased, and their hardenability is relatively low. Also, plain carbon steels lose most of their strength at high temperatures, become too brittle at low temperatures and are subject to corrosion in most environments. These limitations can be eliminated by adding the appropriate alloying elements, as will be discussed in the following section.

2.4   ALLOY STEELS

Alloy steels are usually defined as those which contain more than any of the following amounts: 1.65 percent manganese, 0.60 percent silicon, 0.60 percent copper, or specified amounts of one or more other alloying elements. Steels that do not contain more than a 5 percent total of combined alloying elements are called low-alloy steels and are designated as shown in Table 2.1. Alloy additions may be intended to enhance mechanical, physical or chemical properties; hardenability, weldability or other fabrication characteristics; or some other attribute of the steel. The commonly used alloying elements and their effect on the steel properties are given in Table 2.6. Alloy steels whose compositions have been standardized are given designation numbers as discussed in Section 2.1. Many of the alloy steels listed in the AISI–SAE designation system are also available as 'H-steels' and are used where hardenability is a major requirement. These steels are supplied to meet the hardenability standards specified by the customer and, therefore, slightly broader variations in composition are permitted.

Since there are more elements to be kept within specified limits, better quality control is required and consequently alloy steels are expected to be more expensive than plain carbon steels.

A large variety of the standard alloy steels are available as sheet and strip, either hot rolled or cold rolled. Regular quality is the description used for the basic or standard quality level to which alloy steel products are produced. Steels for this quality are killed and are usually produced to a fine grain size. Hot rolled regular quality alloy steel sheet and strip are normally available as rolled or as heat treated. Standard heat treated conditions include annealed, normalized and normalized and tempered. While regular quality alloy steel bars may contain some surface inperfections, cold working quality bars are made from steels which are produced under closely controlled steelmaking practice

Table 2.6 Effects of major alloying elements in steel

Element	Amount	Effect
Aluminum	Small	Deoxidizes, restricts grain growth, aids surface hardness in nitriding
Boron	0.001–0.003%	Increases hardenability
Carbon	less than 1.5%	Increases strength and hardness, decreases ductility and toughness
Chromium	0.5–2.0%	Increases hardenability and strength
	4.0–18.0%	Increases corrosion and oxidation resistance
Cobalt	up to 12%	Improves cutting tool life at high temperatures, increases hardness
Copper	0.1–0.4%	Improves corrosion resistance and strength
Manganese	0.25–0.40%	Combines with sulfur to prevent brittleness
	more than 1%	Improves hardenability, strength and hardness
Molybdenum	0.2–5%	Improves hardenability, strength and hardness, enhances creep strength
Nickel	2.0%–5.0%	Increases toughness especially at low temperatures
	12.0–20.0%	Improves corrosion and oxidation resistance
Silicon	0.2–0.7%	Deoxidizes, increases strength
	2.0%	Improves elastic properties for springs
	more than 2.0%	Decreases losses in magnetizing steel with alternating current
Sulfur	0.08–0.15%	Improves machinability
Tungsten	up to 20.0%	Increases hardness at room and high temperatures, improves hardenability
Vanadium	up to 5.0%	Increases strength while retaining ductility both at room and high temperatures

and are subject to special inspection standards. This will ensure better surface quality and more uniform chemical composition than is expected in regular quality bars. Aircraft quality and magnaflux quality apply to cold finished alloy steel bars for critical or highly-stressed aircraft parts and other similar applications. Alloy steels are used for castings that have to meet requirements that cannot be met by plain carbon steel such as hardenability, strength or corrosion resistance.

EX steels

From Table 2.6 it is apparent that two or more alloying elements may produce similar effects. Thus it is possible to obtain steels with almost identical properties but with different chemical compositions. Some alloying elements are much more expensive than others, and some may be in short supply in certain countries, which can affect the cost of the different alloy steels. This fact should be kept in mind when selecting an alloy steel for a given application. In most cases the best steel to use is the cheapest one that can be satisfactorily heat treated to have the desired properties. An example of what can be achieved is the series of EX steels that have been designed by SAE to reduce the need for expensive alloying elements. In many EX steels, some of the Ni and Cr content is substituted with Mn which is less expensive.

High-strength low-alloy steels

In many specifications, acceptance for a given steel is based more on physical or mechanical properties than on chemical composition, as in the case of ASTM specifications. This makes it possible for the producer to supply a cheaper proprietary material that will meet the desired properties as in the case of high-strength low-alloy (HSLA) steels. These steels have yield strengths in excess of about 300 MPa (43 ksi or

43 000 lb/in^2) and usually contain from 0.05 to 0.33 percent C, and 0.2 to 1.65 percent Mn with small additions of Cr, Mo or Ni. The constituents of these steels are insufficient to be hardened to martensite by quenching, which is advantageous because they can be welded without becoming brittle. At the same time, the alloying elements are sufficient to make HSLA steels stronger and exhibit higher toughness at low temperatures than plain carbon steels. Also, HSLA grades which contain appropriate amounts of phosphorus, copper, silicon, chromium or molybdenum are more resistant to atmospheric corrosion than plain carbon steels and are sometimes known as weathering steels. HSLA steels can be used for the manufacture of bridges, buildings, automobiles and railroad cars with substantial cost and weight savings. The range of strength and ductility of HSLA steels is shown in Fig. 2.7 and the mechanical properties of some HSLA steels are given in Table A.2. Dual-phase steels are similar to HSLA steels but can be treated to have a microstructure of hard martensite in a soft matrix of ferrite and can be strengthened by working. Dual-phase steels have become important sheet metal materials in the automobile and similar industries.

Another group with even higher strength than the above steels is the ultrahigh-strength steels with yield strengths of about 1400 MPa (200 ksi) or higher (Fig. 2.7). The mechanical properties of some ultrahigh-strength steels are given in Table A.3. These steels differ widely in composition and in the way that their optimum strength is obtained. They include some low, medium and high alloy steels, maraging steels and some stainless steels. Maraging steels develop tough and ductile low-carbon martensite upon cooling from austenitizing temperature. In the quenched condition, these steels can be cold worked and can be further strengthened by precipitation hardening. With their high strength and toughness, maraging steels are particularly useful in the manufacture of large structures with critical strength requirements.

Stainless steels

Stainless steel is the generic name for more than 70 types of corrosion-resistant steels which contain a minimum of 12 percent chromium. Increasing chromium content gives additional corrosion resistance. Stainless steels can be divided into five groups: (a) austenitic; (b) martensitic; (c) ferritic; (d) precipitation hardening; and (e) duplex stainless steels. The names of these categories reflect the microstructure of the steel and this depends on the balance between the alloying elements present and the heating and cooling cycle to which the steel has been subjected. Carbon, nickel, nitrogen and manganese are among the alloying elements that enhance the retention of austenite while chromium and the strong carbide formers enhance the retention of ferrite. The martensitic steels contain balanced amounts of austenite and ferrite stabilizers so that, upon heating, the structure becomes austenitic and, upon cooling at a suitable rate, it becomes martensitic. Precipitation hardening stainless steels have either austenitic, martensitic or semi-austenitic microstructures. Hardening is accomplished by precipitation of titanium or copper in the martensitic microstructures, aluminum in the semi-austenitic microstructures and carbides in the austenitic microstructures. The range of strength and ductility covered by stainless steels is shown in Fig. 2.7 and the composition and properties of some of them are given in Table A.4.

The ferritic steels are used in applications which require corrosion resistance but can tolerate lower strength, as in automobile body trim. The martensitic stainless steels are used in applications where hardness and wear resistance are required, as in the case of surgical instruments, razor blades and cutlery. The austenitic stainless steels are nonmagnetic and are used in applications where good corrosion and oxidation resistance should be combined with high-temperature strength. Many of the austenitic grades show excellent performance at temperatures up to about 725°C (1340°F) and retain most of their ductility and impact strength down to at least −190°C (−315°F). Austenitic grades are the most expensive of the stainless steels and should not be specified where the less expensive ferritic or martensitic grades would be adequate.

Problems with basic stainless steel grades usually arise due to the localized loss of corrosion resistance, sensitization, where the level of chromium in solution drops below 12 percent. This local chromium depletion is usually caused by the formation of chromium carbides along grain boundaries when the material is heated in the critical range of about 480 to 815°C (900 to 1500°F). Sensitization, which is a form of intergranular corrosion, can be prevented either by reducing the carbon content of the steel, usually below 0.1 percent, as in the case of types 304L and 316L, or by adding stabilizing elements like titanium or niobium which have a higher affinity to carbon than chromium, as in the case of types 321 and 347. Rapidly cooling through the critical temperature range also retards the carbide formation. Extra-low carbon stainless steels and stabilized varieties are commonly recommended when welding is involved in fabrication.

Tool steels

Tool steels constitute a class of high-carbon alloy steels that are used to form or cut other materials. To perform its requirements efficiently, a tool steel is required to have excellent wear and abrasion resistance, high hardness at moderate or elevated temperatures, and high toughness. The American Iron and Steel Institute (AISI) has an identification and type classification of tool steels which is based on the end use, common properties or manner of heat treatment. The compositions and properties of some of these steels are given in Table A.5.

High-speed steels retain enough hardness to cut metals at rapid rates that generate tool temperatures up to about 630°C (1166°F), and yet return to their original hardness when cooled to room temperature. Of the different compositions four grades, T1, M1, M2 and M10 make up a large proportion of the tool steels used for production machining and are the most readily available.

2.5   CAST IRONS

Cast irons are a group of materials that are basically ternary alloys of iron, carbon and silicon which contain more than 2 percent carbon and from 1 to 3 percent silicon. Wide variations in properties can be achieved by varying the balance between carbon and silicon, by alloying with various metallic or nonmetallic elements and by varying melting, casting and heat treating practices. The group of cast irons includes gray, white, ductile and malleable irons. White and gray irons derive their names from their respective fracture surface color. Ductile irons exhibit measurable ductility, in contrast with white and gray irons which exhibit brittle behavior in tension. Malleable iron is initially cast as white iron, then malleabilized by heat treatment to give it the required ductility.

In gray cast irons, the silicon content is high enough to cause the excess carbon to take the form of graphite flakes during solidification. This form of excess carbon is the basis for many of the desirable properties of gray iron, such as high fluidity, high damping capacity, low notch sensitivity and good machinability. Gray irons cannot be classified in terms of their chemical compositions since these depend largely on sectional thickness of castings. For this reason ASTM Specifications classifies gray cast irons in terms of tensile strength expressed in ksi while British Standards designates the grade in terms of tensile strength expressed in MPa. The range of strength covered by gray cast iron is shown in Fig. 2.7 and the compositions and properties of some grades given in Table A.6, Appendix A. Gray cast iron is the least expensive of all cast metallic materials and should always be considered first when a cast alloy is being selected. Another material should be chosen only when the mechanical or physical properties of gray cast iron are inadequate. In white cast irons almost all the excess carbon is present as iron carbide, which makes them very hard but brittle. They are most frequently used for their excellent wear and abrasion resistance.

Nodular cast irons, also known as ductile irons or spheroidal-graphite (SG) cast irons, contain graphite in the form of tiny spheres instead of flakes as in gray irons. The composition of unalloyed nodular irons is similar to that of gray irons but the spheroidal graphite is produced by the addition of one or more of the elements Mg, Ce, Ca, Li, Na and Ba to the molten metal. Most of the specifications for standard grades of ductile iron are based on mechanical properties while the composition is loosely specified. For example, ASTM designates the ductile iron grades in terms of tensile strength in ksi, the yield strength in ksi and elongation in percent while British Specifications specifies the grade in terms of tensile strength in MPa and elongation in percent. The range of strength and ductility of nodular cast iron is shown in Fig. 2.7 and some examples of composition and properties are given in Table A.6. Many ductile irons are used in the as-cast condition but in some cases castings are given a ferritizing anneal or are normalized to produce a pearlitic matrix. Hardening treatments may also be given to produce bainitic or martensitic microstructures. These treatments affect the properties in a similar way to steels. Ductile irons can also be alloyed with nickel, molybdenum, copper or chromium to improve their strength, hardenability or corrosion resistance. Nodular cast iron is competing with steels in many applications, as in the case of crankshafts of motor car engines for example.

Malleable cast iron is produced by annealing white iron to transform some or all of the iron carbide into graphite nodules. Malleable irons, like ductile irons, have good ductility and toughness and can also be heat treated to achieve similar structures. Malleable and ductile irons are used for similar applications and the choice between them is based on availability and economy rather than properties. However, in thin-section castings and in applications requiring higher modulus of elasticity malleable irons are preferred. On the other hand ductile irons are preferred in castings where low solidification shrinkage is required to avoid hot tears or where sections are too thick to permit solidification as white cast iron.

Compacted graphite (CG) cast irons have microstructures that are intermediate between those of gray and ductile irons. These structures can be obtained by magnesium, titanium and rare earths additions in a similar manner to ductile irons. The amount of added elements in this case is critical since undertreatment would result in gray iron while overtreatment would result in ductile iron. The mechanical properties of CG irons are also intermediate between those of gray and ductile irons with the tensile strength being similar to that of high-strength gray iron while the ductility is about 1 to 6 percent.

Thermal conductivity and damping capacity are similar to those of gray iron while fatigue and impact strengths, although not as good as those of ductile iron, are substantially better than those of gray iron. This combination of properties makes CG irons suitable for applications where neither gray iron nor ductile iron is entirely satisfactory; examples include disc-brake rotors and diesel-engine heads.

2.6   LIGHT NONFERROUS METALS AND ALLOYS

Nonferrous metals and alloys are relatively more expensive than steels and are usually considered in the category of special materials to meet special needs. They are selected only when the relatively high cost per unit weight can be justified by one or more of their special properties, such as light weight, high strength/weight ratio, high electrical and thermal conductivity, or corrosion resistance.

Aluminum and its alloys

Among the commercial metals, aluminum is second only to iron in production and consumption on a weight as well as on a volume basis. Pure aluminum, specific gravity 2.7, and its alloys are the most important light nonferrous metallic materials and their selection is usually based on one or more of the following considerations:

1. High strength/weight ratio which allows the design and construction of strong lightweight structures. This is particularly advantageous for portable equipment, vehicles and aircraft applications.
2. Resistance to corrosion in atmospheric environments, in fresh and salt waters, and in many chemicals and their salts. Aluminum has no toxic salts, which makes it suitable for processing, handling, storing and packaging of foods and beverages.
3. High electrical and heat conductivity, especially when light weight is important.

Aluminum alloys can generally be divided into wrought and casting alloys depending on the method of production. Wrought alloys constitute about 78 percent, castings about 14 percent and ingots about 8 percent of the total production.

Flat rolled products comprise about two-thirds of the total aluminum mill products and are produced in either the hot or cold rolled condition. Rod and bar are produced by either hot rolling or hot extrusion and could be finished by cold working, while wire is generally produced by cold drawing. Tubular products are usually produced by extrusion from ingot but can also be made by forming and welding of sheet. Shapes which are long products with cross-sections other than rod or tube are produced by extrusion and can be solid, hollow or semihollow. Standard structural shapes are included in this category. Most aluminum forgings are produced in closed dies and are intended for engineering products designed to perform specific functions. Products produced by impact forming are included in this category.

The AA designation system used to describe wrought aluminum alloys in the US was discussed in Section 2.1 and the designation system used to describe the temper is given in Table 2.4. The range of strength and ductility covered by aluminum alloys is shown in Fig. 2.8 and the composition and properties of representative alloys are given in Table A.7 in Appendix A. When using aluminum and its alloys for the design of engineering components account must be taken of the fact that their modulus of elasticity is about

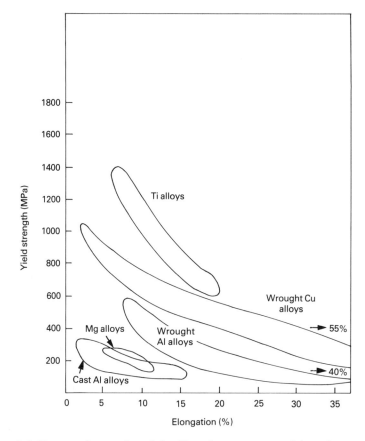

Figure 2.8 Ranges of strength and ductility of some commercial nonferrous alloys.

one-third that of steels. Unlike steels, these materials do not exhibit a clearly defined fatigue limit. It is customary, therefore, to quote an endurance limit.

Magnesium and its alloys

Magnesium, with a specific gravity of only 1.74, is the lightest metal available for use by the engineer. Aluminum is the most commonly added alloying element and is used in amounts of up to about 10 percent. Zinc, rare earths, thorium, zirconium and manganese are also used as alloying elements as shown in Table A.8. Although magnesium alloys are more expensive and weaker than aluminum alloys (Fig. 2.8) their low density gives a cost/volume ratio that is competitive with aluminum alloys. Most magnesium alloys also have specific tensile strengths (strength/density) that are comparable with other common structural materials. Magnesium alloys are thus used in a wide variety of applications in the aerospace and aircraft fields. Other industrial used include storage tanks, portable tools and moving parts in printing and textile equipment.

The Mg-Al-Zn alloys (AZ) are the largest group of magnesium alloys and contain up to 10 percent aluminum and not more than 1.5 percent zinc. These alloys can be precipitation hardened and are generally the strongest of the magnesium alloys, as shown in Table A.8. Magnesium alloys are very easy to machine. The casting alloys have excellent castability and, in die-castings, wall thicknesses of about 1 mm are possible. Notches and other stress raisers, however, should be reduced to a minimum in view of the relatively low impact values of some of the alloys. The modulus of elasticity of magnesium alloys is about 65 percent that of aluminum alloys and about 22 percent that of steels. Consequently it is necessary to stiffen sections made of these alloys if deflections are to remain the same as those in sections made of steels or aluminum alloys. However, based on specific stiffness (modulus of elasticity/density), magnesium alloys compare favorably with steels and aluminum alloys.

Titanium and its alloys

Titanium has a specific gravity of 4.5 and is considered as one of the light metals. Titanium alloys are also among the strongest of the light alloys, especially at elevated temperatures (540°C or 1000°F), which gives them the highest strength/weight ratio among structural alloys. Titanium alloys also have excellent corrosion resistance to atmospheric and sea environments as well as to a wide range of chemicals.

Like iron, pure titanium is allotropic and exists as close packed hexagonal (CPH) alpha-phase, at temperatures up to about 885°C (1625°F) and then transforms to body-centered cubic (BCC) beta-phase, which is stable up to the melting point (1668°C or 3034°F). This allotropic transformation is affected by the addition of alloying elements. For example, aluminum and carbon stabilize the alpha-phase while copper, chromium, iron, molybdenum and vanadium stabilize the beta-phase. The mechanical properties of titanium alloys cover a wide range of strength and ductility (Fig. 2.8) and are closely related to the allotropic phases, e.g. beta-phase is much stronger but more brittle than alpha-phase.

The composition and mechanical properties of some commercial purity titanium and its alloys are given in Table A.9. The commercial purity titaniums are used where corrosion resistance and ductility are the primary requirements. Among these applications are heat exchangers, desalination plants and reactor vessels for chemical processing. Alpha-alloys contain alloying elements like Al, Sn, Nb and Zr in amounts of up to about 10 percent. They are nonheat treatable with a good combination of strength, toughness and weldability. They are also stable in the range −250 to 527°C (−420 to 980°F). The alpha-alloys are used extensively for aircraft components, where an operating temperature of up to 477°C (890°F) is required. Examples are castings, tailpipes and compressor blades. Chemical processing equipment that is required to work at temperatures of up to 477°C (890°F) is also made from these alloys. Beta-alloys are very strong but their lack of toughness limits their use industrially. Alpha-beta alloys combine the characteristics of the alpha and beta phases, so that they are stronger than the alpha alloys but less tough and more difficult to weld. The alpha-beta alloys can be strengthened by solution treatment and ageing and are used in aircraft structural members and skins which operate at temperatures of up to 315°C (about 600°F). The Ti-6Al-4V alloy is the most widely used titanium alloy as it combines attractive properties with inherent workability. It is also used for castings that must exhibit superior strength.

2.7   HEAVY NONFERROUS METALS AND ALLOYS

Copper and its alloys

Copper and copper alloys are considered as one of the major groups of engineering materials. They are widely used because of their excellent thermal and electrical conductivities, high corrosion resistance, high ductility and strength, and color, in addition to ease of fabrication. Copper and its alloys can be classified into the following categories: (a) commercially pure coppers; (b) dilute copper alloys; (c) brasses; (d) bronzes; (e) copper nickels; and (f) nickel silvers.

Commerical purity coppers contain less than 0.7 percent impurities and are used mainly as electrical conductors when in the form of wire and as heat exchangers when in the form of tubes. The dilute copper alloys contain small amounts of various alloying elements that modify one or more of the basic properties of copper. The other major alloys and brasses (Cu + Zn), tin bronzes (Cu + Sn), phosphor bronzes (Cu + Sn + P), aluminum bronzes (Cu + Al), silicon bronzes (Cu + Si), manganese bronzes (Cu + Zn + Sn + Mn), copper nickel (Cu + Ni) and nickel silver (Cu + Ag). The alloys can be broadly divided into wrought alloys and cast alloys. In the wrought class, single-phase alloys are usually strengthened by solid solution or strain hardening, and multiphase alloys are strengthened by precipitates or second-phase dispersions. The range of strength and ductility of copper alloys is shown in Fig. 2.8 and the composition and properties of selected wrought alloys are listed in Table A.10, Appendix A. The cast alloys offer a wider range of structures and are more tolerant of impurities than wrought alloys, e.g. alloys containing a high percentage of lead are not suited for hot working, but can be easily cast.

The major markets for copper and its alloys are:

1. Building construction applications including electrical wire and hardware, plumbing and heating and air-conditioning systems.
2. Electrical and electronic products.
3. Industrial machinery and equipment including valves and fittings, heat exchangers for chemical and marine applications and parts for vehicles and railroad applications.
4. Consumer and general products including electrical appliances, fasteners, coinage and jewelry.

Zinc and its alloys

Zinc is a relatively inexpensive metal and is the fourth most widely used metallic material after iron, aluminum and copper. Almost half of the zinc produced in the world is used to assist the control of corrosion of iron and steel by galvanizing. The major product forms of zinc-coated products are sheet, wire and tube in addition to post-fabrication coated products like fasteners and structures. About 13 percent is used as an alloying element in brasses and die-casting alloys represent more than 25 percent of the total consumption. Zinc is also used in dry batteries, photoengraving and zinc-base chemicals.

The major die-casting alloys are known in the US as Zamak alloys 3, 5 and 7 and are based on the Zn–4 percent Al composition. In Britain, alloys A and B are similar in composition and properties to alloys 3 and 5. The composition and properties of alloys 3 and 5 are given in Table A.11. Alloy 3 has better ductility and retains its impact strength better at elevated temperatures, while alloy 5 is harder and stronger and has better castability. Alloy 7 is a high-purity version of alloy 3 and thus has similar mechanical properties but it exhibits better casting and finishing characteristics.

Zinc die-casting alloys usually shrink slightly after casting. When close dimensional limits are necessary, a stabilizing anneal in dry air at about 100°C (212°F) for six hours is given before machining. Alloys 3, 5 and 7 can be strengthened by ageing and at room temperature, the impact strength of zinc alloy die-castings is much higher than that of either aluminum or magnesium die-castings or iron sand-castings. The above zinc die-casting alloys are not intended for high-temperature service and will suffer significant creep under load at temperatures of about 75°C (165°F). Higher service temperatures can be achieved by increasing the aluminum content or by adding small amounts of chromium and titanium. Some wrought zinc alloys exhibit superplastic behavior, i.e. elongations higher than 500 percent. An example is Zn–22 percent Al which can be formed into complex shapes in one step (see Section 3.6).

Nickel and its alloys

Nickel is an important metal that is used as an alloying element in stainless and alloy steels, copper-base alloys as well as other families of nonferrous alloys. Commercially, pure nickel is resistant to caustics, high-temperature halogens, salts and other oxidizing halides as well as foods. Nickel is also used as the base metal for a number of alloys which can be divided into:

1. Binary systems such as Ni-Cu, Ni-Si and Ni-Mo.
2. Ternary systems such as Ni-Cr-Fe and Ni-Cr-Mo.
3. Complex systems such as Ni-Cr-Fe-Mo-Cu.

Monel alloys are based on the binary Ni-Cu system and have an average composition of 67 percent Ni, 28 percent Cu and 5 percent Fe. These alloys resist corrosion from most chemicals and are competitors to stainless steels in the chemical, marine, power equipment, food service, petroleum and paper industries. Inconel alloys are based on the ternary Ni-Cr-Fe system and are used primarily for their oxidation resistance and strength at elevated temperatures. Additions of molybdenum and copper to these alloys make these alloys highly resistant to pitting, intergranular corrosion and chloride-ion stress-corrosion cracking in addition to retaining their strength and oxidation resistance at elevated temperatures.

The nickel-base superalloys, which have been developed for service at temperatures in the range 1075 to 1225°C (1967 to 2237°F), are basically Ni-Cr alloys with additions of Co, Mo, Ti or Al and depend mostly on precipitation hardening for strengthening. The composition and properties of representative superalloys are given in Tables A.12 and A.13. Several nickel-chromium alloys are also used in cryogenic applications as they have outstanding strength, toughness and ductility at low temperatures.

Cobalt and its alloys

Cobalt is used as an alloying element in many important steels and as a base metal for a number of important superalloys. These superalloys contain 20–30 percent chromium for oxidation resistance and many of them contain tungsten, carbon, nickel, aluminum and minor additions of manganese, silicon, titanium and molybdenum. Most of these alloying elements form hard carbides in a fairly hard matrix and give excellent abrasion and wear resistance. The composition and properties of representative cobalt-base superalloys are given in Tables A.12 and A.13. Cobalt is also a significant element in low-expansion alloys, e.g. 53–54.5 percent Co, 36.5–37 percent Fe and 9–10 percent Cr alloy, which has an exceedingly low coefficient of expansion, at times negative, over the range from 0 to 100°C (32 to 212°F). Cobalt–chromium-base alloys are used in dental and surgical applications because they are not attacked by body fluids.

Refractory metals and alloys

Refractory metals are a group of metals with melting points above 1630°C (2966°F). The most important metals in this group are tungsten, molybdenum, tantalum and niobium. At temperatures above 1075°C (1967°F) these metals and their alloys generally have higher strengths than other high-temperature alloys. However, refractory metals are generally susceptible to oxidation at elevated temperatures. Consequently they must be protected with high-temperature coatings. Also these refractory metals and their alloys involve difficult problems in fabrication, especially when the product is in the sheet form. The most advanced applications of refractory metals and their alloys are in the manufacture of rocket motors and turbojet engines. Some relevant properties of representative refractory metals and alloys are given in Table A.14 in Appendix A.

Low-melting materials

Tin, lead, bismuth, antimony, cadmium and indium and their alloys can be grouped together as low-melting materials. The common characteristics of these materials, besides their low melting points, are low strength and hardness, high ductility and relatively high corrosion resistance. These characteristics give most of these materials wide use as bearing materials and soldering alloys. About 40 percent of total world production of tin is used in coating steel sheets (tinplate) to provide corrosion resistance, non-toxicity and white color. The second largest use of tin is in solders where it is alloyed with lead. Tin is also an important alloying element in bronzes. In the metallic form, lead finds an important application in building construction as pipe, sheet and solder. Lead is also used in important alloys like antimonial lead for storage batteries, bearing alloys, foil and collapsible tubes. Lead is also used as shielding against X-rays and gamma radiation. The major uses for bismuth, antimony, cadmium and indium are as alloying elements, especially to the low-melting point alloys.

The precious metals and alloys

The precious metals include gold, silver and the platinum group metals (platinum, palladium, rhodium, ruthenium, iridium and osmium). The precious metals, except for silver, are the most expensive known metals. In addition to their use in jewelry and coins, precious metals and their alloys are used extensively as catalysts. For example Pt-Rh alloy is used in automobile pollution-control equipment and in the production of nitric acid from ammonia and air. Precious metals are also used in the electrical and electronic industries. For example, silver, gold, rhodium and palladium are used for wiring in solid-state electronics while electrical contacts are frequently made from gold or silver alloys. Electrochemical applications of precious metals include the use of platinum and Pt-Pd alloys as insoluble anodes for electrolytic protection and for electroplating. Pt-Rh and Au-Pt alloys are used for the spinnerets employed in manufacturing rayon and glass fibers. Pt-Rh alloys are also used in thermocouples in combination with platinum. In dentistry, gold and platinum castings are used as inlays, crowns and bridges and silver is an important constituent of several dental amalgams. Certain compounds of precious metals, especially those of silver, platinum and gold, have applications as pharmaceutical agents. For example silver nitrate is an anti-infective and some gold compounds are used in the treatment of rheumatoid arthritis.

2.8 DESIGN AND SELECTION CONSIDERATIONS FOR METALLIC MATERIALS

Discussions in this chapter have illustrated the very wide range of properties and characteristics of metallic materials. One of the major issues that should be considered when designing structural components is the deflection under the expected service loads. This deflection is not only a function of the applied forces and geometry of the component, but also of the stiffness of the material used in making it. As the stiffness of the material is difficult to change, either the shape or the material has to be changed in order to achieve a large change in the stiffness of a component. The following example illustrates how the stiffness can influence the selection of materials for a given application.

It is required to select a structural material for the manufacture of the tie rods of a suspension bridge. A representative rod is 10 m long and should carry a tensile load of 50 kN without yielding. The maximum extension should not exceed 18 mm. Which one of the materials listed in Table 2.7 will give the lightest tie rod?

Table 2.7 Candidate materials for suspension bridge tie rods

Material	Y.S. MPa	E GPa	Specific gravity	Area based on yield strength (mm²)	Area based on deflection (mm²)	Mass (kg)
ASTM A675 grade 60	205	212	7.8	244	131	19
ASTM A572 grade 50	345	211	7.8	145	131	11.3
ASTM A717 grade 70	485	211	7.8	103	131	10.2
Maraging steel grade 200	1400	211	7.8	36	131	10.2
Al 5052 H38	259	70.8	2.7	193	392	10.6
Cartridge brass 70% hard temper	441	100.6	8.0	113	276	22.1

For the present case, calculations of the area will be carried out twice:

1. Area based on yield strength $= \dfrac{\text{load}}{YS}$

2. Area based on deflection $= \dfrac{\text{load} \star \text{length}}{E \star \text{deflection}}$

The larger of the two areas will be taken as the design area and will be used to calculate the mass.

The results show that steel A717 grade 70 and maraging steel grade 200 give the least mass. As the former steel is more ductile and less expensive it will be selected.

The load-carrying capacity of a component can be related to the yield strength, fatigue strength, or creep strength of the material depending on loading and service conditions. These properties are structure sensitive and can be considerably changed by changing the chemical composition of the alloy, method and conditions of manufacture as well as heat treatment. It should be noted, however, that increasing the strength of most metallic materials causes their ductility and toughness to decrease, which could adversely affect the performance of the component in service. This subject will be discussed in more detail in Chapters 8 and 11.

Electrical and thermal conductivities of metallic materials can greatly affect the design and impose severe limitations on material selection in many applications. For example, electrical quality copper and aluminum may represent the only possible materials for the manufacture of a component where electrical conductivity is a primary requirement. Corrosion resistance and specific gravity requirements could impose similar limitations. However, judicious design and careful selection can, in many cases, overcome these limitations.

The design and material selection for a given component are also influenced by manufacturing considerations. The majority of metallic components have either cast or wrought microstructures. Wrought microstructures are usually stronger and more ductile than cast microstructures and this is one of the reasons why about 80 percent of metallic materials are produced in the wrought form. Wrought alloys are available in many shapes and size tolerances. Tolerances are relatively wider for hot-worked products and this could cause difficulties in further processing, as in the case of automatic machining. The poorer surface quality of hot-worked products can also cause difficulties, as in the case of deep drawing of sheets and drawing of wires. Cold-worked products are usually supplied with narrower tolerances, but the presence of residual stresses in them can cause unpredictable size changes during machining. Under these conditions stress relief treatments may be necessary. In some cases, shape requirements can severely limit the range of materials available for selection, as in the case of thin foil and fine wire.

Weldability of an alloy is a function of its composition (Section 3.7) and should be considered when selecting materials for welded structures. Machinability can also be an important parameter in cases where large amounts of material have to be removed by machining (Section 3.8). In some cases machinability can be improved by heat treatment, as in the case of normalizing low-carbon steels. Machinability can also be improved by the addition of alloying elements like sulfur or lead, but this is usually accomplished at the expense of strength and ductility.

Economics represent another important parameter in design and material selection. The design should be based on the material that will perform the required function at the least overall cost. The material selected may be the most cost effective because it has a low initial cost and yet provides good service life or because it provides the possibility of making a component of low operating and maintenance costs. Among metallic materials, plain carbon steels and cast irons are the least expensive and the most widely used materials. Other metals and alloys are only used because they have special characteristics that are not matched by these materials.

2.9   REVIEW QUESTIONS AND PROBLEMS

2.1   What are the main advantages and limitations of plain low-carbon steels? How are these limitations overcome in practice?

2.2 What are the advantages and limitations of nodular cast irons in comparison with plain carbon steels and gray cast irons?

2.3 How does hardness differ from hardenability?

2.4 Which of the strengthening methods increase both the strength and ductility of the material?

2.5 In Fig. 2.1, metallic materials are shown to constitute a large proportion of the motor car. List some of the important metallic components and suggest the alloys that are best suited for their manufacture.

2.6 Compare the use of aluminum, magnesium and titanium alloys for a small aircraft wing structure.

2.7 What are the main material requirements for the following components: motor car exhaust manifold; coil for electrical resistance heater; kitchen knife blade; and railway line? Suggest suitable materials for these components.

2.8 What are the advantages of using gray cast iron instead of steel in making machine tool frames?

2.9 Suggest the strengthening mechanism used for an aluminum alloy that will serve at 150°C (302°F).

2.10 Use Fig. 2.5 to compare oil, water and brine as quenching media. What are the conditions that affect the selection of a quenching medium for a given part?

2.11 What is meant by austempering? What are its advantages and limitations? Compare austempering with martempering.

2.12 Compare the use of pure aluminum and copper for overhead electrical cables.

2.13 What are the distinguishing features of superalloys?

BIBLIOGRAPHY AND FURTHER READING

Ashby, M. F. and Jones, D. R. H., *Engineering Materials 1, An Introduction to their Properties and Applications*, Pergamon Press, London, 1980.

Ashby, M. F. and Jones, D. R. H., *Engineering Materials 2, An Introduction to Microstructures, Processing and Design*, Pergamon Press, London, 1988.

Boyer, H. E. and Gall, T. L., *Metals Handbook Desk Edition*, ASM, Ohio, 1985.

Farag, M. M., *Materials and Process Selection in Engineering*, Applied Science Publishers, London, 1979.

Flinn, R. A. and Trojan, P. K., *Engineering Materials and Their Applications*, 2nd ed., Houghton Mifflin, Boston, 1981.

Jastrzebski, Z. D., *The Nature and Properties of Engineering Materials*, 3rd ed., John Wiley and Sons, New York, 1987.

Keyser, C. A., *Materials Science in Engineering*, Charles Merrill, Columbus, 1986.

Lyman, T. Ed., 'Properties and selection of metals' in *Metals Handbook 8th Ed.*, vol. 1, ASM, Ohio, 1961.

Materials Databook, Institute of Metals, London, 1988.

Materials Data Source Book, Institute of Metals, London, 1987.

Ray, M. S., *The Technology and Applications of Engineering Materials*, Prentice Hall, New Jersey, 1987.

Reed-Hill, R. E., *Physical Metallurgy Principles*, Van Nostrand, Princeton, New Jersey, 1964.

Sims, T. *et al. Superalloys*, 2nd ed., John Wiley & Sons, New York, 1987.

Stainless Steels, Institute of Metals, London, 1985.

Thorton, P. A. and Colangelo, V. J., *Fundamentals of Engineering Materials*, Prentice Hall, London, 1985.

Van Vlack, L. H., *Elements of Materials Science and Engineering*, 5th ed., Addison Wesley, Reading, Mass., 1985.

Chapter 3

Processing of Metallic Materials

3.1   CLASSIFICATION OF MANUFACTURING PROCESSES

Manufacturing can be defined as the act of transforming materials into usable and saleable end products. This means that the products must function satisfactorily and provide value for money. Cost, rate of production, availability and quality of product are important criteria which should be considered when selecting a manufacturing process. With the increasing sophistication of many products and the increasing range of available materials, new manufacturing technologies have been developed to produce more sound and accurate components economically. The different manufacturing processes that are normally used in processing metallic materials can generally be related as shown in Fig. 3.1.

Most metallic materials are found in nature in the form of ores which have to be reduced or refined and then cast as ingots of convenient size and shape for further processing. Casting processes involve pouring the liquid metal, or alloy, into a mold cavity and allowing it to solidify. Powdered materials can be shaped by pressing in suitably shaped dies and then sintering, heating, to achieve the required properties. An important advantage of casting and powder metallurgy processes is that parts of complex shapes can be obtained in one step.

Bulk forming processes are generally used to change the shape of metallic materials by plastic deformation. The deformation can be carried out at relatively high temperatures, as in the case of hot-working processes, or at relatively low temperatures, as in the case of cold-working processes. The basic bulk deformation processes are forging, rolling, extrusion, swaging, and drawing of rod, wire and tube. These processes are called primary working when applied to ingots to break down their cast structure into wrought structure and to change their shape to slabs, plates or billets. Secondary working involves further processing of the products from primary working into final or semifinal shapes.

Sheet-metal working processes are normally carried out at room temperature and usually involve the change of sheet form without greatly affecting its thickness. The basic sheet-metal working processes include shearing, bending, stretch forming, bulging, deep drawing, spinning and press forming. Other sheet-metal forming operations have been developed for the manufacture of certain components and special materials.

Heat treatment is normally carried out to control the structure and properties of the material. By proper control of temperature and cooling rate, the material can be softened to permit further processing or hardened to increase its mechanical strength. Heat treatment can also be used to remove internal stresses, to control grain size, or to produce

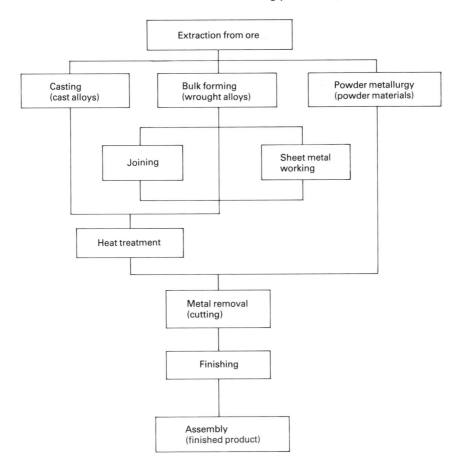

Figure 3.1 Types and usual sequence of manufacturing processes that are normally used in processing metallic materials.

a hard surface on a ductile interior. With the proper heat treatment, a less expensive material could replace a more expensive material or less costly processing could be employed.

Material removal or cutting processes are normally used to remove the unwanted material in the form of chips by using cutting tools which are mounted in machine tools. The traditional basic machine tools are the lathes, boring machines, shapers and planers, milling machines, drill presses, saws, broaches and grinding machines. The productivity in cutting processes can be improved by using machining centers, which are single machines that can perform the functions of several basic machine tools. When cutting very hard metals or when machining intricate shapes and delicate parts, nontraditional or chipless processes can be used. These cutting methods include ultrasonic, electrical discharge, electrochemical, chemical milling, abrasive jet, electro-arc, plasma-arc, electron beam and laser cutting.

In many cases, products are manufactured as separate units and then assembled either by fastening, as in the case of temporary or semipermanent joints, or by joining, as in the case of permanent joints. Exampes of fastening methods include screws, bolts and rivets, while joining methods include welding, brazing, soldering and adhesive bonding. Assembly by press and shrink fitting is also used in some applications.

Finishing processes are normally used to control the quality of the surface and to make it ready for service. Cleaning, deburring and polishing, anodizing, tinning, galvanizing, plating and painting are among the frequently used finishing processes. In addition to controlling the appearance of the surface, many of the finishing processes provide some protection against corrosion.

A brief description of the various manufacturing processes that are commonly used in processing metallic materials will be given in the following sections of this chapter. The processing of polymeric materials will be discussed in Chapter 4, the processing of ceramic materials will be discussed in Chapter 5 and the processing of composite materials will be discussed in Chapter 6. The effect of manufacturing processes on design will be discussed in Chapter 12 and the economics of manufacturing will be discussed in Chapter 18.

3.2   DIMENSIONAL ACCURACY AND SURFACE FINISH

Interchangeability of manufactured components is an essential requirement in mass production of both consumer and producer goods. This is because it allows ease of assembly and repair of manufactured systems and also makes it possible to standardize products and methods of manufacture. For components to be interchangeable, they should be fabricated to accurate and reproducible dimensions. The accuracy of industrially

Table 3.1 Approximate values of surface roughness and tolerance that are normally obtained with different manufacturing processes

Process	Typical tolerance ($\pm$)		Typical surface roughness (R_a)	
	mm	in $\times 10^{-3}$	μm	μin
Sand casting	0.5–2.0	20–80	12.5–25	500–1000
Investment casting	0.2–0.8	8–30	1.6–3.2	63–125
Die-casting	0.1–0.5	4–20	0.4–1.6	16–63
Powder metallurgy	0.2–0.4	8–16	0.8–3.2	32–125
Forging	0.2–1.0	8–40	3.2–12.5	125–500
Hot rolling	0.2–0.8	8–30	6.3–25	250–1000
Hot extrusion	0.2–0.8	8–30	6.3–25	250–1000
Cold rolling	0.05–0.2	2–8	0.4–1.6	16–63
Cold drawing	0.05–0.2	2–8	0.4–1.6	16–63
Cold extrusion	0.05–0.2	2–8	0.8–3.2	32–125
Flame cutting	1.0–5.0	40–200	12.5–25	500–1000
Sawing	0.4–0.8	15–30	3.2–25	125–1000
Turning and boring	0.025–0.05	1–2	0.4–6.3	16–250
Drilling	0.05–0.25	2–10	1.6–6.3	63–250
Shaping and planing	0.025–0.125	1–5	1.6–12.5	63–500
Milling	0.01–0.02	0.5–1	0.8–6.3	32–250
Chemical machining	0.02–0.10	0.8–4	1.6–6.3	63–250
EDM and ECM	0.02–0.10	0.8–4	1.6–6.3	63–250
Reaming	0.01–0.05	0.4–2	0.8–3.2	32–125
Broaching	0.01–0.05	0.4–2	0.8–3.2	32–125
Grinding	0.01–0.02	0.4–0.8	0.1–1.6	4–63
Honing	0.005–0.01	0.2–0.4	0.1–0.8	4–32
Polishing	0.005–0.01	0.2–0.4	0.1–0.4	4–16
Lapping and super-finishing	0.004–0.01	0.16–0.4	0.05–0.4	2–16

available manufacturing processes varies widely and although some of them are capable of producing parts to high levels of accuracy, their ability to duplicate a specific dimension is still limited. This is caused by changes in temperature during manufacture, wear of tools, deflections and vibrations in workpieces and machines, as well as human errors. Deviations from exact sizes are called tolerances and may be defined as the permissible variations in dimensions. Another important factor which affects the manufacture of parts is surface finish. Generally, manufacturing processes that yield closer tolerances also result in better surface finish, as shown in Table 3.1. Experience shows that the manufacturing cost increases exponentially with tighter tolerances, and better surface finish. It is, therefore, important that no better surface finish or closer tolerance than is needed be specified for a given surface.

Manufacturing of a component is normally based on production drawings which specify the material, heat treatment, dimensional tolerances, surface finish, etc. This information is easier to communicate by reference to published standards. The American National Standards Institute, Inc. (ANSI), The British Standards Institution (BSI), Deutscher Normenausschuss (DNA) and International Standards Organization (ISO) are among the bodies which issue widely used standards.

Tolerances, fits and allowances

Tolerances may be specified according to three different systems: bilateral, unilateral and limiting dimensions. Bilateral tolerance is specified as a plus or minus deviation from the nominal size, e.g. 76.2 ± 0.1 mm (3.000 ± 0.004 in). The dimension of the part could vary between 76.3 and 76.1 mm (3.004 and 2.996 in). The total tolerance in this case is 0.2 mm (0.008 in). In unilateral tolerance, the deviation is in one direction from the nominal size, e.g.

$$76.2^{+0.2}_{-0.0} \text{ mm} \qquad (3.000^{+0.008}_{-0.000} \text{ in})$$

In this case, the total tolerance is also 0.2 mm (0.008 in), but the dimension of the part could vary between 76.4 and 76.2 mm (3.008 and 3.000 in). These numbers show that in order to obtain the same maximum and minimum dimensions with the two systems, different nominal sizes must be used. Bilateral tolerances are usually specified for nonmating parts and unilateral ones for mating parts. This is because the unilateral system permits changing the tolerance while still retaining the same allowance or type of fit. With the bilateral system, this is not possible without also changing the nominal size of one or both of the mating parts.

The maximum and minimum dimensions that result from the specified tolerance system are called limiting dimensions. One of the basic reasons for specifying limiting dimensions for a given part is to be assured that the different components of an assembly will fit properly and function well in service. For example, a shaft must be free to move within its bearings which requires some clearance, yet the clearance must not be too great, in order to avoid vibrations. The clearance, allowance, in this case is defined as the intentional difference in the sizes of the mating parts. When the mating parts are not permitted to move relative to each other, negative clearance or interference may be specified. ANSI has established eight classes of fit as a guide to specifying tolerances for typical applications, as shown in Table 3.2. In the ANSI system, the hole is always considered basic, because the majority of holes are produced using standard-size drills

Table 3.2 ANSI recommended allowances and tolerances (d in inches)

Class of fit	Description	Clearance (+) Interference (−)	Hole tolerance	Shaft tolerance
1	Loose fit. Accuracy not essential	$+0.0025\sqrt[3]{d^2}$	$+0.0025\sqrt[3]{d}$	$-0.0025\sqrt[3]{d}$
2	Free fit. For running and sliding fits where speeds are over 600 rev/min and pressures are 4.1 MPa (600 lb/in^2) or higher	$+0.0014\sqrt[3]{d^2}$	$+0.0013\sqrt[3]{d}$	$-0.0013\sqrt[3]{d}$
3	Medium fit. For running and sliding fits under 600 rev/min and pressures less than 4.1 MPa (600 lb/in^2)	$+0.0009\sqrt[3]{d^2}$	$+0.0008\sqrt[3]{d}$	$-0.0008\sqrt[3]{d}$
4	Snug fit. Zero allowance where no movement under load is intended and no shaking is acceptable. The tightest fit that can be assembled by hand	$+0$	$+0.0006\sqrt[3]{d}$	$-0.0004\sqrt[3]{d}$
5	Wringing fit. Zero to negative allowance. Assemblies are selective and not interchangeable	-0	$-0.0006\sqrt[3]{d}$	$+0.0004\sqrt[3]{d}$
6	Tight fit. Slight negative allowance. Interference fit for parts that should not separate in service and are not to be disassembled. The fit can only transmit light loads	$-0.00025\,d$	$+0.0006\sqrt[3]{d}$	$+0.0006\sqrt[3]{d}$
7	Medium force fit. Interference fit and requires considerable pressure to assemble. Assembly can be facilitated by heating external member or cooling internal member. The tightest fit that is allowed on cast parts	$-0.0005\,d$	$+0.0006\sqrt[3]{d}$	$+0.0006\sqrt[3]{d}$
8	Heavy force and shrink fit. Considerable negative allowance. Use for permanent shrink fits on steel members	$-0.001\,d$	$+0.0006\sqrt[3]{d}$	$+0.0006\sqrt[3]{d}$

while the shaft can be produced equally readily to any size. The following example illustrates the use of the table. For a basic hole size of 3 in (76.2 mm) and a class 2 fit, the dimensions would be:

Allowance $= +0.0014\sqrt[3]{3^2} = 0.0029$ in (0.0739 mm)

Tolerance $= +0.0013\sqrt[3]{3} = 0.0019$ in (0.0476 mm)

Hole dimensions: maximum $= 3.0000 + 0.0019$

$= 3.0019$ in (76.248 mm)

minimum $= 3.0000$ in (76.200 mm)

Shaft dimensions: maximum $= 3.0000 - 0.0029$

$= 2.9971$ in (76.126 mm)

minimum $= 2.9971 - 0.0019$

$= 2.9952$ in (76.078 mm)

The BS 4500 ISO system of limits and fits is a comprehensive system which is also widely used. In this system, each part has a basic size and deviations which define the limiting sizes. The amount of deviation determines the class of fit, while the amount of tolerance determines the quality of the fit. The system defines three classes of fit:

1. Clearance fits, which allow sliding or rotation.

2. Transition fits, where the assembly may have either clearance or interference, are specified for accurate location.
3. Interference fits, which assure negative clearance (interference), are specified for rigidity and alignment.

The system caters for diameters ranging from 3 to 3150 mm which are grouped into bands or size ranges. Either a shaft-basis system or a hole-basis system may be used, but the latter system is more widely adopted. In the basic hole, the shaft tolerances occupy one of 28 different positions, each designated by a lowercase letter (or two letters). In practice, values from c to t are adequate for most classes of manufacture. Shafts from c to h have minus deviation giving clearance fits, with c shafts giving extra slack fits and h shafts giving close slide fits. Shafts from k to n have plus deviation giving transition fits, and shafts from p to t also have plus deviation giving interference fits. Holes are designated by capital letters and for most manufacturing processes the H hole only is used.

The ISO system provides 18 grades of tolerance designated IT 01 to IT 18. The smaller the IT number (international tolerance number) the smaller is the actual tolerance. The relationship between the tolerance grade and the type and quality of the manufacturing process is shown in Table 3.3.

Generally, grades 11 to 6 are adequate for most manufacturing work. Figure 3.2 combines the deviations and tolerances for holes and shafts for sliding fits. As holes are more difficult to machine to high accuracy, they are usually allocated a tolerance one grade less, coarser, than the shaft, e.g. H9/f8. The BS 4500 ISO standards provide tables which give the limits of tolerance for the different diameter ranges covered by the system. A sample is shown in Table 3.4. As an example, the limiting dimensions of a shaft and bearing assembly with a nominal diameter of 25 mm and a normal running fit, can be determined as follows:

From Fig. 3.2, an f shaft may be used for required fit and high quality turning is selected for its manufacture, i.e. an IT number of 7. According to earlier discussion, the hole of the

Table 3.3 Tolerance grades in the ISO system

IT number	Manufacturing process
16	Sand casting and flame cutting
15	Stamping
14	Die-casting or molding
13	Press work and tube rolling
12	Light press work and tube drawing
11	Drilling, rough turning and boring
10	Milling, slotting, planing and shaping
9	Horizontal and vertical boring
8	Turning, different types of machines
7	High quality turning, broaching and honing
6	Grinding and fine honing
5	Machine lapping and fine grinding
4	Gagemaking and fine lapping
3	Good quality gages
2	High quality gages
1	Workshop standards
0	Inspection standards
01	Work of the highest quality

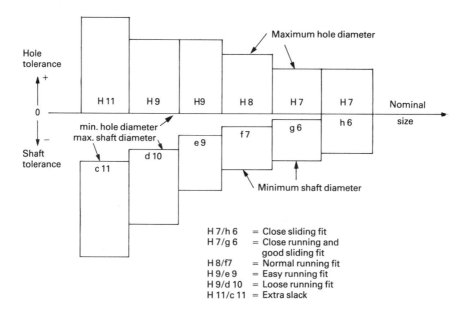

H 7/h 6   = Close sliding fit
H 7/g 6   = Close running and
              good sliding fit
H 8/f7   = Normal running fit
H 9/e 9   = Easy running fit
H 9/d 10   = Loose running fit
H 11/c 11   = Extra slack

Figure 3.2 Sliding fit designations in the ISO system.

Table 3.4 Limits of tolerances for selected holes and shafts (for complete information refer to BS 4500: – ISO limits and fits)

Nominal sizes (mm)		Hole tolerances (+ve) (μm)								Shaft tolerances (−ve) (μm)									
		H7		H8		H9		H11		g6		f7		e9		d10		c11	
Over	Up to and including	U	L	U	L	U	L	U	L	U	L	U	L	U	L	U	L	U	L
–	3	10	0	14	0	25	0	60	0	8	2	16	6	39	14	60	20	120	60
3	6	12	0	18	0	30	0	75	0	12	4	22	10	50	20	75	30	145	70
6	10	15	0	22	0	36	0	90	0	14	5	28	13	61	25	98	40	170	80
10	18	18	0	27	0	43	0	110	0	17	6	34	16	75	32	120	50	205	95
18	30	21	0	33	0	52	0	130	0	20	7	41	20	92	40	149	65	240	110
30	40	25	0	39	0	62	0	160	0	25	9	50	25	112	50	180	80	280	120
40	50	25	0	39	0	62	0	160	0	25	9	50	25	112	50	180	80	290	130
50	65	30	0	46	0	74	0	190	0	29	10	60	30	134	60	220	100	330	140

U = Upper deviation, L = Lower deviation

bearing should be given a tolerance one grade coarser than the shaft, i.e. IT 8. From Table 3.4, the limiting dimensions of 25 H8/f7 is:

Hole diameter   maximum dimension = 25.000 + 0.033 = 25.033 mm
                minimum dimension = 25.000 mm
Shaft diameter maximum dimension = 25.000 − 0.020 = 24.980 mm
                minimum dimension = 25.000 − 0.041 = 24.959 mm

Surface roughness and integrity

Surfaces of manufactured components can never be absolutely smooth or exact. At the same time, the quality of the surface can greatly influence the behavior of components in service, which makes it essential to specify it along with linear and geometric dimensions. The American National Standards Institute (ANSI) has developed standards to define the different surface characteristics, such as profile, roughness, waviness, flaws and lay, as illustrated in Fig. 3.3. Profile is defined as the contour of any section through a surface. Roughness refers to fine surface irregularities which result from manufacturing processes and is expressed in terms of its height, width and length, which is the distance along the measuring direction. Waviness refers to surface irregularities which are of greater spacing than roughness and is measured in terms of the width and height of the wave. Flaws are defects which occur at random locations in the surface. Scratches, cracks, checks, pores, pits, cavities, laps, seams, inclusions and ridges are considered as flaws. Lay is the direction of the predominant surface pattern, which is usually discernible by the eye. The lay, directionality, is characteristic of the finishing process. It is random and uniform after lapping or shot peening, and random but not uniform on surfaces produced by casting or forging. Directional finishes may be periodic, as in the case of turning or grinding, or irregular, as in the case of extrusion or rolling.

Roughness is most commonly specified and is expressed in units of micrometers (μm), nanometers (nm), or microinches (μin). The arithmetic average deviation of the ordinates of profile height of the surface from its center line, R_a, is taken as the standard measure of surface roughness by ANSI in the US and several other countries. According to Fig. 3.3, R_a (also called center-line average CLA, or arithmetic average AA) can be approximately calculated as:

$$R_a = \frac{Y_1 + Y_2 + Y_3 + \ldots + Y_n}{n} \tag{3.1}$$

where R_a = arithmetic average deviation from the centerline
Y = ordinate of the curve of the profile
n = number of readings

The root mean square (RMS) average is sometimes used to describe surface roughness and is calculated as:

$$\text{RMS} = \left[\frac{Y_1^2 + Y_2^2 + Y_3^2 + \ldots + Y_n^2}{n} \right]^{1/2} \tag{3.2}$$

RMS values are about 11 percent larger than R_a values for a sinusoidal profile.

Waviness height and width may also be specified as in Fig. 3.3. If no width is given, it is usually implied that the specified waviness height must be observed over the full length of the surface.

The different characteristics of the surface can be represented in engineering drawings by a check mark with a horizontal extension as shown in Fig. 3.3. Where the symbol is used with a dimension, it affects all surfaces defined by the dimension. Areas of transition, such as chamfers and fillets, should conform with the roughest adjacent finished area unless otherwise indicated. Surface-roughness symbols, unless otherwise specified, apply to the finished surface.

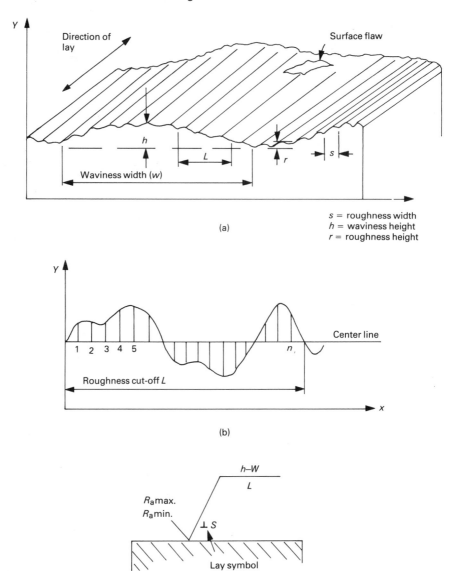

Figure 3.3 Surface finish terminology and symbols. (a) Typical surface highly magnified. (b) Surface roughness measurement. (c) Representation of surface characteristics in drawings. (d) Lay symbols for representation in drawings.

There is a close relationship between surface roughness and tolerances. In general, tolerance must be greater than the distance between the highest peak and the deepest trough of the surface roughness and waviness. This value is usually about 10 times the value of R_a. Typical surface roughness values that are normally obtained with different manufacturing processes are given in Table 3.1.

3.3   MANUFACTURING BY CASTING

In casting, the material is melted, superheated to the proper temperature, poured into a mold and then cooled to solidify in the required form. Alloying elements can be added to molten material to achieve the required chemical composition and degassing agents are added to ensure soundness of the casting. When it is important to refine the grain size or to control the structure of the cast component, nucleating agents are added to the molten material. After removal from the mold, finishing operations and heat treatment may need to be performed before the casting is ready to leave the foundry. Casting is a versatile manufacturing process, since it can be applied to a wide variety of alloys, shapes and product sizes. Tiny parts measuring a few millimeters and weighing a fraction of a gram as well as large parts measuring 10 m or more (30 ft or more) and weighing several tons can be produced by casting. In many cases, casting is the only economic way in which bodies of complex internal shape can be made, and it is the most economical way of forming the more complex high-temperature alloys. Several casting processes are industrially available (see Fig. 3.4) and the differences between them can be expressed in terms of the possible product size, surface finish, tolerance on dimensions, minimum section thickness, production rate, mold or pattern cost and scrap loss. Table 3.5 compares these parameters for some of the commercially available casting processes.

Sand casting, in which a mixture of sand and bonding agents is used to make the mold, is the cheapest and most versatile process as it can be employed for the widest range of alloys and product shapes and sizes, but it is the least accurate, as shown in Table 3.5. As a new mold must be made for each casting, sand casting is a labor-intensive and a relatively slow process.

Shell molding, in which a resin-coated sand is formed into a thin-walled mold by contact with a heated pattern, gives higher accuracy and better surface finish than sand casting. The production rate is also higher than that in sand casting. In addition, the consistency between castings is superior to that obtainable by ordinary sand casting. The metal pattern and the mold materials which have to be used in shell molding are generally more expensive than those used for sand casting. However, the higher productivity and accuracy of shell molding make the process competitive for even moderate quantities, as shown in Table 3.5.

Permanent-mold casting processes have the advantage over sand casting in that the mold can be used for the production of more than one casting as it is usually made of metal, which restricts these processes to the casting of lower melting point metals and alloys. However, when the mold is made of graphite, iron and steel castings can be produced by permanent mold casting. Advantages of permanent-mold casting processes include close tolerances, good surface finish and higher rates of production. In addition, some of these processes can be automated and used for mass production. On the other hand the mold life decreases as the melting point of the alloy being cast increases and mold complexity is often restricted. Permanent-mold casting processes include the low-pressure die-casting

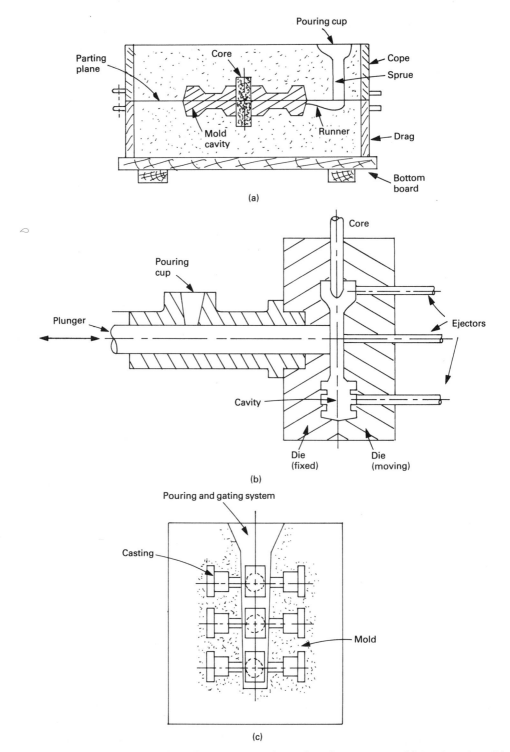

Figure 3.4 Diagrammatic sketches of some commonly used casting processes. (a) Sand casting. (b) Cold chamber pressure die-casting. (c) Investment casting. (d) Centrifugal casting.

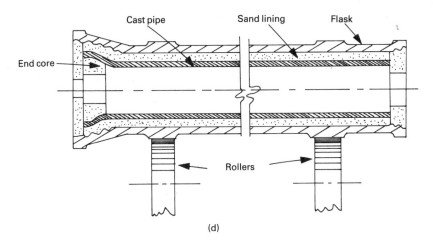

Figure 3.4 – contd.

process, where the molten metal is introduced in graphite dies under moderate pressure but allowed to solidify under atmospheric pressure; and the pressure die-casting process, where the molten metal is forced into metallic dies under pressure and held under pressure during solidification. Pressure die-casting gives higher dimensional accuracy and an improved surface finish which is characteristic of metallic dies. The high price of the metal mold in die-casting makes it economical only when large numbers of the product are to be cast. Other characteristics of die-casting are given in Table 3.5.

Plaster-mold casting processes use gypsum plaster with small additions of other minerals to prevent cracking, to control dimensional changes during baking and to reduce setting time. Investment casting and the Shaw process utilize plaster molds. Investment casting, also known as the lost wax process, is used to produce many castings in one mold and gives the designer an almost unlimited freedom in choosing shapes and materials in sizes varying from a few grams to several kilograms. The ability to hold close tolerances enables final machining processes to be kept to a minimum or eliminated. This latter property enables alloys that are difficult to machine to be made into precise shapes using investment casting. The Shaw process is similar to investment casting but the composition of the mold material provides permeability and good collapsibility to accommodate the shrinkage of the solidifying metal.

Centrifugal casting utilizes the centrifugal force, resulting from the rotation of the mold, to press the solidifying metal against the mold walls. This results in a dense structure, with the lighter impurities tending to be at the inner surface. These impurities can then be removed by light machining. Although round shapes are most common, other symmetrical shapes can be cast. No core is required to form the inner surface of the casting. In addition to large cylinders with thin walls, centrifugal casting can be used to provide forced metal flow from a central gating system into thin intricate mold cavities.

The quality of cast products is affected by the properties of the cast alloy in addition to the process variables discussed above. Fluidity is an important property of metal to be cast and is defined as the ability of a molten alloy to fill a mold cavity completely. Another property is castability which depends on the ease with which an alloy responds to ordinary foundry practice without undue attention to gating, risering, melting and mold condition.

Table 3.5 Characteristics of different casting processes and powder metallurgy

Process	Alloy	Weight kg (lb)	Surface finish μm (μ in)	Tolerance m/m or in/in	Minimum section thickness mm (in)	Porosity rating	Least economical quantity	Relative production rate 1-10
Sand casting	Most	0.2 and up (0.4 and up)	12.5-25 (500-1000)	0.03-0.20	3-5 (0.12-0.20)	Fair	1	1
Shell molding	Most	0.2-10 (0.44-22)	1.6-12.5 (63-500)	0.01-0.03	2-5 (0.08-0.20)	Good	500	4
Gravity die-casting	Nonferrous	0.2-10 (0.44-22)	1.6-12.5 (63-500)	0.02-0.05	3-5 (0.12-0.2)	Very good	500	4-5
Pressure die-casting	Al, Zn, Mg, Cu alloys	0.2-10 (0.44-22)	0.4-1.6 (16-63)	0.001-0.05	1-2 (0.04-0.08)	Excellent	10 000	10
Investment casting	Most	0.1-10 (0.22-22)	1.6-3.2 (63-125)	0.002-0.005	0.5-1.0 (0.02-0.04)	Very good	50	6
Powder metallurgy	Most	0.01-5	0.8-3.2 (32-125)	0.002-0.005	0.8 (0.03)	Variable	5000	8

The shrinkage of the alloy on solidification and cooling to room temperature is also an important parameter in determining the quality of cast products. High shrinkage necessitates more careful design to promote directional solidification, reduce abrupt changes in cross-section, liberalize fillets and permit the best placement of gates and risers, thus avoiding internal shrinkage.

The as-cast structures usually suffer from segregation or coring and in some cases porosity is also present. These defects make cast structures relatively weak and brittle. However, this is not usually serious in unstressed components or if the material is to be subsequently worked. If a stressed component is to be produced by casting, special effort should be made to obtain porosity-free structures and some form of stress relief and homogenization treatment is usually recommended. The porosity rating of products made by different casting processes is given in Table 3.5. Shrinkage cavities can form in the casting due to shrinkage on solidification and can be avoided by using external or internal chills and with sufficient liquid metal supply.

In the cases of mass production and automatic machining operations, the cast material should have highly reproducible properties with respect to hardness, machinability and freedom from defects. These requirements can be more easily achieved in some alloys than in others. The cooling rate during solidification and on cooling to room temperature can have a considerable effect on the structure and mechanical properties. Sensitive materials can become weaker with slower cooling rates, and a section that has been made excessively thick in order to bear more load can be a source of potential weakness. Hot tearing is another serious defect which can develop in a casting if it is restrained from shrinking freely during solidification and subsequent cooling, as shown in Fig. 3.5. Although many factors

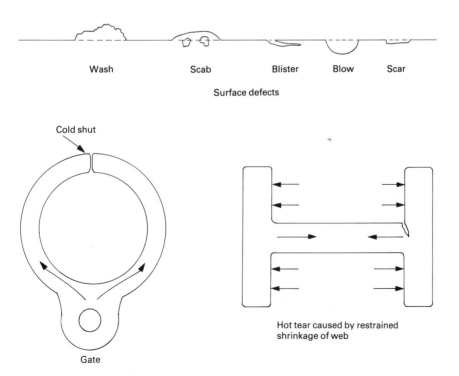

Figure 3.5 Some examples of casting defects.

are involved, coarse grain size and low-melting-point segregates increase the tendency for hot tearing. Proper selection of casting shapes to allow for free contraction is also an important factor in eliminating hot tears, as will be discussed in Chapter 12. Misruns, which result in an incomplete casting, and cold shuts, which are a result of incomplete fusion of two streams of liquid metal, are serious casting defects which can be eliminated by better control of the pouring temperature and pouring rate. Blows, scars, blisters and washes are surface defects which can be overcome by better control of mold material and pouring conditions.

3.4   POWDER METALLURGY PROCESSES

Powder metallurgy is the manufacture of products from powders of metals and metallic compounds. The powders are blended to the required composition and compacted into the desired shape. The green compacts are then sintered in a controlled atmosphere at a temperature below the melting point of the major constituent, for a time sufficient to bond the particles and achieve the required properties. In some applications, the compaction and sintering processes can be combined, as in the case of hot isostatic pressing (HIP). In some applications, powder metallurgy is the only feasible means of processing, as in the case of refractory metals like tungsten, tantalum, and molybdenum. Unusual materials or mixtures like aluminum-graphite mixtures and ceramic-metal composites (cermets) as well as porous components are also important candidates for powder metallurgy techniques. In addition to unusual applications, powder metallurgy techniques can be used to make many machine and structural parts economically. This is because the process lends itself to mass production of small intricate parts of precise dimensions and high tolerances. Although the cost of powder materials is considerably higher than the cost of wrought or cast stock, there is little material waste and often no further machining is required, which could result in an economical finished product. Examples of parts which are successfully produced by powder metallurgy include small gears, pawls, cams and brackets.

The properties of powder metallurgy products are influenced by their porosity in addition to the shape and size of powder particles, manufacturing variables and finishing treatments. Special effort should be made to eliminate porosity in components that need to have high fatigue strength and toughness. Forging of powder preforms gives higher density and greater strength.

Because the powder metallurgy dies are subjected to high pressures and severe wear, they must be made of expensive materials. Production volumes of less than 10 000 parts are normally not economic. Another limitation is usually imposed by the capacity of the production equipment. The cross-sectional area and weight of powder metallurgy products are usually limited to less than about 25 cm^2 (4 in^2) and 10 kg (22 lb) respectively. Table 3.5 compares some of the characteristics of powder metallurgy products with castings. The design rules which should be followed in designing powder metallurgy components will be discussed in Chapter 12.

3.5   BULK FORMING PROCESSES

Bulk forming processes produce wrought structures which are usually stronger and tougher than the cast structures produced by casting processes. The use of wrought metals and alloys is mandatory for many applications, such as critical structural members and high-pressure piping systems. Metallic materials are generally worked for two main reasons. First, to produce shapes that would be difficult or expensive to produce by other means; these may range from sheets or wire to more complex shapes such as I-beams and rails. The second reason is that the mechanical properties of metallic materials are usually improved by mechanical working. The improvement is achieved through grain refinement by recrystallization, directional control of flow lines, homogenization and elimination of segregation, breaking-up and distribution of inclusions, and closing up and welding of porosity.

Metal forming above the recrystallization temperature but below the melting point is called hot working, while forming at relatively low temperatures is cold working. Deforming the material at intermediate temperatures is called warm working. For most metallic materials, the minimum hot-working temperature is about $0.5\ T_m$, where T_m is the absolute melting temperature. Generally, hot-worked materials are soft and ductile while cold-worked materials are strain-hardened and less ductile. Therefore hot-working processes are usually performed in the semifinishing stages of production and cold-working processes in the finishing stages in order to take advantage of the improved strength, the closer tolerances and the improved surface finish that result. Table 3.1 lists the dimensional tolerances and surface finish of some common industrial working processes.

Forging processes

Forging is the oldest known working process and through the ages many related processes have been developed to provide great flexibility, making it economically feasible to forge a single piece or to mass produce thousands of identical parts. Open die forging is used for large components which have to be made in small numbers and, since the tooling costs are low compared with a closed die technique, it is cheaper. Besides being less accurate, open die forging needs skilled labor and has a slow rate of production. Closed die forging gives greater design flexibility, enabling more complex forms with closer tolerances to be made; and with automation, labor costs are reduced and reproducibility is improved. Figure 3.6 gives a schematic representation of closed die forging. The surface finish and dimensional tolerances that are normally achieved in forging are given in Table 3.1.

Many forgings are finished by machining and enough stock on the surface should be provided. As forging dies are expensive, closed die forging is usually not economic for small quantities. In many cases several hundred parts have to be produced before the process becomes economically viable.

Parts made by forging can usually be made by casting, machined by cutting from standard stock, or fabricated by welding. Forging is, in many cases, preferred because the product can be stronger, tougher and more fatigue resistant than the competing cast products. This is because forged components usually have small grain size and fibrous structure with highest strength in the direction of loading, as will be discussed in Section 12.3.

Rolling processes

Rolling is used to produce a large proportion of flat products like plate, sheet and strip, as well as structural shapes like angles, rails and channels. A large quantity of rods and bars are also produced by rolling. From the tonnage point of view, hot rolling is predominant among all manufacturing processes. Because the production is on such a large scale, however, hot-rolled products normally can be obtained only in standard sizes and shapes for which there is sufficient demand to permit economical production. The structure and properties of hot-rolled products can be controlled by the rolling schedule and finishing temperature. Many hot-rolled products are finished by cold-rolling to improve their mechanical properties, surface finish, texture and dimensional accuracy. Examples include bars of all shapes, rods, sheets and strips. Sheet steel less than 1.5 mm (0.05 in) thick is cold rolled in all cases because it cools too rapidly for practical hot rolling. Cold-rolled sheets and strips are commercially available as skin-rolled, quarter-hard, half-hard, full-hard or extra-hard depending on the amount of cold rolling that is given to the material. Cold-rolled sheets and strips are given surface finishes varying from smooth and

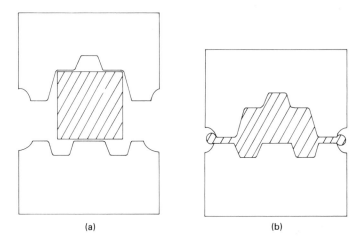

(a)   (b)

Figure 3.6 Schematic representation of closed die forging.

bright as a base for plating to rough matted finish for enameling. The dimensional accuracy and surface finish that can be achieved in industrial hot- and cold-rolled processes are given in Table 3.1.

Metal extrusion

Extrusion involves compressing a billet to force the material to flow through a suitably shaped die to form the product. Figure 3.7 gives some examples of cross-sections of extruded products. Although extrusion can be performed either hot or cold, hot extrusion is employed for many metals to reduce the forces required, eliminate strain hardening and reduce directionality of the product. Hot extrusion is widely used in the manufacture of aluminum, copper, lead, magnesium and alloys of these metals, as complicated shapes of constant cross-sectional area can be produced easily. Steel is more difficult to extrude in view of its high strength and the high temperatures that are required. However, with the development of molten glass lubricants, increasing amounts of steel are being produced by hot extrusion. Soft nonferrous metals and alloys can also be formed by cold extrusion. Examples of cold-extruded products are cans, fire extinguisher cases and thin-wall collapsible tubes. Advantages of cold extrusion are that it is fast, wastes little or no material and gives higher accuracy and better tolerances.

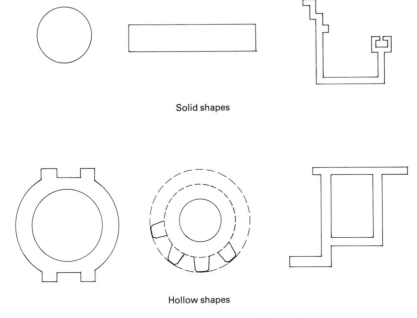

Solid shapes

Hollow shapes

Figure 3.7 Examples of extruded sections.

Extrusions are usually lighter, more sound and stronger than castings. They need no draft or flash to trim and need less machining as they are more accurate than forgings. The use of hydrostatic extrusion makes it possible to cold-extrude many metals difficult to form such as the high-strength super alloys, tungsten and molybdenum. In this process, pressurized liquid is used to push the metal to be extruded through the die.

Wire drawing

Wire and tube drawing are cold-working processes which follow hot rolling or extrusion in the production sequence. The drawing is usually carried out to reduce the size, increase the strength, improve the finish, change the shape or provide better accuracy. Thin wires and tubes can only be produced by cold drawing, as they would cool too fast for practical hot working. In practice, if a large reduction in area is required, several drawing steps may have to be performed. After a number of draws, a work-hardening material like steel may become too brittle to form and intermediate annealing is then required if it is to be drawn further.

3.6   SHEET METAL FORMING PROCESSES

Sheet metal forming processes are usually cold-working operations which have relatively thin sheets as their starting stock material. The products made by sheet metal forming processes include a wide variety of shapes and sizes. Examples are appliance bodies, car bodies, beverage cans and domestic pots and pans. The operations used for sheet metal forming may be classified as shearing, bending, drawing and stretching. The machines and tools used for most of these operations are presses and dies, which is why they are frequently called pressworking operations. Important material requirements that affect the quality of the formed sheet are the uniform elongation, after which necking begins, and the total elongation, which is the sum of uniform elongation and postuniform elongation. Uniform elongation is governed by the strain-hardening exponent, n, whereas postuniform elongation is governed by the strain rate sensitivity index, m. Higher values of n and m indicate higher elongation.

Anisotropy, or directionality, is an important material factor in sheet-metal forming. Anisotropy is the cause of the ears that form in deep-drawn cups, producing a wavy edge. Ears are objectionable because they have to be trimmed off, resulting in waste of material. Yield-point elongation is another material property which affects the quality of many sheet-steel products. Large yield-point elongations are undesirable as they result in stretcher strain marks which can be objectionable in the final product and can cause difficulties in subsequent coating and painting operations.

Grain size is important because of its effect on mechanical properties and surface appearance of formed parts. Coarser grains lead to lower strength and rougher surface appearance. Residual stresses, which result from nonuniform deformation of the sheet during forming, could lead to distortion or stress-corrosion cracking of sheet metal products unless they are properly relieved.

Shearing operations

The shearing process involves cutting sheet-metal by subjecting to shear stresses, usually between a punch and a die. The punch and die may be circular, straight, or any other shape. If the object of the shearing operation is to cut a hole and the material removed is scrap, the process is called punching or piercing. If, on the other hand, the cut piece that is removed from the sheet is the required part and the surrounding material is the scrap, the process is called blanking. Unless special tooling and rigid presses are used, as in the case of fine blanking, the sheared edges are jagged and not square with the surface of the sheet. The quality of the cut surfaces can be improved by shaving, which is a light cut in a second operation, by milling or any other suitable finishing process.

Bending processes

Bending is one of the most common sheet metal forming operations and is used to form flanges, seams, corrugations and similar shapes. Softer tempers of the material allow larger bend angle and smaller bend radius before cracking the sheet. The bendability of the material can also be increased by:

1. Heating the bent area and thus increasing ductility.
2. Better edge preparation and thus removing localized cold working and pre-existing cracking.
3. Reducing stringer-shaped inclusions in the material thus improving its transverse ductility.

Deep drawing

Deep-drawing operations are used to form sheet metal blank into a cylindrical or box-shaped part by means of a punch and die arrangement, as shown in Fig. 3.8. Typical

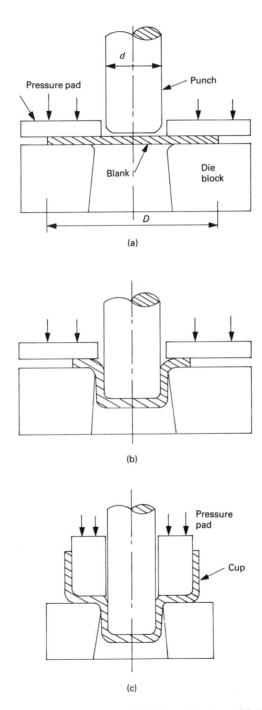

Figure 3.8 Deep drawing of cylindrical cups. (a) Before drawing. (b) During drawing. (c) Re-drawing.

parts produced by this process include beverage cans, containers and motorcar bodies. In addition to the properties of the sheet material, several other parameters can affect the deep-drawing operation, as shown in Fig. 3.8. Increasing the ratio of the blank diameter to punch diameter, decreasing the clearance between the punch and die, decreasing the punch and die corner radii, and increasing the blank holder pressure beyond the optimum value are all known to increase the difficulty of obtaining a sound deep-drawn part. Parts that are too difficult to draw in one operation are generally redrawn, as shown in Fig. 3.8. Practical drawing ratios, $(D-d)/D$, fall in the range of 35 percent to 50 percent for the initial draw and 30, 25 and 16 percent for the first, second and third redraws. If the material becomes too work-hardened to continue the drawing operations, intermediate annealing is performed. If the thickness of the sheet is greater than the clearance between the punch and the die, the thickness will be reduced. This is known as ironing and it produces cups with thinner walls than the base, as in the case of beverage cans and household saucepans.

An important modification of the punch and die drawing process is rubber pad forming, where the workpiece is forced into the die or wrapped over the punch by uniform hydraulic or rubber pad pressure. As either the die or the punch are eliminated, the tooling cost is reduced. Furthermore, one of the surfaces of the sheet is prevented from being damaged or scratched during forming. Because rubber pad forming processes are slower than the conventional drawing operations, they only have an economic advantage for quantities up to several hundreds or thousands of pieces depending on the part size and shape.

High-energy-rate forming is another modification of the drawing operation where the forces needed to form the sheet are released in a very short time. The basic operations are explosive forming, electrohydraulic forming and magnetic-pulse forming, depending on whether the process uses chemical, electrical or magnetic sources of energy. The main advantages of explosive forming is that relatively large parts like radar dishes and rocket motor cases, especially if they are made from materials difficult to form, can be made using inexpensive equipment. Steel plates 25 mm (1 in) thick and 3.5 m (about 12 ft) in diameter have been formed by this method.

Spinning operations

Spinning involves the forming of axisymmetric parts over a rotating form block using rigid tools or rollers. In conventional spinning the displacement of the metal is carried out in several steps and the pressure may be applied manually or mechanically. In shear spinning (also known as roll turning, flow turning, power spinning, or hydrospinning) the diameter of the spun part, unlike conventional spinning, is the same as that of the blank. Examples of parts that are usually manufactured by spinning include radar dishes and rocket motor cases. Figure 3.9 shows some of the shapes that can be produced by spinning. Spinning is generally slower but requires lower tool costs than conventional drawing processes and is only economical for small quantities.

Superplastic forming

Superplastic forming is a relatively recent development and is based on the large uniform plastic strains, no necking, that can be endured by some materials under certain conditions of strain rate and temperature. Resistance to necking during tensile elongation is enhanced by a relatively high value of strain rate sensitivity index, m, about 0.5 or more.

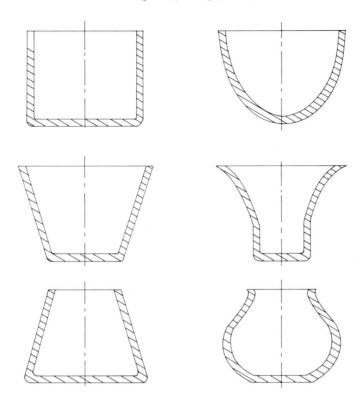

Figure 3.9 Examples of shapes that can be produced by metal spinning.

Several commercial alloy systems exhibit superplastic behavior. These include Ti-6Al-4V, IN 100 nickel-base alloy, Zn-22-Al, AA 7000 aluminum alloy series, and Al-Cu-Zr supral alloys. Conditions that lead to such high strain rate sensitivity are: (a) very fine and stable grain size, less than 10 μm; (b) temperatures in the range of 0.3 to 0.7 the absolute melting temperature; and (c) strain rates of the order of 10^{-4} to 10^{-3}s. Such strain rates are much slower than the strain rates of 0.1–1/s that are normally used in commercial hot-working processes. However, when superplastic forming is combined with diffusion bonding, complex structures like the one shown in Fig. 3.10 can be manufactured in one step, which makes the process competitive.

3.7   FASTENING AND JOINING PROCESSES

Nearly all engineering structures are constructed from various parts that have to be joined together into units. This is done for ease of manufacturing, ease of assembling, convenience of transportation or for economic reasons. Forming a joint between two parts can either be achieved by:

1. Mechanical fastening.
2. Joining which includes welding, brazing and soldering.
3. Adhesive bonding.

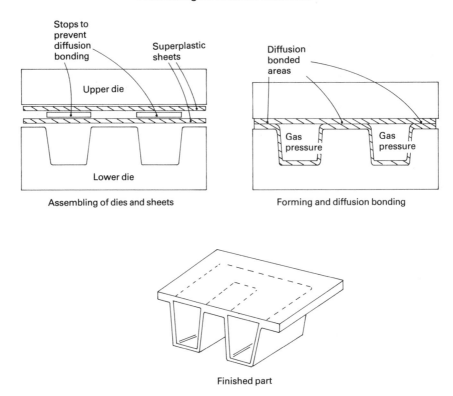

Figure 3.10 Forming of complex structures from sheets using superplastics forming and diffusion bonding.

There is a great diversity of processes within each of the major groups and even within a process or joining method there could be a large number of variations which have been developed to increase productivity, improve reliability, offer better resistance to environment or decrease cost. There are many obvious applications of each of the major groups in which the use of any other group would not be realistic. For example, welding of pressure vessels and earthmoving equipment, fastening of assemblies such as automotive products where disassembly is necessary for servicing, soldering of printed circuit components and adhesive bonding of composite laminates. With new developments, however, the traditional boundaries of process applicability are being gradually removed and it is common now to find more than one process competing for a given application. For example, the attachment of stiffeners to panels to give high-strength, lightweight airframe structures can be carried out by riveting, spot welding, or adhesive bonding, and each of these processes is currently being used in aerospace industry. The different join designs for the various fastening methods will be discussed in Section 12.7.

Mechanical fasteners

Mechanical fasteners include items that provide only temporary joints (nuts and bolts, lock washers and retaining rings) and items that provide permanent joints (rivets and metal stitching). Bolts, nuts and screws are the most common means of temporary fastening. These generally require holes in the components through which the fasteners are inserted. A wide variety of fastener designs is available, with standards specifying thread dimensions, tolerances, pitch, strength and materials used. An ordinary nut can loosen if the forces of vibration overcome those of friction. Introducing a lock washer or a specially designed nut supplies an independent locking feature presenting the nut from loosening.

Rivets are the most commonly used type of permanent mechanical joint. For example, as many as 400 000 rivets can be used in assembling the different parts of an aeroplane. The equipment used in riveting is relatively inexpensive and portable and can be used for on-site assembly. In addition, no postprocessing treatments are normally needed. The materials usually used for making rivets are aluminum alloys, monel metal, brass and steel. The compatibility of the rivet material with that of the parts to be fastened is an important factor that should be considered when selecting the rivet material. Another important factor is the stress concentration at the rivet hole, which frequently leads to fatigue failure. Mechanically joined parts can also suffer crevice corrosion if the mating surfaces are not sealed against the environment.

Metal stitching and stapling can be used to join thin sections of metals and nonmetals at high production rates. No precleaning, drilling, punching or hole alignment is necessary and this reduces the costs and increases production rates. Seaming is another form of mechanical fastening which is achieved by folding the sheets to be joined, as in the case of the tops of beverage cans and food containers.

Welding processes

Welding is defined by the American Welding Society (AWS) as a 'localized coalescence of metals or nonmetals produced by either heating of the materials to a suitable temperature, with or without the application of pressure, or by the application of pressure alone, and with or without the use of filler metal'. The various welding processes have been classified by the AWS as shown in Table 3.6. The different processes have been assigned letter symbols to facilitate their designation. The main factors that distinguish the different processes are: (a) the source of the energy used for welding; and (b) the means of protection or cleaning of the welded metal. The processes given in Table 3.6 cover a wide range which makes it possible to find an efficient and economic means of welding of almost all industrial metallic systems.

Arc welding processes are the most widely used in industry. SMAW process is widely used because of its versatility, portability and low cost. The GTAW process is a strong competitor with SMAW because of its adaptability to certain materials, such as titanium, zirconium, stainless steel and aluminum. Other welding processes which are widely used in industry are: OFW, GMAW and FCAW for manual welding and GMAW, FCAW and SAW for mechanized welding. Resistance welding is a low-cost, high-production process and is frequently used in assembly of motorcar bodies and similar parts made of thin metal sheets. The LBW, EBW, EXW, FRW and USW are more specialized and limited in application. The selection of the welding process for a given application depends mostly on the material to be welded. Carbon steel, the most widely used material, can be welded by most manual or automatic welding processes. Aluminum, on the other hand, is frequently welded with an inert gas process, such as GTAW or GMAW. In addition, applicable codes and standards and the job location influence the selection of the welding process.

In fusion welding processes, the molten filler metal solidifies quite rapidly by heat

Table 3.6 Common welding processes and their recommended use

Process	AWS designation	Recommended use				
		Carbon steels	Low-alloy steels	Stainless steels	Nickel and alloys	Aluminum and alloys
FUSION WELDING						
Arc welding	W					
Shielded metal arc welding	SMAW	a	a	a	a	–
Gas metal arc welding	GMAW	b	b	b	a	e
Pulsed arc	GMAW-P	a	a	a	a	d
Short-circuit arc	GMAW-S	d	d	d	d	–
Gas tungsten arc welding	GTAW	d	d	d	d	e
Flux-cored arc welding	FCAW	b	b	b	e	–
Submerged arc welding	SAW	a	a	a	c	–
Plasma arc welding	PAW	–	–	e	e	g
Stud welding	SW	a	a	a	–	a
Oxyfuel gas welding	OFW					
Oxyacetylene welding	OAW	e	g	g	g	g
PLASTIC WELDING						
Solid state welding	SSW					
Forge welding	FOW	a	–	–	–	–
Cold welding	CW	c	–	–	–	e
Friction welding	FRW	b	b	b	b	b
Ultrasonic welding	USW	g	g	g	g	f
Explosion welding	EXW	e	e	e	e	a
Resistance welding	RW					
Resistance spot welding	RSW	f	f	f	f	f
Resistance seam welding	RSW	f	f	f	f	f
Projection welding	RPW	f	f	f	f	f
OTHER WELDING PROCESSES						
Laser beam welding	LBW	e	e	e	e	f
Electroslag welding	ESW	i	i	i	i	–
Flash welding	FW	a	a	a	a	a
Induction welding	IW	g	–	–	–	–
Electron beam welding	EBW	a	a	a	a	a
BRAZING AND SOLDERING	B&S					
Dip brazing	DB	f	g	g	g	e
Furnace brazing	FB	a	a	a	a	e
Torch brazing	TB	e	e	e	e	e
Induction brazing	IB	e	e	e	f	g
Dip soldering	DS	g	g	g	g	g
Furnace soldering	FS	g	g	g	g	g
Torch soldering	TS	g	g	g	g	g

Key:
a. All thicknesses.
d. Up to 6 mm ($\frac{1}{4}$ in).
b. 3 mm ($\frac{1}{8}$ in) and up.
c. 6 mm ($\frac{1}{4}$ in) and up.
e. Up to 18 mm ($\frac{3}{4}$ in).
g. Up to 3 mm ($\frac{1}{8}$ in).
f. Up to 6 mm ($\frac{1}{4}$ in).
i. 18 mm ($\frac{3}{4}$ in) and up.

conduction into the metal adjacent to the weld. Columnar grains are usually present in the weld bead while the base metal closest to it undergoes a considerable overheating and grain growth. This latter area, the heat affected zone (HAZ), is usually a source of failure in welded components. Thermal contraction of welded metals may cause residual stresses and distortions. Preheating of joints is an effective method of reducing the cooling rate of the weld which reduces distortion and residual stresses. Postwelding heat treatment can be used to relieve internal stresses and to control the microstructure of the weld area. The need for preheating and postwelding heat treatment depends primarily on the weldability of the welded metal. Weldability can be considered to have two components:

1. Fabrication weldability, which is related to the ease with which a material can be welded.
2. Service weldability, which is related to the ability of the process–material combination to form a weld that will perform the intended job successfully.

In general, weldability of steel decreases as hardenability increases, because higher hardenability promotes the formation of microstructures which are more sensitive to cracking. Higher hardenability means more possibility of forming brittle martensite which cannot withstand the shrinkage strains in the weld zone. Hydrogen-induced cracking is also more prevalent in welding of hardenable steels than in welding of low carbon steels. Proper preheat, high heat input and maintenance of adequate interpass temperatures reduce the rate of cooling in the HAZ and this results in a softer, less-sensitive microstructure. The HAZ may also be softened by postweld heat treatment in the range of 480 to 670°C (895 to 1240°F). The carbon equivalent (CE) is often used to estimate the weldability of hardenable carbon and alloy steels. In this approach, the significant composition variables are reduced to a single number, CE, using one of several similar formulas as follows:

$$CE = \%C + Mn/6 + (\%Cr + \%Mo + \%V)/5 + (\%Si + \%Ni + \%Cu)/15 \quad (3.3)$$

Steels with CE less than 0.35 percent usually require no preheating or postheating. Steels with CE between 0.35 and 0.55 percent usually require preheating, and steels with CE greater than 0.55 percent may require both preheating and postheating. In addition to CE, other factors such as hydrogen level, restraint and thickness must be considered simultaneously in relation to a specific application.

In addition to cracking and residual stresses, defects like porosity, slag inclusions, incomplete fusion and incorrect weld profile can also exist in the welded joint. Such defects can be eliminated by following the correct welding procedure and selecting the appropriate technique. Generally, strict quality control and nondestructive testing are essential if welding defects are to be eliminated and high reliability of welded joints is to be maintained.

Welding jigs and fixtures are frequently used in production to reduce distortion, warping and buckling of the welded parts. The use of jigs also increases productivity, reduces costs and results in higher accuracy. Typical dimensional tolerances which may be held on average weldments are:

3 mm for small parts with little welding
6 mm for moderate sized parts with a small amount of welding
9 mm for large parts with a moderate amount of welding
9–12 mm for large parts with a large amount of welding.

Brazing and soldering are joining processes which employ filler materials that have melting points below that of the metal being joined. Brazing filler alloys have melting points above 427°C (800°F), while soldering filler alloys have melting points below this limit. The heat for brazing and soldering may be provided by torch, furnace, induction or hot dipping. The main advantages of brazing and soldering processes are:

1. Dissimilar metals can be joined.
2. Materials of different thicknesses can be joined.
3. No postjoining heat treatment is normally needed and joints are virtually free of internal stresses.
4. Complex assemblies can be joined in several steps by using filler materials with progressively lower melting points.
5. The processes lend themselves to automation.

On the other hand, brazing and soldering processes have the limitations of:

1. Relatively lower strength, especially in the case of the lower melting point alloys.
2. Temperature resistance is limited by the filler material melting point.
3. Inspection of the joint could be difficult.
4. Cost of the joint could be high.

Adhesive bonding

Adhesive bonding has reached a high level of development and reliability and is used widely in manufacturing and construction. The dollar value of adhesive bonded products per year rivals or exceeds the value of welded products. Adhesives have an advantage over other joining methods in that they can be applied to any surface, which enables them to join widely different materials, e.g. metal to glass, metal to metal, metal to polymer and polymer to polymer. Very thin sections can be bonded and, since the operation is generally carried out at temperatures below 200°C (400°F), there is no significant distortion of the parts or change in their original properties. The elimination of rivets and other fasteners results in better structural integrity since no holes are required and smoother surfaces and better appearance are achieved. Because of the inherent plasticity of some adhesives, the load on the interface is more uniformly distributed, which eliminates localized stresses that generally result from mechanical fastening or welding. In addition, adhesive joints prevent fluid leakage and bimetallic corrosion at joints, as well as providing some degree of thermal and electrical insulation.

Structural adhesives are generally based on polymeric systems and many of them are composites containing several components. The main polymeric systems that are widely used for structural adhesives include: epoxies, acrylics, anaerobics, urethane, cyanoacrylates and polyimides. The characteristics of the different adhesive systems will be discussed in Section 4.7. The major factors to be considered in the selection of an adhesive include:

1. The materials to be bonded.
2. Joint design.
3. Required strength.
4. Service temperature and other environmental conditions.
5. Whether disassembly will be required for inspection, maintenance, or replacement.

Surface preparation is an important factor in achieving the optimum performance of adhesive-bonded joints. Rough surfaces promote mechanical adhesion and cleanliness promotes wetting by the adhesive. Adhesives are available in solid, liquid, powder or tape form, which makes them suitable for mass production techniques. Curing cycles can range from 30 min to several days, depending on the adhesive, which could slow down the production rates. Inspection of the quality of the joint can also be difficult once assembled. A serious disadvantage of adhesives is that, because most of them are organic materials, they cannot be used at high temperatures and their strength decreases rapidly as the temperature rises. Most adhesives are not stable above about 315°C (600°F). In addition, adhesive properties tend to degrade with time. The cost of the adhesive can be high and elaborate jigs and fixtures may be needed to apply heat and pressure, depending on the curing cycle. In spite of these limitations there are many applications where the overall economics makes adhesive bonding an attractive alternative to other fastening and joining processes. As an example from the aircraft industry, acrylic structural adhesives have replaced mechanical fasteners in fabricating a glider plane structure. A similar example exists in the automotive industry, where adhesives replaced costly welding and finishing operations in the assembly of a truck cab quarter panels and stiffening ribs. An example from the electrical appliances industry is bonding a strontium ferrite magnet disc to steel in the fabrication of electric generators.

3.8   MACHINING PROCESSES

Components which have been processed by casting, working welding or heat treatment often require further finishing operations before they are ready for use. These finishing operations usually consist of material removal by cutting and various other machining processes. Machining processes of parts on traditional machine tools are generally expensive as they are energy, labor and capital intensive. In addition, a large proportion of the material removed by cutting is normally wasted. In spite of these drawbacks, machining operations are frequently specified in industry. The main reasons for specifying machining processes are:

1. To improve dimensional accuracy and tolerances.
2. To achieve specified surface characteristics or texture on all or part of the surface of the product.
3. To create internal or external surface features that are difficult or not possible to produce by other processes.
4. It may be more economical to manufacture the component by machining.

Basic cutting processes traditionally involve: drilling and reaming, turning and boring, shaping and planing, milling, broaching, filing, sawing, tapping and grinding. There are also some nontraditional machining processes such as chemical milling, electrochemical machining, ultrasonic machining, electric discharge and lasers which have been developed for special metal removal operations. In the basic metal cutting processes, a cutting tool is forced through the workpiece to remove excess material in the form of chips. The main parameters that affect the metal cutting process are:

1. The cutting speed, which is the relative velocity between the cutting tool and the workpiece surface.

2. The feed, which is the width of material removed per revolution or pass of the cutting tool.
3. The depth of cut, which is usually the dimension at right-angles to the feed.

The rate of metal removal increases and the machining time decreases as the cutting speed, feed and depth of cut increase. Figure 3.11 illustrates the cutting action in some cutting operations and Table 3.1 gives a list of the industrial machining processes and the expected dimensional tolerance and surface roughness under normal working conditions.

As machining is relatively expensive it should not be performed unless necessary and tolerances that are closer than necessary should not be specified. The economics of metal cutting can be improved by using high cutting speeds and tools with long lives. If the material to be cut gives discontinuous chips and needs less power for cutting, the

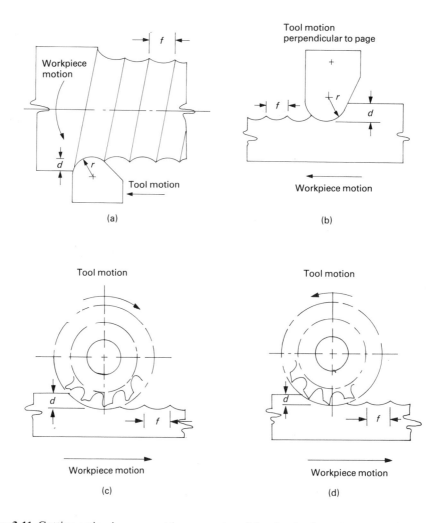

Figure 3.11 Cutting action in some cutting operations (d = depth of cut, f = feed, r = nose radius). (a) Turning. (b) Shaping and planing. (c) Upcut milling. (d) Downcut milling.

economics are further improved. The ease with which one or more of the above factors can be realized for a given material is taken as a measure of its machinability. Thus, a material with good machinability is one which requires less power consumption, causes less tool wear and easily acquires a good surface finish. One of the methods for comparing machinability of materials is to determine the relative power required to cut them using single point tools. Another method is to use the machinability index, which is defined as follows:

Machinability index percent =

$$\frac{\text{Cutting speed of material for 20 min tool life} \times 100}{\text{Cutting speed of SAE 1112 steel for 20 min tool life}} \tag{3.5}$$

In this definition the free machining steel SAE 1112, or AISI B1112, is taken as the standard and its machinability index is arbitrarily fixed at 100 percent. The higher the machinability index, the easier and the more economical it is to finish the material by metal cutting. The machinability of SAE 1112 steel is better than most steels because it contains extra sulfur, resulfurized, and is known as a free-machining steel (see Table 2.1). The extra sulfur is present in the form of manganese sulfide, which could reduce friction at the tool–chip interface and thus increase tool life. The machinability index of some common metallic materials is given in Table 3.7.

When the workpiece material is too hard, too brittle, or its shape is difficult to produce with sufficient accuracy by any of the basic metal cutting processes, abrasive and special metal removal processes are used. Special metal removal processes include chemical

Table 3.7 Machinability index of some common metallic materials

Material	Hardness (BHN)	Machinability index
STEELS		
AISI 1015	121	50
1020	131	65
1030	149	65
1040	170	60
1050	217	50
1112	120	100
1118	143	80
1340	248	65
3140	262	55
4130	197	65
4340	363	45
18-8 stainless steel	150–160	25
CAST IRONS		
Gray cast iron: soft	160–193	80
medium	193–220	65
hard	220–240	50
Malleable iron	110–145	120
NONFERROUS ALLOYS		
Aluminum alloys	35–150	300–2000
Bronze	55–210	150–500
Magnesium alloys	50–75	500–2000
Zinc alloys	80–90	200

milling, electrochemical machining, ultrasonic machining, electric discharge and lasers. In abrasive cutting processes, the tool is made of abrasive particles of irregular geometry whose sharp corners act as small cutting edges. Several abrasive cutting operations have been developed to cover processes ranging from those yielding the finest and smoothest surfaces produced by any machining process, in which very little material is removed, to rough, coarse surfaces that accompany high material-removal rates. The most common example of abrasive cutting is grinding, where a small volume of the workpiece material is removed to yield closely controlled dimensions and tolerances. Excessive temperature rise in grinding can cause tempering and softening of hardened surfaces, thermal cracking, residual stresses or distortion of the workpiece. These effects can be avoided by proper application of cooling fluids, using softer grade grinding wheels, lower depth of cut, lower wheel speeds and higher work speeds. Other finishing operations which are based on abrasive cutting include: honing, which is used primarily for surface finishing of holes; lapping, which is used for flat and cylindrical surfaces; and polishing and buffing, which are used to produce a smooth, lustrous surface finish.

Machining processes which involve metal cutting by chip formation have a number of disadvantages and limitations. In addition to causing distortion and residual stresses, the relatively high forces involved put a definite limitation on the delicacy of the workpiece that can be machined. For example, the production of semiconductor chips would not be possible with any of these processes. Nontraditional machining (NTM) processes have been mainly developed for material removal nonmechanically and without chip formation. The main characteristics of some important NTM processes are given in Table 3.1.

In chemical machining, the material is removed from selected areas of the workpiece by immersing it in a chemical reagent. Chemical milling, chemical blanking, chemical jet machining and chemical polishing are all variations on the basic process. Electrochemical machining (ECM) removes material by anodic dissolution in an electrolyte. The shape of the generated cavity is the mirror image of the tool, which is usually made of copper or brass. Ultrasonic machining employs an ultrasonically vibrating tool to remove material by erosion with abrasive particles suspended in a slurry. The tool forms the reverse image in the workpiece as the slurry abrades the material. Boron carbides, aluminum oxide and silicon carbide are the commonly used abrasives and can cut virtually any material. Ultrasonic machining is best suited for hard, brittle materials, such as hardened steels, ceramics, carbides and precious stones.

Electrodischarge machining (EDM) cuts metals by discharging electric current stored in a capacitor bank across a thin gap between the tool and the workpiece. Thousands of sparks per second are generated and each spark produces a tiny crater by vaporization, thus eroding the shape of the tool into the workpiece. EDM is applicable to all materials that are fairly good electrical conductors and the absence of almost all mechanical forces makes it possible to machine fragile parts without distortion. In addition, fragile tools, such as wires, can be used. EDM is slow compared with traditional methods and produces a brittle, recast layer on the surface which could reduce fatigue life. However, its controllability, versatility and accuracy usually result in superior design flexibility. Significant cost reductions in manufacturing of tools and dies can also be achieved. Laser beam machining (LBM) uses a laser to melt and vaporize materials. The beam can be used to drill microholes as small as 0.005 mm (0.0002 in) with a depth-to-diameter ratio of 50:1 and can be used for any variety of metallic and nonmetallic materials. The final choice of a machining process for a given application depends not only on technical considerations, but also on economic factors. Machining a component can involve considerable time, thus contributing significantly to the overall cost of the product. The economics of manufacturing processes will be discussed in Chapter 18.

3.9   SURFACE TREATMENT

The surface quality of a component plays an important role in determining its behavior in service. This is because wear resistance and corrosion resistance are principally affected by the surface condition of the material. In addition, fatigue resistance is known to be greatly influenced by the surface quality of the component. Surface treatment covers a wide range of finished processes in industry which are usually required to perform one or more of the following functions:

1. Protection against corrosion and oxidation.
2. Protection against abrasion and wear.
3. Provision of a desirable surface appearance and color.

Surface treatments can be broadly classified as:

1. Mechanical finishing processes, as in the case of grinding, honing, lapping and polishing.
2. Metallurgical processes, as in the case of flame hardening, carburizing, carbonitriding and nitriding.
3. Coating treatments which can be further divided into conversion or additive coating treatments. In conversion coatings, the surface to be treated is changed by chemical, electrochemical, thermal, or mechanical means, while in additive coatings a layer of material is applied to the surface to be finished.

The following paragraphs will discuss the different surface treatments in terms of their functional requirements.

Surface treatments for corrosion protection

In many applications corrosion resistance of a component can be achieved by covering its surface with a protective coating. Protective coatings can be metallic, ceramic, or polymeric depending on the required function and service conditions. Metallic coatings can be anodic or cathodic with respect to the substrate. For example, steels may be protected satisfactorily by nickel, chromium, or tin which are cathodic to iron and can also be protected by zinc or cadmium which are anodic to iron. Nickel, chromium and tin are relatively more corrosion resistant in many environments and do not suffer from the disadvantage of forming unsightly and voluminous corrosion products like iron rust. As protection in this case depends on the fact that the steel is no longer in contact with the corrosive environment, the coating must be free of pores. On the other hand, zinc, cadmium and aluminium are anodic relative to iron and corrode preferentially, thus protecting it sacrificially. In the latter cases, iron oxide is replaced by zinc, cadmium or aluminum oxides which are less objectionable and more protective. Metallic coatings can be applied by:

1. Electrodeposition, where the coating metal is deposited from an electrolyte, as in the case of chromium, cadmium, zinc, copper and nickel.

2. Chemical or electroiess plating is an aqueous reduction process in which metal is deposited from solution, often under a catalytic action, as in the case of nickel deposition.
3. Hot dipping involves dipping the metal to be coated in the molten coating metal which is usually zinc, tin, or aluminum.
4. Metal spraying uses a spray gun to spray the molten coating metal, as in the case of zinc and aluminum.
5. Cladding involves bonding the coating metal on the substrate by rolling, normally at high temperature, as in the case of aluminum or stainless steel on carbon steel and aluminum on aluminum alloys.
6. Sintering of powdered coating material on the substrate, as in the case of aluminum and nickel.
7. Vacuum deposition involves evaporating the coating material in vacuo and depositing it on the substrate within the vacuum chamber, as in the case of aluminum.
8. Vapor decomposition of a metal compound vapor can be used to deposit the coating material, as in the case of chromizing.
9. Cementation, where the coating material is diffused at relatively high temperatures into the surface from a gas, liquid or solid, as in the case of chromizing, sheradizing (zinc) and calorizing (aluminum).

Corrosion protection can also be provided through the formation of oxides on the surface using conversion coating methods. Two processes are in commercial use: (a) anodizing, which is an electrolytic process for thickening and stabilizing oxide films on base metals, is most commonly used for aluminum to give an oxide film which can be colored; and (b) phosphating, which is used primarily for steels as an underlayer for paint finishes and finds little application for corrosion protection by itself. Vitreous enamels and ceramic coatings cover the surface to be protected with an adhering, hard and durable layer. Vitreous enamels usually consist of borosilicate matrix in which crystalline opacifiers and pigments are suspended. Not all metals can be satisfactorily covered with vitreous enamels. Steels that are selected for vitreous enameling should be low in metalloids and may contain titanium or very low carbon. Vitreous enamels are hard and smooth with good abrasion resistance, but are susceptible to chipping. In general, thinner coatings are less liable to chipping. Ceramic materials will be discussed in more detail in Chapter 5.

Organic materials are very widely used as coating to metallic surfaces. It has been estimated that over 50 percent of all metallic surfaces for which an impervious, pore-free, adherent and attractive property is required are treated with polymeric paints. As a family, polymeric coatings are relatively inexpensive, provide an opportunity for a variety of pleasing colors and give good resistance to corrosion. However, they do not allow the holding of close tolerances and they have only average resistance to abrasion. Their resistance to elevated temperatures is less than that of metallic or ceramic coatings. In general, a thin layer primer, or underlayer, is first applied to the substrate and then thicker layers of paint are applied. The primer protects the coated metal and adheres to both substrate and paint. To promote adhesion, the surface to be coated should be free of grease and scale. Polymeric coatings can be combined with metallic coatings by first applying an anodic coating to the base metal and then a polymeric finish. Another alternative is to mix the metallic powder with the polymeric coating. Paints containing zinc or aluminum powders can be used to protect steels anodically. Polymeric materials will be discussed in Chapter 4.

Surface treatment for wear and abrasion resistance

Wear resistance of surfaces can be provided either by selecting a material that has the required bulk properties or by surface treatment and coatings. The latter solution offers the attractive possibility of combining widely different properties in a given component and allows the possibility of substituting a cheap material for a more expensive one. Typical applications include gears, cams and crankpins where a soft core and a hard, wear-resistant surface are required. In addition to increasing wear resistance, surface hardening treatments tend to develop residual compressive stresses at the surface which can markedly improve fatigue strength. The different surface hardening techniques which are used to modify the surface structure and its properties can be classified into: selective heating of the surface, altered surface chemistry, deposition of an additional surface layer and selective working of the surface.

Induction and flame hardening are selective heating techniques as heat is applied only to the surface. Hardening depends upon the steel's basic carbon content for hardness. After heating, quenching is normally applied using water spray. Induction and flame hardening are normally applied to medium-carbon steels and cast irons. Surface hardness of 50 to 60 RC and case depths of 0.7 to 6 mm (0.03 to 0.25 in) can be achieved with little distortion in the component. The journals of the crankshafts of internal combustion engines are usually treated by this method.

Carburizing, carbonitriding and cyaniding increase the carbon, and in some cases nitrogen, content of the surface by heating in a carbon-rich atmosphere at temperatures in the austenitic range. These processes are usually used for low-carbon and some alloy steels. The presence of nitrogen in the treated surface improves its hardenability and allows the use of low-cost steels. Surface hardness in the range of 55 to 65 RC and case depths of 0.025 to 1.5 mm (0.001 to 0.060 in) can be achieved after heat treatment. Some distortion usually results during heat treatment. Gear teeth are usually treated by this method to increase their wear resistance while maintaining their toughness.

Nitriding is conducted at lower temperatures than carburizing but requires the presence of chromium, aluminum, vanadium, molybdenum or tungsten in the steel to allow the formation of hard nitride layers. The process can be applied to some alloy steels, stainless steels and high-speed tool steels. The process gives case depths of 0.025 to 0.6 mm (0.001 to 0.033 in) and surface hardness up to 1100 on the Vickers hardness scale (VHN), which is considerably higher than that obtained by ordinary case hardening.

Hard facing techniques involve depositing a relatively thick layer, as much as 6 mm (0.25 in), of a hard alloy on the treated surface. Common techniques of hard facing are flame spraying, plasma spraying and weld deposition using arc welding techniques. These processes involve some diffusion at the interface between the coating and the base metal but mechanical bonding is sometimes desirable. Strong mechanical bonds are obtained by roughening the substrate. Hard facing alloys fall into four classes: tungsten, cobalt, nickel and iron base. Tungsten-base alloys have excellent resistance to abrasive wear. Cobalt-base alloys combine excellent resistance to adhesive wear with heat and corrosion resistance. Depending on their composition, nickel-base alloys are capable of balanced adhesive and abrasive wear resistance. Iron-base alloys are less expensive than other systems and are characterized by either excellent resistance to abrasive wear or excellent resistance to impact, depending on their composition. In addition to metallic alloys, ceramics and refractory carbides can also be used to increase wear resistance. They are usually deposited using hard facing techniques. Hard facing techniques can also be used to repair worn-out machine elements such as shafts and gears.

Selective working of the surface can be used to improve the surface properties of finished components. Shot peening involves hitting the surface repeatedly with steel or cast iron shot. The resulting surface deformation is accompanied with compressive stresses which improve the fatigue strength. The depth of surface deformation can reach 1.5 mm (0.060 in) depending on the process variables. Surface rolling involves cold working the surface using a roller instead of shot and yields similar results. Explosive hardening can also be used to increase the surface hardness with little change in the shape of the component.

3.10 REVIEW QUESTIONS AND PROBLEMS

3.1 Suggest possible manufacturing processes for the following items in an internal combustion engine: (a) pistons; (b) connecting rods; (c) crankshaft.

3.2 Compare the use of powder metallurgy and metal cutting for manufacturing: (a) a gear of 35 mm (1.4 in) outer diameter for a household food processor (20 000 parts are required per year); (b) a bronze sleeve for a 50 mm (2 in) diameter shaft (1000 parts are required per year).

3.3 Suggest the sequence of processing steps for an aluminum beverage can.

3.4 Compare the use of deep drawing and spinning for manufacturing cups of 150 mm (6 in) diameter and 250 mm (10 in) high. Number of parts required is 50 000 per year.

3.5 Suggest the possible methods for making a 3 m (3.3 yards) diameter radar dish.

3.6 Select four methods for making pipes and compare between them in terms of possible materials, sizes, length of pipe and rate of production.

3.7 Why are zinc alloys attractive for making die cast products?

3.8 Give the steps involved in making a round steel wire of 0.5 mm (0.02 in) diameter from $100 \star 100$ mm (4×4 in) square billet. What is the length of the wire that results from a one meter (3.3 ft) billet? [Answer: 50 930 m]

3.9 A steel bridge is to be fabricated from low-carbon steel plates of 25 mm (1 in) thickness. Suggest the methods of cutting the plate to size, edge preparation for the different types of joints that are likely to be used, and the method of welding.

3.10 What is the maximum thickness of sheet which can be cut by blanking on a 1 MN ($2.2 \ 10^5$ lb) press? Diameter of the blank = 500 mm (20 in), shear strength of material = 300 MPa (43 ksi). [Answer: 2.1 mm or 0.08 in]

BIBLIOGRAPHY AND FURTHER READING

Agrawal, S. P. ed., *Superplasticity*, ASM, Ohio, 1985.

Amstead, B. H., Ostwald, P. F. and Begeman, M. L., *Manufacturing Processes*, John Wiley & Sons, New York, 1979.

DeGarmo, E. P., Black, J. T. and Kohser, R. A., *Materials and Processes in Manufacturing*, 6th ed., Collier Macmillan, New York, 1984.

Doyle, L. E. *et al.*, *Manufacturing Processes and Materials for Engineers*, 3rd ed., Prentice Hall, New Jersey, 1985.

Gordon, S., 'Acrylics structural adhesives cut bonding costs', *Mechanical Engineering*, Sept. 1983, pp. 60–65.

Kalpakjian, S., *Manufacturing Processes for Engineering Materials*, Addison Wesley, Reading, Mass., 1984.

Lindberg, R. A., *Processes and Materials of Manufacture*, 3rd ed., Allyn and Bacon, Boston, 1983.

Ray, M. S., *The Technology and Applications of Engineering Materials*, Prentice Hall, London, 1987.

Schey, J. A., *Introduction to Manufacturing Processes*, McGraw-Hill, New York, 1977.

Chapter 4

Polymeric Materials and their Processing

4.1 CLASSIFICATION OF POLYMERS

Polymers are an important and fast-growing class of materials with a wide range of mechanical, physical and chemical properties. They are characterized by their low density, good thermal and electrical insulation, high resistance to most chemicals and ability to take different colors and opacities. Compared with structural metals and alloys, unreinforced bulk polymers have high thermal expansion coefficients, are mechanically weaker and exhibit lower elastic moduli. These drawbacks can be greatly overcome by reinforcing polymers with a variety of fibrous materials, as will be discussed in Chapter 6.

Two of the main reasons for the fast expansion of the industrial use of polymers are: (a) the ease with which they can be manufactured into complicated shapes in one step and with little need for further processing or surface treatment; and (b) their versatility. The basic manufacturing processes for polymeric parts are extrusion, molding, casting and forming of sheet. Polymers can also be machined and joined. In addition, several other specialized processes have been developed for the manufacture of fiber-reinforced composites, as will be discussed in Chapter 6. Because of their unique properties and great versatility, polymers are increasingly replacing metallic and ceramic materials in many applications. One of the key advantages of plastics is the ability to produce accurate components, with excellent surface finish and attractive colors, at low cost and high speed. Some of the major applications are in automotive, electrical and electronic products, household appliances, toys, containers, packaging and textiles. As an example, consider a household bucket. To make the body out of galvanized steel, three pieces (body, bottom and handle fixers) have to be cut out of the sheet, assembled by seaming and then soldered. In addition to being slow and labor intensive, this is also wasteful of materials. If made of plastics, the body is lighter, of attractive color and can be made in one step in injection molding.

Plastics are composed of a mixture of polymeric materials and additives. Polymers, which constitute the major portion of the mixture, are made up of extremely large molecules formed from polymerization of different monomers. The monomers used in forming polymers are organic materials which contain carbon atoms joined with other atoms by primary covalent bonds. Other atoms in typical monomers include hydrogen, nitrogen, oxygen, fluorine, silicon, sulfur and chlorine. An example of such a monomer is ethylene, which contains two carbon and four hydrogen atoms (C_2H_4) and can be made from natural gas. The additives, which have much smaller molecules in comparison with polymers, are intended to give the plastic desirable properties of color, flexibility, rigidity, flame resistance, weathering resistance and processability. In general, additives can be grouped into two main categories. The additives of the first category modify the characteristics of the base polymer by physical means and act as plasticizers, lubricants, impact modifiers, fillers and pigments. The second category of additives achieve their effect by chemical reactions and include flame retardants, stabilizers, ultraviolet absorbers and antioxidants.

Plastics are usually classified into thermoplastics and thermosets. These two classes differ in the degree of their intermolecular bonding. Thermoplastics have little or no cross-bonding between molecules, soften when heated and harden when cooled no matter how often the process is repeated. Thermosets, on the other hand, have strong intermolecular bonding, which prevents the fully cured materials from softening when heated. Whether a plastic is thermosetting or thermoplastic is of great importance to the designer and material selector not only because of the difference in behavior but also because of the difference in manufacturing techniques.

Rubbers are similar to plastics in structure and the difference is largely based on the degree of extensibility or stretching. Rubbers can be stretched to at least twice their original length and, upon release of the stress, return to approximately the original dimensions. Some grades of plastics approach this definition of rubber.

4.2   PARAMETERS AFFECTING THE BEHAVIOR OF PLASTICS

As would be expected, the behavior of plastics as engineering materials is affected by the nature of the polymer molecules and the additives. As with other materials, the structure of plastics affects their behavior in a number of ways.

Effect of structure

The structure of a polymeric material is intimately related to the polymerization mechanism involved in forming it. During polymerization, a monomer, or other small molecule, attaches itself to a growing molecule in order to produce the polymeric molecule. In addition-polymerization, molecules of the monomer are usually subjected to heat and pressure, in the presence of a catalyst, to induce polymerization. The resulting molecules are chain-like with secondary weak bonds between them which allows them to slide easily relative to each other. These polymers are described as linear polymers. When the arrangement of linear molecules is completely random, the structure is amorphous. A certain amount of crystallinity is achieved when adjacent chains become aligned, creating regions of regular structure. The frequency and distribution of these regions are determined by the structure of the molecules and by the processing techniques applied to the material. Increasing crystallinity tends to increase the static and fatigue strengths and softening temperature but decreases ductility, as in the case of high-density polyethylene (HDPE). Amorphous polymers, on the other hand, shrink less in molding, which simplifies the design of molds for large parts.

The optical properties of polymers are also influenced by the extent of crystallinity. Crystalline polymers are essentially two-phase systems, with the denser crystalline phase having a higher refractive index than the amorphous matrix. This difference in refractive index makes crystalline polymers either opaque or translucent. Completely amorphous polymers can be transparent, e.g. acrylic and polycarbonate.

Addition-type polymers can also be polymerized with branched molecules which have side chains protruding from the main chain, thus forming a structure resembling the branched shape of a tree. Polyethylene is an example of a polymer in which branching can occur. With increasing length of side chains, the average distance between the main chains increases. This reduces the secondary bonds and is accompanied by a decrease in crystallinity and lower density and stiffness. When the repeating unit in the polymer is the same as that of the monomer, the resulting polymer is known as a homopolymer. Another type of addition polymer can be made to contain more than one kind of monomer and the resulting structure, called a copolymer, is comparable to a solid solution in metallic materials. The properties of the copolymer depend on the type and arrangement of monomers used in building it. Copolymers tend to be less crystalline than homopolymers and they are, therefore, tougher and more flexible. An example of a copolymer is acrylonitrile-butadiene-styrene (ABS), which is made from acrylonitrile, butadiene and styrene. The members of the ABS group are strong and tough with properties that can be tailored to different requirements by varying the proportions of the constituents and the molecular weight.

In contrast to addition reactions, which are primarily a summation of individual molecules into a polymer, condensation reactions form in steps between two or more different molecules which makes them slower processes. Furthermore, there is usually a by-product which must condense, hence the name of the reaction. This by-product is usually water or some other molecule, such as HCl or CH_3OH. Condensation reactions can produce either a chain-like molecule such as 6/6 nylon, or a network structure such as phenolformaldehyde.

Effect of additives

It has been stated earlier that additives can greatly influence color, flexibility, rigidity, flame resistance, weathering resistance and processability. The effect on mechanical strength, electrical properties and toxicity should be taken into consideration when selecting additives for some applications. Solid fillers and reinforcing materials can have dramatic effects on properties and will be discussed in Chapter 6.

Polymers are often colorless, or milky white, unless dyes or pigments are added. Dyes dissolve in the polymer and can be used to color transparent plastics. As dyes are organic materials, they tend to change color if heated beyond about 200°C (c. 400°F) during processing or service. Pigments do not dissolve in the polymer but are dispersed and, therefore, tend to decrease the transparency of the plastic. Organic pigments are more stable than dyes but are still sensitive to heat. Inorganic pigments are used for opaque objects and are the most stable of the coloring additives. They are not affected by molding temperatures and are resistant to light and weathering.

As polymers are organic materials, they tend to burn when heated or subjected to a flame. Some polymers are more resistant to burning than others, depending on their composition. Polymethylmethacrylate, for example, continues to burn when ignited, whereas polycarbonate extinguishes itself. Other examples of self-extinguishing polymers include PVC, chlorinated polyethers, nylons, polyphenyl oxides, polysulfones and silicones. Flame resistance may be further increased by the addition of flame retardants. These materials are usually added after polymerization as they are often insoluble in polymers. Examples of flame retardants include antimony compounds, phosphorous compounds, chlorinated phosphates, polyphosphonates, halogenated paraffins, aluminum trihydrate, ammonium sulfamate, barium metaborate and zinc borate.

Ultraviolet radiation and oxidation can both cause the primary covalent bonds in polymer chains to break, giving rise to molecular chain fragments known as free radicals. Free radicals can recombine to form cross-links between the polymer chains which increases rigidity and decreases ductility of the material. They can also recombine to form smaller polymer molecules which leads to loss of strength. A certain degree of protection against radiation can be achieved by the addition of suitable ultraviolet absorbers. A typical example of protection against ultraviolet radiation is the addition of carbon black to rubber.

Exposure to high temperatures can cause degradation of polymers as a result of the oxidation. Examples of antioxidant additives include phenols, aromatic amines, aminophenols, aldehydes and ketones.

Hardeners are agents that encourage cross-linking between polymer chains, thus increasing rigidity and hardness. On the other hand, plasticizers are added to improve flexibility by reducing the intermolecular secondary bonds between the polymer chains. Plasticizers are low molecular weight solvents that could evaporate from the finished part with time and cause shrinkage and reduction in ductility. An important use of plasticizers is in PVC, where various grades of flexibility can be achieved by controlling the amount of plasticizer added. The use of plasticizers is restricted to thermoplastics, since the brittleness of thermosets results from cross-linking, which cannot be altered by such additives.

Lubricants are additives which are used to help in the processing of plastics. Internal lubricants act within the plastic to reduce the forces between molecules. Other additives may act as external lubricants to reduce the adhesion of the plastic to the metallic surfaces of the processing machinery and molds.

4.3   GENERAL CHARACTERISTICS OF PLASTICS

As discussed in Section 4.1, the mechanical, physical and chemical properties are among the important factors that should be considered when selecting plastics for a given application. As the behavior of plastics can be very different from the behavior of metals, better understanding of the characteristics of plastics is essential for avoiding failure in service. Plastics should not simply be substituted for other materials, instead the design should make use of their advantages and should avoid their limitations. The following parts of this section will discuss the general characteristics of plastics and compare them to other materials.

Mechanical behavior of plastics

When a polymer is subjected to a continuously applied load, it undergoes both elastic deformation, which occurs instantaneously on application of load, and viscous deformation, which increases with time. The latter component of deformation is usually not fully recoverable on unloading. The recovery of dimensions after load removal is dependent on the loading time, and is generally more rapid after short periods of time at low strains than after long periods of time at high strains. The recovery time is an important parameter in applications involving intermittent loading. This behavior is called viscoelastic and is

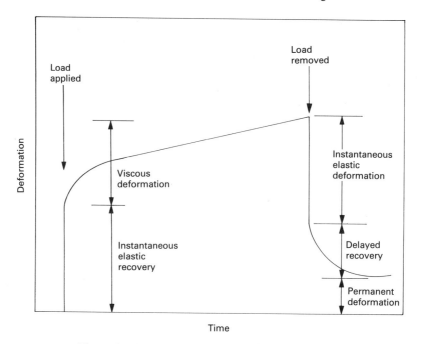

Figure 4.1 Typical viscoelastic behavior of plastics.

illustrated in Fig. 4.1. Generally, increasing the molecular weight of the polymer, by increasing the average chain length and cross-linking, causes the viscous deformation to decrease.

Since the mechanical properties of plastics are time dependent, creep properties should be considered when designing plastic parts that are expected to bear the applied stresses for considerable periods of time. Because of the viscoelastic nature of polymers, it is necessary to establish creep curves at several stress levels in order to define their performance, as shown in Fig. 4.2a. This behavior can be represented as stress versus time at constant strain, i.e. stress relaxation (Fig. 4.2b). A number of polymers show an exponential stress relaxation according to the relationship:

$$\sigma = \sigma_0 \exp(-t/\lambda) \tag{4.1}$$

where σ_0 = stress at time 0,
$\quad\ \sigma$ = stress at time t,
$\quad\ \lambda$ = relaxation time which is defined as the time necessary to reduce the stress to $(1/e)$ of its original value.

To illustrate the use of (4.1) in design, consider a fastener made of a given polymer with a relaxation time of one year at 20°C. The time needed to reduce the original stress exerted by the fastener by 25 percent is calculated as follows:

$$\sigma = 0.75 \star \sigma_0 = \sigma_0 \exp(-t/12)$$

Solving the above equation gives $t = 3.45$ months.

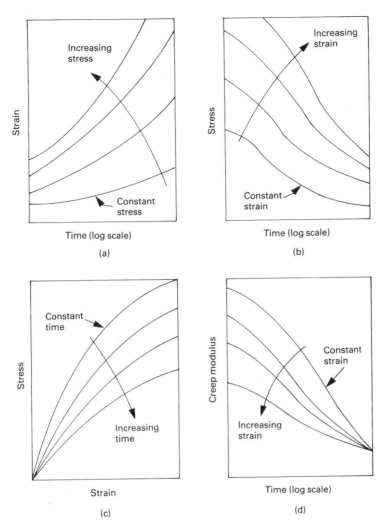

Figure 4.2 Mechanical behavior of plastics. (a) Creep curves at constant stress. (b) Stress relaxation curves at constant strain. (c) Stress–strain curves at different testing speeds. (d) Variation of creep modulus with strain and time.

Source: Farag, M. M., *Materials and Process Selection in Engineering*, Applied Science Publishers, London, 1979.

The viscoelastic behavior of polymers can also be represented as stress versus strain at constant time (Fig. 4.2c). The curves of Fig. 4.2c indicate that the slope of the stress–strain curves for polymers is not constant as in metals but is time dependent, hence the name creep modulus. Creep modulus versus time curves are therefore obtained from the curves of Fig. 4.2c by calculating the slopes at constant strain as shown in Fig. 4.2d.

It should be emphasized that the mechanical behavior of polymers is very sensitive to temperature and in some cases to humidity. For some materials, the creep moduli can change by up to 4 percent for each one degree C change in temperature or each 1 percent change in relative humidity. This factor should be taken into consideration when designing with plastics and when comparing the mechanical properties of different polymers produced by different manufacturers. The dependence of relaxation time (λ) on temperature can be represented by the general relationship:

$$1/\lambda = a\exp(-b/T) \qquad (4.2)$$

where T is absolute temperature
a and b are parameters which depend on the material.

In the case of the fastener discussed above, if the relaxation time is 2/3 year at 50°C for example, then the time needed to reduce the original stress exerted by the fastener by 25 percent at 40°C can be calculated from (4.2) as follows:

$$1/12 = a\exp(-b/293) \text{ and } 1/8 = a\exp(-b/323)$$

Solving simultaneously for a and b gives $a = 6.58$ and $b = 1280$. At 75°C, $1/\lambda = 6.58\exp(-1280/313) = 0.11$.

i.e. λ is about 9 months.

$$\sigma = 0.75\sigma_o = \sigma_0\exp(-t/9)$$

$$t = 2.59 \text{ months.}$$

Although short-term tests have been primarily developed for metallic materials, which do not usually exhibit viscoelasticity, they have been also used for plastics as they are convenient and readily available. Figure 4.3 compares some commonly used plastics on the basis of short-term tensile strength and tensile modulus. Short-term tests can predict the behavior of thermosetting plastics more accurately than the behavior of thermoplastics. This is because the rigid, brittle thermosetting plastics, like melamine and phenolics, exhibit an elastic behavior which extends essentially up to the fracture stress. The soft, flexible thermoplastics, however, exhibit considerable viscoelasticity and, for the short-term results to have any meaning, the test temperature and strain rate must be accurately specified. These results are often of more interest to the producer as a quick measure for material quality than they are to the designer who is usually concerned with the long-term performance of his design.

The impact strength of polymers can be measured in a similar way to metals and is usually presented in joule or ft lb per unit width of specimen or notch depth. The impact strength generally decreases with decreasing temperature and many polymers become very brittle at their glass transition temperatures. Plastics which are considered tough include polycarbonates, PTFE, ABS, nylon, polyethylene and high-impact polypropylene. On the other hand, acrylics and general-purpose polystyrene are among the brittle plastics.

Unlike steels, which can resist an infinite number of fatigue cycles if the applied stress is below their endurance limit, plastics do not have well-defined fatigue limits. Also, because of the viscoelastic nature of plastics, high-frequency cyclic loading can cause temperature rise and further reduction in strength.

Generally, plastics are much softer than many other engineering materials and their hardness is usually expressed by special scales of Rockwell and Shore hardness. Because of their low hardness, plastics are usually susceptible to abrasive damage from other harder materials. However, there are exceptions to this rule and some plastics have considerable

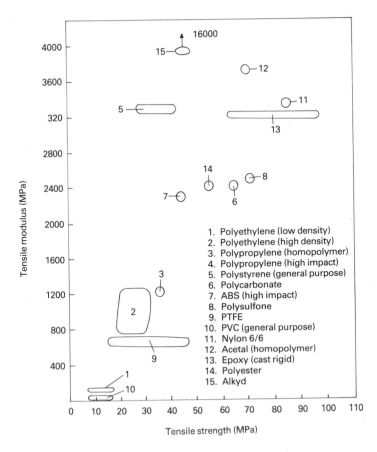

Figure 4.3 Comparison of some commonly used plastics on the basis of tensile strength and tensile modulus.

resistance to abrasion, even compared with metallic materials. Polyurethane is particularly good in this regard.

Physical properties of plastics

Low specific gravity is one of the important features that distinguishes plastics from other materials. The specific gravity of simple hydrocarbon polymers, like polyethylene and polystyrene, ranges between 0.9 and 1.1. Adding heavier atoms, such as chlorine or fluorine, at the expense of hydrogen increases the weight, e.g. the specific gravity of PVC ranges between 1.2 and 1.55 and that of teflon ranges between 2.1 and 2.2. The crystalline form of a polymer is always denser than the amorphous form, due to more efficient packing.

One of the severest limitations on the use of plastics in many applications is their limited resistance to high temperatures. The heat deflection temperature is usually considered as an indication of the suitability of polymers to high-temperature service. According to ASTM standards, this temperature is determined as follows: a sample of the plastic to be tested is shaped as a bar and placed in a liquid medium and loaded to a known

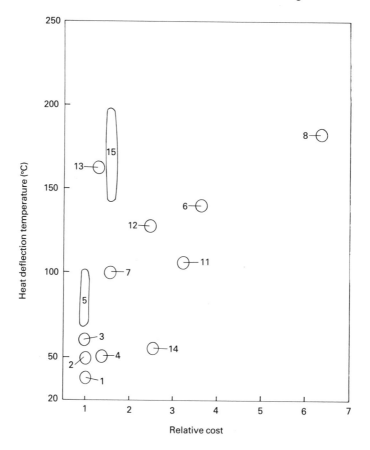

Figure 4.4 Comparison of some commonly used plastics on the basis of heat deflection temperature and cost (refer to Fig. 4.3 for key to numbers).

level. The temperature of the liquid is raised steadily (1.1°C/min or 2°F/min) until the bar has distorted by a predetermined amount (0.25 mm or 0.010 in). The temperature at which this distortion is reached is the heat distortion temperature. The data from such tests give reasonably good indication of the maximum temperature for short-term service for amorphous thermoplastics. For crystalline plastics, however, the deflection temperature usually gives a poor indication of the upper temperature limit for use of the material. Figure 4.4 compares some commonly used plastics on the basis of heat deflection temperature and relative cost.

Generally, the coefficient of expansion of bulk polymers is approximately two to 20 times as great as that of metals and is usually not linear. This means that if a plastic material is to replace a metallic material in a part with close tolerances, due allowance should be made. Fillers may be incorporated in plastics to reduce thermal expansion which could give anisotropic characteristics in the case of fibrous fillers.

Plastics are nonconductors of heat and electricity and some are among the best available insulators. As with mechanical properties, electrical properties vary with time and temperature and sometimes with humidity. The frequency and direction of the electric field, as well as the direction of the applied mechanical stresses in relation to molecular orientation can affect the measured values of electrical properties. Adding radicals and degradation due to radiation or heat, can result in some conductivity.

Plastics are usually transparent or translucent unless additives mask these properties. If the densities of the amorphous and crystalline forms of the polymer are close, or if the material is totally amorphous, there is little dispersion as light passes through the thickness and the material is transparent. Application of high mechanical stress can cause crazing which adversely affects transparency. Transparency can also be reduced in thicker sections and most polymers become dull in thicknesses above 25 mm (1 in). An exception is acrylic which can be obtained in thick sections with excellent optical properties. Among the optically clear polymers are acrylic, allylic and cellulosic. Among the polymers which can be translucent are: epoxy, polycarbonate, polyester, polyethylene, polypropylene, polystyrene, polysulfone, polyurethane, PVC and silicone.

Chemical properties

Generally, plastics can exhibit excellent resistance to many forms of chemical attack and are better than many metals, especially in weak acids or alkalis. They are, however, attacked by strong oxidizing acids. Thermoplastics can also be dissolved by various organic solvents. As the molecular weight increases, solubility in a particular solvent decreases. Cross-linking, even in slight amounts, may make the plastic insoluble. More crystalline polymers exhibit higher chemical resistance. The higher resistance is due to the denser packing of the chain molecules which makes the penetration of a solvent or other chemical substance more difficult. Fuels, fats, oils and even water may cause some plastics to swell and soften. This is of particular importance for materials used in gaskets and seals. Immersing stressed thermoplastics in some liquids could cause stress cracking.

Ageing and weathering of plastics depend on the nature of the environment and the incident radiation. Most plastics will oxidize and degrade if kept for long periods at elevated temperatures in the presence of air. Sunlight is also damaging, since the ultraviolet radiation can cause polymer degradation unless stabilizers are added.

4.4   THERMOPLASTICS

Of the two basic types of plastics, the thermoplastics family is the larger, both in number and in volume used (about 78 percent of the market). Generally, thermoplastics soften into a state of high viscosity when heated for processing. Most thermoplastics are furnished chemically complete, polymerized and in the form of small pellets or ready to mold. Some are also available as liquid resins for casting, laminating or coating. Some commonly-used plastics are compared in Figs 4.3 and 4.4 and typical properties of the major classes of plastics are given in Tables A.15 and A.16 in Appendix A.

Polyethylenes are the most widely used group of plastics and occupy 28 percent of the total world market of plastics. Polyethylenes are relatively cheap, tough, resistant to chemicals, have excellent dielectric strength and are easy to process. Because of their low water absorption and permeability, polyethylenes are widely used in sheet form as moisture barriers. Their largest drawbacks are their poor heat resistance and dimensional stability. The different types of polyethylenes can be classified according to their density into low, medium and high density, as shown in Tables A.15 and A.16. They can be easily processed by the different methods discussed later in this chapter. Typical applications include high-frequency insulators, piping, housewares, blow-molded bottles and containers, battery parts, toys and flexible film packaging.

Polystyrenes are second only to polyethylenes in volume of use and occupy about 17 percent of the total market share of plastics. They are based on the styrene monomer with low crystallinity and considerable branching. Various modifications create a wide range of structures and properties. Polystyrenes are generally rigid, have low cost and have exceptional insulating properties. Although they have reasonable tensile strength (Table A.15) they are subject to creep and their maximum service temperature is limited to about 77°C (c. 170°F) (Table A.16). The impact grades and glass-filled types are used widely for engineering parts and polystyrene foams are highest in volume use of all the plastics. They can be easily processed by common methods. Typical applications include housings for appliances like TV and radios, automobile dashboards, containers, housewares, luggage and furniture parts, fan blades and foam products.

Vinyls are a very widely used group of plastics and occupy about the same market share as polystyrenes. Like polyethylenes and polystyrenes, they are versatile and low in cost. Plasticizers can change the vinyls from a stiff and rigid state to a flexible and rubber-like one (Table A.15). Vinyls have excellent chemical resistance, high abrasion resistance and good electrical resistivity. Their maximum useful temperature is about 77°C (c. 170°F) (Table A.16). The most widely-used vinyls are polyvinyl chloride (PVC) and PVC-acetate copolymer. Vinyls can be easily processed by common methods. Typical applications include rigid PVC for the construction industry such as pipes, conduits, fume hoods and various building components. Flexible PVC is used in communication and low-tension cables, flexible tubing, footwear, imitation leather, upholstery, records, and film and sheet for chemical tanks and other applications.

Polypropylenes are the fourth-largest group of thermoplastics and they occupy about 5 percent of the total market share of plastics. They are similar in many respects to high-density polyethylenes and can be classified into three basic groups: homopolymers, copolymers and reinforced, as shown in Tables A.15 and A.16. Copolymers are produced by adding other types of olefin monomers to improve properties like low-temperature toughness. The mechanical properties can be further improved by adding fillers, glass fibers, or asbestos. The ability of polypropylenes to withstand fatigue loading has made them popular for surgical implant hinges. Although polypropylenes have low creep resistance, reinforcing them with asbestos can raise their maximum service temperature to about 125°C (c. 260°F). They are easily processed by common methods. Typical applications include house wares, luggage and furniture, automotive components, medical devices, piping, mechanical parts, housings, TV cabinets, containers, ropes and film and fibers.

The ABS plastics have a good balance of toughness, hardness and rigidity, which they maintain over the range of temperatures −40 to 107°C (c. −40 to 225°F). ABS have low water absorption, hence good dimensional stability. They also exhibit high abrasion resistance. As shown in Tables A.15 and A.16, ABS polymers are available in medium-, high- and very high-impact grades, as well as heat-resistant, transparent, flame retardant and expanded grades. ABS plastics are readily processed by all manufacturing methods for thermoplastics. ABS chromium-plated components are in wide use as a replacement for metals, e.g. in automobiles and appliances. Other applications include piping and fittings, appliance housings, telephone housings, refrigerator components, fume hoods and ducts, automotive components and tool handles.

Acrylic plastics are mostly based on methyl methacrylate polymers, PMMA, modified by copolymerization or blending with other monomers. They have high optical clarity, can acquire glossy surface and are available in brilliant transparent colors. Acrylics are strong, hard and stiff, but regular grades are brittle (Tables A.15 and A.16). High-impact grades are produced by blending with rubber stock. Acrylics are available as cast sheets, rods, tubes, or blocks and as powders for injection molding and extrusion. They can be processed by machining, molding and thermoforming. Molding compounds can be extruded or injection molded. Typical applications include transparent enclosures, optical uses such as lenses, signs, displays, window glazing, lighting fixtures, protective goggle lenses, reflectors, pump parts, control knobs and tool handles.

Polyamides (nylons) have basically linear molecular structures with a relatively high degree of crystallinity, which gives them good mechanical properties, as shown in Table A.15. They also have a very high abrasion resistance and good frictional characteristics. Nylons also have excellent electrical properties and chemical resistance. Their major disadvantage is their relatively high cost (Table A.16) and moisture absorption which results in dimensional changes. Nylon 6/6 is the most widely used of the nylons and types 6/10, 6/11 and 6/12 are less moisture absorbent than other types. Nylons are processed mainly by molding and extruding, but small parts can be processed by powder sintering. Polyamides are used in applications where precision is not the first requirement and where high wear resistance and low friction are required. Typical applications include bearings, gears, housings, bushings, tubing, pump housings, rollers, fasteners, zippers and electrical parts.

Acetals are highly crystalline and among the strongest and stiffest thermoplastics (Tables A.15 and A.16). Their excellent creep resistance and low moisture absorption give them excellent dimensional stability and they retain most of their properties in hot water. Acetals also have excellent resistance to vibration fatigue. They have low coefficient of friction and good abrasion resistance. They are useful for continuous service up to about 105°C (c. 220°F). Acetals are processed mainly by molding or extruding but some parts are blow molded. Typical applications include mechanical parts such as gears, bushings, bearings, cams, rollers, impellers, latches, wear surfaces and housings where high performance is required over a long period.

Polycarbonate is a linear, noncrystalline, transparent plastic. Polycarbonate is tough and has excellent electrical resistivity as well as outdoor stability. Its maximum service temperature is about 140°C (c. 285°F) and has good creep resistance under load. Although it has negligible moisture absorption, it is easily attacked by organic solvents. Polycarbonate has very low and uniform mold shrinkage and can be easily processed by all thermoplastic manufacturing techniques. Typical applications include safety helmets, safety shields, window glazing, windshields, lenses, load-bearing electrical parts, electrical insulators, battery cases, tool housings, medical apparatus and parts requiring dimensional stability.

Fluoroplastics are composed basically of linear molecules with fluorine replacing some or all of the hydrogen atoms. They are highly crystalline with high molecular weight. As a class fluoroplastics rank among the best plastics for chemical resistance and for high-temperature performance up to about 260°C (500°F). Polytetrafluoroethylene (PTFE) is the most widely used fluoroplastic because of its high service temperature. PTFE is usually fabricated by extruding or compacting the resin and then sintering. Typical applications include linings for chemical processing equipment, chemical pipes, pump parts, valves, coatings for home cookware, insulation for high-temperature wire and cable, gaskets and low-friction surfaces.

Polyesters, thermoplastic type, have good mechanical and electrical properties combined with excellent dimensional stability. Their chemical resistance is good except against strong acids or bases. Polyesters are not suitable for outdoor service or in water above 52°C (c. 125°F). They can be processed by molding and extrusion and typical applications include switch housings, tapes, gears, cams, bushings, pumps, automotive parts and similar load-bearing parts. Polyesters are also available as thermosetting formulations.

Polyurethanes are tough, with excellent abrasion and impact resistance. Their electrical and chemical properties are also good, but exposure to ultraviolet rays produces brittleness, reduces strength and causes yellowing. They can be made in solid moldings and/or flexible foams. Polyurethanes are also available in thermosetting formulations.

Cellulosics are a family of cellulose derivatives and the four common ones are cellulose acetate, ethyl cellulose, cellulose propionate and cellulose acetate butyrate. Generally they are tough and hard with good optical properties. Cellulosics can be processed by many thermoplastic processing techniques. Typical applications include tool handles, pens, knobs, safety goggles, lighting fixtures, steering wheels, toys and packaging.

Several groups of thermoplastics have been developed to withstand temperatures in excess of 200°C (c. 390°F) for extended periods. These include polyimide, polysulfone, polyphenylene sulfide and polyarylsulfone (Tables A.15 and A.16). In addition to their high temperature resistance, these plastics have high strength and tensile modulus and excellent resistance to solvents. Their major disadvantage is the difficulty in processing them. Their high price limits their use to specialized applications where resistance to high temperature is important.

4.5   THERMOSETTING PLASTICS

Although thermosetting plastics are fewer in number and their manufacturing processes are more limited than thermoplastics, their special characteristics make them indispensable in some applications and at present they occupy about 14 percent of the total market share of plastics. The presence of the network of strong covalent bonds, which are developed due to cross-linking between molecules, in thermosetting plastics makes them react quite differently to temperature and mechanical stresses than thermoplastics. Thermosetting plastics generally resist higher temperatures and are harder but more brittle than thermoplastics. They also have better dimensional stability, creep resistance, chemical resistance and electrical properties. Thermosetting plastics are generally more difficult to process than thermoplastics and, once cured, i.e. cross-linked, they will not return to their original state. Before curing, thermosetting plastics are composed of a resin system and fillers. Sometimes reinforcements are also added. Thermosetting plastics are usually classified according to the resin component, which consists of a polymer, curing agents, hardeners, inhibitors and plasticizers. The resin component influences the dimensional stability, heat and chemical resistance, electrical properties and flammability. The fillers usually have a major effect on mechanical properties, especially when in the form of fibers. The behavior of fiber-reinforced materials will be discussed in Chapter 6. Figures 4.3 and 4.4 compare some thermoplastics with thermosetting plastics and Tables A.15 and A.16 list some properties of selected thermosetting plastics.

Phenolics (phenol formaldehyde) are the most widely used group of thermosetting plastics and occupy about 6 percent of the total market share of plastics. Although brittle, phenolic resins have high resistance to heat, water and chemicals. They are relatively cheap and are readily molded with good stiffness and impact resistance, but their color is usually limited to black or brown, Phenolics can be classified as: general purpose with wood flour fillers, shock resistant with paper or fabric fillers, heat resistant with mineral or glass fillers, electrical grade with mineral fillers, chemical grade with no fillers and rubber phenolics. Phenolics can be processed by compression, transfer or injection molding, as well as by extrusion. Typical applications include motor housings, pulleys, wheels, appliance connector plugs, ignition parts, condenser housings, electrical insulators, handles and knobs.

Aminos, urea and melamine, are the second-largest group of thermosetting plastics and occupy about 4 percent of the total market of plastics. Their properties depend on composition but they are generally hard and rigid as well as abrasion and chip resistant. Urea molds faster and costs less, but melamine has harder surface and higher heat and chemical resistance. Typical applications of urea are in electrical and electronic components and typical applications of melamine are in dinnerware and handles.

Epoxies are one of the most versatile plastics and are used in high-performance applications where their high cost is justified. They are available in a wide variety of forms, both liquid and solid, and are cured into the finished product by catalysts. The epoxy resins give good chemical resistance and electrical properties in addition to excellent bonding properties. They also have exceptionally high strength, especially when reinforced with glass fiber or other similar fibers, as will be discussed in Chapter 6. Liquid epoxies are processed by casting, and powders by molding. Some formulations can be cured without heat or pressure. Typical uses of epoxies include adhesives, components requiring high strength and thermal insulation, encapsulation of electronic components and tools and dies.

Polyester thermosetting resins are copolymers of a polyester and, usually, styrene. Unreinforced polyesters have limited use and the majority of products are glass reinforced either as moldings or laminates. When reinforced, polyesters exhibit excellent balance of mechanical, electrical and chemical properties. They can be made in a large number of colors and give off no volatiles during curing. Polyesters have low moisture absorption. They are widely used to produce 'fiberglass' boat hulls, swimming pools, chairs, automobile body panels and other high-strength components.

Polyurethane thermosetting plastic can be flexible or rigid, depending on formulation. It has excellent toughness and abrasion resistance. It is particularly suitable for foamed parts, in either flexible or rigid types.

Silicones are semiorganic polymers composed of monomers in which oxygen atoms are attached to silicon atoms together with radicals. Silicones are available as liquids, elastomers or rigid solids depending on the type of radicals present. Silicones can be used in a wide range of temperatures from -73 to $+260°C$ (-100 to $+500°F$). They have excellent electrical properties and resistance to water and certain chemicals. They are also compatible with body tissues. Silicones are premium plastics and are used in critical or high-performance applications, such as aircraft, aerospace and electronics.

4.6 ELASTOMERS

Elastomers are a large family of polymers with the characteristic ability to undergo very large elastic deformations without rupture. They are soft and have a low elastic modulus and a low glass-transition temperature. The term elastomer is derived from the words

elastic and mer, and indicates a family of rubber-like materials. All rubbers are elastomers, but not all elastomers are rubbers. An elastomer is defined as being capable of
recovering substantially in shape and size after the load has been removed. Rubber is
defined as being capable of recovering quickly from large deformations.

Rubbers are generally hydrocarbon polymers with sulfur or other additives cross-
linking the polymers to each other, producing a thermoset-like structure. The rubber
becomes stronger but less extensible as the amount of cross-linking between sulfur and
carbon atoms increases. When rubber is stretched, the polymer chains tend to straighten
and become aligned, which increases crystallinity and strength. Table A.17 lists some
properties of selected rubbers which will be discussed below.

Natural rubber is a homopolymer of the isoprene monomer. A number of grades of
rubber are made from the raw material by adding fillers like carbon black, silica and
silicates. The soft grades have excellent resilience and good abrasion resistance, with low
hysteresis and heat build-up under repeated loading. They have low resistance to oil, heat,
ozone and sunlight. Additives like carbon black improve the resistance to ultraviolet
radiation. Major uses of natural rubber include conveyor belts, tire products, sound
damping and gaskets.

Styrene butadiene rubbers (SBR) are copolymers of butadiene and styrene and are
the most widely used group of rubbers because of their low cost and their use in automobile
tires. A wide range of properties can be obtained by changing the butadiene to styrene
ratio, and the grades containing more than 50 percent styrene are usually considered as
plastics. SBRs have excellent impact and abrasion resistance but poor chemical resistance.
Their major uses include automobile tire treads, conveyor belts, gaskets and sound
damping.

Neoprene (chloroprene) is a synthetic rubber which is chemically and structurally
similar to natural rubber. It has excellent resistance to oils, chemicals, sunlight, weathering, ageing and ozone and retains its properties at temperatures up to 115°C (c. 240°F). In
addition, it has excellent resistance to permeation by gases, but it is more expensive. Major
applications include heavy-duty conveyor belts, V-belts, footwear, brake dia-
phragms, bridge mounts and motor mounts.

Butyl rubbers consist of copolymers of isobutylene and a few percent of isoprene and
are one of the lowest-priced synthetics. Because of their excellent dielectric strength and
impermeability, they are widely used for cable insulation, encapsulating components and
similar electrical applications. Other applications include coated fabrics, high-pressure
steam hoses, inner tube linings and machine mounts.

Some silicones can be considered as rubbers in view of their extensibility. Silicone
rubbers are the most stable of all rubbers with excellent resistance to high and low
temperatures, oils and chemicals. They maintain their electrical properties over the range
of temperatures −73 to +270°C (c. −100 to +520°F). All silicone rubbers can be classified
as high-performance, high-price materials and their major uses include seals, gaskets,
O-rings, encapsulation of electronic components and insulation for wire and cable.

4.7   ADHESIVES

One important application of polymers is as adhesives. Adhesives with strengths higher
than the strength of some metals have been developed and are being used to replace

welding and riveting in many applications. The use of adhesives covers the range from aircraft and missile skins to food and beverage cans. This is because:

1. Adhesive bonding can be effectively used to combine different materials while maintaining their integrity.
2. Stresses are uniformly distributed over the bonded area.
3. Adhesive joints provide sealing against the environment.
4. The fatigue behavior of adhesive bonded structural assemblies is excellent.

There are many types of adhesive and they are usually classified according to their chemical composition into:

1. Natural resins such as starches and dextine, and animal glues. Natural adhesives can be applied in solution or as emulsions. Evaporation of the vehicle, which is usually water, results in hardening. Such adhesives do not usually resist water and are relatively weak.
2. Inorganic materials such as silicates.
3. Polymeric materials such as thermoplastic resins, thermosetting resins and rubbers (elastomers). This group is most widely used in engineering applications and will be discussed in more detail here.

The mechanism of adhesion in many cases could be a combination of mechanical locking and van der Waals bonding between the adhesive and substrate. Mechanical locking can be enhanced by increasing the roughness of the surfaces by grit-blasting or rough machining. Surface cleanliness and the presence of chemical compounds on the surface are also important in the case of adhesive bonding. Most adhesives develop their maximum load-carrying capacity under shear stressing and some excel in tension. In all cases, however, adhesives are weak in peel and tear loading. In general, good joint design profits from maximizing shear and tensile/compressive stresses while minimizing those of peel and tear, as discussed in Sections 3.7 and 12.7. The factors that affect the selection of an adhesive for a given joint include the service temperature, the chemical environment, the expected life of the joint and the nature of the surfaces to be joined. The method of application and curing treatment can also play an important role in the selection of adhesives.

There is a very large number of polymers that can be used as structural adhesives. Many of these are composite systems with several components and may be available as liquids, pastes, solids, pellets, cartridges or films. Table A.18 gives the properties and method of preparation of some commonly used adhesives.

Epoxies consist of an epoxy resin plus a hardener and are available in both one-part, hot cure, and two-part, cold or warm cure, form. They allow great versatility in formulation since there are many resins and hardeners in this group.

Cyanoacrylates are one-part adhesives which consist of liquid monomers that polymerize through reaction with moisture on the surfaces to be bonded. They set very quickly, in a few seconds, but they require close-fitting surfaces because they do not fill large voids. Exposure to temperatures above 80°C (c. 175°F) in high humidity can cause bond failure. Most consumer-oriented superglues are of this type.

Anaerobics are one-part polyester acrylic adhesives which harden when in contact with metal surfaces and when air is excluded. Rubber-modified anaerobics remove odor, flammability and toxicity, while speeding the curing operation. They can bond almost anything, including oily surfaces, and are often used to secure, seal and retain threaded or similarly close-fitting parts.

Toughened acrylics are fast curing and give high strength and toughness. They are supplied in two parts, resin and catalyst, which can be applied as premixed or applied separately. In the latter case the resin is applied to one surface and the catalyst to the other, Toughened acrylics tolerate minimal surface preparation and bond well to a variety of materials including oily metals and many plastics. They often contain volatile, flammable monomers and special vapor extraction systems may need to be installed if used on a large scale.

Polyurethanes are either one-part or two-part adhesives which cure fast at room temperature. They provide strong resilient joints with high impact resistance. Their fast cure and sensitivity to moisture in the uncured state usually necessitate machine application. They are weak in creep and should only be considered for nonstructural applications.

Hot melts are thermoplastic adhesives which are solid at room temperature but melt when heated. They are generally applied as heated liquids and form a bond as they cool. They are generally used in nonstructural applications.

4.8   PROCESSING OF PLASTICS BY MOLDING

There are many methods of manufacturing plastic parts which can be considered as molding processes. These processes usually employ the following sequence of steps:

1. Plastics in the form of powder, pellets, or granules are usually heated above the softening point.
2. The molten plastic is forced or placed into a mold which determines the dimensions of the molded part.
3. The material is then allowed to harden, by curing or freezing, and is then ejected from the mold.

In some molding processes the ejected part is ready for use. With others, trimming and other finishing processes are necessary to make it ready for use. The pressure-molding processes, as opposed to gravity processes, are the most widely used because they are faster and give closer dimensional control. Pressure-molding processes cover compression molding, transfer molding and injection molding. These processes are similar in some ways to die-casting processes used for metals. It should be noted, however, that molten plastics are more viscous and much less conducting to heat than metals. These differences are reflected in the design and operation of the plastics-molding machines. Because of the very low thermal conductivity of plastics, the heating process is slow. Attempting to increase the production rate by increasing the wall temperature of the heating chamber could cause degradation of the surface layers of the plastic and poorer performance of the molded part. Pressure-molding processes will be discussed in the following sections. Gravity molding, usually called casting, will be discussed in Section 4.11.

Compression molding

Compression molding is usually used for thermosetting plastics. It involves feeding the premeasured amount of partially cured material into the mold and then squeezing it with a ram to fill the cavity, as shown in Fig. 4.5. Under the action of both heat and pressure the molding material becomes plasticized, fills the mold to the shape of the cavity, cures and hardens. The heat required for curing is supplied through the mold walls. When the

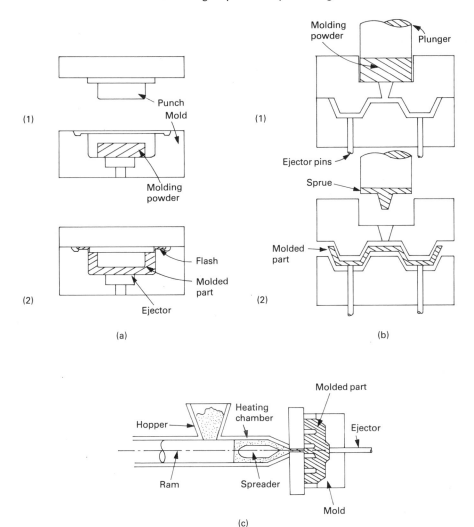

Figure 4.5 Processing of plastics by molding. (a) Compression molding. (b) Transfer molding. (c) Injection molding.

material is supplied as loose powder or pellets, several minutes are needed for each cycle. The cycle time can be reduced by preheating the material to about 140°C (*c*. 280°F) before placing it in the mold. Cold pressing the material in the shape of a preform can also greatly reduce the cycle time. Compression molding is similar to forging, with the same problem of flash formation at the parting plane and the need for its removal. Another time-saving process is to cold press the material to the required form and then bake it in an oven to cure it. This latter process is similar to the powder metallurgy techniques described in Section 3.4.

The most commonly used thermosetting materials for compression molding are various types of phenolics and aminos (ureas and melamine). A range of filler materials can be used. Other important materials are the alkyds and polyesters.

The size of articles which can be compression molded is limited by the size and force capacity of the available press. Compression molding is most suitable for simple shapes with uniform wall thicknesses, preferably in the range 3 to 6 mm (1/8 to 1/4 in). Thicker walls are undesirable because of long curing time, and thicknesses of about 12 mm (1/2 in) represent a practical upper limit. Very intricate parts with undercuts, small holes, or sidedraws may not be suitable for compression molding. Normally a draft of at least one degree per side is needed. All materials shrink to varying degrees during the molding process and the greater the shrinkage the more tolerance should be allowed. The range of expected molding shrinkage per unit length in the commonly used materials is:

Phenolics	0.001 to 0.010
Alkyds	0.000 to 0.006
Urea-formaldehyde	0.004 to 0.008

Typical parts produced by this process include door knobs, handles, fittings and housings. Fiber-reinforced materials may also be formed by compression molding.

Transfer molding

Transfer molding is closely allied to compression molding and is used mainly for thermo-setting plastics. In this process, the material is first heated and compressed in a chamber, then forced through an orifice into the mold cavity, as shown in Fig. 4.5. Considerable work is done on the material as it flows to the mold cavity, and this results in homogenizing its consistency and raising its temperature. The mold is also heated so that the temperature of the molded material is further raised to continue the cross-linking process. The mold remains closed until the curing reaction is completed, then it is opened to remove the part.

Transfer molding was developed to overcome some of the design limitations on compression molded parts. Cycle times are shorter but the mold costs are higher than comparable compression molds. Transfer molding can be used to produce more intricate sections, thinner walls, closer tolerances and more uniform density than compression molding. Typical parts include electrical and electronic components.

Injection molding

Injection molding is a versatile production process for converting thermoplastics and thermoset materials into molded parts of relatively intricate shapes at fast rates and with good dimensional accuracy. This process is used to produce more thermoplastic products than any other process. Injection molding is essentially similar to the hot-chamber die-casting process described in Section 3.3. Viscosity of the melt is an important parameter as it affects the material flow and the required injection pressure. Raw material is fed by gravity into a pressure chamber ahead of a ram, or screw (Fig. 4.5). As the ram advances, the polymer is forced into a heating chamber and then through a nozzle to the split-mold cavity. In this process the mold remains cool, for thermoplastics, so that the material solidifies as soon as the mold is filled. In this case, the complete cycle requires only a few seconds. The fast cooling of the molten plastic can cause internal stresses to exist in the molded part and could cause warping and poorer performance. Decreasing the cooling rate by increasing the mold temperature reduces the internal stresses but lengthens the cycle time and increases the cost of production. A compromise has to be achieved between the amount of internal stresses and production cost, depending on the shape and expected use of the part.

Another factor which affects the performance of injection molded parts is the difference in crystallinity between thin and thick sections and between the surface layers and bulk of the section. Normally the surface layers and thin sections will cool faster and, therefore, show lower crystallinity. As discussed earlier, materials with lower crystallinity have less strength and lower modulus of elasticity, but are more ductile and tougher.

Practically all thermoplastics can be injection molded. Thermosetting plastics may also be injection molded, but the mold has to be heated to allow the completion of cross-linking. In this case, the machine setting must ensure that the material does not cure before it fills the mold. Typical injection molded parts include gears for instruments and small machines, screw caps for bottles, battery cases, trash cans, cups and containers, various electrical and communications components, toys and pipe fittings. Parts that need to be made from different color polymers can be injection molded in several steps. The high cost of molds limits the process to high-production volumes.

In reaction injection molding (RIM), a mixture of two or more reactive fluids is forced under high pressure into the mold cavity. Chemical reaction takes place in the mold cavity and the material solidifies. Major applications of RIM are in automotive parts, bumpers and fenders, thermal insulators, refrigerators and freezers.

4.9   EXTRUSION OF PLASTICS

Extrusion is a process used to produce continuous shapes from the raw material which can be in the form of granules, pellets, or powder. The raw material is fed from a hopper into a screw chamber where it is preheated, compressed and then forced through a heated die onto a conveyor belt, as shown in Fig. 4.6. The material may be heated either through internal friction during extrusion and/or by external means such as a heated chamber. The process blends, compounds, homogenizes and extrudes the plastic at temperatures in the range 130 to 400°C (c. 270 to 750°F). As the material leaves the die it must be cooled rapidly to maintain the extruded shape. Cooling is effected by air blast, water spray, or submersion in water. The rate and uniformity of cooling is important for dimensional control because of shrinkage and distortion. There is usually a significant difference in size between the die orifice and the final dimensions of the product. This is due to the stretching of the plastic immediately after leaving the die.

Extrusion can be used to produce sheets (thicker than 0.75 mm or 0.030 in), films (thinner than 0.75 mm or 0.030 in), rods, filaments, tubes and a variety of profiles. Extrusion is also used to coat metallic wires, cables, strips and roll-formed shapes with plastics. Dual extrusion can be used to combine two dissimilar plastics into a single profile. Simplicity of sectional profile is always desirable. Balanced wall thickness is usually of particular importance, especially for rigid materials. Wall thicknesses that can be extruded range between 0.50 and 6.25 mm (c. 0.020 and 0.250 in). Sink marks, due to shrinkage, are likely to occur on flat surfaces opposite an adjoining leg or rib.

Most thermoplastics and elastomers as well as some thermosets can be extruded. The most commonly extruded materials are thermoplastics and these include rigid vinyl, flexible vinyl, polystyrene, ABS, polypropylene and polyethylene. Other plastics that can be extruded include nylon, polycarbonate, polysulfone and acetal. Because the extruded material is normally pulled as it comes out of the die, the polymer chains tend to be aligned in the drawing direction. This results in directionality of the properties and the material is stronger in the length direction than in the width direction, especially in the case of filaments, thin sheets and films. While directionality may be desirable in the case of filaments, it may be unacceptable in films. Directionality can be reduced in the latter case by stretching the film in both the length and width directions as it leaves the die.

Although some thermosets can be extruded in the form of tubes and wire coatings, strict control of temperatures and speeds is necessary to avoid premature, or insufficient, curing of the material within the extruder.

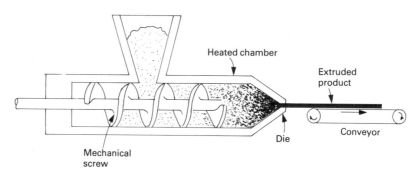

Figure 4.6 Extrusion of plastics.

4.10   THERMOFORMING

Thermoforming processes include many shaping processes which are applied to previously shaped plastics with the help of heat. In some cases the term is limited to sheet forming, while thermoforming of tubes is called blow molding. In general, all thermoforming processes are used for thermoplastics which soften when heated and then harden when cooled. Most thermoplastics which are available in the sheet or tube forms are suitable for thermoforming. Materials which are widely used for thermoforming include polystyrene, ABS, PVC, acrylic, cellulosics, polycarbonate, high- and low-density polyethylene and polypropylene.

Thermoforming of sheets

There is a variety of techniques for thermoforming of sheet or film which are all based on the application of heat and air pressure or vacuum, as shown in Fig. 4.7. Large, thin-walled parts of uniform thickness with generous radii, draft and flowing contours are well suited for these processes. Although the side in contact with the mold can be produced with relatively high gloss, the side away from the mold normally has very little gloss. If rigidity is a problem, ribs, ridges or metallic inserts should be introduced. These processes cannot form parts with openings or holes since the pressure cannot be maintained. These features should be introduced by secondary operations like blanking or drilling. Typical applications of sheet thermoforming include liners for appliances like refrigerators and freezers, baths and panels for showers, sinks, covers and housings, advertising signs and packaging.

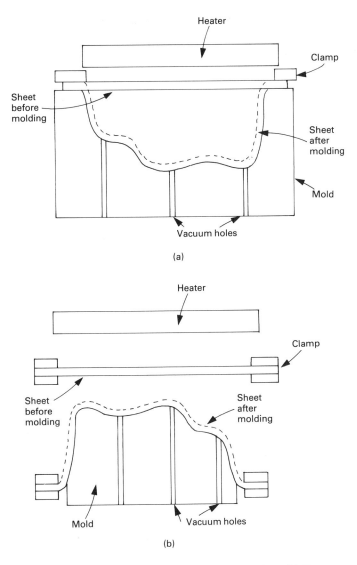

Figure 4.7 Thermoforming of plastics. (a) Straight vacuum forming. (b) Drape vacuum forming.

Blow molding

In blow molding, a tubular piece of thermoplastic is heated and then pressurized internally and expanded into a cavity of relatively cool split mold, as shown in Fig. 4.8. Blow molding is usually combined with extrusion or injection molding in automated processes. In this case a hollow tube of molten plastic, parison, is produced by an extruder and is then pinched at one end. The parison is then placed into the open mold. The mold is closed and the parison is expanded, blown, with an air nozzle to fill the mold. The part cools off after contact with the mold walls and is then ejected. In some cases, the extruder is replaced by an injection molding machine in the production of parison.

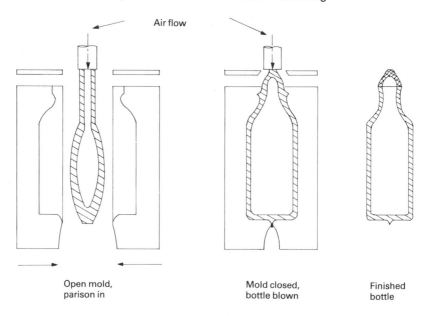

Figure 4.8 Blow molding of plastics.

Typical products include one-piece thin-walled containers and bottles, storage tanks, laboratory ware, waste baskets and toys. In general, the shape should be designed with generous corner radii. Blow molded parts tend to have nonuniform wall thickness. Those areas of the parison which touch the cool mold wall first will have the thickest sections. Deep narrow grooves are likely to result in unacceptable thinning of the sheet and should be avoided. Narrow bottle necks and handles should be located on the parting plan of the mold.

4.11   CASTING OF PLASTICS

Some thermoplastics and some thermosets can be processed by casting. These materials can be cast under gravity in relatively cheap molds, which could be rigid or flexible. Molds can be made from a variety of materials including wood, lead, plaster, plastics, rubber or metal. In some castings, the top of the mold is left open and this side of the part is simply the horizontal free surface of the liquid. Other casting molds are made in two or three pieces and are filled through a pouring system that is later trimmed off. The mold materials are cheap and the process involves no expensive machinery, but it is slow. Casting is used to form parts which could be prohibitively difficult if produced by other methods or where the production volume is low. Casting lends itself to multiple pouring which could be used to produce parts containing different materials and embedments. Most reinforced plastic parts can be made by pouring liquid resin over glass-mat reinforcement that has been placed in the mold.

Casting of thermoplastics involves mixing a monomer, catalyst and additives, heating the mixture and then pouring into the mold under atmospheric pressure. A polymerization reaction ensues and the material is formed. The part is ready to be taken out of the mold after polymerization is completed at ambient pressure. Acrylics and nylons can be used to produce large parts of contoured form, free from voids, and several inches thick. Parts weighing from 0.5 to 680 kg (1 to 1500 lb) have been cast. Minimum recommended thickness is about 6 mm (1/4 in), although 1.5 mm (1/16 in) can be achieved. Normal casting tolerances are ±1 percent. Dimensions with better tolerances are achieved by subsequent machining. Typical parts include large gears, bearings and wheels. Intricate shapes can be formed with flexible molds, which are then peeled off.

Casting of thermosets is of similar nature. Degassing may be necessary for improved integrity. Epoxies, phenolics, polyesters and polyurethanes are among the thermosetting materials which are cast. A mixture of monomer, catalyst and additives is poured into the mold. Curing can take place at ambient pressure. Both clear and filled casting resins are used in a variety of items including bowling balls, tools, gears, cams and wheels, as well as pots and containers.

Rotational molding, or casting, permits the manufacture of completely enclosed hollow moldings. A hollow thin-walled sectional mold is charged with a predetermined amount of liquid plastic solution and then closed. The closed mold is rotated in two or more planes and at the same time heated to gel or fuse the film of plastic which has been evenly distributed on the interior walls of the mold. The mold is then cooled and opened to remove the molded article. The advantage is that loss in scrap and trimming is minimized. Applications include soft vinyl toys and motor car parts. As an example of the latter application, the arm rest is made by first molding a vinyl skin to give a leather grained surface and then blowing vinyl foam into the mold after the skin surface has gelled.

Another process which is related to casting, is encapsulation which involves casting the plastic around an electrical component or a system, thus embedding it in plastic. The plastic serves as an electrical insulation and also as a structural member supporting the system.

4.12   FASTENING AND JOINING OF PLASTIC PARTS

Plastics can be fastened and joined by a variety of methods including mechanical means, solvents, adhesives and different types of welding.

Mechanical fastening

Press and snap fits can be used to assemble plastic parts by employing molded-in ribs and mating slots. These fits require close dimensional tolerances. The plastic must be flexible and tough to allow one part to be forced into the other without cracking. In addition, conditions which could lead to creep at the joint should be avoided. An example of a snap fit is shown in Fig. 12.2(d). In some applications, parts can be assembled by molding them with integral and mating threads. Threads are usually not desirable in molded parts as they require complex molds and longer time to remove from the mold. Both factors contribute to an increased production cost. Metal screws can be used for fastening plastic parts. Self-tapping screws which are made of plastic can be adequate if the parts will not be disassembled frequently. If the parts are intended for frequent disassembly, threaded metal inserts should be molded in. Rivets are easily installed and are of low cost, but they are not as accurate as threaded connections.

Solvents and adhesives

Solvents can be used to bond thermoplastic parts. The surfaces to be bonded are first softened by wetting with an appropriate solvent and then clamped together to achieve a chemical bond between them. This process is applicable to soluble thermoplastics like acrylics, polystyrenes and some vinyls. The strength of the joint is comparable to that of the parent material.

Adhesives can be used to join both thermosets and thermoplastics. Adhesive bonding is the preferred method for joining thermoset parts. When designing adhesive-bonded joints, the area of contact should be maximized, as will be discussed in Section 12.6. Some plastics are difficult to join using either adhesives or solvents because of their inherent chemical resistance. Such materials include acetals, tetrafluoroethylene, polyethylene and polypropylene.

Welding plastic parts

Several methods of welding can be used to join plastic parts and are usually called after the source of energy used for welding. Most thermoplastics can be welded by applying enough heat energy to cause melting of the surfaces to be welded. The surfaces are then pressed together to allow fusion and to ensure a good bond. The sources of energy can be ultrasonic vibrations, friction, hot tool, induction heating or hot gas. Thermosets do not melt when heated and cannot, therefore, be welded in this fashion.

Ultrasonic welding uses the frictional heat generated by ultrasonic vibrations to fuse the interface of thermoplastic parts and join them. Most ultrasonic welding is used for polystyrene and can be used only with relatively rigid thermoplastics. Continuous welds up to 250 mm (10 in) are possible. The joint design of the two parts to be welded should encourage rapid heating and proper alignment. A similar process is spin welding, where the frictional heat is generated by rotating one part relative to the other. Ping-pong balls and similar parts are produced by spin welding.

Hot plate welding is performed by first lightly holding the surfaces to be welded against a heated metal plate until they melt and then pressing them together until they cool. The equipment is simple and cheap, but consistent results are difficult to obtain.

Thermoplastics can be joined by induction heating. An electromagnetic material, which consists of a dispersion of finely divided metal particles in a thermoplastic matrix, is placed between the two surfaces to be welded. The assembly is then put in a high-frequency alternating magnetic field which heats the metal particles. Heat from the metal particles causes the thermoplastic to melt and fuses the surfaces to be welded.

In gas welding, the heat source can be hot air or other gases. Usually an inert gas, such as nitrogen, is preferred to air. This is because oxygen in the air could oxidize and degrade the welded surfaces. The heat of the hot gas melts the surfaces to be joined. Pressure is then applied to allow fusion to take place at the interface and to ensure a good bond. Plastics like PVC, polypropylene, polyethylene, acrylics and ABS can be welded in this manner.

4.13   FINISHING OF PLASTIC PARTS

Finishing operations represent an important step in the manufacture of plastic parts. Incorrect finishing could damage or distort the processed part or could add unnecessary expense to the cost of production. In designing a plastic part the finishing processes, as well as the functional requirements, should be considered.

Machining of plastics

Plastics can be easily machined using conventional processes and machine tools. By using appropriate tools and techniques, plastics can be cut, machined, punched, pierced, folded, bent and ground. The main considerations that should be kept in mind when machining plastics are:

1. Plastics do not dissipate the heat generated in cutting.
2. Thermosets are sensitive to thermal gradients during cutting and the machined surface may char if the cutting conditions are too severe.
3. Thermoplastics may soften and clog the tools if overheated in cutting.

Plastics are best cut with tools that have positive rake angle (up to 20 degrees) and high relief angles. Small depth of cut, small feed, relatively high speeds, sharp tools, sufficient cooling and proper support of the workpiece are therefore required. Residual stresses may develop in thermoplastics during machining. To relieve residual stresses, parts are annealed at temperatures ranging from 80 to 160°C (175 to 315°F) and then cooled slowly and uniformly.

Machining of plastics is usually performed either as a finishing process or as the main manufacturing process. Generally, it is desirable to reduce secondary processing of molded parts. However, machining of molded parts may be necessary to introduce features that are difficult or impossible to incorporate in the molding process. Examples include undercuts and holes at different angles. Machining can be selected as the main manufacturing process when only a few parts are needed, when the accuracy required is too high to be achieved by other processes, when the shape of the part permits economic production by machining form solid, or when the nature of the material does not permit molding. Under these conditions, the part design should take into account the available basic shapes which will be used as the stock material. Both thermoplastics and thermosets are available in the form of sheet, rod, tubing, billets and various other standard mill shapes. Thermosetting plastics are also available as laminates with fiber or fabric reinforcement. Laminates are graded as either hot-punch or cold-punch. In the former case, the material must be heated before punching to avoid cracking and chipping. Practically all thermoplastics are also available as flat or tubular films and many are available as foams of different densities and rigidity.

Decorating operations

Decorating operations consist largely of painting various features on molded parts or basic shapes such as sheets. Plastics vary considerably in their ability to take and hold various paint finishes. For example, cellulose acetate and polystyrene can be painted easily and successfully while polyethylene and polypropylene must have their surfaces oxidized. Polyolefines and some thermosets can be painted with difficulty only by the use of special pretreatment. Among the techniques used for painting plastic parts are spray painting, wipe-in painting, silk screening and printing. Thermoplastics may also be decorated by hot stamping, where a heated tool is pressed on the surface to produce the desired depression. A foil of the desired color could be placed between the tool and the surface to create the required effect. Engraving gives a similar effect to hot stamping and can be used to create a two-color effect on laminates made from layers of different colors.

Another technique for decorating thermoset moldings is the use of impregnated foils or papers which are placed in the mold with the plastic to be molded. The foils are impregnated with the same plastic and decorated by printing prior to placing in the mold. An example is the decorated melamine tableware. Similar techniques can be used for thermoplastics. In this case, the foil is placed between two thermoplastic films which are fused to the surface during molding.

Metallic coatings

Vacuum metallizing and electroplating are the two main methods of producing metallic coatings on plastic surfaces. In vacuum metallizing, the plastic surface is first coated with a lacquer to promote adhesion of the metallic film. The part is then placed in a vacuum chamber where it is coated with a vaporized metal, which is usually aluminum. The surface is finally given a top coat of lacquer to protect the delicate, mirror-like, metallic film. Baking of the lacquer films may be necessary in some cases to harden them and to remove volatiles which could cause blistering of the coating. Colors other than the silvery color of aluminum can be given by tinting the top coat of lacquer. In electroplating, the plastic parts, which are normally nonconducting, are first given a conductive coating. This will allow the plastic parts to be treated like metallic parts.

4.14   DESIGN AND SELECTION CONSIDERATIONS FOR POLYMERS

The increasing use of plastics in engineering applications can be explained on the basis of economics or performance. Although the price of most plastics may appear high in comparison with other materials, when compared on a weight basis, conversion of this cost to volume basis shows plastics to be in a favorable position. The ease of manufacture and the need for little or no finishing further adds to the economic advantage of plastics. From the performance point of view, there are many applications where plastics outperform other materials. Their light weight, corrosion resistance, low coefficient of friction, low thermal and electrical conductivity, optical properties and decorative appeal are among the factors that explain the increasing use of plastics. Specific examples where plastics successfully replaced metals in the motorcar include bumpers, front and rear ends, radiator end tanks and door handles. In most cases, plastics are not direct substitution for metals. This is because plastics have widely different properties and employ widely different processing techniques. The design should be changed to make use of the advantages of plastics and to avoid their limitations.

Plastics can be divided roughly into two categories. The high-volume, reasonably priced materials like polyethylene, polystyrene and polypropylene are at one end of the scale while the high-performance, high-price materials like silicones, polycarbonates and fluoroplastics are at the other end. Glass reinforced and otherwise modified grades fall between the two ends of the scale. In addition to cost, availability of the selected plastic in the required formulation is a prime concern in selection. For example, many molding resins are formulated solely for injection molding and cannot be used for the manufacture of a relatively small number of parts.

When the part is to carry loads, i.e. a structural part, it should be remembered that the strength and stiffness of plastics vary significantly with temperature. Room temperature data cannot be used in design calculations if the part is going to be used at any other temperature. Long-term properties, i.e. creep behavior, cannot be predicted from short-term properties. Many of the commonly used engineering plastics have a notched impact strength less than 5.4 J/cm (10 ft lb/in). From the design point of view, this means that they are brittle and the effect of stress raisers must be considered. For example, one type of acetal has an unnotched Izod impact strength of greater than 54 J/cm (100 ft lb/in) and a notched Izod impact of only about 1 J/cm (c. 2 ft lb/in). A complicating factor is that many plastics achieve their impact resistance through the addition of plasticizers. With time, these additives may vaporize, leaving the aged plastic relatively brittle.

4.15   REVIEW QUESTIONS AND PROBLEMS

4.1 Give the main reasons for mixing additives with polymers when forming plastics.

4.2 Explain what is meant by crystallinity in polymers. How does increasing crystallinity affect strength, ductility, transparency and dimensional changes on solidification?

4.3 What are the attractive features of plastics when compared with metals?

4.4 Compare the processing of metals and plastics by extrusion. Explain the reasons for the differences between the two processes.

4.5 Low elastic modulus is one of the main limitations in using plastics for load-bearing components. Suggest design features that will overcome this limitation.

4.6 In Fig. 2.1, plastics are shown to constitute an important part of the motor car. List some of the parts that are made of plastic and suggest the possible methods of manufacturing them. Suggest other parts that are now being made of other materials and can be advantageously substituted with plastics.

4.7 What are the advantages and disadvantages of plastic gears? How can the performance of such gears be improved?

4.8 What are the main material requirements for the following components: disposable teaspoon; flexible electrical insulator for low-tension cables; TV cabinet; and motor car battery casing? Suggest suitable polymeric materials and methods of manufacturing them.

4.9 Recommend suitable plastics and manufacturing processes for the following applications: (a) body of a telephone set; (b) 2-liter (0.5 gal) lubricating oil container; (c) bag for frozen vegetables; (d) safety shield for a mechanical press; and (e) hard hat for wearing on a building site.

4.10 Compare polypropylene and aluminum as materials for making toothpaste tubes.

4.11 Why is it more desirable for lettering on a plastic part to be raised above the surface?

BIBLIOGRAPHY AND FURTHER READING

Ashby, M. F. and Jones, D. R. H., *Engineering Materials 1, an Introduction to their Properties and Applications*, Pergamon Press, London, 1980.
Ashby, M. F. and Jones, D. R. H., *Engineering Materials 2, an Introduction to Microstructures, Processing and Design*, Pergamon Press, London, 1988.
Baer, E., *Engineering Design for Plastics*, Robert E. Krieger, New York, 1975.
Bee, J. V. and Garrett, G. V., *Materials Engineering*, Pergamon Press, London, 1986.
Billmeyer, F. W., *Textbook of Polymer Science*, 3rd ed., John Wiley & Sons, Sussex, 1984.

Brown, R. L. E., *Design and Manufacture of Plastic Parts*, John Wiley & Sons, New York, 1980.

Brydson, J. A., *Plastic Materials*, 2nd ed., Van Nostrand Reinhold, New York, 1970.

Crawford, R. J., *Plastics Engineering*, Pergamon Press, London, 1987.

DeGarmo, E. P., Black, J. T., Black, J. T. and Kosher, R. A., *Materials and Processes in Manufacturing*, 6th ed., Macmillan, New York, 1984.

Flinn, R. A. and Trojan, P. K., *Engineering Materials and their Applications*, 2nd ed., Houghton Mifflin, Boston, 1981.

Jastrzebski, Z. D., *The Nature and Properties of Engineering Materials*, 3rd ed., John Wiley & Sons, New York, 1987.

Kalpakjian, S., *Manufacturing Processes for Engineering Materials*, Addison-Wesley, Massachusetts, 1984.

Keyser, C. A., *Materials Science in Engineering*, 4th ed., Charles E. Merrill, Columbus, Ohio, 1986.

Kroschwitz, J. Ed., *Polymers: an Encyclopedic Source Book of Engineering Properties*, John Wiley & Sons, New York, 1987.

Ray, M. S., *The Technology and Applications of Engineering Materials*, Prentice Hall, New Jersey, 1987.

Rodriguez, F., *Principles of Polymer Systems*, 2nd ed., McGraw-Hill, London, 1983.

Thornton, P. A. and Colangelo, V. J., *Fundamentals of Engineering Materials*, Prentice Hall, New Jersey, 1985.

Chapter 5

Ceramic Materials and their Processing

5.1 CLASSIFICATION OF CERAMIC MATERIALS

Ceramics cover a wide variety of materials of widely different compositions and properties. The characteristic feature of all ceramic materials is that they are inorganic compounds of one or more metals with a nonmetallic element. The nonmetallic elements are usually oxygen as in the case of aluminum oxide, carbon as in the case of silicon carbide and nitrogen as in the case of silicon nitride. The crystal structures of ceramics are complex, since they have to accommodate more than one element of widely different atomic size. The interatomic forces generally alternate between ionic and covalent bonds, which leave relatively few free electrons in ceramic structures as compared with metals. This makes ceramic materials generally insulators to heat and electricity. The strong ionic and covalent bonds give ceramics their high hardness, stiffness and stability. At the macrostructural level, ceramic materials can have one of the following types of structures:

1. Noncrystalline structures, or glasses, where only the short-range arrangement of the atoms exists but do not have the long-range repetitious crystalline pattern.
2. Crystalline structures where all the material is arranged according to a three-dimensional long-range order. Crystalline ceramics generally have higher stability than glasses, and on the average have higher melting points and hardness.
3. Crystalline material bonded together by a noncrystalline or glassy matrix. The glassy phase is usually the weaker of the two phases and reducing its volume fraction increases the strength of the material.

The American Ceramic Society has classified ceramics into the following groups: whitewares, glass, refractories, structural clay products and enamels. These groups will be discussed in this chapter with the exception of enamels, which will be discussed in Chapter 8.

5.2 GENERAL CHARACTERISTICS OF CERAMICS

Mechanical properties of ceramics

The most noticeable characteristic of ceramics is that they are hard and brittle at room temperature. The mechanical properties of ceramics, like those of other materials, are influenced by their structure. Variations in mechanical behavior result from the various

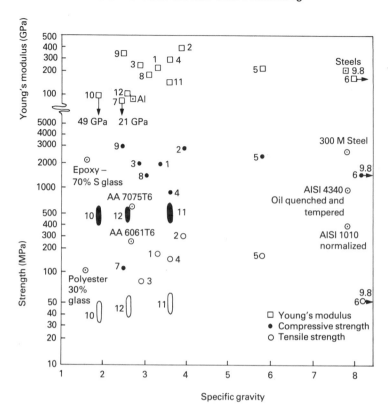

Figure 5.1 Comparison of some ceramic materials on the basis of specific gravity, tensile strength, compressive strength and Young's modulus. Some metals and composites are included for comparison.
1 = Alumina 85%. 2 = Alumina 99.5%. 3 = Beryllia 98%. 4 = Magnesia. 5 = Zirconia. 6 = Thoria. 7 = SiC (Silicate bond). 8 = SiC (Self-bonded). 9 = Boron Carbide. 10 = Cordierite. 11 = Zircon. 12 = Steatite.

combinations of covalent, ionic and van der Waals bonds that exist within the structures. While the coordination arrangement of metals is not altered by slip, new neighbors normally result from slip in ceramics. In the latter case, slip has to be achieved by breaking the strong ionic and covalent bonds, which is more difficult than breaking the metallic bonds. The difficulty of slip in ceramic materials make them brittle but very strong. Such high strength can be achieved provided no porosity or structural defects are present. In practice, however, structural irregularities such as microcracks, grain boundaries and microporosity are difficult, if not impossible, to eliminate. While in ductile materials the stress concentrations around these structural defects may be relieved by plastic flow of the material, this is not possible in brittle materials and fracture occurs at stresses much lower than the theoretically possible strength. Once started, the fracture propagates readily under tension, because the stress concentration is intensified as the crack proceeds. Under compression, a crack will not be self-propagating as loads can be transferred across it; therefore, brittle materials are usually much stronger in compression than in tension. The ratios between tensile strength, modulus of rupture and compressive strength can be roughly taken as 1:2:10. Figure 5.1 compares some mechanical properties of commonly used ceramics with some metallic materials and plastics and Table A.19, Appendix A, gives numerical values of some properties.

Certain ceramic materials, including Al_2O_3, BeO, B_4C, and SiC, exhibit very high values of tensile strength when fabricated in the form of thin fibers or whiskers. These materials can be used as the reinforcing phase in fiber-reinforced composites, as will be discussed in Chapter 6.

Physical properties of ceramics

Ceramic materials generally have higher melting points than commonly used metals, and their thermal conductivities fall between those of metals and polymers. The thermal conductivity of refractories depends on their composition, crystal structure and texture. Simple crystalline structures, as in silicon carbide, usually have higher thermal conductivity. The variation of thermal conductivity with temperature usually depends on whether the material is crystalline or noncrystalline. For example, fireclay bricks show an increase of thermal conductivity with rising temperature, whereas the more crystalline forsterite and some high aluminas show a decrease in thermal conductivity with rising temperature.

Thermal shock resistance is a function of thermal conductivity and expansion coefficient in brittle materials. Ceramics generally have lower thermal expansion coefficients than do metals and polymers, but there is a wide range of variation between different types and grades. Ceramics with a lower thermal expansion coefficient and higher thermal conductivity usually exhibit better thermal shock resistance.

When building refractory linings, allowance for expansion should be made otherwise compressive stresses will arise which could reduce the service life. For example, when a stabilized zirconia lining, which is constrained from expanding, is heated from room temperature to 400°C the compressive stresses can be calculated as follows:

Thermal expansion coefficient is $9.8 \star 10^{-6}/°C$.

Modulus of elasticity is 210 GPa.

Increase per unit length of unconstrained material due to thermal expansion $= 9.8 \star 10^{-6} (400 - 25) = 3.68 \star 10^{-3}$.

In a constrained material, compressive stresses that will correspond to a strain equal to the expansion is $3.68 \star 10^{-3} (210\,000) = 772$ MPa, which is more than one third the compressive strength (see Table A.19).

Ceramics are generally nonconducting to electricity and are considered to be dielectric. Most porcelains, aluminas, quartz, mica and glasses have volume electrical resistivity values greater than 10^{15} ohm cm and dielectric constants of up to 12 (1 kHz a.c.). The electrical resistivity of many ceramics decreases with the introduction of impurities in their structures.

Magnetic ceramics include a number of spinel structures commonly referred to as ferrites, e.g. $MgFe_2O_4$, Fe_3O_4, $CoFe_2O_4$, and $CuFe_2O_4$. Ferrites find widespread applications in magnetic recording tape and disks, electron beam deflection coils, transformer cores and computer memory parts.

The specific gravities of most ceramic materials range from 2 to 3, which makes them comparable to those of light metals.

Chemical resistance of ceramics

The strong ionic and covalent bonds in ceramics make them chemically very stable even at elevated temperatures. Almost all ceramic materials are resistant to chemical attack, except by hydrofluoric acid and, to some extent, by hot caustic solutions. They are not affected by organic solvents. Glasses are among the most chemically stable materials, although their relative performance in various environments may vary considerably between different grades, as will be discussed in Section 5.6.

5.3   REFRACTORY CERAMICS

The most widely used common refractories are the alumina-silica types, with compositions ranging from nearly pure silica to nearly pure alumina, together with some impurities, such as iron and magnesium oxides. The structure usually consists of a glassy matrix binding the crystalline constituents. Silica bricks are made of at least 95 percent SiO_2. The refractory bricks based on silica are less expensive, but are slightly acidic, i.e. give acidic water solutions. Slags high in CaO and MgO, which are used to refine molten metals by removing phosphorus and sulfur, are basic and react with silica to form low-melting-point materials that erode the bricks. Under such basic slag conditions, basic refractories, such as those based on MgO and CaO, are used. Magnesia is susceptible to thermal shock, and less stable when in contact with most metals at temperatures above about 1700°C (c. 3100°F) in reducing atmospheres.

Alumina is one of the most important commercial oxides. At moderately high temperatures its strength is high, it is chemically stable and it can be used either in oxidizing or reducing atmospheres up to about 1930°C (c. 3500°F) for short periods. Pure alumina refractories can withstand higher temperatures than those containing binders. Alumina ceramics are successfully used in applications like cutting-tool inserts. However, their relatively high thermal expansion coefficient, coupled with high modulus of elasticity, makes them too prone to thermal shock to be used in applications like gas turbines. Table A.19, Appendix A, lists numerical values of typical properties.

Beryllia combines good electrical insulation with relatively higher thermal conductivity, which makes it useful for special electronic applications. Beryllia is stable to about 1700°C (c. 3100°F) in air, reducing atmospheres and in vacuum. Its resistance to chemical attack at high temperatures makes beryllia suitable as a crucible material for melting high-purity metals and alloys. However, it is expensive and difficult to work with and its dust particles are toxic. Table A.19 gives typical properties of beryllia.

Pure zirconia suffers an inversion in crystal structure at about 1000°C (c. 1830°F), which causes a 7 percent contraction in volume with accompanying cracking. When stabilized with about 5 percent CaO, zirconia can be used at elevated temperatures and can be repeatedly heated above 2230°C (c. 4040°F) without reversion. Zirconia is not wetted by most metals, and it is used for crucibles for melting platinum, palladium, ruthenium and rhodium. It is also used for refractory bricks away from the slag line, as slags react severely with it. Partially stabilized zirconia (PSZ) ceramics exhibit good thermal shock resistance, low coefficient of friction, excellent wear resistance and a thermal expansion coefficient similar to that of steel. These characteristics make PSZ good candidates for applications in several parts of advanced diesel engines. PSZ is also used in applications where wear and thermal shock resistance are needed, as in the case of high-temperature metal-extrusion dies. Table A.19 gives some properties of zirconia refractories.

Thoria has the highest melting point of all oxides, 3327°C (6021°F) and is stable under most conditions. Thoria crucibles have been used for melting titanium. Thoria is sensitive to thermal shock due to its high coefficient of thermal expansion and low thermal conductivity. It is also expensive and radioactive. Table A.19 lists some properties of thoria.

Carbides, as a family, contain materials with the highest melting points of all engineering materials. Hafnium and tantalum carbides both have melting points of 3944°C (7131°F). However, carbides cannot be used unprotected at high temperatures because of their poor oxidation resistance. The only exception is silicon carbide, which can be used at temperatures up to about 1680°C (*c*. 3050°F). Silicon carbide combines high thermal conductivity, low thermal expansion and high thermal shock resistance, which allows it to be used as a structural material for parts which are subjected to high mechanical and thermal stresses. As their density is only about one third that of metallic superalloys, parts made from SiC have additional advantages where rotary and oscillating motion is involved. A SiC turbine rotor for an exhaust-operated turbocharger has performed satisfactorily in an experimental diesel engine. Boron carbide is best known for its extreme hardness and its abrasion resistance. Table A.19 lists some properties of commonly used carbides.

Most nitrides are relatively brittle and have poor resistance to oxidation. The most widely used refractory nitrides are boron nitride (BN) and silicon nitride (Si_3N_4). Cubic boron nitride is best known as the synthetic diamond material, borozon, and is stable in air up to 1730°C (*c*. 3140°F) and has a hardness approaching that of diamond. It also has a thermal conductivity five times that of copper at room temperature, which makes it useful for very small heat-sink devices. Boron nitride is anisotropic and its thermal expansion parallel to the direction of pressing is 10 times that in the perpendicular direction. Silicon nitride has outstanding resistance to wetting or reaction with molten nonferrous metals and is used for crucibles and boats for melting and refining semiconductor materials like germanium. It has excellent thermal shock resistance, which makes it useful for thin, thermocouple protection tubes and radiant heat shields. Its excellent erosion resistance at high temperatures makes it useful for rocket nozzle inserts.

5.4 WHITEWARES

Whitewares include the various types of pottery, china, tile and porcelain. The raw materials used in making whitewares are clay, quartz and feldspar. Oxides, coloring additives and fluxes may also be added. The clays used are generally composed of hydrated aluminum silicate, alumina and silica in different ratios. Clays become plastic when mixed with water and act as a bonding material on drying. The finer the particle size, the stronger and more plastic the clay. However, plasticity and shrinkage increase together, and shrinkage is usually limited to 12 percent for control of dimensions and warping. Shrinkage may be reduced by the addition of nonplastic, nonshrinking materials such as fired clay, ground flint or similar minerals. The second raw material involved in making whitewares is quartz, or sometimes flint. The addition of quartz reduces shrinkage and improves the component strength in the dry and fired condition. The third constituent of whitewares is feldspar, which is a class of alkali-alumina-silicate mineral. In the unfired body the feldspar helps in reducing shrinkage, and during firing it acts as a flux which dissolves the clay and then the quartz. In addition to feldspar, other fluxing ingredients like $CaCo_3$, $BaCO_3$, $Na_2O.Al_2O_3.2SiO_2$, ZnO and Na_2CO_3 may also be added.

Besides their household applications in tiles and sanitary ware, whitewares have many industrial applications in electrical insulation. Electrical whitewares (porcelains) are conventionally divided into low-voltage insulators, high-voltage insulators and high-frequency insulators. Low-voltage insulators are suited for applications involving up to about 500 V and may be used in the unglazed or glazed condition. Cordierite ($2MgO.2Al_2O_3.5SiO_2$) is widely used in low-voltage applications and Table A.20 lists its relevant properties. High-voltage whitewares are suitable for voltages higher than 500. These grades are vitrified bodies that are usually glazed for weathering resistance. Zircon porcelain is used extensively in high-voltage applications and consists principally of zircon with some clay and alkaline fluxes. Zircon porcelain has superior properties to normal porcelain, as shown in Table A.20. High-frequency applications require low power loss, high dielectric strength and high resistivity. Steatite whiteware is usually used for such applications, although oxide ceramics are also used. Steatite is essentially composed of talc and a little clay. It has excellent electrical properties and a low cost, but its poor thermal shock resistance and narrow firing temperature range limit the size and shape of its products. Table A.20 lists some of the properties of steatite whiteware.

5.5 CLAY PRODUCTS

Clay products form one of the most important classes of structural materials. Structural clay products may be classified as follows: brick for building, paving, or firebrick; tile for roofing, floors, walls and load-bearing; and pipe for sewers, drains and conduits. The minerals commonly found in clays are mostly kaolinite ($2SiO_2.Al_2O_3.2H_2O$) and other hydrated silicates of alumina. Silica forms 40 to 80 percent of the raw materials used in making structural clay products other than firebrick. In firebrick the silica content may be as high as 98 percent. The alumina content ranges from 10 to 40 percent, except in silica brick. Iron oxide, which usually constitutes less than 7 percent, determines the color of the clay and of the fired product. Lime normally constitutes less than 10 percent of the clay. Alkalis form less than 10 percent but are of great value as fluxes.

The essential requirements for building brick are sufficient strength in crushing and bending, durability, a proper suction rate and a pleasing appearance if it is to be used as a facing brick. The requirements for tiles are generally higher than those for brick, and better clays and more accurate processing methods are employed. Wall tiles are usually finished by glazing. Clay pipes used for sewers should successfully withstand the action of acids and should be impervious. They should be reasonably straight, smooth and free from cracks and blisters.

5.6 GLASS

Glasses include a large family of noncrystalline inorganic materials of widely different composition and properties. Silica is the most important constituent of glass, but other oxides are added to achieve certain characteristics. The constituents of glass can be grouped into three categories:

1. Glass formers such as silica and boron oxide.
2. Modifiers such as sodium and potassium which are added to lower melting temperature and improve processability.
3. Intermediate oxides which do not form glasses by themselves but affect the properties. An example is lead oxide which is added to improve the refractive index of the glass.

Table A.21 gives typical compositions and properties of some commercial glasses.

Fused silica and 96 percent silica glass have high softening temperature and low thermal expansion coefficient, which makes them resistant to heat and thermal shock. They also have very good chemical durability and electrical resistance. However, high cost and fabrication difficulties limit their applications to highly specialized products like furnace sight glasses, high-temperature thermocouple sheaths and space and astronomical applications.

Soda-lime glass is the most widely used glass in terms of tonnage produced and variety of applications, because it is the least expensive and the easiest to fabricate. It is used for windows, bottles, jars, electric bulbs and fluorescent tubes. However, its resistance to high temperature and thermal shock is poor and its resistance to chemical attack is only fair.

Lead glasses are somewhat more costly than soda-lime glasses, but have excellent electrical resistivity and high refractive index. They are composed basically of silica, potash and lead oxide. Glasses of this type are used for neon sign tubes and other applications requiring good electrical insulation. With higher lead content, these glasses are used for electric capacitors and for absorption of X-rays, gamma rays and other forms of higher radiation. Lead glasses are also used for optical purposes where they are commonly called flint glass. Like soda-lime glasses, however, these glasses are not resistant to high temperatures or thermal shock.

Borosilicate glasses, pyrex, are made by replacing most of the soda in soda-lime glass by boric oxide. This reduces the thermal expansion coefficient, which improves the thermal shock resistance. Their resistance to heat and chemical attack is also better than soda-lime and lead glasses, but they are also more expensive and difficult to fabricate. Among its uses are laboratory glassware, industrial pipe lining, sight glasses, boiler-gage glasses and domestic cookware. More recent applications include both flat-plate and tubular-type solar energy collectors.

Aluminosilicate glass is also resistant to thermal shock and stands higher temperatures than borosilicate glass. However, it is more expensive and more difficult to fabricate. Uses include halogen lamps for automobile head lamps, range-top cookware and, when coated with electrically conducting film, resistors for electronic circuitry.

In addition to the above classification, which is based on chemical composition, glasses can also be classified according to their unique characteristics or special applications as:

1. Optical glass is made free of bubbles and other defects which could interfere with the passage of light. Most of the major types of glass can be produced to optical quality.
2. Colored glass is made by adding coloring agents to a base glass. Soda-lime and lead glasses have the widest range of colors.
3. Opal glass is either translucent or opaque. Translucence makes some of these glasses suitable for light fittings.
4. Photochromic glass darkens when exposed to UV and fades when the source is removed. The rate of fading depends on the composition. Photosensitive glass also responds to UV and if then heated it changes to opal. If exposed through a mask, an image can be permanently printed on the glass. The exposed area is attacked by acid

enabling intricate patterns to be formed. This technique can be used in the production of semiconductor integrated circuits and in printing.

5. Sintered glass parts can be fabricated from any composition by a technique similar to powder metallurgy techniques. The powdered glass is mixed with a liquid to form a thick paste which is pressed or cast to the desired shape and then fired. The properties of sintered glass are similar to those of the original melt.

6. Foam or cellular glass is produced by mixing glass powder with a foaming agent and heating the mixture until the glass melts and the foaming agent gives off gas which is trapped in the form of tiny bubbles. Because of its cellular structure this glass is a good thermal and acoustic insulator.

7. Glass fibers can be made from different compositions depending on the intended application. Common types are those used for thermal and acoustic insulation and for reinforcing plastics. Optical fibers are made from ultrapure materials and are being increasingly used in telecommunications, medical applications, blind-spot inspection and computer links.

8. Thermally tempered glass is considerably stronger and more impact resistant than ordinary glass because of the compressive surface stresses introduced by heat treatment. Thermal strengthening can be induced in many glass compositions of medium to high thermal expansion coefficient. Uses include windows of automobiles and aircraft, industrial piping and household applications. Chemically strengthened glass also contains compressive surface stresses, but these are introduced by changing the chemical composition of the surface. Chemically strengthened glasses are usually based on the soda-aluminosilicate composition and are generally stronger than thermally tempered glass. However, they are more expensive and their maximum service temperature is limited to about 300°C (c. 570°F).

9. Laminated glass is a sandwich structure of one or more glass layers and one or more clear plastic layers. Major uses include security glass, ballistic glass and windshields for motorcars, trains and aeroplanes.

10. Glass ceramics, pryocerams, are a family of fine-grained, crystalline materials made from special glass compositions by controlled crystallization. Glass ceramics have much greater impact strength than commercial glasses and ceramics. Their thermal expansion coefficient varies from negative to small positive values, depending on the composition, and this leads to excellent thermal shock resistance and good dimensional stability. Glass ceramics are useful up to about 1100°C (c. 2020°F), with excellent corrosion and oxidation resistance. They are electrical insulators and are suitable for high-temperature, high-frequency applications.

5.7   PROCESSING OF CERAMIC PRODUCTS

The raw materials used for making ceramic parts are usually in the form of particles or powder. After mixing and blending the appropriate ingredients, processing is carried out either in a dry, semi-dry, or liquid state and either cold or hot condition.

Casting

Slip casting consists of suspending powdered raw materials in liquid, to form a slurry or slip that is poured into porous molds, usually made of gypsum. The mold absorbs the liquid,

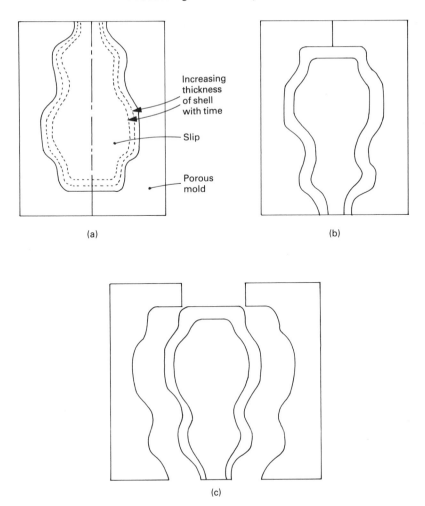

Figure 5.2 Processing of hollow ceramic products by clip casting. (a) Pouring slip in mold and waiting for shell of required thickness to form. (b) Removing excess slip. (c) Removing green product out of mold in preparation for drying and firing.

leaving a layer of solid material on the mold surface. For hollow components, the excess slip is removed after the desired shell thickness has formed (Fig. 5.2). The process is especially economical for short production runs. Products made by slip casting are liable to large shrinkage on firing because the green density obtained by this method is only about 70 percent of the theoretical density. Large and complex parts can be easily produced by slip casting. The main disadvantages of this process are poor dimensional accuracy and slow rates of production.

Molding

Molding processes in ceramics are similar to the injection molding of plastics. The ceramic is mixed with a thermoplastic resin and heated to provide the fluidity needed for the mixture to flow into the mold cavity. The resin is later burned off before firing. Sections as thin as 0.5 mm and as thick as 8 mm can be injection molded and complex shapes with good dimensional accuracy can be rapidly produced. The main limitation is high tooling costs.

Jiggering

Jiggering is used for circular or oval shapes like dinnerware and porcelain electric insulators. In jiggering, the plastic ceramic body is rotated while a profiling tool forms the surface by cutting away excess material (Fig. 5.3). This process can be adapted for automatic machines and high rates of production.

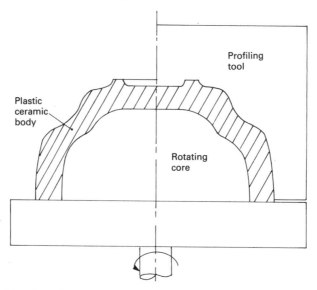

Figure 5.3 Forming of ceramic shapes of circular or oval cross-section by jiggering.

Extrusion

Extrusion consists of forcing a plastic ceramic body through a forming die and then cutting the product to length. Most clay ceramic products such as brick and rain-tile are made with auger extruders. The advantages of extrusion are low tooling costs and high production rates, but it is difficult to produce parts with thin walls and the parts must be symmetrical.

Pressing

Pressing can be done with dry, plastic or wet raw materials. In dry pressing, ceramic mixtures with liquid levels up to 5 percent by weight are pressed into shape under high pressure in a metal die. This method is widely used for manufacturing nonclay refractories, electrical insulators and electronic ceramic parts. Small parts can be produced to close tolerances by this method. Semidry pressing and wet pressing, in which water content is from 5 to 15 percent and from 15 to 20 percent respectively, use lower pressures and less expensive dies. In isostatic pressing, dry ceramic powder with small amounts of binder are pressed and then sintered to densities up to about 95 percent of the theoretical densities.

The process is widely used for high-grade oxide ceramic components of complex shapes. Hot pressing combines the pressing and firing operations and is used with isostatic or uniaxial pressing techniques. Components produced by hot pressing have high densities and strength, e.g. aluminum oxide cutting tools with a porosity of less than 1 percent.

Drying and firing

After forming the plastic ceramic mass into the required shape by one of the above processes, it is dried to remove the water. If drying is carried out too quickly, cracking or warping may occur. After drying, the part is fired or sintered into a permanent product. In whiteware products, firing takes place at temperatures between about 900 and 1450°C (c. 1650 and 2640°F). During firing, vitrification takes place and the resulting liquid glassy-phase fills the pores, thus forming the matrix. This holds the unmelted particles together. In high-grade refractories, the sintering process produces a crystalline bond instead of the glassy-phase bond resulting from vitrification. The density of the material after firing is usually in the range of 98.5 to 99.5 percent of the theoretical density.

Finishing

Dimensional accuracy and surface finish of the fired parts can be improved by grinding and lapping operations. Ultrasonic and laser machining can also be used to produce cross holes, blind holes and similar features in fired ceramic products.

5.8   FORMING OF GLASS PRODUCTS

Glass can be cast, rolled, drawn, pressed like metals and, in addition, it can be blown. Casting is accomplished by pouring the liquid glass into a mold and cooling slowly. Centrifugal casting techniques, similar to those used for metals, can also be used to produce parts with axial symmetry, such as the funnels at the back of the television tubes. Rolling is widely used for flat products like window glass and plate glass. The temperature of the glass should be controlled to achieve the right viscosity which would allow it to be rolled into a sheet. The resulting sheet is then slowly cooled in an annealing furnace. Plate glass is normally produced by floating the molten glass on a bath of molten tin. The sheet then passes through an annealing furnace, which has smooth rollers to avoid harming the surface finish. This process results in mirror-smooth surfaces that need no further grinding or polishing.

Drawing of glass tubing is accomplished in a way similar to rolling, in which glass of the proper viscosity flows directly around a ceramic tube or mandrel pulled by asbestos-covered rollers. Air blowing keeps the tube from collapsing after it leaves the mandrel. Pressing is similar to die-forging of metals and is accomplished by placing a gob of molten glass into a metal mold, pressing it, then transferring it to an annealing furnace. The press-and-blow method involves feeding a gob into a mold, pressing, then blowing the glass to take the final shape. This process is similar to the blow molding of polymers, shown in Fig. 4.8. The products are then annealed to remove the internal stresses caused during cooling. This process is used to make containers and similar parts. Glass fibers are generally produced by pulling molten glass through heated platinum orifices. Fiber diameter is controlled by changing the orifice size and the drawing speed.

Glass processing often involves removing material from the surface. This can be accomplished by cutting, using carborundum or diamond abrasive wheels. Another method is to cover the surface with an etchant-resistant wax, leaving the areas to be marked uncovered. The part is then immersed in a solution containing hydrofluoric acid. Coatings are also frequently applied to glass surfaces during or after processing. A thin film of an inorganic oxide, usually tin or titanium oxide, can be deposited on glass to improve lubricity, thereby improving the scuffing resistance of the surface. Such films are not apparent to the unaided eye, although surface reflection is increased. Iridescent effects are possible if the film is applied thickly enough.

The quality of glass products varies widely, depending on the end use. The main requirements for flat products are flatness, transparency, freedom from stresses and freedom from bubbles and inclusions. For containers, accuracy of volume is usually important, while in chemical ware control of chemical composition is necessary to avoid corrosion. In optical glasses, the refractive index is most important, while in electrical applications the electrical resistivity is normally specified.

5.9   DESIGN CONSIDERATIONS FOR CERAMIC PRODUCTS

Designing ceramic products needs special considerations in view of their brittleness and relatively low mechanical and thermal shock resistance. If the same configuration is used when a ceramic material is substituted for a metallic alloy, the ceramic part could fail in

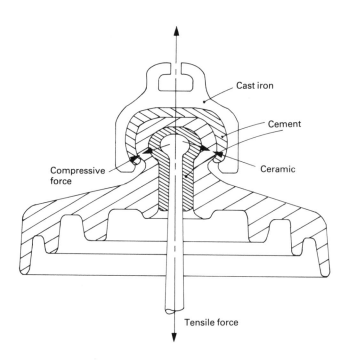

Figure 5.4 Selecting design features to convert tensile loads on a component into compressive forces.

service or even during assembly. Special designs should be developed to make use of the advantages of ceramics and to avoid their limitations. As the ratios between tensile strength, modulus of rupture and compressive strength are usually in the range of 1:2:10, every effort should be made to load ceramic parts in compression and to avoid tensile loading. An example of a design which converts the general tensile loading on a component into a compressive force on the ceramic material is shown in Fig. 5.4.

Design tools like finite element analysis can be used to eliminate tensile stresses and to ensure uniform distribution of stresses on loading. As ceramics are brittle, they are sensitive to stress concentration. Features like sharp corners, notches and unstrengthened holes should, therefore, be avoided. Press fits and shrink fits permit successful attachment of ceramics to steel and allow prestressing the ceramic part in compression which increases its load-carrying capacity. Variability of properties in addition to lower toughness and thermal shock resistance results in lower reliability of ceramic parts in comparison with metallic parts.

The dimensional changes which take place in ceramic parts on drying and firing should be taken into consideration when making the design. Large flat surfaces can cause warping and large changes in thickness can lead to nonuniform drying and cracking. Dimensional tolerances should be generous to avoid the need for machining, which is usually difficult and expensive. Shrinkage can be minimized by appropriate selection of material composition and processing method. When high dimensional accuracy is essential, grinding or lapping should be specified. In cylindrical grinding, accuracies of ± 0.1 mm (± 0.004 in) can be held, and parallel grinding gives an accuracy of ± 0.0003 mm per mm (± 0.0003 in per in). Experiments on the use of ceramics in diesel engine applications show that ceramic parts with smoother surface finish perform better than those with rougher finish, especially under conditions of sliding friction under high contact stresses. On coated ceramics, the coating thins out on sharp corners, so they should be avoided if uniform thickness of the coating is required.

5.10   REVIEW QUESTIONS AND PROBLEMS

5.1   In Fig. 2.1, ceramics are shown to be needed in making some parts of the motor car. List some of the components that are made of ceramics and discuss why ceramics were selected for their manufacture. Suggest other parts that are now being made of other materials and can be advantageously substituted with ceramics.

5.2   What are the main material requirements for the following components: ceramic high-tension insulator; motor car windshield; the glass part of an electric bulb; cut-glass decanter; ceramic cutting tool? Suggest suitable materials and manufacturing processes for the above components.

5.3   Compare high-density polyethylene and glass for making bottles for pharmaceutical products.

5.4   What are the attractive features of ceramics in comparison with metals?

5.5   In general, how do the properties of ceramics differ from those of thermosetting plastics?

5.6   Compare the use of plastics and ceramics for making plates and dishes for the following purposes: (a) household use; (b) camping; and (c) food service on airplanes.

5.7   Compare the use of glass and plastics for laboratory ware for chemistry experiments.

5.8   Why should sharp corners and large changes in thickness be avoided when designing ceramic parts?

5.9 Ceramic materials are now being considered for several parts of the internal combustion engine. What are the main difficulties that have to be overcome before ceramics are widely used in such applications?

5.10 What are the main advantages and limitations of ceramic cutting tools in comparison with high-speed steel tools?

BIBLIOGRAPHY AND FURTHER READING

Clauser, H. R., *Industrial and Engineering Materials*, McGraw-Hill, New York, 1975.

Ichinose, N., *Introduction to Fine Ceramics*, John Wiley & Sons, Sussex, 1987.

Jastrzebski, Z. D., *The Nature and Properties of Engineering Materials*, 3rd ed., John Wiley & Sons, New York, 1987.

Kalpakjian, S., *Manufacturing Processes for Engineering Materials*, Addison Wesley, Reading, Mass., 1984.

Kingery, W. D. *et al.*, *Introduction to Ceramics*, 2nd ed., John Wiley & Sons, New York, 1976.

Norton, F. H., *Elements of Ceramics*, Addison Wesley, Reading, Mass., 1957.

Onoda, G. Y. and Hench, L. L., *Ceramic Processing Before Firing*, John Wiley & Sons, Sussex, 1978.

Ray, M. S., *The Technology and Applications of Engineering Materials*, Prentice Hall, New Jersey, 1987.

Thornton, P. A. and Colangelo, V. J., *Fundamentals of Engineering Materials*, Prentice Hall, New Jersey, 1985.

Van Valck, L. H., *Elements of Materials Science and Engineering*, 5th ed., Addison Wesley, Reading, Mass., 1985.

Chapter 6

Composite Materials and their Processing

6.1 INTRODUCTION

A composite material can be broadly defined as an assembly of two or more chemically distinct materials, having a distinct interface between them and acting in concert to produce a desired set of properties. Composite materials can be classified as being either natural, as in the case of wood, or synthetic, as in the case of glass fiber reinforced plastics (GRP). The different materials in a composite may be present as macroscopically distinct phases, as in the case of coatings, sandwich materials and laminates, or they may be mixed on a microscopic scale, as in the case of dispersion, particulate and fiber composites. In the latter group the matrix, the continuous phase, can be used to identify the composite, e.g. metal-matrix or polymer-matrix composite. The overall behavior of composite materials depends on:

1. Properties of the constituents.
2. Size and distribution of the constituents.
3. Volume fraction of the constituents.
4. Shape of the constituents.
5. Nature and strength of bond between the constituents.

Polymer-matrix composites are more widely used than metal-matrix composites. This is because a wider variety of reinforcing materials and better established manufacturing techniques are available for polymer reinforced polymers. The demanding mechanical and chemical conditions that are usually encountered in metal-matrix composites allow only a narrow choice of reinforcing materials and the techniques are often more complex and costly. The bulk of fiber-reinforced composites consists of plastics reinforced with continuous fibers. The reinforcing phase for polymers may be in the form of bundles of very fine glass fibers, or other fibers such as graphite, aramid polymers, boron or SiC. Applications of glass reinforced plastics (GRP) are varied and accordingly glass fibers are available in several commercial forms. Continuous strand fiber and roving, which resemble ribbon or tape, are used for filament winding processes. Chopped fiber, 6 to 25 mm (1/4 to 1 in) long, is utilized in manufacturing molded GRP shapes with randomly oriented fibers. Woven fabrics and nonwoven mats are also important and are produced as sheets, cloth and roving. The manufacture of composite parts will be discussed in Section 6.7.

Most composite materials have been developed to improve mechanical properties such as strength, stiffness, creep resistance and toughness, but a few examples have been developed to achieve certain physical or chemical characteristics. Some properties of composite materials depend only on the amount of each phase (volume fraction) and are insensitive to the microstructural geometry. These structure-insensitive properties may be determined by suitable weighted averages of the properties of each of the individual phases. Density and heat capacity are examples of such properties. The variation of structure-insensitive properties with the amount of phases can be represented by the rule-of-mixtures as:

$$P_c = f_1 \star P_1 + f_2 \star P_2 + \ldots \tag{6.1}$$

where P_c, P_1 and P_2 are the properties of the composite, phase 1 and phase 2 respectively, and f_1 and f_2 are the volume fractions of phases 1 and 2 respectively; $f_1 + f_2 + \ldots = 1$.

Composite-material properties which are sensitive to the geometry of the phases as well as to their volume fractions are called structure-sensitive properties. Examples are elastic modulus, strength and thermal and electrical conductivities. The variation of these properties with the volume fraction of constituent phases does not follow a unique relationship, but usually falls between an upper limit and lower limit depending on the arrangement of phases. This is particularly true in the case of fiber-reinforced composites. The upper limit is represented by the rule of mixtures of (6.1), while the lower limit is represented by the relationship:

$$P_c = \frac{P_1 \star P_2}{(f_1 \star P_2) + (f_2 \star P_1)} \tag{6.2}$$

The upper limit represents the case where the fibrous constituents are arranged in parallel, while the lower limit is for the case where they are arranged perpendicular to the direction of measurement, as shown in Fig. 6.1. The properties of the composite can be made more isotropic by orienting some of the fibers in the transverse direction (cross ply). However, this increased isotropy is achieved at the expense of reduced longitudinal properties. Anisotropy is not always undesirable and the load-carrying efficiency of structures can be increased by tailoring the material to provide greater local strength and stiffness where it is most needed.

6.2   DISPERSION-STRENGTHENED COMPOSITES

Dispersion composites are characterized by microstructures consisting of fine particles, of sizes less than $0.1\ \mu m$ ($4\ \mu in$) and volume fractions are usually less than 15 percent, dispersed in a matrix. This type of composite is usually produced to improve the mechanical strength of the matrix, which is the load-carrying constituent. The fine dispersions impede the motion of matrix dislocations, thus increasing the resistance to plastic deformation. The main variables in determining the effectiveness of a dispersion are: 1. size d; 2. volume fraction f; and 3. interdispersion separation D_p. The relationship between these parameters is:

$$D_p = \frac{2d^2}{3f}(1-f) \tag{6.3}$$

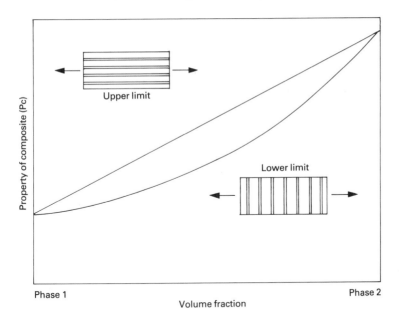

Figure 6.1 Variation of structure-sensitive properties with volume fraction for composite materials.

Typical ranges for these parameters in dispersion-strengthened composite materials are: $f = 0.01$ to 0.15, $d = 0.01$ to $0.1\ \mu m$ (0.4 to 4 μin), and $D_p = 0.01$ to $0.3\ \mu m$ (0.4 to 12 μin). The strengthening effect of dispersions is inversely related to D_p.

Dispersion-strengthening has an advantage over precipitation-hardening because the hard, dispersed phases function as dislocation obstacles at high temperatures, whereas the precipitates go back into solution. Industrial examples of dispersion-strengthened composites are the aluminum-Al_2O_3 system, known as sintered aluminum powder (SAP), and the nickel-3 percent THO_2 system, known as TD nickel. For high-temperature stability, the dispersed phase must not coarsen. This means that it should have low solubility and low rate of diffusion in the matrix, low interfacial energy and low reactivity with the matrix, high melting point and high negative heat of formation.

Several techniques have been developed for processing dispersion composites. Surface oxidation is the industrial process used for making SAP, and it relies on the formation of a thin, adherent refractory oxide film on the aluminum powder. The thickness of the oxide film is of the order of 100 Å. The oxidized powders are then shaped by conventional powder metallurgy techniques. Internal oxidation is another industrial process and is based on the preferential oxidation of the solute element in a dilute solid solution. This process results in the formation of a fine oxide dispersion throughout the metal matrix. This requires that the solute oxide has a higher negative heat of formation than the matrix oxide. Copper containing 3.5 percent by volume of Al_2O_3 in fine dispersion is produced by this method. A third process for manufacturing dispersion composites is the coprecipitation process which is based on the chemical precipitation of a nickel salt around very fine thorium oxide particles. This nickel salt is decomposed to nickel oxide and then reduced in hydrogen to produce a nickel powder with a dispersion of thorium oxide; the resulting material is called TD nickel.

6.3   PARTICULATE-STRENGTHENED COMPOSITES

Particulate-strengthened composites can be considered as an extension of dispersion-reinforced composites in that particle size is larger, more than $1 \, \mu m$ ($40 \, \mu in$) and the volume fraction is greater, more than 20 percent. Both the matrix and particles share the applied load in particle reinforcement. Strengthening in this case is similar to dispersion strengthening as the particles restrict the deformation of the matrix. The yield strength (σ_c) of composites that are strengthened with undeformable particles is inversely proportional to the square root of the interparticle spacing D_p, as follows:

$$\sigma_c \propto \frac{1}{\sqrt{D_p}} \qquad (6.4)$$

The ratio of D_p/d and the ratio of the elastic properties of the particle and matrix can also affect the strength of the composite. Generally, the elastic modulus of particulate composites falls between the upper and lower limits given by (6.1) and (6.2) respectively. Positive deviations from (6.2) signify matrix constraint. Beyond the elastic portion of the stress–strain curve of inorganic particulate composites, the softer matrix begins to deform plastically while the harder particles usually remain elastic until fracture. This behavior imparts strength to the composite but drastically lowers its ductility below that of the matrix phase alone. Typical composites of this type are WC-Cu, WC-Co, TiC-Ni, TiC-Ni-Mo and other ceramic-metal combinations, called cermets. The widest use for cermets is for cutting tools. It has been reported (*Advanced Materials and Processes*:5/ 1987) that ceramic-particulate reinforced aluminum (6061 T6, 2124 T6 and 7091 T6) composites are being considered for connecting rods and pistons of motor car engines. At reinforcement levels of 20–30 percent, these materials are said to be about one-third the weight of steel with lower thermal expansion and equivalent wear resistance.

Cermet-type composites can be prepared by impregnation, where a green porous compact of the particles is prepared by cold or hot pressing and then impregnated with the liquid matrix material. Conventional powder metallurgy techniques involving either solid state or liquid state sintering are also suitable for the preparation of inorganic particulate composites.

The use of particulate fillers in polymeric materials is widespread. Inorganic particulate fillers may be used to raise the elastic modulus, increase the surface hardness, reduce shrinkage and eliminate crazing after molding, improve fire resistance, improve color and appearance, modify the thermal and electrical conductivities and raise the viscosity for ease of molding. Also, the use of inorganic fillers can greatly reduce the cost without necessarily sacrificing other desirable properties. An important application of this type of composite is in automotive tires, where particles of silica and carbon black are added to enhance the strength of the rubber matrix. The elastic modulus of rubber varies with the volume fraction of carbon particles f according to the relationship:

$$E_c = E_o(1 + 2.5f + 14.1f^2) \qquad (6.5)$$

where E_c and E_o are the elastic moduli of the filled and unfilled rubber respectively.

A different type of behavior is achieved in particulate composites when the particles are softer than the matrix. An example is the addition elastomer particles to cross-linked epoxy in order to improve its crack resistance and toughness. The low molecular weight carboxy-terminated butadiene-acrylonitrile copolymers (CTBN) are soluble in epoxy at relatively high temperatures. On cooling, CTBN precipitates in the epoxy matrix causing a marked increase in toughness. Polyesters can also be made tougher using a similar mechanism.

6.4   MECHANICS OF FIBER REINFORCEMENT

Fibrous composites cover a wide variety of materials in which a matrix is used to bind together the fibers and to protect their surfaces from damage or chemical attack. Furthermore, the matrix separates the individual fibers and prevents brittle cracks from spreading across the composite. Glass fiber reinforced plastics (GFRP) account for over three-quarters of total fiber composite production. Although many types of plastics can be used as the matrix for glass reinforced plastics, polyester and epoxy resins are the most widely used. Table A.22 lists the properties of selected fiber reinforced composites.

The strength of fibrous composites is determined by:

1. Strength of the fibers.
2. Orientation of the fibers in relation to the external load.
3. Whether the fibers are continuous or chopped.
4. Properties of the matrix.
5. Compatibility of the constituents, e.g. differences in thermal expansion coefficient.
6. Strength and nature of the bond between the fibers and the matrix.

In many cases, the matrix can be considered as a medium that transfers and distributes the load to the fibers. The bond between the fiber and matrix must be strong enough to prevent interfacial separation or fiber pull-out under axial loads. Extensive chemical reaction between the fibers and the matrix can damage the fibers and reduce their strength. Such reactions can take place during the processing of the composite or while in service and measures should be taken to prevent them.

Elastic modulus (continuous fibers)

When both the fibers and matrix are deforming elastically, the elastic modulus of the composite E_c in the direction of fibers can be represented by the rule of mixtures:

$$E_c = E_f \star f_f + E_m \star f_m \qquad (6.6)$$

where E_f and E_m are the elastic moduli of the fibers and matrix respectively.
f_f and f_m are the volume fractions of the fibers and matrix respectively; $f_f + f_m = 1$.

To illustrate the use of (6.6), the elastic modulus of a polyester-60 percent E-glass fiber reinforced composite is:

$$E_c = 0.6 \, E_g + 0.4 \, E_m$$

Taking $E_g = 73.5$ GPa (Table A.23) and $E_m = 2.8$ GPa gives:

$$E_c = 45.22 \text{ GPa.}$$

When the composite is loaded at 90° to the direction of fibers, the elastic modulus of the composite, E_{ct}, can be represented by the lower limit relationship of the rule of mixtures (6.2) thus:

$$E_{ct} = \frac{E_f \star E_m}{(E_f \star f_m) + (E_m \star f_f)} \qquad (6.7)$$

For the polyester-60 percent E-glass composite discussed above, the elastic modulus at 90 degrees to the direction of fibers is:

$$E_{ct} = \frac{73.5 \star 2.8}{(73.5 \star 0.4) + (2.8 \star 0.6)} = 6.62 \text{ GPa}$$

This value is less than 15 percent of the elastic modulus in the direction of fibers.

When the composite is subjected to shear stresses parallel to the fiber direction, the shear modulus of the composite, G_c, is also represented by the lower limit relationship (6.2) thus:

$$G_c = \frac{G_f \star G_m}{(G_f \star f_m) + (G_m \star f_f)} \qquad (6.8)$$

where G_f and G_m are the shear moduli of the fibers and matrix respectively.

Assuming $f_f = 0.5$ and a fiber modulus of a hundred times the shear modulus of the matrix, which is not unreasonable in the case of fiber reinforced polymers, the ratio E_c to G_c can be shown to be about 25:1. This large difference between the longitudinal elastic modulus and shear modulus in a composite leads to relatively low stiffness in bending of short composite beams.

Under conditions where the matrix is strained beyond its elastic limit while the fibers are still elastic, the composite modulus, E'_c, usually called secondary modulus, is given by the relationship:

$$E'_c = E_f \star f_f + \left(\frac{d\sigma_m}{d\varepsilon_m} \right) f_m \qquad (6.9)$$

where $(d\sigma_m / d\varepsilon_m)$ is the slope of the stress–strain curve of the matrix at the strain under consideration (Fig. 6.2).

Tensile strength (continuous fibers)

When the fibers are continuous and when the externally applied load is in the direction of fibers, the strains in the matrix and in the fibers are the same, assuming that no slippage takes place. Under these conditions, stresses are distributed according to the rule of mixtures:

$$\sigma_c = \sigma_f \star f_f + \sigma_m \star f_m \qquad (6.10)$$

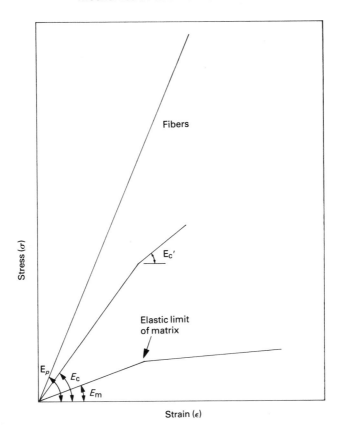

Figure 6.2 Effect of fiber and matrix properties on the stress—strain diagram of fiber-reinforced composites.

where σ_c, σ_f, and σ_m are the stresses in the composite, the fibers, and the matrix respectively.

From Fig. 6.2, it can be shown that:

$$\frac{\sigma_f}{\sigma_m} = \frac{E_f}{E_m} \tag{6.11}$$

Equation (6.11) shows that the ratio between the stresses carried by the fibers and matrix within the elastic range is the same as the ratio between their elastic moduli. For example, if the polyester-60 percent E-glass fiber composite discussed above is used to make a structural member of cross-sectional area 300 mm² to carry a load of 40 kN, then the stress in the fibers is:

$$\sigma_f = \frac{40\,000 \star 73.5}{300 \star 45.22} = 216.7\,\text{MPa}$$

The stress in the matrix is:

$$\sigma_m = \frac{40\,000 \star 2.8}{300 \star 45.22} = 8.26 \text{ MPa}$$

The load carried by the fibers $= 0.6 \star 300 \star 216.7 = 39$ kN

The load carried by the matrix $= 0.4 \star 300 \star 8.26 = 1$ kN

The above numbers show that the fibers carry most of the load.

Continuous fiber composites, containing ductile matrix and brittle fibers, fail when σ_f reaches the fracture stress of the fibers. This is true only when f_f is greater than the critical value required for effective fiber strengthening, which is so in most practical cases.

The alignment of fibers in relation to the applied loading can greatly affect the behavior of the composite. When the angle between the fibers and applied load is small, less than about 8°, the strength of the composite is unaffected. With larger angles of misorientation, the strength decreases precipitously and can be estimated from the relationship:

$$\sigma_\phi = \sigma_c \star \sin\phi \star \cos\phi \tag{6.12}$$

where ϕ is the angle of misorientation, σ_ϕ is the strength of composite in the ϕ orientation, and σ_c is the strength of composite in the fiber direction.

When fibrous composites are loaded at 90° to the fiber direction, they fail at stresses as low as the matrix strength or even lower. This anisotropy can be reduced by using the cross-plied arrangement in which some fibers are oriented at 90° to the main direction. A more isotropic composite can be obtained if alternate layers are rotated through 45°. However, the composite remains weak in the thickness direction.

Discontinuous fibers

The behavior of composites containing discontinuous fibers cannot be described by the above relationships unless their aspect ratio, defined as length/diameter, is greater than a critical value which is determined by several factors including:

1. Relative strength of the fiber and matrix.
2. Fiber shape and end geometry.
3. Interfacial bond strength.
4. .Volume fraction of fibers.

As a rough estimate, the critical aspect ratio, A_c, can be expressed as:

$$A_c = \frac{l_c}{d_f} = \frac{\sigma_{fu}}{2\tau_m} \tag{6.13}$$

where l_c is critical fiber length, d_f is fiber diameter, σ_{fu} is fiber strength, and τ_m is matrix shear strength.

For GFRP, the value of Ac in (6.13) is of the order of 70. With a glass fiber diameter of 0.01 mm, the critical fiber length is about 0.7 mm. This value is important when manufacturing GFRP composites by the spray-on-mold technique, as will be discussed in Section 6.7.

When the aspect ratio is less than the critical value, an effective volume fraction, which is less than the actual value, can be substituted for f_f in the appropriate rule of mixtures in order to anticipate the composite behavior. The smaller the fiber aspect ratio, the lower the effective fiber strength or volume fraction.

6.5 MATERIALS FOR FIBER REINFORCEMENT

There is a wide range of materials which can be used in making fibers for reinforcing composites. From a materials point of view, fibers can be divided into metallic fibers or wires, nonmetallic fibers and whiskers. In selecting a certain fiber for a given application the following factors should be considered:

1. The matrix should wet the fibers to reduce the probability of voids at the interface. Fibers are frequently precoated with a thin film of a suitable material to improve wettability.
2. Extensive detrimental reactions should not take place between the matrix and the fiber materials. These reactions are usually prevented by fiber coatings. For example, glass fibers are coated with resins, while alumina fibers or whiskers do not require coating when used with the epoxy resin matrix. Metals generally do not wet ceramic whiskers, and metallic coatings are required. In the case of carbon fibers, the bond is mainly mechanical in nature and is determined by the difference in thermal expansion coefficients between the fibers and the matrix and by interlocking of the matrix with surface irregularities of the fibers.
3. The difference between the thermal expansion coefficients of the fibers and the matrix should not be so large as to result in excessive thermal stresses on cooling or heating the composite. Thermal stresses can adversely affect the properties of the composite and may cause warping of the finished component. In severe cases, matrix cracking can occur if the local tensile stress exceeds the matrix tensile strength. The differences in thermal expansion coefficients can be used to advantage in hot pressed ceramic matrices with metallic fibers. The fiber material is chosen to have a higher thermal expansion coefficient than the matrix, and on cooling after hot pressing the matrix is placed under compressive stresses in a similar way to prestressed concrete.

Glass fibers

Glass fibers, in their various forms, are the most widely used form of reinforcement for plastics because of their low cost and availability. Glass fibers are manufactured by the drawing of molten glass as individual fibers, then coating with size. The size protects the fiber surface, acts as a lubricant and provides a chemical link with the matrix. Fiber diameters of about 0.01 mm (0.0004 in) are usually available. Glass fibers can be supplied in a variety of forms. Continuous-strand roving is supplied as untwisted, continuous strands, bundled and wound into a cylindrical package. Roving may be obtained with surface treatment for use with different resins. Woven roving is a heavy fabric which is usually used in lay-up processes. Reinforcing mats are often made of randomly-oriented chopped fibers held together by a resin binder. Chopped fibers, 3 to 50 mm (1/8 to 2 in) are usually mixed with the resin for compression or injection molding, as will be discussed in Section 6.7. E-glass, which is alumina borosilicate glass, is the most commonly used in plastic reinforcement. S-glass was developed for higher strength, but it is more expensive then E-glass. Table A.23, Appendix A, gives some properties of glass fibers.

Carbon fibers

Carbon and graphite fibers were developed to meet the need of the aircraft industry for light materials with superior strength and stiffness. The scope of use of this material has now expanded and carbon fiber reinforced plastics (CFRP) are used in several nonaerospace areas, particularly in sports, leisure equipment and industrial applications. Carbon fibers are produced in a number of ways from a variety of precursors and their properties are strongly influenced by the manufacturing technique employed. Most high-performance fibers are manufactured from polyacrylonitrile (PAN) precursor first by oxidation at 200–250°C (c. 390–480°F), then carbonization in a nonoxidizing atmosphere at 1000°C (c. 1830°F); graphitization is then performed in a nonoxidizing atmosphere at 2500–3000°C (c. 4530–5430°F). The surface of the resulting fibers is then chemically treated to promote good adhesion with matrix materials. Carbon fibers are usually classified into:

1. High-modulus, HM, or type I.
2. High-strength, HS, or type II.
3. General-purpose grade, A, or type III. This type is produced at lower temperatures, achieving intermediate performance at lower cost.

Table A.23 gives some characterstics of carbon fibers. Carbon fibers can be used with a variety of matrix materials, but epoxy resins represent the major part of CFRP products. Other matrix materials include: (a) thermosets, such as polyesters, vinyl esters, phenolics and polyimides; (b) thermoplastics, such as nylons, polycarbonates, polyesters and polysulfones; and (c) metals, such as aluminum, magnesium, tin and lead.

Carbon fibers are available in the form of continuous filaments which are grouped in tow bundles or rovings containing 3000 to 12 000 fibers. Larger tows containing 40 000 to more than 300 000 fibers are also available. Most fibers are sized with a small amount of resin binder to maintain their integrity during handling. Rovings are also available as preimpregnated tapes (prepregs) containing about 40 percent by weight of resin. Other forms include chopped fibers and woven fabrics which include unidirectional and bidirectional fabrics. Unidirectional fabrics contain carbon fibers in the warp direction only and the weft is formed with glass, polyester or aramid (Kevlar). Hybrid mixtures containing two types of fibers in the warp, e.g. carbon–glass and carbon–Kevlar, are also used to combine the attractive properties of the different reinforcements. Bidirectional fabrics utilizing carbon in both warp and weft directions are available in different styles of weave, e.g. plain, satin and twill.

Kevlar fibers

DuPont's Kevlar, an aramid, is lighter than carbon fibers and is available in different crystalline structures. More alignment of molecules results in higher crystallinity and higher elastic modulus. Kevlar 29 has an elastic modulus of about 62 000 MPa (9 million lb/in^2) with about 3.6 percent elongation. Kevlar 49 has an elastic modulus of 117 000 MPa (17 million lb/in^2) with about 2.5 percent elongation. Both types have strengths of about 3400 MPa (500 000 lb/in^2). Unlike carbon and glass fibers, which have similar elastic moduli in tension and compression, Kevlar exhibits a much lower modulus in compression than in tension. Table A.23 gives some properties of aramid fibers.

Kevlar's coefficient of thermal expansion is very slightly negative. It also has good dielectric and vibration damping properties. These characteristics are useful in the manufacture of printed circuit boards. It also exhibits better impact resistance in comparison with carbon and glass fibers and is used in army helmets and rigid armor for aircraft, ships, tanks and mobile shelters. Kevlar is available in tow bundles and woven fabrics in a similar way to carbon fibers. Chopped aramid fibers can also be used in bulk molding compounds. Kevlar can be used with a wide range of resin systems, including epoxies, unsaturated polyesters, phenolics and polyurethanes.

Metal wires

Metal wires are reasonably strong, relatively ductile and exhibit more consistent behavior than brittle fibers. They are usually manufactured by wire drawing in many steps, which makes their production much more laborious compared with the one-step drawing of glass fibers. The finer the metal wire the more expensive it becomes. Steel wires are normally used for reinforcing rubber and concrete. Other metallic fibers include tungsten, beryllium, molybdenum and rene 41, Table A.23.

Nonmetallic fibers

Nonmetallic fibers cover a wide variety of materials including boron, Al_2O_3, SiO_2, SiC and Si_3N_4. Many of these materials are also available as whiskers. Boron fibers are produced by vapor deposition of boron on a tungsten wire and they range from 0.1 to 0.15 mm (4 to 6 mil) in diameter. Typical properties of nonmetallic fibers are given in Table A.23.

6.6 LAMINATED COMPOSITES

Laminated composites, or layered materials, consist of two or more different layers bonded together. The layers can differ in material, orientation or form. Clad metals are an example of metal–metal laminates where the outer layers are usually selected to give corrosion resistance or decorative appearance, while the inner layers are usually selected to give high strength. Other possible laminate combinations include metal–plastics as in the case of soft-packaging materials, elastomer–fabrics as in the case of V-belts and motorcar tires and composite–metal as in the case of FRP faced honeycomb core sandwich materials.

Fiber reinforced laminates

Composites which are reinforced with unidirectional fibers are known to have directional properties and are generally much stronger and stiffer in the longitudinal direction than in the transverse direction, as shown in Fig. 6.1. On the other hand, the torsional modulus reaches a maximum when the angle between the fibers and the principal stress axis is 45°. To reduce this anisotropy and to be able to use composites in multidirectional stress systems, it is necessary to build up laminar layers with different fiber orientations, as shown in Fig. 6.3, so that the finished component or laminate is capable of carrying loads in several directions. An individual layer in the laminate is designated by the angle between its fibers and the axis of the component or laminate. Considering the laminate in Fig. 6.3, the lay-up is as follows: $0/ + 45/ - 45/90/ - 45/ + 45/0$. Balancing the individual layers around the neutral axis will avoid warping of the laminate. The properties of a given fiber reinforced laminate in a certain direction can be determined on the basis of volume fraction and the properties of the different layers in that direction.

Sandwich materials

Sandwich materials can be classified as laminated composites. These materials usually consist of a thin facing material and a low-density core (Fig. 6.4). Sandwich materials combine high section modulus with low density, e.g. an aluminum-faced, honeycomb sandwich structure beam is about one-fifth the weight of a solid aluminum beam of equivalent rigidity. The facing material in a sandwich structure carries the major applied load and therefore determines the stiffness, stability and strength of the composite. Examples of possible facing materials are aluminum, stainless steel, magnesium, titanium, plastics and fiber reinforced materials. The core forms the bulk of a sandwich structure. Therefore, it is usually of light weight but must also be strong enough to withstand normal shear and compressive loads. The core can be in the form of foam or cells. Foam cores are usually made from plastics, especially polystyrene, urethane, cellulose acetate, phenolic, epoxy and silicone. Foamed inorganic materials like glass, ceramics and concrete can also be used. Cellular cores can have corrugated or honeycomb structures and are usually made from metal foils joined by welding, brazing or adhesive bonding. Other core materials are glass reinforced plastics, ceramics or paper. Plastic cored sandwiches faced with steel or aluminum have been shown to be weight-saving, cost-effective substitutes for automotive sheet steel. Thermal and sonic insulation characteristics of these sandwiches provide secondary benefits, along with possibilities of using lower-capacity forming presses.

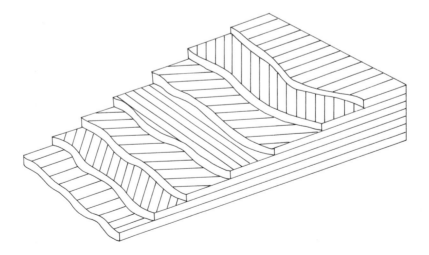

Figure 6.3 Building up fiber-reinforced laminates with different fiber orientation to reduce anisotropy.

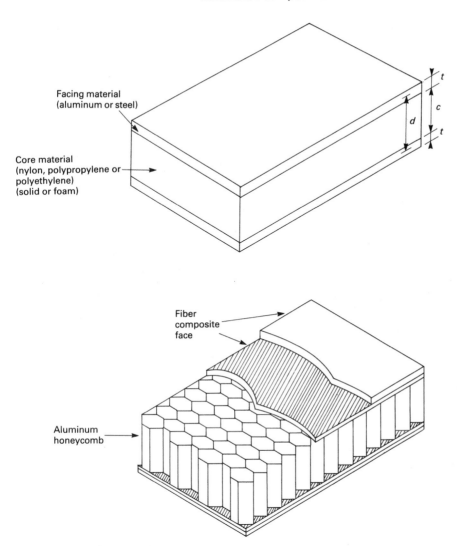

Figure 6.4 Examples of sandwich materials.

However, sandwich materials suffer from lower in-plane strength, lower dent resistance, limited joining capability and more difficulties in recycling.

An important use of sandwich materials is in stiffness-limited applications, where they provide weight reduction and cost savings. In applications where a structural member is replaced by a sandwich material, the flexural rigidity of the sandwich should be at least equal to that of the original member. The flexural rigidity in bending, D_1, of the symmetrical laminated composite (sandwich beam) shown in Fig. 6.4 can be calculated as:

$$D_1 = E_f \star \frac{b \star t^3}{6} + E_f \frac{b \star t \star d^2}{2} + E_c \frac{b \star c^3}{12} \qquad (6.14)$$

where E_f and E_c = moduli of elasticity of the face and core materials respectively,

b = beam width,
t = face thickness,
c = core thickness, and
$(d + t)$ = laminate thickness = $c + 2t$; i.e. $d = c + t$.

When t is thin compared with the sandwich thickness and E_c is small compared with E_f, the second factor of (6.14) becomes dominant.

If the composite laminate is used to replace a sheet material, the equivalent flexural stiffness of the laminate, D_1, should be equal to the flexural rigidity of the sheet, D_s, if it is to behave in a similar manner in bending. In such a case:

$$D_1 = D_s = (b \star t_s^3 \star E_s)/12 \qquad (6.15)$$

where t_s is sheet thickness and E_s is elastic modulus of sheet material.

Equations (6.14) and (6.15) can be used to calculate the required thicknesses in preparation for comparing weight and cost of different sandwich and sheet materials.

To illustrate the use of the above equations, consider the case of a steel plate of thickness t_s = 10 mm which is to be replaced by a steel-polyethylene sandwich panel of the same width and stiffness. If the thickness of the steel facing is 0.5 mm calculate the total thickness of the laminate and the mass ratio. From (6.14) and (6.15):

$$D_1 = D_s = \frac{b \star 10^3 \star E}{12} = \frac{b \star 0.5 \star d^2 \star E}{2}$$

d = 18.26 mm
Laminate thickness = $d + 0.5$ = 18.76 mm
Core thickness $c = d - t$ = 18.26 − 0.5 = 17.76 mm
Mass of steel plate/(m^2) = 78 kg
Mass of laminate/(m^2) = 7.8 + 3.55 = 11.35 kg
Mass ratio of steel plate to laminate = 6.87

Composite structures

Composite structures are often built of different types of composite materials which are arranged to take advantage of their capabilities, as illustrated in Fig. 6.5. The floor beam is subjected to bending stresses in service. The high strength and stiffness capabilities of boron-epoxy composites were used in the flanges which are expected to bear most of the tensile and compressive stresses. The titanium-faced sandwich is used in the web and is expected to bear most of the shear stresses. The rotor blade is subjected to torsional stresses, which will be mainly borne by the 45° fiber reinforced laminates; and to bending stresses, which will be carried by the unidirectional composites and the 45° laminates.

6.7   MANUFACTURING OF COMPONENTS MADE OF COMPOSITE MATERIALS

Manufacture of reinforced-plastic components

Thermoplastics which are reinforced with short-chopped, randomly oriented fibers are easily fabricated using conventional techniques. Injection molding and extrusion are

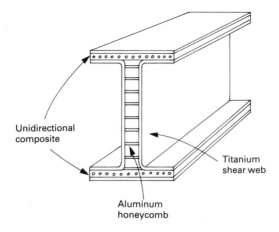

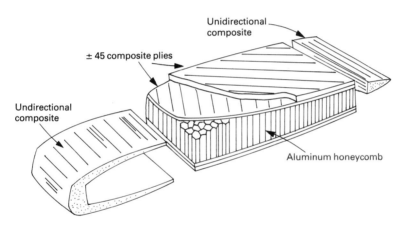

Figure 6.5 Examples of composite structural members.

Source: McCullough, R. L., *Concepts of Fiber-resin Composites*, Marcel Dekker, New York, 1971.

widely used. However, since glass reinforced thermoplastics set more quickly than nonreinforced resins, the mold designer has to make due allowance when sizing thin areas and gating systems. For more details on shape design of molded components refer to Section 12.3.

Composites based on thermosetting plastics are processed using specially developed methods like:

1. Contact molding which employs single surface molds as in the case of hand lay-up, spray-up and filament winding.
2. Compression-type molding as in the case of sheet molding, bulk molding, preform molding and cold molding.
3. Resin-injection molding which is similar to the process used for nonreinforced materials and reinforced thermoplastics.
4. Pultrusion, which is a modification of the extrusion process.

Contact molding processes are used for liquid resin systems and various forms of fibrous reinforcement. A mixture of the resin, liquid monomer and catalyst is applied in the liquid form to the mold. The reinforcing phase is usually placed in the mold in the form of a mat, woven-roving cloth or a preformed fibrous shape. The reinforcement is saturated with the liquid resin, which is then cured by heating. A male or female single surface mold is used, thus providing only one smooth surface. Molds are inexpensive and can be made of wood, plaster, sheet metal or plastic. A thin pigmented gel coat about 0.4 to 0.5 mm (c. 0.015 to 0.020 in) can be sprayed on the mold surface at first to enhance surface finish. Hand lay-up is used for manufacturing low to medium production volumes of large, relatively thin-walled parts, such as boat hulls. Spray-up uses a continuous strand glass roving which is fed through a combination fiber chopper and spray gun. The chopped roving and liquid resin are sprayed on the mold surface to build up the required wall thickness. Curing under pressure can be achieved by placing pressure bags over the surface of the molding.

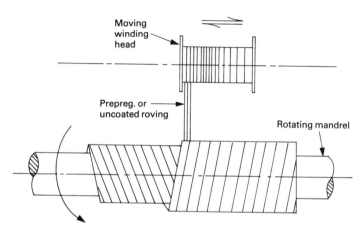

Figure 6.6 Filament winding process used in making reinforced axisymmetric components.

Filament winding is used for manufacturing highly reinforced, high-strength parts, which are usually axisymmetric in shape. Continuous strands, roving or woven tapes are drawn through a resin bath and wound onto a rotating mandrel (Fig. 6.6). The wound part is then cured by heating. Prepreg, in which the fibers have been previously coated with resin and stored at low temperature, can also be used for filament winding. Examples of filament wound products include tubes and storage tanks.

High-pressure laminates are made by stacking layers of resin-impregnated sheets (prepregs) then applying heat and pressure to bond the layers together. The pressures applied range from 8 to 14 MPa (1200 to 2000 lb/in²). In addition to making flat and curved laminated sheets, this method can be used to produce laminated tubing by winding the impregnated sheet under pressure on a mandrel and then curing the resin. Because of the high pressures used, the size of parts made by this method is limited by the sizes of available presses.

Compression molding processes, sometimes referred to as matched-metal molding, employ either bulk molding compound (BMC) or sheet molding compound (SMC). BMC is a prepared mixture of polyester resin, chopped glass fibers of length ranging from 3 to

25 mm (1/8 to 1 in), filler, catalyst and other additives. It is supplied in bulk form or as extruded 'rope'. The required amount of the mixture is placed in a heated mold, which is then closed, subjected to pressure and cured. SMC is similar to BMC, but the reinforcement is longer glass fibers, 25 to 75 mm (1 to 3 in), and the compound is supplied in sheet form. The sheet is first cut to a shape similar to that of the mold and is then pressed and heated in the mold for curing.

In preform molding, a mixture of chopped glass roving and weak binder is sprayed on a pattern resembling the product. The preform is then placed in a heated metal mold and saturated by a mixture of resin, catalyst, fillers and pigment. Curing then takes place under pressure. Cold press molding is similar to preform molding, except that curing takes place at room temperature and under low mold pressure. This allows the use of inexpensive mold materials like plastics and plaster. Preform molding gives more consistent and better-looking products than those produced by lay-up or spray-up processes.

Resin-injection molding is a matched-mold process which uses low-cost molds. The reinforcement is laid up dry in the mold and liquid catalyzed resin is pumped in until the mold is filled and the pressure reaches about 170 kPa (c. 25 lb/in^2). Curing takes about 15 min. This process may be considered as intermediate between spray-up and the faster pressure molding methods. The process is suitable for medium-volume production and can be used for manufacturing complex parts for motor car and truck bodies. Tailoring composites by placing extra reinforcement in highly stressed areas makes it possible to consolidate several smaller parts into one large structure, thus reducing assembly operations and cost.

Pultrusion is used for reinforced thermosetting plastics to make shapes similar to extruded shapes. The reinforcing fibers are immersed in liquid-resin bath and then pulled through a long heated die to cure the resin (Fig. 6.7). Rubber-faced rolls, which are shaped to the required form of the pultrusion, pull the cured and finished shape from the die. The pulling speed ranges from 50 to 5000 mm/min (2 to 200 in/min), depending on the resin compound, shape and thickness of the section, mold temperature and mold length. The most commonly pultruded materials are glass reinforcement and polyester resins. Standard pultruded shapes are available in the form of: (a) solid bar of round, rectangular and dogbone cross-section; (b) hollow shapes of round and rectangular cross-section; (c) structural shapes, like channels and I-beams; and (d) sheet of flat and corrugated cross-section. Epoxy resins can also be used in pultrusions, but they require longer curing time and do not release cleanly from the dies.

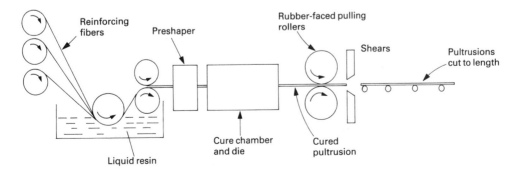

Figure 6.7 Steps of manufacture of composite shapes by pultrusion.

Manufacture of reinforced-metal components

Fiber reinforced metal–matrix composites can be prepared in a variety of ways which can be broadly classified into:

1. Processes based on solid-state diffusion.
2. Liquid-phase infiltration.
3. *In situ* processes.

Selection of the appropriate process is not only a function of the required shape and properties of the composite, but also of the fiber-matrix combination.

In solid-state diffusion processes, the matrix and fibers must be intimately mixed as it is difficult for the matrix material to move between tightly packed fibers. The fibers are usually coated by a layer of appropriate thickness of the matrix material using vapor deposition, coating with molten metal or electroplating. This layer protects the fibers and ensures an even distribution of the constituents. This method is easier to adopt in the cases of large volume-fractions of fine fibers. The loosely assembled coated fibers are then compacted and sintered. Solid state diffusion is relatively slow and partial melting of the matrix increases the diffusion rate and tolerates less intimate contact between fibers and matrix material. Secondary operations such as heat treatment, further shaping, machining or finishing may be needed before the composite part is ready for use. This method has been used for the composite systems of: Al-steel, Al-W, Ni-W, Co-W, Cr-Al$_2$O$_3$, Ni-Al$_2$O$_3$ and Ni-Si$_3$N$_4$.

Liquid-phase infiltration involves injecting the molten matrix material between the fibers and allowing it to solidify. Shrinkage on solidification and trapped gases can cause cavities in the composite. Pressure infiltration or vacuum infiltration can be used to overcome this problem. Since the infiltration technique is rapid, contact time between the melt and fibers can be reduced, which minimizes undesirable chemical reactions. These reactions can be further reduced by coating the fibers with an intermediate material which is wetted by the matrix but does not dissolve in the molten matrix. Composites prepared using this method can be made ready for use using the secondary operations described above. Systems which have been prepared by liquid-phase infiltration include: Ni-Al$_2$O$_3$, Cu-W, Al-glass, and Al-Al$_2$O$_3$. The latter system is reported by Chrysler as a possible candidate for connecting rods in motor car engines.

In situ composite materials are prepared by simultaneous solidification of the constituents under controlled conditions. The shape, size and distribution of the reinforcement depend on the solidification rate and temperature gradient across the solid–liquid interface. This method can be used to manufacture complex shapes in one step with little or no secondary processing, as in the case of turbine blades. A major limitation of this method is that the constituents have to be metallurgically compatible. High-temperature *in situ* composite systems include: Ta-Ta$_2$C, Nb-Nb$_2$C, nickel-base eutectics and cobalt-base eutectics.

6.8 DESIGNING WITH COMPOSITES

Earlier discussion has shown that the characteristics of composite materials are different in many respects from those of common metallic materials. This means that major design and manufacturing changes have to be made when replacing a metallic material by a composite material. As composite materials are usually more expensive on a cost per unit weight basis, weight saving and efficient use of the material are necessary if the finished part is to be competitive from the economic point of view. Weight saving is possible when the relevant specific property (property/density) of the composite material is better than that of the material being replaced. Figure 6.8 compares the specific strength (strength/density) and the specific elastic modulus (elastic modulus/density) for some metallic and composite materials.

Efficient use of composite materials can be achieved by tailoring the material for the application. For example, to achieve maximum strength in one direction in a fibrous composite material, the fibers should be well aligned in that direction. In applications where a composite material is subjected to tensile loading, the important design criterion is the tensile strength in the loading direction. Under compression loading, failure by buckling becomes an important parameter. For a structural component of a certain configuration, the buckling stress is proportional to the elastic modulus of the component material. In such cases the important design criterion is the composite elastic modulus.

The desirable feature of tailorability is illustrated by the use of carbon fibers, which exhibit a negative coefficient of thermal expansion, in building dimensionally stable structures. The negative coefficient of thermal expansion of carbon fibers is balanced by the positive coefficient of thermal expansion of the epoxy matrix and the resulting composite has a zero coefficient of thermal expansion. Space structures and finely-tuned optical equipment are typical applications for such composites.

An important factor that should be borne in mind when designing with composite materials is that their high strengths are obtained only as a result of large elastic strains in the fibers. In some structures these strains can cause unacceptable deflections. An example is the case of an aircraft wing where a strain of about 1.7 percent could cause the wing tips to become vertical. This indicates that composite materials should not simply be substituted for other materials without making the necessary modifications in design and construction.

Differences between the fatigue behavior of metals and composite materials should also be taken into account when designing with composites. Unlike steels, which show an endurance limit or a stress below which fatigue failure does not occur, composite materials suffer fatigue damage even at relatively low stress levels. While fatigue crack propagation in metals is usually limited to a single crack which progressively grows with each cycle, fibrous composites may have many cracks which can be growing simultaneously. Cracks may propagate through the matrix, be arrested by a fiber, or move along a fiber-matrix interface. This makes fatigue failure of fibrous composites a more gradual, noncatastrophic event than for metals. Fatigue loading of fibrous composites is also accompanied by a progressive reduction in modulus with increasing number of cycles. In addition, cyclic creep, which is the increase in strain at the minimum fatigue stress, can also occur in chopped-fiber reinforced plastics. This latter weakness can be overcome by using continuous fibers which are oriented in the principal stress direction.

In spite of their high strength, fibrous composites can exhibit low fracture toughness. This is because the conditions required for high strength, e.g. short critical aspect ratio and strong interfacial bond between fibers and matrix, are in conflict with the conditions required for high fracture toughness, e.g. long critical aspect ratio and weak interfaces parallel to the fibers. In general, a compromise between strength and toughness will be required for a given application. Several hybrid composites have been developed to achieve such compromise. For example, Kevlar fibers, which have good impact resistance but relatively low compressive strength, complement graphite fibers, which have roughly four times the compressive strength but are less tough. Because each fiber retains its own identity, hybrid technology is becoming increasingly important as a way of selectively utilizing the dominant properties of each fiber and reducing the overall cost of the composite. As an example, mixing carbon and glass fibers in a composite and placing the carbon fibers at critical locations, high-performance light-weight structures with minimal amount of the expensive carbon fibers can be achieved.

6.9   SELECTION AND USE OF COMPOSITE MATERIALS

Composite materials offer an extremely wide variety of properties and their use in industry ranges from jet engine turbine blades to golf clubs. Although composites are usually used because of their attractive mechanical properties and light weight, the ability to change

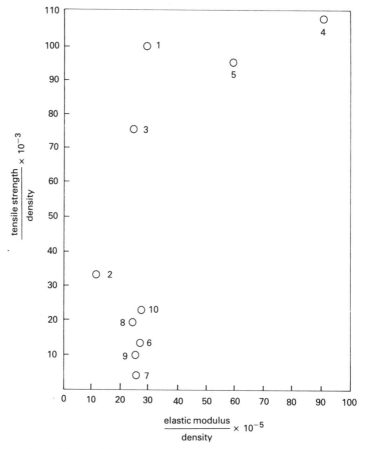

Figure 6.8 Comparison of the specific strength and the specific elastic modulus for some metallic and composite materials.
1 = Epoxy-70% S glass fibers. 2 = Epoxy-70% S glass fabric. 3 = Epoxy-73% E glass fibers. 4 = Epoxy-63% carbon fibers. 5 = Epoxy-62% aramid fibers. 6 = AISI 4340 steel (oil quenched and tempered). 7 = AISI 1010 steel (normalized). 8 = AA 2014 aluminum alloy (T6). 9 = AA 6061 aluminum alloy (T6). 10 = titanium alloy (Ti−6 Al−4 V).

and tailor their physical and chemical properties should not be overlooked. In spite of their attractive characteristics, composite materials still have serious challenges to meet. A major advantage of traditional materials, such as steels and aluminum alloys, is that most design experience and production facilities are metal-oriented. Major production lines use metal forming presses and metal cutting machine tools. Components are usually assembled by fasteners and fusion welding. Computer-controlled flexible machining systems represent a high degree of sophistication achieved over a long history of development in metal-based production technology. In contrast, much composite production is still carried out by hand lay-up, although numerically controlled machines are being increasingly used. Because of the random orientation of short fibers, dough-molding and sheet molding are not suitable for high-strength parts. All these processes are comparatively slow. This is acceptable in aircraft and aerospace industries, where high strength/weight is very important and production runs are short, but poses severe problems in the automotive and similar mass-production industries. Another difficulty is met where composites have to be joined to conventional materials like steel and aluminum. Special joint design and bonding materials have to be used to allow for the mismatches in stiffness and thermal expansion. Some of the above difficulties can be overcome by reducing the number of separate units and molding composites in one complex part. This lowers both the fabrication and secondary operation costs.

Another factor which affects the selection of composite materials is legislation. For example, the Corporate Average Fuel Economy legislation in the US, which required that by 1985 new cars should average 29 miles to the US gallon, initiated major development programs to introduce lightweight materials. Another example is legislation on crash padding in car interiors which led to a large increase in the amounts of plastics employed in European cars. On the other hand, legislation on side impact resistance has restricted the design and introduction of all plastic doors and other structural members.

Uses of composites in aerospace and aircraft industry

The outstanding values for strength/weight and elastic modulus/weight of some composite materials (Figs. 6.8 and 8.2), make them prime candidates for the aerospace and aircraft industries. In such applications, lifting a greater payload with a given amount of power can be achieved by reducing the weight of the structure. An added advantage of weight saving is that it may permit reduction in some structures like wing area, which would lead to further weight reduction. Composites with high specific strength at relatively low temperatures are usable for the fuselage and landing gear. As an example, the Boeing 767 employs several types of FRP composites in several locations, as shown in Fig. 6.9 and Table 6.1. The weight saving that resulted from using composites instead of traditional materials came to more than 900 kg (c. 2000 lb). In addition, composites that retain their strength and creep resistance at high temperatures are usable for the propulsion system. In addition to their use in the jumbo jet category, advanced FRP composites have been used even more extensively for both primary and secondary structural parts in military aircraft and helicopters.

Uses of composites in automotive industry

Motor cars represent an important market for composite materials and several types of FRP have been developed to meet the needs of the automotive industry. These materials

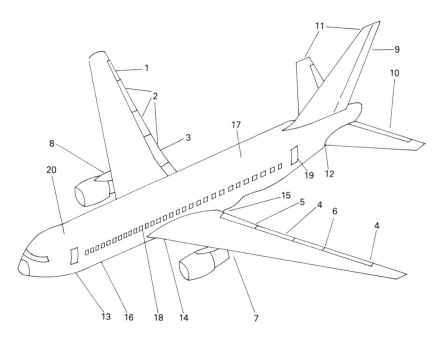

Figure 6.9 Examples of FRP composites used in the Boeing 767. See Table 6.1 for key to numbers. (Reference: *Design Engineering*, October 1981, p. 47.)

Table 6.1 Examples of FRP composites used in the Boeing 767

Location	Part	Composite system
Wing	[a]1. Fixed panels – wing T.E.	Hybrid, Kevlar-carbon
	2. Fairing – T.E. flap linkage	Hybrid, Kevlar-carbon
	3. Inboard and outboard spoilers	Carbon-epoxy
	4. Outboard and inboard ailerons	Carbon-epoxy
	5. Inboard flap – debris protection	Kevlar-epoxy
	6. Outboard flaps – L.E. and T.E.	Kevlar-epoxy
	7. Fairings – engine pylon	Kevlar-epoxy
	8. Fairing – thrust reverser	Kevlar-epoxy
Tail	9. Rudder	Carbon-epoxy
	10. Elevator	Carbon-epoxy
	11. Fixed T.E. tip – empennage	Hybrid, Kevlar-carbon
	12. Seal plates – stabilizer	Hybrid, Kevlar-carbon
Landing gear	13. Nose landing gear doors	Hybrid, glass-carbon
	14. Body landing gear doors	Hybrid, Kevlar-carbon
Body	15. Fairing – wing/body	Hybrid, Kevlar-carbon
	16. Cargo liner	Kevlar-epoxy
	17. Lavs, closets and partitions	Kevlar-epoxy
	18. ECS ducts	Kevlar-epoxy
	19. Emergency escape system	Kevlar-epoxy
	20. Outboard stowage bins and center stowage supports	Kevlar-epoxy

[a]Numbers refer to parts in Fig. 6.9.

provide some of the advantages of high-performance composites used in aerospace applications, but some aspects of performance are traded for reduced cost and speed of manufacturing. Composites used in the automotive industry are usually made of filled or unfilled polymers reinforced with milled, chopped, continuous or combinations of chopped and continuous fibers. Thermosetting polymers such as polyester, vinylester, epoxy and urethane are typically used as matrix materials. The primary reinforcing fiber is E-glass. The uses of composites in motor cars can be grouped into the following areas:

1. Panels, as in the exterior body panels of the Chevrolet Corvette. These panels are made of 25 mm (1 in) chopped fiberglass reinforced polyester sheet molding compound (SMC). More recently, flake glass reinforced urethane reaction injection molding (RIM) was used for the exterior body panels of the Pontiac Fiero. Advantages of composite panels include corrosion resistance, dent resistance and design freedom. Similar parts which can be made of composite materials include headlamp and taillamp housings, hoods, fenders and instrument panels.
2. Semistructural parts, e.g. the tailgate of the 1980 Oldsmobile station wagon. This part has significant load-bearing requirements in the open position and is made of two SMC pieces which are bonded with a urethane adhesive. Bumpers can be treated similarly and are more economical than the traditional rolled steel bumpers.
3. Structural parts, as in the case of the Chevrolet Corvette rear leaf spring. This is a single leaf made of unidirectional glass reinforced epoxy and was used to replace a 10-leaf steel spring and provide 80 percent weight reduction. Composites have also been shown to be feasible materials for rear axles and drive shafts.
4. High-temperature applications of composites in and around the engine are still in the development stages. The development of low-cost high-temperature polymers would increase the potential for using composites in these areas.

Uses of composites in sports and recreational industry

Sports and recreational goods have made extensive use of FRP materials. For example, vaulting poles were first made of 60 percent E-glass reinforced epoxy which was later increased to 70 percent. Substituting S-glass for E-glass allowed a 25 percent increase in stiffness without adding weight. Carbon fibers or carbon-glass hybrids are widely used for golf-club shafts, ski poles, fishing rods, tennis rackets, badminton rackets, squash rackets and many parts of gliders and sailing yachts. The use of FRP in making tennis rackets will be discussed in Chapter 24.

Other applications of composite materials

Composite materials are being increasingly used in chemical processing where corrosion resistance is an important requirement. Typical uses include chemical and fuel tanks, piping, ducting, fume collection hoods and duct systems, process pump and valve bodies, impellers, electroplating racks and photographic processing equipment. Household applications include furniture, lawn furnishings, patio covers, fans and blowers, microwave oven trays, sinks, bath tubs and shower units. Composites are also used in business machine housings, keyboards and printer heads. Electrical and electronic industries also make use of composite materials in printed circuit boards and other electronic components, electrical pole line hardware, cross arms and shatterproof street lighting globes.

Examples of marine applications include pleasure, commercial and military boat hulls and superstructures, in addition to masts, spars, marker buoys and floating docks.

6.10   REVIEW QUESTIONS AND PROBLEMS

6.1   What are the attractive features of fiber reinforced composites?

6.2   Compare the use of galvanized steel sheets and glass fiber reinforced plastics in making $1 \, m^3$ ($35 \, ft^3$) water storage tanks.

6.3   How do the properties of metallic sheets differ from those of fiber reinforced laminates?

6.4   In Fig. 2.1, composites are shown to be used in making some parts of the motor car. List some of these parts and suggest methods of manufacturing them. Suggest other parts that are now being made of other materials and can be advantageously substituted with composites.

6.5   Apply the rule of mixtures to calculate the elastic modulus of polystyrene containing 50%, 60% and 70% by volume of E-glass fibers both in the direction of the fibers and at right-angles to the fibers. [Answer: $E_c = 38.4$, 45.4, and $52.4 \, GPa$ parallel to the fibers, $E_c = 6.3$, 7.7, and $9.96 \, GPa$ at right-angles to the fibers].

6.6   What are the main limitations on the maximum volume fraction of the reinforcing fibers in fiber reinforced composites?

6.7   Compare the use of aluminum and gfrp for use in making lawn furnishings.

6.8   Compare the use of enameled steel and gfrp for making bath tubs.

6.9   Explain the increasing use of carbon fiber reinforced plastics in sports equipment.

6.10 Compare the use of steel sheet and gfrp for making motor car bodies.

6.11 What are the main advantages of substituting carbon fiber reinforced plastics for aluminum alloys in aircraft structures?

BIBLIOGRAPHY AND FURTHER READING

Bracke, P., Schurmans, H. and Verhoest, J., *Inorganic Fibers and Composite Materials*, Pergamon Press, New York, 1984.

Brown, R. L. E., *Design and Manufacture of Plastic Parts*, John Wiley & Sons, New York, 1980.

Demmler, A. W., 'Kevlar and carbon composites compared', *Automotive Engineering*, vol. 93, No. 2, 1985, pp. 44–52.

Farag, M. M., *Materials and Process Selection in Engineering*, Applied Science Pub., London, 1979.

Harris, B., *Engineering Composite Materials*, Institute of Metals, London, 1986.

Lockwood, P. A., 'Composites for industry', *ASTM Standardization News*, December, 1983, pp. 28–31.

McCullough, R. L., *Concepts of Fiber-resin Composites*, Marcel Dekker, New York, 1971.

Parratt, N. J., *Fiber-reinforced Materials Technology*, Van Nostrand Reinhold, London, 1972.

Riegner, D. A. and Hau, J. C., 'Fatigue considerations for FRP composites', *SAE paper 820698* (p. 109), 1982.

Thompson, J. M., Bauer, A. and Brodowsky, D., 'A plastic suspension part?', *SAE paper 820802*, 1982.

Thornton, P. A. and Colangelo, V. J., *Fundamentals of Engineering Materials*, Prentice Hall, New Jersey, 1985.

Trebilcock, N. and Epal, J. N., 'Light truck FRP leaf spring development', *SAE paper 810325*, 1981.

Van Vlack, L. H., *Elements of Materials Science and Engineering*, Addison Wesley, Reading, Mass., 1985.

Chapter 7

Failure of Components in Service

7.1   CAUSES OF FAILURE OF ENGINEERING COMPONENTS

The service behavior of a material is governed not only by its inherent properties but also by the stress system acting on it and the environment in which it is operating. Causes of failure of engineering components can be classified into the following main categories:

1. Design deficiencies. Failure to evaluate correctly working conditions due to the lack of reliable information on loads and service conditions is a major cause of inadequate design. Incorrect stress analysis, especially near notches, and complex changes in shape could also be a contributing factor.
2. Poor selection of materials. Failure to identify clearly the functional requirements of a component could lead to the selection of a material that only partially satisfies these requirements. As an example a material can have adequate strength to support the mechanical loads but its corrosion resistance is insufficient for the application.
3. Manufacturing defects. Incorrect manufacturing could lead to the degradation of an otherwise satisfactory material. Examples are decarburization and internal stresses in a heat-treated component. Poor surface finish, burrs, identification marks and deep scratches due to mishandling could lead to failure under fatigue loading.
4. Exceeding design limits and overloading. If the load, temperature, speed, voltage, etc., are increased beyond the limits allowed for by the factor of safety in design, the component is likely to fail. As an example, if an electrical cable carries a higher current than the design value, it overheats and this could lead to melting of the insulating polymer and then short-circuit. Subjecting the equipment to environmental conditions for which it was not designed also falls under this category. An example here is using a freshwater pump for pumping sea water.
5. Inadequate maintenance and repair. When maintenance schedules are ignored and repairs are poorly carried out, service life is expected to be shorter than anticipated in the design.

Anticipating the different ways in which a product could fail is an important factor that should be considered when selecting a material or a manufacturing process for a given application. The possibility of failure of a component can be analyzed by studying on-the-job material characteristics, the stresses and other environmental parameters that will be acting on the component and the possible manufacturing defects that can lead to failure. This technique is called failure analysis and can be carried out in the following ways:

1. By an environment profile that provides a description of the expected service conditions. These include operating temperature and atmosphere, radiation, presence of contaminants and corrosive media, other materials in contact with the component and the possibility of galvanic corrosion, and lubrication conditions.
2. By fabrication and process flow diagrams that provide an account of the effect of the various stages of production on the material properties and of the possibility of quality control. Certain processes can lead to undesirable directional properties, internal stresses, cracking, or structural damage, which could lead to unsatisfactory component performance and premature failure in service.
3. By failure logic models that describe all the possible types of failure and the conditions that can lead to them. Failure due to chemical causes occurs when corrosion is so excessive that it becomes hazardous for the component to remain in service. Electrical failures occur in insulating materials when the applied voltage exceeds the breakdown voltage of the material, or when flashover takes place. Mechanical failures are more varied and represent a serious threat to all load-bearing components and will be discussed in more detail in this chapter.

7.2   TYPES OF MECHANICAL FAILURE

Generally, a component can be considered to have failed when it does not perform its intended function with the required efficiency. The general types of mechanical failure encountered in practice are:

1. Yielding of the component material under static loading. Yielding causes permanent deformation which could result in misalignment or hindrance to mechanical movement.
2. Buckling, which takes place in slender columns when they are subjected to compressive loading, or in thin-walled tubes when subjected to torsional loading.
3. Creep failure, which takes place when the creep strain exceeds allowable tolerances and causes interference of parts. In extreme cases failure can take place through rupture of the component subjected to creep. In bolted joints and similar applications failure can take place when the initial stressing has relaxed below allowable limits, so that the joints become loose or leakage occurs.
4. Failure due to excessive wear, which can take place in components where relative motion is involved. Excessive wear can result in unacceptable play in bearings and loss of accuracy of movement. Other types of wear failure are galling and seizure of parts.
5. Failure by fracture due to static overload. This type of failure can be considered as an advanced stage of failure by yielding. Fracture can be either ductile or brittle.
6. Failure by fatigue fracture due to overstressing, material defects or stress raisers. Fatigue fractures usually take place suddenly without apparent visual signs.
7. Failure due to the combined effect of stresses and corrosion, which usually takes place by fracture due to cracks starting at stress concentration points, e.g. caustic cracking around rivet holes in boilers.
8. Fracture due to impact loading, which usually takes place by cleavage in brittle

materials, for example in steels below brittle-ductile transition temperature and plastics below glass transition temperature.

Of the above types of mechanical failure, the first four do not usually involve actual fracture, and the component is considered to have failed when its performance is below acceptable levels. On the other hand, the last four types involve actual fracture of the component, and this could lead to unplanned load transfer to other components and perhaps other failures. This can be avoided by careful design and the selection of the appropriate factor of safety as discussed in Chapter 9.

7.3 FRACTURE TOUGHNESS AND FRACTURE MECHANICS

Fracture toughness can be qualitatively defined as the resistance of a material to the propagation of an existing crack. Such cracks cause high local stresses at their tips and these depend on the geometry of the flaw and the geometry of the part. The ability of a particular flaw or stress concentrator to cause fracture depends on the fracture toughness of the material. Therefore, to predict the fracture strength of a part both the severity of the stress concentration and the fracture toughness of the material must be known. Quantitative prediction of the fracture strength can be made using fracture mechanics techniques and in the simple case of glass, Griffith showed that:

$$\sigma_f = \sqrt{\left(\frac{2 \star E \star \gamma_s}{\pi \star a}\right)} \tag{7.1}$$

where σ_f = fracture stress in MPa (or lb/in^2)

a = crack length for edge cracks and 1/2 crack length for center cracks. This is measured in m (or in),

E = Young's modulus in MPa (or lb/in^2),

γ_s = energy required to extend the crack by a unit area in J/m^2 (or in lb/in^2).

For glass, γ_s is simply equal to the surface energy. However, this is not the case with metals due to the plastic deformation which occurs at the tip of the propagating crack. In the latter case, the fracture toughness is proportional to the energy consumed in the plastic deformation. Because it is difficult to measure this energy accurately, the parameter called the stress intensity factor, K_I, is used to determine the fracture toughness of most materials. The stress intensity factor, as the name suggests, is a measure of the concentration of stresses at the tip of the crack under consideration. For a given flawed material, catastrophic fracture occurs when the stress-intensity factor reaches a critical value, K_c. The relationship between stress intensity factor, K_I, and critical intensity factor, K_c, is similar to the relationship between stress and tensile strength. The value of K_I is the level of stress at the crack tip and is material independent. On the other hand, K_c is the highest value for K_I that the material can withstand without fracturing and is material and thickness dependent. The reason why K_c is thickness dependent is that lateral constraint imposed on the material ahead of a sharp crack in a thick plate gives rise to a triaxial state of stress which reduces the apparent ductility of the material. Thus the fracture strength is less for thick plates compared with thinner plates, even though the inherent properties of the material have not changed. As the thickness increases, K_c decreases and reaches a minimum constant value, K_{Ic}, when the constraint is sufficient to give rise to plane-strain conditions, as shown in Fig. 7.1. The thickness, t, at which plane-strain conditions occur is related to the fracture toughness, K_{Ic}, and yield strength of the material, YS, according to the relationship:

$$t = 2.5\left(\frac{K_{Ic}}{YS}\right)^2 \qquad (7.2)$$

The critical stress intensity factor for plane-strain conditions, K_{Ic}, is found to be a material property which is independent of the geometry. In an expression similar to Griffith's, the fracture stress σ_f can be related to the fracture toughness, K_{Ic}, and the flaw size, $2a$:

$$\sigma_f = \frac{K_{Ic}}{Y\sqrt{\pi a}} \qquad (7.3)$$

where Y is a correction factor which depends on the geometry of the part, i.e. thickness, width W and the flaw size $2a$ for center crack and a for edge crack. For the case of thick plates, as a/W decreases to 0, i.e. plane-strain, Y decreases to 1. From (7.3) it can be shown that the units of K_{Ic} are MPa $\sqrt{m}$ or lb/in^2 $\sqrt{in}$. The values of Y for different geometries are given in Section 11.5.

As K_{Ic} is a material property, the designer can use it to determine the flaw size that can be tolerated in a part for a given applied stress level. Conversely, the designer can determine the stress level that can be safely used for a flaw size that may be present in a part. As an example, consider a wide plate containing a crack of length $2a$ extending

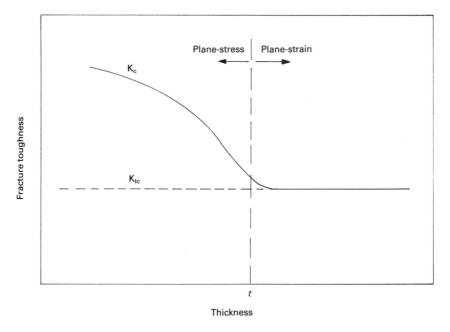

Figure 7.1 Effect of thickness on fracture toughness behavior.

through the thickness. If the fracture toughness of the material is 27.5 Mpa√m and the yield strength is 400 MPa, calculate the fracture stress σ_f and compare it to the yield strength σ_y for different values of crack lengths. Assume $V = 1$. Using (7.3), σ_f can be calculated for different crack lengths. The results are given in the following table:

a (mm)	1	2	4	6	8	10
σ_f (MPa)	490.6	346.9	245.3	200.3	173.5	155.2
σ_f/σ_y	1.23	0.87	0.61	0.50	0.43	0.39

With the smallest crack, the yield strength is reached before catastrophic failure occurs. However, longer cracks cause fracture before yielding.

Fracture toughness data are available for a wide range of materials and some examples are given in Section 8.5. Fracture toughness data can also be easily established for new materials using standardized testing methods, e.g. ASTM Standard E399. Because of possible anisotropy of microstructure, it is important to orient the test specimen to correspond to the actual loading conditions of the part in service. Fracture toughness, like other material properties, is influenced by several factors including strain rate or loading rate, temperature and microstructure, as discussed in more detail in Section 8.5. Also, increasing the yield and tensile strengths of the material usually causes a decrease in K_{Ic}. The use of fracture mechanics in design will be discussed in Section 11.5.

Fracture toughness is widely accepted as a design criterion for high-strength materials where ductility is limited. In such cases, the relationship between K_{Ic}, applied stress and crack length governs the conditions for fracture in a part or a structure. This relationship is shown schematically in Fig. 7.2. If a particular combination of stress and flaw size in a structure reaches the K_{Ic} level, fracture can occur. Thus there are many combinations of stress and flaw size that may cause fracture in a structure made of a material having a particular value of K_{Ic}. The figure shows that materials with higher K_{Ic} values tolerate larger flaws at a given stress level or higher stress levels for a given flaw size.

Figure 7.2 also shows that if a material of known K_{Ic} is selected for a given application, the size of the flaw that will cause fracture can be predicted for the anticipated applied stress. If the design stress of a part is taken as 0.5YS, the critical flaw length would be (a_1). Therefore, provided that no defect of size greater than (a_1) is present, failure should not occur on loading. If in a proof test the part is loaded to a stress above the expected service stress and the test was successful, then a flaw of size greater than a_2 could not have existed. During service life crack growth of the order of (a_1–a_2) could be tolerated before failure. From (7.3) and Fig. 7.2, it can be shown that the maximum allowable flaw size is proportional to $(K_{Ic}/YS)^2$, where K_{Ic} and YS are measured at the expected service temperature and loading rate. Thus the ratio (K_{Ic}/YS) can be taken as an index for comparing the relative toughness of structural materials. Higher values of (K_{Ic}/YS) are more desirable as they indicate tolerance to larger flaws without fracture, as will be discussed in Section 8.5. The sensitivity of the nondestructive testing techniques used to detect manufacturing defects that approach the critical size in the part or structure is determined by the value of the allowed flaw size.

As indicated earlier, high-strength materials usually have low fracture toughness which allows the above analysis, known as linear elastic fracture mechanics (LEFM), to be successfully used to predict their behavior. In many applications, however, especially

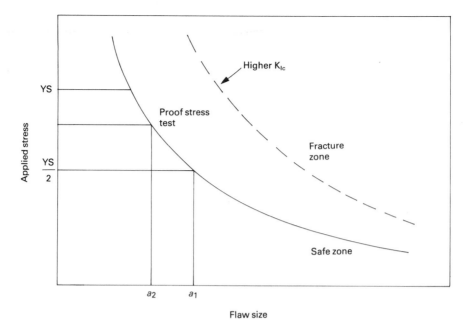

Figure 7.2 Schematic relationship between stress, flaw size and fracture toughness.

when low- and medium-strength structural materials are used, the section sizes are of insufficient thickness to cause plane-strain conditions under normal service temperatures and strain rates. For these cases, the linear elastic analysis used to estimate K_{Ic} values does not hold, since general yielding and large plastic deformations take place and plane-stress conditions prevail. In such cases, elastic-plastic fracture mechanics EPFM rather than LEFM must be used. At present, three methods are available for the evaluation of tough materials under plane-stress conditions where fracture is accompanied with considerable plastic deformation:

1. Crack-opening displacement, COD, which gives a measure of the prefracture deformation at the tip of a sharp crack.
2. J integral, which is an average measure of the elastic-plastic stress/strain field ahead of a crack.
3. R-curve analysis, which relaxes the K_c, COD and J to crack extension in a given material under plane-stress conditions.

The crack-opening displacement (COD) concept is based on the assumption that crack extension, i.e. fracture, takes place when the material at the crack tip has reached a maximum permissible plastic strain. The crack tip strain or the crack tip opening displacement (CTOD) can be related to COD, which is a measurable quantity. It can be shown that this criterion is equivalent to the K_{Ic} criterion in the case of LEFM. LEFM is often used to compare strong materials on the basis of their critical crack size (a_c) at yield stress:

$$a_c = A \left(\frac{K_{Ic}}{YS} \right)^2 \tag{7.4}$$

where A is a constant.

In the same way, a critical crack size a_c can be determined from COD measurements as:

$$a_c = B\left(\frac{COD_c}{e_y}\right) \tag{7.5}$$

where B is a constant, COD_c is critical COD, and e_y is strain at yield.

The values of the constants A and B in (7.4) and (7.5) depend on the crack geometry and loading configuration.

Although COD testing has prospects, further research is still needed to develop the concept to a stage where it can give unique toughness parameters for high-toughness materials.

The J integral offers potentials for application to fracture problems where the stresses are close to or above yield. For elastic behavior, the J integral is identical with the energy release rate per unit crack extension and can be shown to be:

$$JIC = \frac{(1 + v^2)K_{Ic}^2}{E} \tag{7.6}$$

where v is Poisson's ratio and E is elastic constant.
It can also be shown that:

$$J = C \star COD \star YS \tag{7.7}$$

where C is a constant which depends on the stress state. $C = 1$ for plane-stress conditions.

Equations 7.6 and 7.7 show the general compatibility between K_{Ic}, COD and J. This indicates that the J integral could offer a means of extending fracture mechanics concepts from linear elastic to general yielding or plastic behavior.

When a ductile part containing a crack is subjected to an increasing load, localized yielding is first observed at the crack tip, which then takes a rounded shape, called crack blunting. At a certain load corresponding to a critical value of J, JIC, crack growth initiates. Beyond this value, increasing J causes further increase in crack length, as shown in Fig. 7.3. The curve relating J to crack extension is called J-resistance curve or J-R curve. Similarly, K_c-R and CTOD-R curves respectively relate K_c and CTOD to crack extension. K_c in this case is the critical stress-intensity level for fracture under plane stress conditions. As discussed earlier, the value of K_c does not only depend on the material but also on the thickness and constraint. For ductile materials, the values of J and K_c have to be increased to several times the JIC and K_{Ic} values before final instability occurs. This corresponds to an important margin of safety between initiation and instability. Therefore, knowing the R-curve for the material at the required thickness, temperature and rate of loading would be useful in design. Under these conditions, the relationship between fracture toughness, applied stress and crack length can be represented schematically in Fig. 7.4.

When a part or structure is subjected to fatigue loading or stress corrosion, a relatively harmless defect could develop into a dangerous crack in service. Under these conditions, crack growth can occur at lower levels of stress intensity than K_{Ic}, as shown in Fig. 7.5. The relationship between fatigue crack growth per cycle, da/dN, and the stress intensity range corresponding to the load cycle applied to the part or structure, is discussed in Section 7.5. If the relationship between da/dN and the stress intensity range is known for a

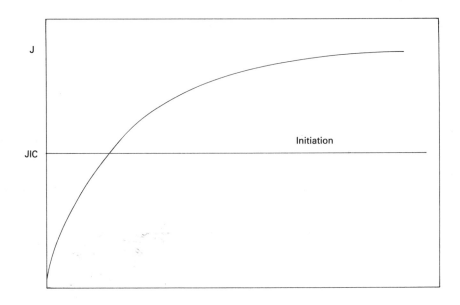

Figure 7.3 Typical J-resistance curve.

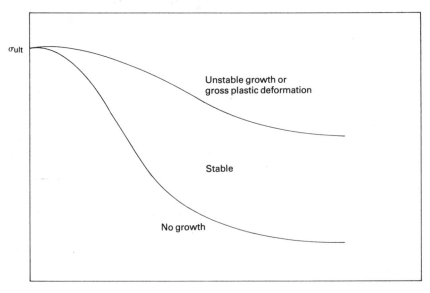

Figure 7.4 Schematic representation of the relationship between applied stress, crack length and fracture toughness for a ductile material.

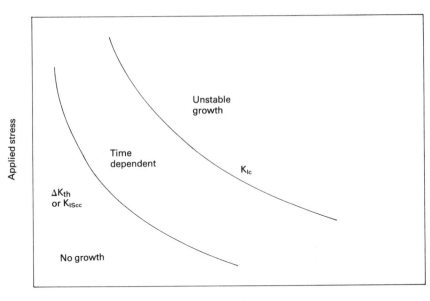

Figure 7.5 Effect of fatigue loading or stress-corrosion on crack growth.

given material, it can be used to calculate fatigue life of components and structures made of this material.

7.4 DUCTILE AND BRITTLE FRACTURES

Machine and structural elements often fail in service as a result of either ductile or brittle fracture. The terms ductile and brittle are usually used to indicate the extent of macroscopic or microscopic plastic deformation which precedes fracture. The terms ductile and brittle are also related to toughness, which is a measure of the energy needed for fracture. Service failures which occur solely by ductile fracture are relatively infrequent and may be a result of errors in design, incorrect selection of materials, improper fabrication techniques, or abuse. The latter condition arises when a part is subjected to load and environmental conditions which exceed those of intended use. As an example of ductile fracture, consider the case of an aluminum ladder 3 m long and made of four T-sections and hollow cylindrical rungs, as shown in Fig. 7.6. The ladder failed when a man weighing 90 kg climbed half way up when it was leaning against a wall at an angle of 15°. Although this was the first time for the man to use the ladder, his wife, who weighs 60 kg, had used it many times before. As a result of failure, T-sections S2 and S3 suffered severe plastic deformation and buckling caused by bending while T-sections S1 and S4 cracked just under the rung where the man was standing, Fig. 7.6. Investigation showed that large reduction in area accompanied the fracture and chemical analysis showed that the T-sections were made of AA 6061 alloy. The hardness of the alloy was in the range 25 to 30 RB in most areas but was about 20 RB in section S2. These hardness values correspond to T4 temper condition of the AA 6061 alloy. It is expected that the weakest section S2, which was on the tension side during loading, has yielded causing the loads to be redistributed and section S3

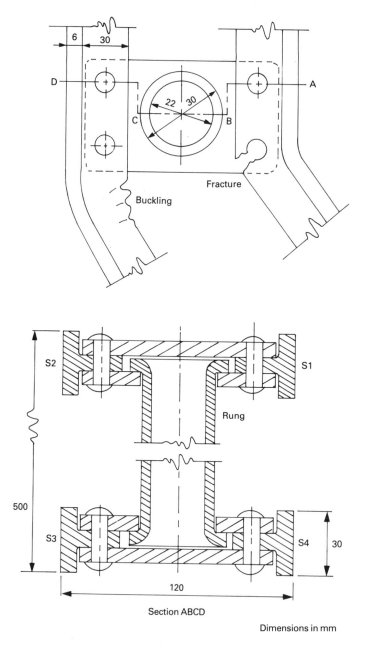

Figure 7.6 Failure of aluminum ladder.

to yield. This, in turn, caused sections S1 and S4 to be overloaded in tension. As failure is caused by overload during normal use, it is recommended that a stronger material be used. It would be sufficient to change the temper condition from T4 to T6. The AA 6061 T6 has a hardness of 45–55 RB and yield strength about twice that of the AA 6061 T4.

Brittle fractures are usually initiated at stress raisers, such as large inclusions, cracks or surface defects and sharp corners or notches. The single most frequent initiator of brittle fracture is the fatigue crack, which accounts for more than 50 percent of all brittle fractures in manufactured products. Brittle fractures are insidious in character because they may occur under static loading at stresses below the yield strength and without warning. Once started, the brittle fracture will run at high speed, reaching 1200 m/s in steel, until total failure occurs, or until it runs into conditions favorable for its arrest. The risk of occurrence of brittle fracture depends on the notch toughness of the material under a given set of service conditions. A characteristic feature of brittle fracture surfaces is the chevron pattern, which consists of a system of ridges curving outwards from the center line of the plate, as shown schematically in Fig. 7.7. These ridges, or chevrons, may be regarded as arrows with their points on the center line and invariably pointing towards the origin of the fracture, so providing an indication of its propagation pattern. This feature is useful in the analysis of service failures.

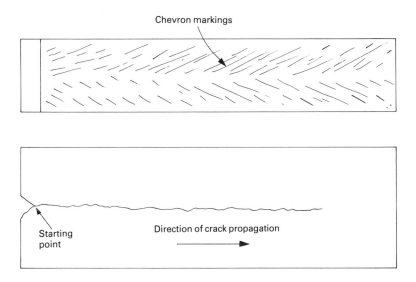

Figure 7.7 Chevron patterns in brittle fracture.

The temperature at which the component is working is one of the most important factors that influence the nature of fracture. Brittle fractures are usually associated with low temperature, and in some steels conditions may exist where a difference of a few degrees, even within the range of atmospheric temperatures, may determine the difference between ductile and brittle behavior. This sharp ductile–brittle transition is only observed in body-centered cubic (BCC) and close-packed hexagonal (CPH) metallic materials and not in face-centered cubic (FCC) materials, as illustrated schematically in Fig. 7.8. The most widely used tests for characterizing the ductile-to-brittle transition are the Charpy, ASTM standards A23 and A370, and Izod. The temperature at which the material behavior changes from ductile to brittle is called the ductile–brittle transition temperature, T_c, and may be taken as the temperature at which the fractured surfaces exhibit 50 percent brittle fracture appearance. In V-notch Charpy experiments the transition temperature can be set at a level of 20.3 J (15 ft lb) or at 1 percent lateral contraction at the notch. The transition temperature based on fracture appearance always

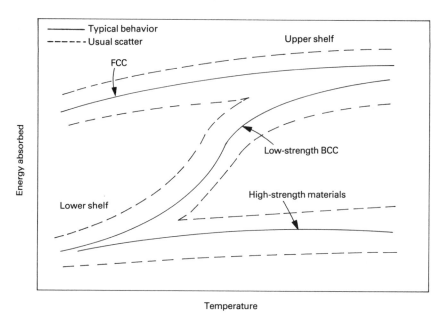

Figure 7.8 Schematic representation of the effect of temperature on energy absorbed in fracture.

occurs at a higher temperature than if based on a ductility or energy criterion. Therefore the fracture appearance criterion is more conservative.

The rate of change from ductile to brittle behavior depends on the strength, chemical composition, structure and method of fabrication of the material. The state of stress and the speed of loading also influence the nature of fracture. A state of triaxial tensile stresses, such as those produced by a notch, can be the cause of brittle fracture. The notches in a component can be due to sharp changes, processing defects, or corrosion attack. Materials which behave normally under slowly applied loads may behave in a brittle manner when subjected to sudden applications of load, such as shock or impact. The ductile-to-brittle transition also shifts to higher temperatures as the rate of loading increases. In the case of steels, the shift in transition temperature depends on the strength and can be as high as 68°C (155°F) in steels of yield strength of 280 MPa (*c.* 40 ksi). The shift in transition temperature between static and impact loading decreases with increasing strength and becomes negligible at yield strengths of about 900 MPa (*c.* 130 ksi).

In applying the Charpy V-notch results to industrial situations it should be borne in mind that the shock conditions encountered in the test may be too drastic. Many industrial components operate successfully in extreme cold without special consideration for notch toughness values or transition temperature. However, where stress concentration and rate of strain are high and service temperatures are low, special design and fabrication precautions should be taken and materials with low transition temperatures should be selected. The design and fabrication precautions that should be taken include:

1. Abrupt changes in section should be avoided and thickness should be kept to a minimum.
2. Welds should be located clear of stress concentrations and of each other and should be easily accessible for inspection.

3. Whenever possible, welded components should be designed on a fail-safe basis, as discussed in Chapter 9.

A useful relation between plane-strain fracture toughness (K_{Ic}) and the upper-shelf Charpy V-notch impact energy (CVN) was suggested for steels of yield strengths (YS) higher than about 770 MPa (*c.* 110 ksi) by Rolfe, S. T. and Barson, J. M.: 'Fracture and fatigue control in structures', Prentice Hall, NJ, 1977, as:

$$\left(\frac{K_{Ic}}{YS}\right)^2 = \frac{5}{YS}\left(CVN - \frac{YS}{20}\right) \tag{7.8}$$

where K_{Ic} is in (ksi in$^{1/2}$), YS in ksi, and CVN in ft lb.

7.5 FATIGUE FRACTURE

Fatigue fractures account for about 80 percent of part failures in engineering and occur in materials when they are subjected to fluctuating loads. Generally, fatigue fractures occur as a result of cracks which usually start at some discontinuity in the material, or at other stress concentration locations, and then gradually grow under repeated application of load. As the crack grows, the stress on the load-bearing cross-section increases until it reaches a high enough level to cause catastrophic fracture of the part. This sequence is reflected in the fracture surfaces which usually exhibit smooth areas, which correspond to the gradual crack growth stage, and rough areas, which correspond to the catastrophic fracture stage, as shown schematically in Fig. 7.9. The smooth parts of the fracture surface usually exhibit beach marks which occur as a result of changes in the magnitude of the fluctuating fatigue load. Another feature of fatigue fractures is that they lack macroscopic plastic deformation and, in this respect, they resemble brittle fractures.

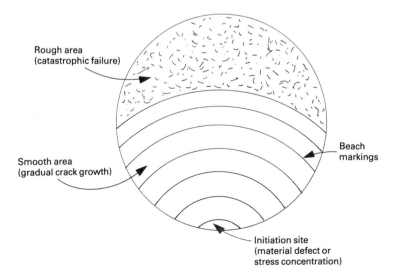

Figure 7.9 General appearance of a fatigue fracture surface.

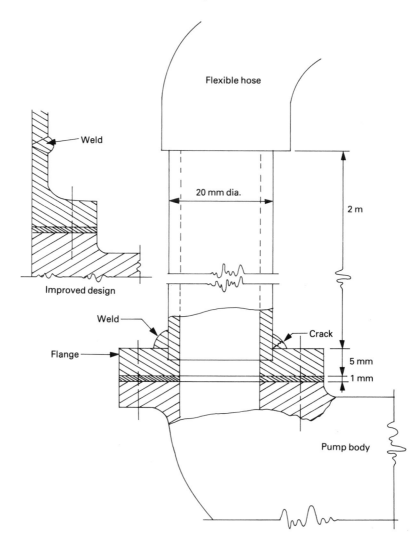

Figure 7.10 Failure of pressure line of a hydraulic pump.

As an illustration of fatigue fracture, consider the following example. The steel pressure line of a hydraulic pump in a power generation unit started leaking at the exit line flange assembly, shown in Fig. 7.10. The source of leaking was found to be a crack in the fillet weld. Investigation of the working conditions showed that although the pressure in the line was within the design limits, excessive vibrations existed in the 2 m-long tube which was not sufficiently supported by the flexible hose at its end. This caused the line to act as a cantilever beam with maximum forces at the flange. It was concluded that the crack in the fillet weld took place as a result of fatigue loading caused by the vibrations in the line. The corrective action taken was to change the design to move the weld from the area of high stress concentration, as shown in Fig. 7.10. The line was also adequately supported at the point where it joined the flexible hose in order to minimize vibrations.

The simplest type of fatigue loading is the alternating tension-compression without a

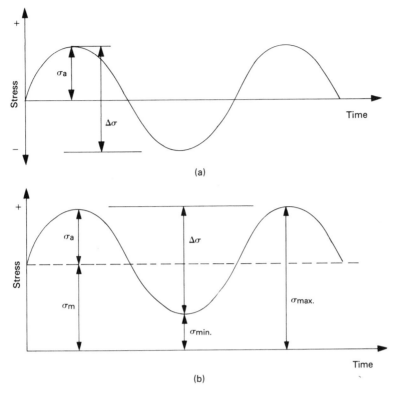

Figure 7.11 Types of fatigue loading. (a) Alternating stress, R = − 1. (b) Fluctuating stress.

static direct stress, Fig. 7.11. In this case the stress ratio defined as $R = \sigma_{min}/\sigma_{max}$, is −1. If a static mean stress is superimposed on the alternating stress, then the stress varies between the limits of:

$\sigma_{max} = \sigma_m + \sigma_a$, and $\sigma_{min} = \sigma_m - \sigma_a$,

as shown in Fig. 7.11. A special case is the pulsating stress with $R = 0$.

Under actual service conditions, parts may be subjected to more than one form of load, e.g. alternating torsion with static tension. Many other combinations are known to be met in different applications. However, most of the available fatigue test results are for the simple alternating stresses, i.e. $R = -1$. Such results are usually presented as S–N curves, as shown in Fig. 7.12. In this case S is the alternating stress and N is the number of cycles to failure. Some materials, such as steels and titanium, exhibit a well-defined fatigue or endurance limit below which no fatigue fracture occurs. Other materials, such as aluminum alloys and plastics, do not have such a limit and their S-N curves continue to decrease at high numbers of cycles. For these materials, fatigue strength is reported for a specified number of cycles. As this number of cycles is not standardized, the reported fatigue strength values are subject to large variations depending on whether the strength is taken at $N = 10^6$, 10^7 or 10^8 cycles. Therefore, it is necessary to specify the number of cycles for which the strength is reported. When the material does not exhibit a well-defined fatigue or endurance limit, it is only possible to design for a limited life.

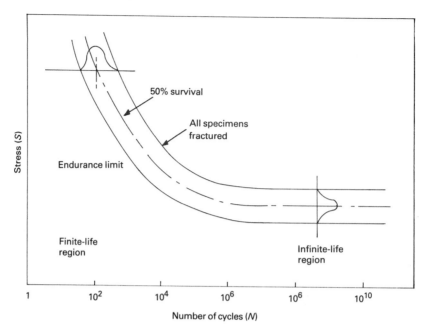

Figure 7.12 Representation of fatigue test results on S–N curve.

The endurance limit, or fatigue strength, of a given material can usually be related to its tensile strength, as shown in Table 7.1. The endurance ratio, defined as (endurance limit/tensile strength), can be used to predict fatigue behavior in the absence of endurance limit results. As the Table shows, the endurance ratio of most ferrous alloys varies between 0.4 and 0.6.

Although most available S–N curves and endurance limit values are based on laboratory experiments and controlled test conditions, fatigue results always show larger scatter than other mechanical properties. The standard deviation for endurance limit results is usually in the range of 4 to 10 percent, but in the absence of statistical values an 8 percent standard deviation can be assumed. The reported endurance results can be taken to correspond to 50 percent survival reliability.

An important limitation of S–N curves and the endurance limit results is that they are usually determined for relatively small specimens under controlled conditions and simple loading systems. In addition, these results do not distinguish between crack initiation life and crack propagation life. These disadvantages limit their use in designing large structural components where crack-like defects can exist in the material or as a result of manufacturing. Under such conditions, it is the rate of fatigue crack propagation that determines the fatigue life of the part.

Crack initiation

Even if the nominal stresses acting on the part are below the elastic limit, local stresses may exceed the yield stress as a result of stress concentration or material discontinuity. As a result, cyclic plastic deformation takes place on favorably oriented slip planes leading to local strain hardening, residual stresses, intrusions and extrusions on the surface and eventual crack nucleation.

Table 7.1 Comparison of static and fatigue strengths of some engineering materials

Material	Tensile strength		Endurance limit		Endurance ratio
	MPa	ksi	MPa	ksi	
FERROUS ALLOYS					
AISI 1010 Normalized	364	52.8	186	27	0.46
1025 Normalized	441	64	182	26.4	0.41
1035 Normalized	539	78.2	238	34.5	0.44
1045 Normalized	630	91.4	273	39.6	0.43
1060 Normalized	735	106.6	315	45.7	0.43
1060 Oil Q, tempered	1295	187.8	574	83.3	0.44
3325 Oil Q, tempered	854	123.9	469	68	0.55
4340 Oil Q, tempered	952	138.1	532	77.2	0.56
8640 Oil Q, tempered	875	126.9	476	69	0.54
9314 Oil Q, tempered	812	177.8	476	69	0.59
302 Annealed	560	81.2	238	34.5	0.43
316 Annealed	560	81.2	245	35.5	0.44
431 Quenched, tempered	798	115.7	336	48.7	0.42
ASTM 20 gray cast iron	140	20.3	70	10.2	0.50
30 gray cast iron	210	30.5	102	14.8	0.49
60 gray cast iron	420	61	168	24.4	0.40
NONFERROUS ALLOYS					
AA 2011 T8	413	59.9	245	35.5	0.59
2024 Annealed	189	27.4	91	13.2	0.48
6061 T6	315	45.7	98	14.2	0.31
6063 T6	245	35.5	70	10.2	0.29
7075 T6	581	84.3	161	23.4	0.28
214 As cast	175	25.4	49	7.1	0.28
380 Dia-cast	336	48.7	140	20.3	0.42
Phosphor bronze, annealed	315	45.7	189	27.4	0.60
hard drawn	602	87.3	217	31.5	0.36
Aluminum bronze, quarter hard	581	84.3	206	29.9	0.35
Incoloy 901, at 650°C (1202°F)	980	142.1	364	52.8	0.37
Udimet 700, at 800°C (1472°F)	910	132	343	49.7	0.38
REINFORCED PLASTICS					
Polyester – 30% glass	123	17.8	84	12.2	0.68
Nylon 66 – 40% glass	200	29	62.7	9.1	0.31
Polycarbonate – 20% glass	107	15.5	34.5	5	0.32
40% glass	131	19	41.4	6	0.32

It can be shown that the local stress range at the site of crack nucleation, $\Delta\sigma_{max}$, can be related to the stress-intensity factor range, ΔK_I, by the following relationship:

$$\Delta\sigma_{max} = \Delta\sigma \star K_t = \frac{2}{\sqrt{\pi}} \frac{\Delta K_I}{\sqrt{r}} \qquad (7.9)$$

where r is the notch-tip radius, $\Delta\sigma$ is the range of applied nominal stress, and K_t is the stress concentration factor.

Experience shows that $(\Delta K_I/\sqrt{r})$, is the main parameter that governs fatigue-crack initiation in a benign environment. In the case of steels, there is a fatigue-crack-initiation threshold, $(K_I/\sqrt{r})_{th}$, below which fatigue cracks do not initiate. The value of this threshold increases with increasing strength and with decreasing strain hardening exponent.

Crack propagation

The performance of most parts and structures under fatigue loading is more dependent on their resistance to crack propagation than to crack nucleation. This is because microcracks are known to nucleate very early in the lives of parts and notched high-strength materials may have propagating cracks effectively throughout their service lives. Initially the crack propagates along the slip plane along which it nucleated, stage I, and then turns on to a plane perpendicular to the direction of the maximum tensile stress, stage II. Stage I may account for more than 90 percent of the life of a smooth ductile part under light loads or may be totally absent in a sharply notched highly-stressed part.

Experience based on experimental data shows that the fatigue-crack-propagation behavior is controlled primarily by the stress-intensity-factor range, ΔK_I, and can be divided into three regions, as shown in Fig. 7.13. In region 1, fatigue cracks grow extremely slowly or not at all and are related to the fatigue-crack-propagation threshold, ΔK_{th}, which corresponds to the stress-intensity-factor range below which cracks do not propagate under cyclic loading. In the case of steels this threshold is primarily a function of the stress ratio, $R = (\sigma_{min}/\sigma_{max})$, and is essentially independent of chemical composition or mechanical properties. The crack growth in the intermediate second region can be represented by a power law, which is usually called the Paris relationship:

$$da/dN = C(\Delta K_I)^n \qquad (7.10)$$

where a is crack length, N is number of cycles, da/dN is crack growth per cycle, and C and n are experimentally determined constants which depend on material properties and environment. Generally, C decreases with increasing elastic constant, yield strength and K_{Ic}. The constant n also decreases with increasing K_{Ic} which indicates that tougher materials offer higher resistance to fatigue crack propagation. On the other hand n

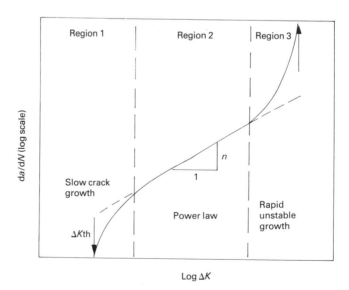

Figure 7.13 Schematic illustration of the effect of stress–intensity-factor range (K_I) on fatigue-crack growth rate (da/dN).

increases with increasing yield strength, which indicates that stronger materials tend to have faster growing cracks. For the case of austenitic stainless steels, C is $3.0 \star 10^{-10}$ and n is 3.25 when ΔK_I is expressed in (ksi in$^{1/2}$). The third stage is characterized by increase in growth rate leading to rapid unstable growth as K_c or K_{Ic} is approached.

The use of the above principles to select materials and to design parts that will resist fatigue fracture will be discussed in Sections 8.4 and 11.6 respectively.

7.6   WEAR FAILURES

Wear is a surface phenomenon which involves a progressive loss of material and reduction of dimensions over a period of time. Mechanical parts that are subjected to sliding or rolling contact are bound to suffer some degree of wear. Examples of such parts are bearings, gears, guides, brakes and clutches. The amount of wear which the part can suffer before it fails to function efficiently depends on the accuracy and tolerances involved. Wear takes place by several mechanisms and can be classified thus:

1. Adhesive wear occurs between two surfaces that slide against each other under pressure. The process involves adhesion, plastic deformation and fracture of the asperities. Scoring, scuffing and galling represent severe types of adhesive wear.
2. Abrasive wear is caused by the removal of material from a surface when it moves relative to hard particles or relative to the hard projections on a mating surface. When hard particles are involved, both sliding surfaces may be abraded or the particles may be embedded in one surface and abrade the other surface. Abrasion wear normally predominates under lubricated conditions.
3. Erosive wear is abrasive wear involving loss of surface material by contact with fluid that contains foreign particles. Although the foreign particles are usually solid, liquid-impingement erosion can be caused by liquid droplets carried in a fast-moving stream of gas.
4. Chemical or corrosive wear is caused by the combined effects of chemical or electro-chemical reaction and mechanical action. The combined effects may be mutually enhancing and even mild conditions could lead to severe failure.
5. Erosion-corrosion takes place when there is relative motion between the surface and a corrosive fluid. Under these conditions, the wear rate can be directly related to the rate of relative movement.
6. Surface fatigue is a special type of surface damage where parts of the surface are detached under externally applied cyclic stresses. This is an important source of failure in rolling-contact systems such as ball bearings and railway lines. Surface fatigue damage can also take place on a surface in contact with a liquid when the flow conditions give rise to repeated formation and collapse of vapor bubbles at the surface. This action imposes large repetitive contact stresses that can cause surface fatigue resulting in pitting or spalling. This type of damage is called cavitation erosion.

In practice, determining the cause of wear may be difficult because failure may have resulted from the combined effect of different types of wear modes. In addition, as wear progresses there may be a change in the predominant wear mode. An important method of combating wear is lubrication. Many kinds of surface films can act as lubricants, preventing cold welding of asperities and thus reducing friction and wear. Lubricants may be solid, liquid or gas. Liquid lubricants have the advantage of combining cooling with lubricating action. In systems which depend on lubricants to combat wear, failure of the lubricant can lead to scuffing, galling, or even seizure. Most lubricant failures occur due to chemical decomposition, contamination or change in properties due to excessive heat. In many cases in practice, more than one of the above causes can be involved in lubricant failures.

7.7   CORROSION FAILURES

Corrosion may be defined as the unintended destructive chemical or electrochemical reaction of a material with its environment. Metallic, polymeric and ceramic materials are susceptible to attack from different environments and while the corrosion of metals is electrochemical in nature, the corrosion of other materials usually involves chemical reaction. The nature, composition and uniformity of the environment and the attacked surface can greatly influence the type, rate and extent of corrosion. In addition, externally imposed changes and changes that occur as a result of the corrosion process itself are known to influence the type and rate of corrosion. Corrosion frequently leads to failure of engineering components or renders them susceptible to failure by some other mechanism. The rate and extent of corrosive attack that can be tolerated in a certain component depend on the application. For example, in many structural applications some uniform corrosion can be allowed, while in food-processing equipment even a minute amount of metal dissolution is not tolerated.

In the case of metallic materials, where corrosion takes place by electrochemical attack, the corroding metal is the anode in the galvanic cell and the cathode can be another metal, a conducting nonmetal or an oxide. Atmospheric corrosion of metals is probably the most commonly encountered form of corrosion. When a metal is exposed to the atmosphere, its surface is covered with a thin layer of condensed or absorbed water, even at relative humidity less than 100 percent, and this layer can act as the electrolyte. The presence of industrial contaminants in the atmosphere increases the corrosion rate. Examples are dust, sulfur dioxide and ammonium sulfate. Sodium chloride is also an impurity, which is present in marine atmospheres and increases the corrosion rate.

Corrosion can take place evenly over the entire surface or it can be concentrated at certain locations. Uniform corrosion commonly occurs on surfaces of uniform composition and microstructure. Pitting corrosion occurs when one area of the surface becomes anodic with respect to the rest of the surface due to segregation of alloying elements or inclusions in the microstructure. Surface deposits that set up local concentration cells, dissolved halides that produce local anodes by rupture of the protective oxide film or mechanical ruptures in protective organic coatings are also common sources of pitting corrosion. Localized changes in the corrodent are also known to cause localized attack. Intergranular attack is another type of localized corrosion which takes place at grain boundaries when they become more susceptible to corrosion than the bulk of the grains. Intergranular attack is often strongly dependent on the mechanical and thermal treatment given to the alloy. For example, unstabilized stainless steels are susceptible to intergranular corrosion when heated in the temperature range 550 to 850°C (1000 to 1550°F). In this temperature range (sensitizing range) chromium combines with carbon to form chromium carbides which precipitate at the grain boundaries and this depletes the neighboring areas of chromium. In many corrosive environments the chromium-depleted areas are attacked.

Table 7.2 Position of some metallic materials in the galvanic
series based on sea water

Protected, noble or cathodic end

Platinum

Gold

Graphite

Titanium

Silver

Chlorimet 3 (61 Ni, 18 Cr, 18 Mo)
Hastelloy C (62 Ni, 17 Cr, 15 Mo)
Inconel 625 (61 Ni, 21.5 Cr, 9 Mo, 3.6 Nb)

Incoloy 825 (21.5 Cr, 42 Ni, 3 Mo, 30 Fe)
Type 316 stainless steel (passive)
Type 304 stainless steel (passive)
Type 410 stainless steel (passive)

Monel alloy 400 (66.5 Ni, 31.5 Cu)

Inconel alloy 600 (passive) (76 Ni, 15.5 Cr, 8 Fe)
Nickel 200 (passive) (99.5 Ni)

Leaded tin bronze G, 923, cast (87 Cu, 8 Sn, 4 Zn)
Silicon bronze C65500 (97 Cu, 3 Si)
Electrolytic tough pitch copper C11000 (99.9 Cu, 0.04 O)
Red brass C23000 (85 Cu, 15 Zn)
Aluminum bronze C60800 (95 Cu, 5 Al)
Admiralty brass C44300, C44400, C44500 (71 Cu, 28 Zn, 1 Sn)

Chlorimet 2 (66 Ni, 32 Mo, 1 Fe)
Hastelloy B (60 Ni, 30 Mo, 6 Fe, 1 Mn)

Inconel 600 (active)
Nickel 200 (active)

Naval brass C46400 to C46700 (60 Cu, 39.25 Zn, 0.75 Sn)
Muntz metal C28000 (60 Cu, 40 Zn)

Tin
Lead

Type 316 stainless steel (active)
Type 304 stainless steel (active)

Lead-tin solder (50 Sn, 50 Pb)

Cast irons
Low carbon steels

Aluminum alloy 2117 (2.6 Cu, 0.35 Mg)
Aluminum alloy 2024 (4.5 Cu, 1.5 Mg, 0.6 Mn)

Aluminum alloy 5052 (2.5 Mg, 0.25 Cr)
Aluminum alloy 3004 (1.2 Mn, 1 Mg)
Aluminum 1100, commercial purity aluminum (99 Al min, 0.12 Cu)

Galvanized steel
Zinc

Magnesium alloys
Magnesium

Corroded, anodic, least noble end

Dissolving chromium carbides by solution heat treatment at 1060 to 1120°C (about 1950 to 2050°F) followed by water quenching eliminates sensitization. Susceptibility of stainless steels to sensitization can also be reduced by reducing the carbon content to less than 0.03 percent as in the case of extra-low carbon grades, e.g. 304L, or by adding sufficient titanium and niobium to combine with all the carbon in the steel, e.g. 347 or 321 stainless steels.

When dissimilar metals are in electrical contact in an electrolyte, the less noble metal becomes the anode in the galvanic cell and is attacked to a greater extent than if it were exposed alone, while the more noble metal becomes the cathode and is attacked to a lesser extent than if it were exposed alone. The severity of this galvanic corrosion depends on the separation of the two metals in the galvanic series, Table 7.2. In most cases, metals from one group can be coupled with each other without causing a substantial increase in the corrosion rate. Another factor which affects the severity of galvanic corrosion is the relative areas of the anodic metal to the cathodic metal. Thus a steel rivet in a copper plate will be more severely corroded than a steel plate containing a copper rivet. Galvanic corrosion can also take place between two different areas of a structure, which is made of the same metal and immersed in the same electrolyte, if the contact areas are at different temperatures. For a steel structure in contact with dilute aerated chloride solution, the warmer area is anodic to the colder area while for copper in aqueous salt solution, the warmer area is cathodic to the colder area. If a structure, which is made of the same material is in contact with two different concentrations of an electrolyte, concentration-cell corrosion will take place. This type of attack is known to take place in buried metals as a result of their being in contact with soils that have different chemical compositions, especially with respect to the concentration of sodium chloride, sodium sulfate and organic acids. Differences in water content or degrees of aeration can also be detrimental. Corrective action in such cases usually involves coating of the buried metal in asphalt, enclosing in a concrete trough and/or adopting cathodic protection.

A crevice at a joint between two metallic surfaces or between a metallic and nonmetallic surface will also provide conditions for concentration-cell corrosion. This is called crevice corrosion. Some alloys are susceptible to selective leaching where the less corrosion-resistant element is removed by corrosion. Common examples include dezincification, where zinc is removed from brasses, and graphic corrosion, where iron is removed from gray cast irons.

Plastics do not corrode in the same way as metals since they are electrical insulators. Instead, chemical reaction, dissolution or absorption can take place depending on the type of polymer and nature of the solution.

The deleterious effects of the different types of corrosion can be eliminated or at least reduced by adopting one or more of the following preventive measures:

1. Selecting the appropriate material, as will be discussed in Section 8.6.
2. Observing certain design rules, as will be discussed in Section 12.10.
3. Using protective coatings, as will be discussed in Section 8.9.
4. Using corrosion inhibitors.
5. Using galvanic protection.

Discussion of the last two methods of combating corrosion is beyond the scope of this book and the reader is referred to the specialized texts given in the bibliography at the end of this chapter.

7.8   STRESS CORROSION AND CORROSION FATIGUE

Stress corrosion cracking (SCC) occurs in some alloys as a result of the combined effect of tensile stresses and chemical attack. The stresses involved in SCC can be due either to normal service loads or to residual stresses resulting from manufacturing and assembly processes. Examples of manufacturing processes which could lead to residual stresses include casting, welding, cold forming and heat treatment. Normally, a threshold stress is required for SCC to occur and shorter lives are expected with higher stresses. This threshold stress may be as low as 10 percent of the yield stress and is not usually a practical design stress. Susceptibility of an alloy to SCC is often a function of the content of major alloying elements, such as nickel and chromium in stainless steels and zinc in brasses. Increasing the strength is also known to increase the susceptibility to SCC. Another important factor that affects the occurrence of SCC is the environment. The presence of certain ions, even in small concentrations, can be detrimental to some alloys but not to others. For example, stainless steels crack in chloride environments but not in ammonia-containing environments, whereas brasses crack in ammonia-containing environments but not in chlorides. Examples of harmful ions include gaseous HCl, H_2S or hydrogen which cause SCC in high-strength low-alloy steels and halides in aqueous solutions, gaseous H_2O or fuming nitric acid which cause SCC in high-strength aluminum alloys. In addition to the damaging ions, other environmental variables are also known to affect SCC of alloys. Some of these variables are temperature, pH, electrochemical potential and aeration.

In the presence of certain chemicals and under the influence of stress, some plastics can fail by gradual cracking. This is known as environmental stress cracking. For example, polyethylenes suffer environmental stress cracking in detergents and oils.

Stress corrosion cracking may be reduced or prevented by using one or more of the following methods:

1. Lowering the stress below the threshold value by eliminating residual stresses and reducing externally applied stresses.
2. Eliminating the critical environmental species.
3. Selecting the appropriate alloy. For example, carbon steels, rather than stainless steels, are often used in the construction of heat exchangers used in contact with seawater. This is because carbon steels are more resistant to SCC although they are less resistant to general corrosion than stainless steels.
4. Applying cathodic protection.
5. Adding inhibitors to the system.
6. Applying protective coatings.
7. Introducing residual compressive stresses in the surface by processes like shot-peening.

Corrosion fatigue is caused by the combined effects of fluctuating stresses and corrosive environment. Unfavorable environments cause fatigue cracks to initiate in fewer cycles and increase the crack growth rate, thus reducing the fatigue life. For example, the fatigue strength of smooth samples of high-strength steel in salt water can be as little as 10 percent of that in dry air. Under salt water conditions the smooth surface is attacked, creating local stress raisers that make the initiation of fatigue cracks much easier. Salt water also increases crack growth rate in steels. Another example is high-chromium alloys which retain only 30 to 40 percent of their normal fatigue strength when tested in seawater.

On the other hand, austenitic stainless steels and aluminum bronzes retain about 75 percent of their normal fatigue strength when tested in seawater. Under corrosion fatigue conditions, the frequency of the stress cycle, the shape of the stress wave, the stress ratio as well as the magnitude of the cyclic stress and the number of cycles affect the fatigue life. Generally, corrosion fatigue strength decreases as the stress frequency decreases because this allows more time for interaction between the material and environment. This effect is most important at frequencies of less than 10 Hz. As in the case of stress corrosion, the temperature, pH and aeration of the environment affect the corrosion fatigue life.

Many of the methods used to reduce or eliminate SCC can also be used to combat corrosion fatigue. Among the possible methods are:

1. Reducing the applied stress by changing the design and eliminating tensile residual stresses.
2. Introducing residual compressive stresses in the surface.
3. Selecting the appropriate materials.
4. Using corrosion inhibitors.
5. Applying protective coatings.

7.9   ELEVATED-TEMPERATURE FAILURES

The effect of service environment on material performance at elevated temperature can be divided into three main categories:

1. Microstructural effects, such as grain growth and overaging.
2. Chemical effects, such as oxidation.
3. Mechanical effects, such as creep and stress rupture.

While oxidation and creep can directly lead to failure of a part in service, the microstructural changes can lead to weakening of the material and therefore can indirectly lead to failure. Unlike room temperature service, elevated-temperature service can change the structure of the material and consequently the mechanical behavior. Many of the strengthening mechanisms that are effective at room temperature become ineffective at elevated temperatures. Generally, nonequilibrium structures change during long-term high-temperature service and this leads to lower creep strength. Thus, materials which depend on their fine grains for strengthening may lose this advantage by grain growth and materials which have been strain-hardened by cold working may recover or anneal. Structures which have been precipitation hardened to peak values may overage and steels which have been hardened and tempered may overtemper.

Oxidation

Many materials, metallic and nonmetallic, combine with oxygen during service, especially at elevated temperatures. In metals, oxidation often starts rapidly and continues until an oxide film or scale is formed on the surface. After this stage, the rate of further oxidation depends on the soundness of the oxide film. If the oxide film is porous and allows continuous access of oxygen to the metal surface oxidation will continue until all the material is oxidized. Examples of such metals include sodium and potassium. On the other hand, if the oxide film is dense and impervious it provides protection against further oxidation, as in the case of aluminum oxide on aluminum and chromium oxide on chromium. The degree of protection increases as the thickness of the oxide film increases and oxidation practically stops after a critical thickness is reached. In such cases, for a given service temperature and atmosphere, the oxide film thickness, T, is given by the formula:

$$T = K\log(ct + 1) \qquad (7.11)$$

where K and c are constants and t is exposure time.

In many cases the composition and characteristics of oxide films can be changed by adding alloying elements to the base metal. For example, chromium, aluminum and silicon are added to iron to modify its normally porous oxide layer and make it more protective. Materials for elevated temperature service can also be protected against oxidation by applying protective coatings.

Most plastics and rubbers oxidize in the presence of oxygen. Rubbers are especially susceptible to oxidation and the process is called ageing. The reaction of oxygen with rubber initially reduces elasticity and increases hardness. This is because oxygen diffuses into the structure and provides additional cross-linking. As ageing proceeds, the rubber degrades and eventually loses most of its strength. The rate of ageing depends on the temperature, type of atmosphere, material composition and method of manufacture.

Creep

A major factor which limits the service life of components in service at elevated temperatures is creep. Creep is defined as the time-dependent deformation which occurs under stress. Creep occurs as a result of the motion of dislocations within the grains, grain boundary rotation and grain boundary sliding. It is sensitive to grain size, alloying additions, microstructure of the material and service conditions. When creep reaches a certain value, fracture occurs. Creep fracture (also called stress rupture) usually takes place at strains much less than the fracture strains in tension tests at room temperature. In most practical cases, creep fracture occurs by the nucleation and growth of voids on grain boundaries which are mainly perpendicular to the maximum principal tensile stress axis.

In most practical cases the strain which is suffered by a component under creep conditions can be divided into the stages shown schematically in Fig. 7.14. Following an initial instantaneous deformation, creep takes place at a decreasing strain rate during the primary or transient stage. This is followed by the secondary creep or steady-state stage where the strain rate is constant under constant stress conditions. At the end of the steady-state stage, tertiary creep starts and the strain rate increases rapidly with increasing strain and fracture finally occurs. Tertiary creep can be caused by:

1. Reduction of the cross-sectional area of the component due to cracking or necking.
2. Oxidation and other environmental effects which reduce the cross-sectional area.
3. Microstructural changes that weaken the material such as coarsening of precipitates.

Under certain conditions, some materials may not exhibit all the above stages of creep. For example at high stresses or high temperatures the primary stage may not be present, with secondary creep or even tertiary creep starting soon after load application. Another example is the case where fracture occurs before the tertiary stage is reached as in the case of some low-ductility cast alloys. Creep ductility is an important factor in materials

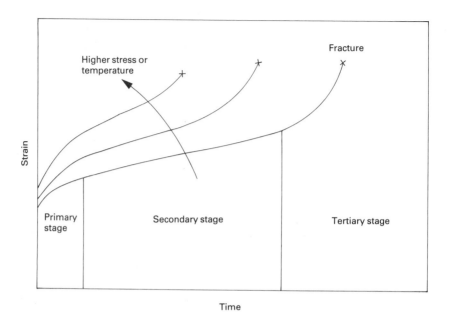

Figure 7.14 Schematic creep curve under tensile loading.

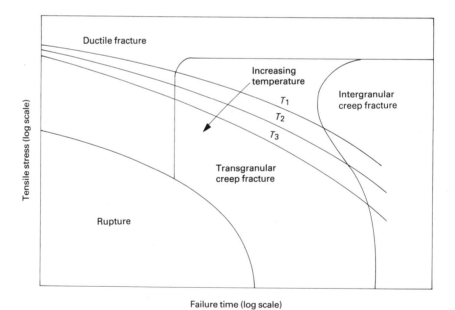

Figure 7.15 Stress–rupture diagram showing the dependence of creep life on stress and temperature. The change of failure mechanism is also shown.

selection. Although the permissible creep strain in practice is usually of the order of 1 percent, selecting materials with higher creep ductility means a higher safety margin. Creep-fracture behavior of a given material can be represented graphically using stress-rupture curves similar to those schematically shown in Fig. 7.15. These curves can be used for design purposes and for alloy-development research. The diagrams show the dependence of creep life and failure mechanism of a given alloy on stress and temperature.

Combined creep and fatigue

In many high-temperature applications in practice, the applied loads are cyclic and could lead to a combined creep-fatigue failure. Under these conditions, the life of a component is determined by the initiation and growth of a creep or a fatigue crack. The growth rate of such a crack is usually affected by the maximum stress-intensity factor, the minimum/maximum stress-intensity ratio (R-ratio), the frequency of the cyclic load, the wave shape of the cyclic load, the hold or sustained-load time, the environment, the material and the operating temperature. At high load frequencies and/or relatively lower temperatures, crack growth is independent of the frequency or temperature. This is because the material just ahead of the crack does not suffer any time-dependent processes, such as oxidation or creep relaxation. Under these conditions, the mechanism of crack growth is essentially the same as room temperature fatigue. At low frequencies and/or relatively high temperatures, crack growth is affected by time-dependent processes. A mixture of the two extreme cases of behavior is expected at intermediate temperatures and load frequencies.

Thermal fatigue

Another form of elevated-temperature failures is thermal fatigue. Stresses and strains induced in a component due to thermal gradients can cause failure if repeated a sufficient number of times. Faster changes in temperature, lower thermal conductivity of the material, higher elastic constant, higher thermal expansion coefficient, lower ductility and thicker component sections often account for shorter service life. Although ceramic materials are particularly prone to thermal fatigue, metallic components whose service conditions cause the surface to change temperature more rapidly than the bulk, e.g. gas-turbine blades, are known to fail after relatively short service life (10 000–100 000 cycles). Thermal fatigue cracks usually initiate on the surface and progress normal to the surface.

In high-temperature applications, the environment plays an important role in determining the performance of components. Selecting the material that will resist the environment, controlling the environment, or protecting the surface is essential for prolonged service. Examples of aggressive environments are those which contain vanadium compounds, sulfur compounds or salt. A vacuum environment may be more harmful than air if some of the alloy constituents evaporate at high temperatures.

7.10   FAILURE ANALYSIS

When a component fails in service, it is important that the source of failure be located in order to identify the responsible party and to avoid similar failures in future designs. Due to the complexity of most failure cases, it is useful to follow a systematic approach to the analysis. An important step is to gather background information about the function, source, fabrication, materials used and service history of the failed component. Site visits involve locating all the broken pieces, making visual examination, taking photographs and selecting the parts to be removed for further laboratory investigation. Macroscopic, microscopic, chemical analysis, nondestructive and destructive tests are normally used to locate possible material and manufacturing defects. Presence of oxidation and corrosion products, temper colors, surface markings, etc., can also provide valuable clues towards failure mode identification. Based on the gathered information it should be possible to identify the origin of failure, direction of crack propagation and sequence of failure. Presence of secondary damage not related to the main failure should also be identified. The final step in failure analysis usually involves writing a report to document the findings and to give the conclusions.

Identification of failure mode is not only important for determining the cause of failure, but also a powerful tool for design review. The following discussion gives a brief review of some of the analytical techniques which have been developed for systematic identification of failure modes. Reference should be made to the original publications for more details.

The materials failure logic model (MFLM) proposed by Marriott and Miller (see bibliography) is very similar to the fault tree analysis described in Section 13.5. The MFLM is based on the assumptions that: (a) material failure can be modeled as a logic sequence of elementary go-no/go events; and (b) each material failure mechanism can be characterized by a logic expression which serves to identify that mechanism regardless of context. As an example to illustrate this model, consider the failure of a welded low-alloy steel pressure vessel which failed during commissioning at less than operating load. The failure event can be described as:

$$F = A.B.(C_1 + C_2).D.E.G.H. \tag{7.12}$$

where A = low alloy steel
$\quad\quad B$ = heat treatment defect resulting in brittle structure
$\quad\quad C_1$ = welding defect
$\quad\quad C_2$ = residual stress from welding
$\quad\quad D$ = presence of corrosive environment
$\quad\quad E$ = high residual stresses as a result of inappropriate postweld heat treatment
$\quad\quad G$ = failure of nondestructive tests to detect initial defect
$\quad\quad H$ = failure to detect incorrect heat treatment of material
$\quad( ).( )$ = Boolean AND operator
$\quad( )+( )$ = Boolean OR operator

In this case, either of the following logic events could have been sufficient to cause failure:

$$1.\; F_1 = A.B.C_1.G.H. \tag{7.13}$$

This means that the initial defect, combined with the brittle structure, constituted a major risk.

$$2.\; F_2 = A.B.C_2.D.E.H. \tag{7.14}$$

This means that stress corrosion cracking is likely to lead to crack growth even in the absence of initial defect.

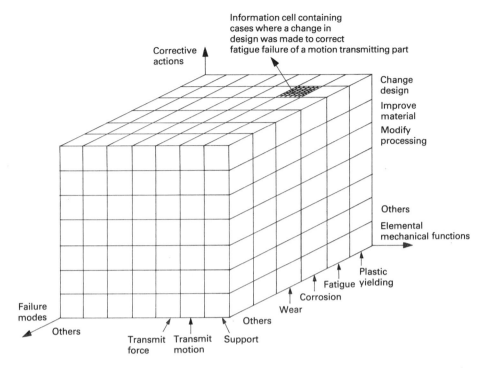

Figure 7.16 Part of failure experience matrix which can be used to store failure information.

In addition to its use in failure analysis, the MFLM can be interfaced with a computer aided design system to aid the designer in the identification of potential failures.

Collins *et al.* (see bibliography) introduced the failure-experience matrix as a means of storing failure information for mechanical systems. The matrix is three dimensional, as shown in Fig. 7.16, with the axes defined as follows:

1. Failure modes. This axis covers the different types of failure, e.g. fatigue, corrosion, wear, etc.
2. Elemental mechanical functions. This axis covers all the different functions that are normally performed by mechanical elements. Examples include supporting, force transmitting, shielding, sliding, fastening, liquid storing, pumping, damping, etc.
3. Corrective actions. This axis gives any measure or combination of steps taken to return a failed component or system to satisfactory performance. Examples of corrective actions include design change, change of material, improved quality control, change of lubricant, revised procurement specifications, change of vendor, etc.

This system can be computerized and could be of help to engineers in designing critical components. If the function of the component is entered, the system will give the most likely modes of failure and the corrective actions needed to avert it.

Another technique which was proposed by Weiss (see bibliography) uses expert systems for failure analysis. A logic program is written using LOGLISP language which is a combination of logic, i.e. predicate calculus and resolution, and LISP, which is the language usually used for artificial intelligence. The principal ingredients for failure analysis are symptom–cause relationships and facts and rules about the system under consideration. For example, the presence of a neck is a symptom of a failure due to tensile overload. Symptoms can be related to loading, e.g. tension, torsion, bending and fatigue; and failure mode, e.g. neck, dimples, shearlip, beach marks or cleavage. Based on the observed symptoms, the expert system program gives the possible causes of failure. Introducing more than one symptom for a given failure reduces the number of possible causes of failure. For a realistic failure analysis case, the expert system needs to interface with a material data base and probably a finite element stress analysis program.

7.11   REVIEW QUESTIONS AND PROBLEMS

7.1 Would you use AISI 1050 steel for manufacturing a component that will serve at $-50°C$ ($-58°F$)? If not suggest substitute materials.

7.2 The wall thickness of a steel tank is measured monthly and the loss in thickness is approximately the same each month, 50 mg/dm^2 per day. What is the useful life of the tank if the initial thickness is 10 mm and the minimum safe thickness is 6 mm? [Answer: 17 years]

7.3 A welded stainless steel tank exhibited considerable pitting corrosion near the welded joints. What do you think caused this corrosion? Suggest three ways of eliminating this type of failure.

7.4 Ti-6 Al-4 V and aluminum 7075 T6 alloys are widely used in making lightweight structures. If the available NDT equipment can only detect flaws larger than 3 mm in length, can you safely use either of the above alloys for designing a component that will be subjected to 400 MPa? Use the information in Table 8.3. [Answer: yes for Ti-6 Al-4 V; no for AA 7075 T6]

7.5 Explain the difference between alternating stress and fluctuating stress cycles. Which one of these loading modes is encountered in the motor car rear-axle and the connecting rod of an internal combustion engine?

7.6 Why is fatigue failure a potentially serious problem in many welded steel structures? What are the best ways of avoiding such failure?

BIBLIOGRAPHY AND FURTHER READING

Boyed, G. M., ed., *Brittle Fracture in Steel Structures*, Butterworths, London, 1970.
Boyer, H. E., ed., *Metals Handbook Desk Edition*, ASM, Ohio, 1985.
Colangelo, V. J. and Heiser, F. A., *Analysis of Metallurgical Failures*, John Wiley & Sons, New York, 1987.
Collins, J. A., *Failure of Materials in Mechanical Design*, John Wiley & Sons, New York, 1981.
Collins, J. A., Hagan, B. T. and Bratt, H. M., 'Failure experience matrix', *Trans. ASME, J. Eng. Ind.*, vol. 98, 1976, pp. 1074–9.
Cook, N. H., *Mechanics and Materials for Design*, McGraw-Hill, New York, 1985.
Farley, J. M. and Nickols, R. W., *Non-destructive Testing*, Pergamon Press, London, 1988.
Fontana, M. G., *Corrosion Engineering*, 3rd edn., McGraw-Hill, New York, 1986.
Marriott, D. L. and Miller, N. R., 'Materials failure logic models', *Trans. ASME, J. Mech. Design*, vol. 104, 1982, pp. 628–34.
Metals Handbook, 8th ed., vol. 10, *Failure Analysis and Prevention*, ASM, Ohio.
Parker, A. P., *The Mechanics of Fracture and Fatigue*, E. & F. N. Spon, London, 1981.
Peterson, R. E., *Stress Concentration Design Factors*, John Wiley & Sons, New York, 1974.
Scott, D. ed., *Wear: Treatise on Materials Science and Technology*, vol. 13, Academic Press, New York, 1979.
Unterweiser, P. M., ed., *Case Histories in Failure Analysis*, ASM, Ohio, 1979.

Weiss, V., 'Towards failure analysis expert systems', *ASTM Standardization News*, April 1986, pp. 30–4.

Whyte, R. R., ed., 'Engineering Progress through Trouble', *The Institution of Mech. Eng.*, London, 1975.

Wulpi, D. J., *Understanding how Materials Fail*, ASM, Ohio, 1985.

Chapter 8

Functional Requirements of Engineering Materials

8.1 INTRODUCTION

The characteristics that are usually considered when selecting a material for a given application can be classified into the following categories:

1. Mechanical behavior including yield strength, tensile strength, elongation percent, reduction in area percent, hardness, toughness, fatigue strength and stiffness. Resistance to abrasion and erosion are also related to mechanical behavior. Mechanical stability includes creep resistance, loss of ductility and dimensional stability. Although the mechanical properties at service temperature are important, satisfactory performance at other temperatures must also be taken into account. Room temperature properties after extended service at elevated temperature can be important for applications where intermittent shutdowns are encountered as in the case of boilers and jet engines.
2. Chemical properties which include corrosion, oxidation and sulfidation resistance.
3. Physical characteristics including electrical, magnetic and thermal properties. Density is also included in this category.
4. Processability which includes castability, workability, weldability and machinability. The ability of the material to acquire good surface finish and resistance to galling and seizing can also be included in this category.

   In selecting materials for a given application it is useful to classify them according to the major function they are expected to perform in service. In this chapter, materials that are usually used in certain applications will be compared according to their functional requirements.

8.2 SELECTION OF MATERIALS FOR STATIC STRENGTH

Static strength can be defined as the ability to resist a short-term steady load at moderate temperatures. This resistance is usually measured in terms of yield strength, ultimate tensile strength, compressive strength and hardness. When the material does not exhibit a well-defined yield point, the stress required to cause 0.1 or 0.2 percent plastic strain (the proof stress) is used instead. For most ductile wrought metallic materials, the tensile and compressive strengths are very close and in most cases only the tensile strength is given. However, brittle materials like ceramics are generally stronger in compression than in tension and both properties are usually given in such cases. Although many engineering materials are almost isotropic, there are important cases where significant anisotropy exists. In the latter cases the strength depends on the direction in which it is measured. The degree of anisotropy depends on the nature of the material and its manufacturing history. Anisotropy in wrought metallic materials is more pronounced when they contain elongated inclusions and when processing consists of repeated deformation in the same direction. Composites reinforced with unidirectional fibers also exhibit pronounced anisotropy, as was discussed in Chapter 6. Anisotropy can be useful if the principal external stress acts along the direction of highest strength.

The level of strength in engineering materials may be viewed either in absolute terms or relative to similar materials. For example, it is generally understood that high-strength steels have tensile strength values in excess of 1400 MPa (c. 200 ksi), which is also high strength in absolute terms. Relative to light alloys, however, an aluminum alloy with a strength of 500 MPa (c. 72 ksi) would also be designated as high-strength alloy even though this level of strength is low for steels. From the design point of view it is more convenient to consider the strength of materials in absolute terms. From the manufacturing point of view, however, it is important to consider the strength as an indication of the degree of development of the material concerned, i.e. relative to similar materials. This is because highly developed materials are often complex, more difficult to process and relatively more expensive. Figure 8.1 gives the strength of some materials both in absolute terms and relative to similar materials. In a given group of materials, the medium-strength members are usually more widely used because they generally combine optimum strength, ease of manufacture and economy. The most developed members in a given group of materials are usually highly specialized and, as a result, they are produced in much lower quantities. The low-strength members of a given group are usually used to meet requirements other than strength. Requirements such as electrical and thermal conductivities, formability, corrosion resistance or cost may be more important than high strength in some applications.

The load-carrying capacity of a given component is a function of both the strength of the material used in making it and its dimensions. This means that a lower strength material can be used in making a component to bear a certain load provided that its cross-sectional area is increased proportionally. However, the designer is not usually completely free in choosing the strength level of the material selected for a given part. Other factors like space limitations, weight limitations and cost could limit his choice. Space limitations can usually be solved by using stronger material which will allow smaller cross-sectional area and smaller total volume of the component. It should be noted, however, that reducing the cross-sectional area below a certain limit could cause failure by buckling due to increased slenderness of the part. As an example, consider the case of a load of 50 kN which is to be supported on a cylindrical compression element of 200 mm length. As the compression element has to fit with other parts of the structure, its diameter should not exceed 20 mm. Weight limitations are such that the mass of the element should not exceed 0.25 kg. Which of the materials given in Table 8.1 is most suited for making the compression element? Table 8.1 shows the calculated diameter of the compression element when made of different materials. The diameter is calculated on the basis of strength and on the basis of buckling. The larger value for a given material is used to calculate the mass of the element. The results in Table 8.1 show that only epoxy-62 percent Kevlar satisfies both the diameter and weight limits.

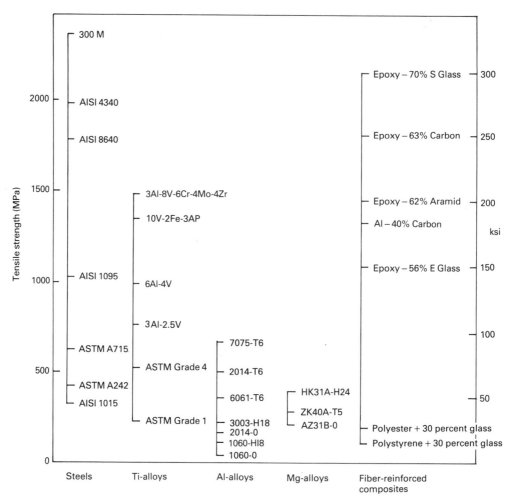

Figure 8.1 Comparison of some engineering materials on the basis of tensile strength.

Frequently, higher strength members of a given group of materials are also more expensive. Using a stronger but more expensive material could result in a reduction of the total cost of the finished component. This is because the amount of material used would be less and consequently processing cost could also be less. Weight limitations are encountered with many applications including aerospace, transport, construction and portable appliances. In such cases, the strength/density, or specific strength, becomes an important basis for comparing the different materials. Figure 8.2 compares the materials of Fig. 8.1 on the basis of specific strength. The figure shows a clear advantage of the fiber-reinforced composites over other materials.

8.3   SELECTION OF MATERIALS FOR STIFFNESS

When a load is placed on a beam, the beam is bent and every portion of it is moved in a direction parallel to the direction of the load. The distance that a point on the beam moves

Table 8.1 Comparison of compression element materials

Material	Strength (MPa)	Elastic modulus (GPa)	Specific gravity	Diameter based on strength (mm)	Diameter based on buckling[a] (mm)	Mass based on larger dia. (kg)	Remarks
STEELS							
ASTM No. A675 grade 45	155	211	7.8	20.3	15.75		Reject (1)
ASTM No. A675 grade 80	275	211	7.8	15.2	15.75	0.3	Reject (2)
ASTM No. 717 grade 80	550	211	7.8	10.8	15.75	0.3	Reject (2)
ALUMINUM							
AA 2014 T6	420	70.8	2.7	12.3	20.7		Reject (1)
PLASTICS AND COMPOSITES							
Nylon 6/6	84	3.3	1.14	27.5	44.6		Reject (1)
Epoxy–70% glass	2100	62.3	2.11	5.5	21.4		Reject (1)
Epoxy–62% Kevlar	1311	82.8	1.38	7.0	19.9	0.086	Accepted

[a]Assuming that the ends of the compression element are not constrained, the Euler formula, see Equation (11.16), can be used to calculate the minimum diameter that will allow safe use of the compression member without buckling.
Reject (1) = Material is rejected because it violates the limits on diameter.
Reject (2) = Material is rejected because it violates the limits on weight.

(deflection) depends on its position in the beam, the type of beam and the type of supports. For example, a beam which is simply supported at both ends suffers maximum deflection, y in its middle when subjected to a concentrated central load L. In this case the maximum deflection, y, is given by:

$$y = \frac{L \star l^3}{48 \star E \star I} \tag{8.1}$$

where l is the length of the beam, E is Young's modulus of the beam material, and I is second moment of area of the beam cross section with respect to the neutral axis.

The deflection of the beam under load can be taken as a measure of its stiffness and (8.1) shows that this is a function of both E and I. Putting the above discussion in general terms, the stiffness of a component can be defined as its resistance to elastic deflection and is a function of both shape and modulus of elasticity of the material. The stiffness of a component may be increased by increasing its second moment of area, which is computed from the cross-sectional dimensions, and selecting a high-modulus material for its manufacture. An important characteristic of metallic materials is that their elastic moduli are very difficult to change by changing the composition or heat treatment. On the other hand, the elastic moduli of composite materials can be changed over a wide range by changing the volume fraction and orientation of the constituents, as discussed in Chapter 6. Table 8.2 gives representative values of the modulus of elasticity of some engineering materials. When a metallic component is loaded in tension, compression, or bending the Young's modulus, E, is used in computing its stiffness. When the loading is in shear or torsion the modulus of rigidity, G, is used in computing stiffness. The relationship between these two elastic constants is given by:

$$G = \frac{E}{2(1 + \mu)} \tag{8.2}$$

where μ is Poisson's ratio.

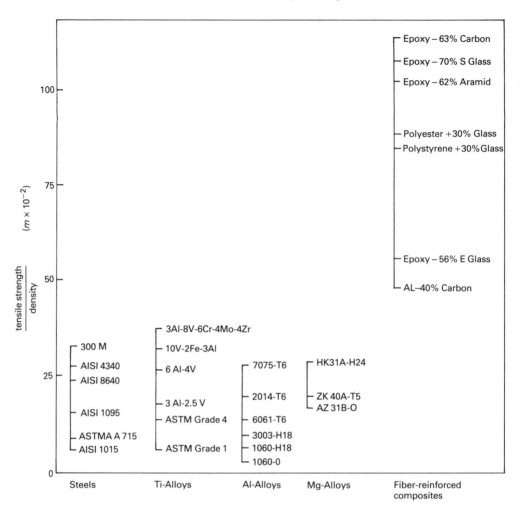

Figure 8.2 Comparison of some engineering materials on the basis of specific tensile strength.

Table 8.2 Comparison of stiffness of selected engineering materials

Material	Modulus of elasticity E (GPa)[a]	Density P (mg/m³)[b]	$\dfrac{E}{P} \times 10^{-5}$	$\dfrac{E^{1/2}}{P} \times 10^{-2}$	$\dfrac{E^{1/3}}{P}$
Steel (carbon and low alloy)	207	7.825	26.5	5.8	35.1
Aluminum alloys (average)	71	2.7	26.3	9.9	71.2
Magnesium alloys (average)	40	1.8	22.2	11.1	88.2
Titanium alloys (average)	120	4.5	26.7	7.7	50.9
Epoxy–73% E glass fibers	55.9	2.17	25.8	10.9	81.8
Epoxy–70% S glass fibers	62.3	2.11	29.5	11.8	87.2
Epoxy–63% carbon fibers	158.7	1.61	98.6	24.7	156.1
Epoxy–62% aramid fibers	82.8	1.38	60	20.6	146.6

[a]To convert GPa to million lb/in² multiply by 0.145. [b]To convert mg/m³ to lb/in³ multiply by 0.036.

Section shape	Formula for I	Value of I for different geometries	
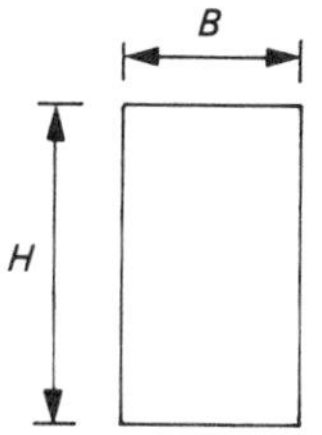 (rectangle, dimensions B, H)	$\dfrac{BH^3}{12}$	$H/B = 1$ $H/B = 2$ $H/B = 3$ $H/B = 4$	$I = 833$ $I = 1650$ $I = 2511$ $I = 3333$
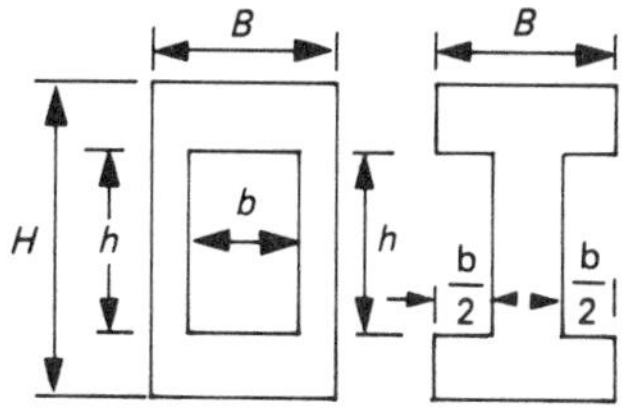 (hollow box and I-section, dimensions B, H, b, h)	$\dfrac{BH^3 - bh^3}{12}$	$H = 19 \quad B = 10$ $h = 15 \quad b = 6$ $H = 21 \quad B = 8$ $h = 17 \quad b = 4$	$I = 4028$ $I = 4536$
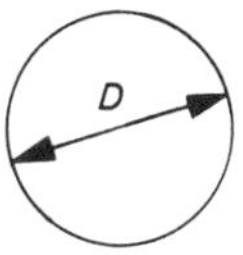 (solid circle, diameter D)	$\dfrac{\pi D^4}{64}$	$D = 11.29$	$I = 796$
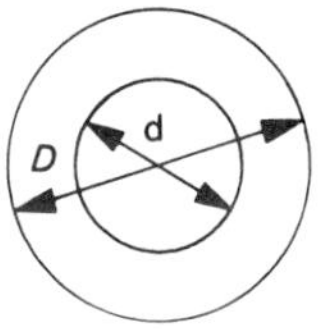 (hollow circle, diameters D, d)	$\dfrac{\pi(D^4 - d^4)}{64}$	$D = 20 \quad d = 16.5$	$I = 4300$

Figure 8.3 Effect of shape on the value of second moment of area (I) of a beam in bending. Cross-sectional area is the same in all cases and equals 100 units of area.

In materials where the initial part of the stress-strain relationship is not linear, the tangent modulus, or secant modulus, is used.

The importance of stiffness arises in complex assemblies where differences in stiffness could lead to incompatibilities and misalignment between various components, thus hindering their efficiency or even causing failure. Using high-strength materials in attempts to reduce weight usually comes at the expense of reduced cross-sectional area and reduced second moment of area. This could adversely affect stiffness of the component if the elastic constant of the new strong material does not compensate for the reduced second moment of area. Another solution to the problem of reduced stiffness is to change the shape of the component cross-section to achieve higher second moment of area, I. This can be achieved by placing as much as possible of the material as far as possible from the axis of bending. Figure 8.3 gives the formulas for calculating I for some commonly used shapes and the values of I for a constant cross-sectional area.

In applications where both the stiffness and weight of a structure are important, it becomes necessary to consider the stiffness/weight or specific stiffness of the structure. In the simple case of a structural member under tensile or compressive load, the specific stiffness is given by E/P, where E is the Young's modulus of the material and P is density. In such cases, the weight of a beam of a given stiffness can be easily shown to be proportional to P/E. This shows that the weight of the component can be reduced equally by selecting a material with lower density or higher elastic modulus. When the component is subjected to bending, the dependence of the weight on P and E is not as simple. From (8.1) and Fig. 8.3 it can be shown that the deflection of a simply-supported beam of square cross-sectional area is given by:

$$y = \frac{L \star l^3}{4 \star E \star b^4} \tag{8.3}$$

where b is breadth or width of beam.

The weight of the beam, w, can be shown to be:

$$w = l \star b^2 \star p = \frac{l^{5/2}}{2} \star \left(\frac{L}{y}\right)^{1/2} \star \frac{P}{E^{1/2}} \tag{8.4}$$

This shows that for a given deflection y under load L, the weight of the beam is proportional to $(P/E^{1/2})$. As E in this case is present as the square root, it is not as effective as P in controlling the weight of the beam. It can be similarly shown that the weight of the beam in the case of a rectangular cross-section is proportional to $(P/E^{1/3})$, which is even less sensitive to variations in E. This change in the effectiveness of E in affecting the specific stiffness of structures as the mode of loading and shape changes is illustrated in Table 8.2.

Another selection criterion which is also related to the elastic modulus of the material and cross-sectional dimensions is the elastic instability, or buckling, of slender components subjected to compressive loading. The compressive load, L_b, that can cause buckling of a strut is given by Euler formula as:

$$L_b = \frac{\pi^2 \star E \star I}{l^2} \tag{8.5}$$

where l is the length of the strut.

Equation 8.5 shows that increasing E and I will increase the load-carrying capacity of the strut. As the buckling can take place in any lateral direction, an axially symmetric cross-section can be considered. For a solid round bar of diameter D the second moment of area, I, is given as:

$$I = \frac{\pi \star D^4}{64} \tag{8.6}$$

The use of the resistance to buckling as a selection criterion was illustrated in the example of Section 8.2.

The weight of a strut, w, is given by:

$$w = l \star \frac{\pi \star D^2}{4} \star P = \frac{2 \star l^2 \star L_b^{1/2}}{\pi^{1/2}} \left(\frac{P}{E^{1/2}}\right) \tag{8.7}$$

Equation 8.7 shows that the weight of an axisymmetric strut can be reduced by reducing P or increasing E of the material. However, reducing P is more effective as E is present as the square root. In the case of a panel subjected to buckling, it can be shown that the weight is proportional to $(P/E^{1/3})$.

8.4 SELECTION OF MATERIALS FOR FATIGUE RESISTANCE

In many engineering applications, the behavior of a component in service is influenced by several other factors besides the properties of the material used in its manufacture. This is particularly true for the cases where the component or structure is subjected to fatigue loading. Under such conditions, the fatigue resistance can be greatly influenced by the service environment, surface condition of the part, method of fabrication and design details. In some cases, the role of the material in achieving satisfactory fatigue life is secondary to the above parameters, as long as the material is sound and free from major flaws. For example, if the component has welded, bolted, or riveted joints, the contribution of crack initiation stage (see Section 7.5) is expected to be small and most of the fatigue life is determined by the crack propagation stage. Experience shows that crack propagation rate is more sensitive to continuum mechanics considerations than to material properties. When design and manufacturing details are optimized, the materials aspects should then be considered. Generally, the presence of microstructural stress raisers must be avoided. Fatigue strength of metallic materials generally increases with increasing tensile strength, but the higher the strength, the higher the notch sensitivity of the material and the greater the need to eliminate coarse second-phase particles and produce a more refined, homogeneous structure. Meeting these needs could require expensive metallurgical processes or the addition of expensive alloying elements.

A measure of the degree of notch sensitivity of the material is usually given by the parameter q:

$$q = \frac{K_f - 1}{K_t - 1} \tag{8.8}$$

where K_f = the ratio of the fatigue strength, in the absence of stress concentrations, to the fatigue strength, with stress concentration.

K_t = the stress concentration factor which represents the severity of the notch and is given by the ratio of maximum local stress at the notch to average stress.

The value of q can be considered as a measure of the degree of agreement between K_f and K_t. Thus, as q increases from 0.0 to 1.0, the material becomes more sensitive to the presence of stress concentrations. Generally, increasing the tensile strength of the material makes it more notch sensitive and increases q. The value of q is also dependent on component size, and it increases as size increases, so stress-raisers are more dangerous in large masses.

Steels are the most widely used structural materials for fatigue applications as they offer high fatigue strength and good processability at a relatively low cost. Steels have the unique characteristic of exhibiting an endurance limit which enables them to perform indefinitely, without failure, if the applied stresses do not exceed this limit. As shown in Table 7.1, the endurance limit is roughly equal to 0.4–0.6 of the ultimate tensile strength of steels. However, with steels of UTS greater than about 1100 MPa (c. 160 ksi), the scatter in fatigue strength becomes wide and the endurance ratio can change over a wide range. The optimum steel structure for fatigue resistance is tempered martensite, since it provides maximum homogeneity. Steels with high hardenability give high strength with relatively mild quenching and hence low residual stresses, which is desirable in fatigue applications. Normalized structures give better fatigue resistance than pearlitic structures obtained by annealing.

Inclusions in steel are harmful as they represent discontinuities in the structure that could act as initiation sites for fatigue cracking. Therefore, free machining steels should not be used for fatigue applications. However, if machinability considerations make it essential to select a free machining grade, the leaded steels are preferable to those containing sulfur or phosphorus. This is because the rounded lead particles give rise to less structural stress concentrations than the other angular and elongated inclusions. By the same token, cast steels and cast irons are not recommended for critical fatigue applications. In rolled steels, the fatigue strength is subject to the same directionality as the static properties.

Unlike ferrous alloys, the non-ferrous alloys, with the exception of titanium, do not normally have an endurance limit. In such cases, the fatigue strength is usually taken as the stress which causes fracture after a given number of loading cycles, usually 10^7 or 10^8 cycles. Aluminum alloys usually combine corrosion resistance, light weight and reasonable fatigue resistance. The endurance ratio of aluminum alloys is more variable than that of steels, Table 7.1, but an average value can be taken as 0.35. Generally, the endurance ratio is lower for as-cast structures and precipitation hardened alloys. Fine-grained inclusion-free alloys are most suited for fatigue applications. Copper alloys, like most other non-ferrous alloys, have no endurance limit and the endurance ratio can vary over a wide range, as shown in Table 7.1. Some bronzes and Cu-Ni alloys have good fatigue strengths which can be improved by cold working.

The viscoelasticity of plastics makes their fatigue behavior more complex than that of metals. In addition to the set of parameters that affect the fatigue behavior of metals, the fatigue behavior of plastics is also affected by the type of loading, small changes in temperature and environment and method of sample fabrication. Because of their low thermal conductivity, hysteretic heating can build up in plastics causing them to fail in thermal fatigue or to function at reduced stress and stiffness levels. The amount of heat generated increases with increasing stress and test frequency. This means that failure of plastics in fatigue may not necessarily mean fracture. In flexural fatigue testing by constant amplitude of force, ASTM D671 sets an arbitrary level of stiffness – 70 percent of the original modulus – as failure. Some unreinforced plastics such as PTFE (polytetrafluoroethylene), PMMA (polymethylmethacrylate), and PEEK (polyetheretherketone) have fatigue endurance limits. At stresses below this level, failure does not occur. Other plastics, usually amorphous materials, show no endurance limit. In many unreinforced plastics the endurance ratio can be taken as 0.2.

The failure modes of reinforced materials in fatigue are complex and can be affected by the fabrication process when differences in shrinkage between fibers and matrix induce internal stresses. There is a growing body of practical experience, however, and some fiber-reinforced plastics are known to perform better in fatigue than some metals, as shown in Table 7.1. The advantage of fiber-reinforced plastics is even more apparent when compared on a per weight basis. For example, because of its superior fatigue properties, glass fiber reinforced epoxy has replaced steel leaf springs in several motor car models. Generally, fiber reinforced crystalline thermoplastics exhibit well-defined endurance limits, while amorphous-based composites do not. The higher strengths, higher thermal conductivity and lower damping account for the superior fatigue behavior of crystalline polymers. As with static strength, fiber orientation affects the fatigue strength of fiber reinforced composites. In unidirectional composites, the fatigue strength is significantly lower in directions other than the fiber orientation. Reinforcing with continuous uni-directional fibers is more effective than reinforcing with short random fibers.

8.5   SELECTION OF MATERIALS FOR TOUGHNESS

Toughness can be defined as the ability of the material to absorb energy and deform plastically before fracture. The amount of energy absorbed during both deformation and fracture is a measure of the material's toughness. As the amount of deformation preceding fracture is a measure of ductility and the force needed to cause fracture is a measure of strength, it becomes clear that toughness is a function of both strength and ductility of the material. The stress required to fracture tough materials generally corresponds to that required to produce yielding and considerable plastic deformation across the whole cross-section. In brittle materials, however, fracture occurs at a lower level of stress and at a much higher speed. Toughness is measured using two main techniques: notch toughness and fracture toughness. Notch toughness is the resistance of the material to impact loading and is typically measured by the amount of energy needed to fracture the Charpy or Izod V-notch impact specimens, as given in ASTM E23 specifications. Notch toughness values cannot be used directly in design calculations but are usually used to compare different materials. Fracture toughness, on the other hand, is defined as the resistance of a material to the propagation of an existing crack, as discussed in Section 7.3. The concepts of fracture mechanics are used to estimate the fracture toughness of materials and to evaluate the load-carrying capacity of flawed structures or components.

There is a close relationship between toughness and other mechanical properties. Within a given class of materials, there is an inverse relationship between strength and toughness, as shown in Fig. 8.4 and Table 8.3. Generally, the toughness of a material is influenced by its chemical composition and microstructure. For example, steels become less tough with increasing carbon content, larger grain size and more brittle inclusions. The grain size of steels is affected by the elements present, especially those used for deoxidizing. Small additions of aluminum to steel are known to promote fine grain size, which improves the toughness. Fully-killed fine-grained steels also have lower transition temperatures and are normally selected for applications where brittle fracture may occur. Fine grains can also be obtained in steels by using alloying elements, by controlling the rolling practice, or by normalizing treatment. A thoroughly deoxidized steel grade has fewer nonmetallic inclusions and gives better toughness. When brittle inclusions are elongated, their influence on ductility is more pronounced in the transverse and through-thickness direc-tions.

The method of fabrication can also have a pronounced effect on toughness, and experience has shown that a large proportion of brittle fractures originate from welds or their vicinity. This can be caused by the residual stresses generated by the welding process, reduction of toughness of the heat affected zone, or by defects in the weld area. The rate of load application also influences the toughness. Materials which are tough under slowly applied load may behave in a brittle manner when subjected to shock or impact loading.

Decreasing the operating temperature generally causes a decrease in toughness of most

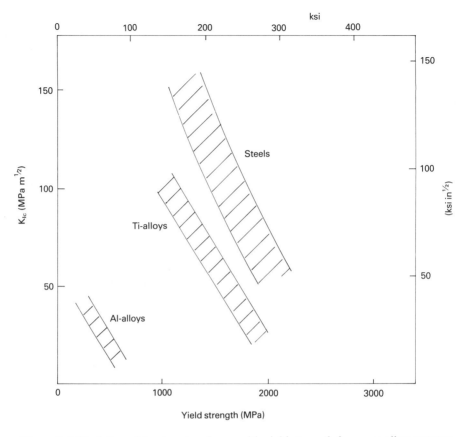

Figure 8.4 Variation of fracture toughness with yield strength for some alloy systems.

engineering materials. This is particularly important in the case of body-centered cubic materials (BCC) as they tend to go through a ductile-to-brittle transition as the temperature decreases. All carbon and most alloy steels are of the BCC group and their behavior is illustrated in Fig. 8.5. When using BCC materials, it is important that the service temperature be higher than the ductile-to-brittle transition temperature. This is easy to apply in the case of low-strength steels as the transition from ductile to brittle behavior (upper and lower shelf energies respectively) is distinct and takes place over a narrow temperature range. However, as the strength of steel increases, the transition becomes less distinct and the difference between the upper and lower shelf energies becomes smaller, which makes it difficult to specify the transition temperature. Under these conditions, the concepts of fracture toughness can provide a useful tool for design and materials selection.

Under cryogenic temperatures most BCC materials become too brittle to use. In such cases, austenitic steels and nonferrous materials of face-centered cubic (FCC) structures become the only possible materials, as they do not suffer ductile-to-brittle transition. Several aluminum, titanium, copper and nickel-base alloys are available for cryogenic applications.

An important aspect of selecting materials for toughness is the likelihood of detection of a crack before it reaches a critical size. As larger cracks can be more easily discovered, it

Table 8.3 Comparison of toughness and strength of some engineering materials

	Yield strength (MPa)	(ksi)	K_{Ic} (MPa m$^{1/2}$)	(ksi in$^{1/2}$)	K_{Ic}/YS (m$^{1/2}$)	(in$^{1/2}$)
STEELS						
Medium carbon steel	260	37.7	54	49	0.208	1.30
ASTM A533B Q&T	500	72.5	200	182	0.400	2.51
AISI 4340 (T260°C)	1640	238	50	45.8	0.030	0.19
AISI 4340 (T425°C)	1420	206	87.4	80	0.062	0.388
Maraging 300	1730	250	90	82	0.052	0.328
ALUMINUM ALLOYS						
AA 2024-T651	455	66	24	22	0.053	0.333
AA 2024-T3	345	50	44	40	0.128	0.80
AA 7075-T651	495	72	24	22	0.048	0.306
AA 7475-T651	462	67	47	43	0.102	0.642
TITANIUM ALLOYS						
Ti-6 Al-4 V	830	120	55	50	0.066	0.417
Ti-6 Al-4 V- 2 Sn	1085	155	44	40	0.04	0.258
Ti- $ Al- 4 Mo-2 Sn-0.05 Si	960	139	45	40	0.047	0.288
PLASTICS						
PMMA	30	4	1	0.9	0.033	0.225
Polycarbonate	63	8.4	3.3	3	0.052	0.357
CERAMICS						
Reaction bonded Si3N4	450	63.3	5	4.6	0.011	0.07
Al203	262	36.9	4.5	4.1	0.017	0.11
SiC (self bonded)	140	19.7	3.7	3.4	0.026	0.173

follows that materials tolerating larger critical cracks are more advantageous. A comparison of materials on the basis of their crack tolerance should be based on similar loading efficiencies, i.e. same ratio of service stress to yield stress. It was shown in Section 7.3 that the maximum tolerable crack is proportional to (K_{Ic}/YS), where K_{Ic} and YS are the fracture toughness and yield strength of the material at the service temperature and loading rate. Higher values of (K_{Ic}/YS) indicate tolerance to larger cracks. The materials listed in Table 8.3 are compared on the basis of this parameter. The values in the table show that a material may exhibit a good crack tolerance even though its fracture toughness is modest. In general, therefore, the material selected for a given application must have such combination of K_{Ic} and YS that the critical crack length is appropriate for that application and the available NDT techniques.

Although unreinforced plastics generally have lower impact strength than most metallic materials, numerous techniques have been developed to improve their toughness. Examples of such techniques include:

1. Alloying the plastic with a rubber phase or with another higher impact plastic. Examples of this method include toughened nylons, which are alloyed with polyolefin or other polymeric modifier; and thermoplastic polyester, which is alloyed with thermoplastic elastomers (TPE). It has been reported that alloys of polycarbonate and polybutylene (PC/PBT) are being used in several motor car bumpers. An alloy of nylon and ABS combines the characteristics of both crystalline and amorphous polymers which results in a combination of high flow rate, high-temperature warp resistance, good surface appearance, chemical resistance and toughness.

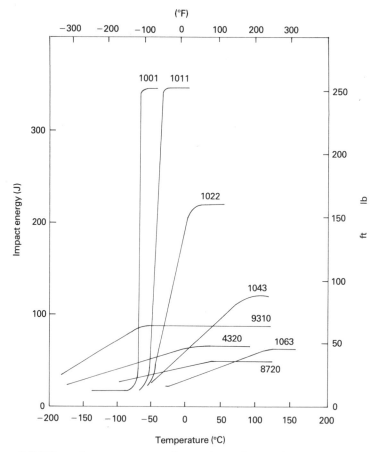

Figure 8.5 Effect of temperature on the notch toughness of some AISI-SAE steels.

2. Copolymerization to create a tougher chemical structure. This approach is used to produce less notch-sensitive polycarbonates which retain their ductility at lower temperature.
3. Incorporating high-impact resistance fibers. For example PET (polyethylene terephthalate), nylon and polyethylene fibers have been used to replace a portion of the glass fibers in injection molded polyesters for automotive components. Advanced composites for aircraft use S-2 glass fibers or hybrids of glass with graphite or aramid fibers in places where improved toughness or damage resistance is needed.

In general, thermoplastic-matrix composites are tougher than those with thermoset-matrix, which is one of the reasons why the former are being developed to replace current epoxies. On the other hand, the toughness of thermoset composites can be improved by incorporating rubber modification. This, however, often reduces stiffness.

The fracture of ceramics is dependent on critical flaw size which is a function of fracture toughness (K_{Ic}). With careful processing, the average flaw size can be reduced to about 30 μm, but this may still be larger than the critical flaw size. In addition, a single flaw in the material which is larger than the critical flaw size is sufficient to cause fracture. This is why toughness data for ceramic materials are often inconsistent and strength and toughness do not always respond in the same manner to changes in microstructure or interfacial properties. Table 8.3 lists typical toughness values of some ceramic materials. An important technique to improve the toughness of ceramics like ZrO_2, Al_2O_3, and Si_3N_4 is to induce a phase transformation in the region of applied stress within the material. This absorbs energy at the tip of the advancing crack, arresting its propagation and significantly increasing both strength and toughness. Another technique is to introduce fibers to increase the toughness as a result of fiber debonding, crack deflection, or fiber pullout. Internal stresses due to differences in thermal expansion between matrix and fibers in a composite can also provide a toughening effect. Because ceramics are sensitive to surface damage, surface modification techniques are also being developed as a means of improving their toughness.

8.6   SELECTION OF MATERIALS FOR CORROSION RESISTANCE

Although corrosion resistance is usually the main factor in selecting corrosion-resistant materials, it is often difficult to assess this property for a specific application. This is because the behavior of a material in a corrosive environment can be dramatically changed by seemingly minor changes in the medium or the material itself. The main factors that can affect the behavior of material can be classified as: (a) corrosive medium parameters; (b) design parameters; and (c) material parameters.

Corrosive medium parameters

Corrosive medium parameters include:

1. Chemical composition and presence of impurities.
2. Physical state whether solid, liquid, gas or combinations.
3. Aeration, oxygen content and ionization.
4. Bacteria content.

In the case of metallic materials, the most significant factor controlling the probability of atmospheric corrosion is whether or not an aqueous electrolyte is provided by condensation of moisture under prevailing climatic conditions. Hot, dry or cold, icy conditions give less attack than wet conditions. Clean atmosphere is less aggressive than industrial or marine atmospheres containing sulfur dioxide and salt respectively. Direction of exposure to the sun, wind and sources of pollution can also affect the rate of atmospheric corrosion.

In buried structures, increasing the porosity of the soil and the presence of water increase the rate of corrosion. In addition to allowing continuing access of oxygen to the corroding surface, porosity also encourages the activity of aerobic bacteria which can lead to local variation in aeration, consumption of organic protection systems and production of H_2S. In general, dry, sandy or chalky soils of high electrical resistance are the least corrosive while heavy clays and saline soils are the most corrosive.

The rate of corrosion of under-water structures is affected by the amount of dissolved oxygen as well as the amount of dissolved salts and suspended matter. Since oxygen enters the water by dissolution from the air, its concentration can vary with depth and flow rate. Soft fresh water is generally more corrosive than hard water, which precipitates a protective carbonate on the corroding surface. In sea water, the presence of chloride ions increases the electrical conductivity and, therefore, the rate of corrosion. The presence of organic matter, such as bacteria or algae, in water can decrease the rate of corrosion in the covered areas but produce regions of local deaeration where accelerated attack occurs. Increasing the water temperature generally increases the rate of attack. In chemical plants, the rate of attack depends on several factors including the temperature, concentration, fluid velocity, degree of aeration, purity and applied stresses. In general, attack is severest where protective or oxide films are disrupted or become locally unstable.

Design parameters

The design parameters which affect the rate of corrosive attack include stresses acting on the material in service, operating temperature, relative motion of medium with respect to the material, continuity of exposure of the material to the medium, contact between the material and other materials, possibility of stray currents, and geometry. Combating corrosion by design will be discussed in Section 12.10.

Material parameters

From the corrosion resistance point of view, it is convenient to divide engineering materials into metallic and nonmetallic materials. This is because of the difference in the atomic structure which is reflected in the electrical conductivity and, therefore, the mechanism of corrosion. In the case of metallic materials, corrosion takes place by electrochemical processes due to the presence of free electrons in their structure. Nonmetals, on the other hand, are generally nonconducting to electricity and have ionic or covalent bonds which are chemically very stable. The chemical stability of the bonds is usually reflected in the bulk properties, as will be discussed later. There are, however, some main parameters which affect the corrosion resistance of materials. These include chemical composition and presence of impurities, nature and distribution of microstructural constituents, surface condition and deposits and processing history. Generally, the corrosion resistance improves in pure metals as their purity increases. An example is the localized attack in commercially pure aluminum due to the presence of iron impurities. Another example is arc-melted zirconium being more corrosion resistant than induction-melted zirconium because the former is more pure. The general behavior of some commonly used engineering materials under different corrosive conditions is shown in Tables 8.4 and 8.5.

Carbon steels and cast irons are used in large quantities because of their useful mechanical properties and low cost. These materials, however, are not highly corrosion resistant, with the exception of resistance to alkalis and concentrated sulfuric acid. Mild steel has adequate resistance to scaling in air up to about 500°C (c. 930°F), but this temperature is reduced in the presence of sulfur in flue gases. The addition of chromium in amounts of about 3 percent increases the resistance to both oxidation and sulfide scaling. Chromium additions also improve resistance to atmospheric corrosion. Nickel is also added to improve the resistance to sodium hydroxide.

Stainless steels represent a class of highly corrosion-resistant materials and have widespread applications in engineering. It should be remembered, however, that stainless steels do not resist all corrosive environments. For example, when subjected to stresses in chloride-containing environments, stainless steels are less resistant than ordinary structural steels. Stainless steels, unless correctly fabricated and heat treated, can also be more susceptible to localized corrosion such as intergranular corrosion, stress-corrosion cracking

Table 8.4 Relative corrosion resistance of some uncoated metallic materials

Material	Industrial atmosphere	Fresh water	Sea water	Acids H_2SO_4 5–15% concentration	Alkalis 8%
Low-carbon steel	1	1	1	1	5
Galvanized steel	4	2	4	1	1
Gray cast iron	4	1	1	1	4
4–6% Cr steels	3	3	3	1	4
18-8 stainless steel	5	5	4	2	5
18-35 stainless steel	5	5	4	4	4
Monel (70% Ni-30% Cu)	4	5	5	4	5
Nickel	4	5	5	4	5
Copper	4	4	4	3	3
Red brass (85% Cu-15% Zn)	4	3	4	3	1
Aluminum bronze	4	4	4	3	3
Nickel silver (65% Cu-18% Ni-17% Zn)	4	4	4	4	4
Aluminum	4	2	1	3	1
Duralumin	3	1	1	2	1

Key:
1. Poor–rapid attack.
2. Fair–temporary use.
3. Good–reasonable service.
4. Very good–reliable service.
5. Excellent–unlimited service.

Source: Farag, M. M.: *Materials and Process Selection in Engineering*, Applied Science Publishers, London 1979.

Table 8.5 Relative chemical stability of selected polymeric materials

Material	Water absorption[a]	Weak acids	Strong acids	Weak alkalis	Strong alkalis	Organic solvents
THERMOPLASTICS						
Fluoroplastics	0	5	5	5	5	5
Polyethylene	0.01–0.02	5	2	5	5	5
Polyvinylidene chloride	0.04–0.10	5	3	5	5	3
Vinylechloride	0.45	5	3	5	3	2
Polycarbonate	–	5	3	5	3	2
Acrylics	0.03	3	3	5	3	1
Polyamides (nylon)	1.50	3	1	5	3	3
Acetals		2	1	2	1	5
Polystyrene	0.04	3	1	3	1	1
Cellulose acetate	3.80	3	1	3	1	2
THERMOSETS						
Epoxy	0.10	5	3	5	5	3
Melamine	0.30	5	1	5	1	5
Silicones	0.15	5	3	3	2	3
Polyesters	0.01	3	1	2	1	2
Ureas	0.60	3	1	2	1	2
Phenolics	0.07–1.00	2	1	2	1	3

[a] After 24 h of immersion (weight percent).
　See Table 8.4 for rating numbers.

Source: Farag, M. M., *Materials and Process Selection in Engineering*. Applied Science Publishers, London 1979.

and pitting than ordinary structural steels. As discussed in Section 2.4, stainless steel is the generic name for more than 70 types of steels which are characterized by containing more than 12 percent chromium. Increasing the chromium content gives additional resistance. Stainless steels can be divided into austenitic, martensitic, ferritic, precipitation hardening and duplex stainless steels, as discussed in Section 2.4.

The corrosion resistance of stainless steels is usually attributed to the presence of a thin film of hydrous oxide on the surface of the metal. The condition of the film depends on the composition of the stainless steel and on the treatment it receives. To give the necessary protection, the film must be continuous, nonporous, self healing and insoluble in the corrosive medium. In the presence of such oxide film the stainless steels are passive and have solution potentials approaching those of noble metals. When passivity is destroyed, the potential is similar to that of iron. Exposing stainless steels to mildly oxidizing corrosive agents causes them to become active and increasing the oxygen concentration causes them to regain passivity. When the passive film is destroyed locally, stainless steels can fail catastrophically by localized mechanisms such as pitting, crevice corrosion, intergranular corrosion or stress corrosion cracking. Chromium plays an important role in forming the passive film on the stainless steel surface. The presence of nickel in high-chromium steels greatly improves their resistance to some nonoxygenating media. It is also an austenite stabilizer. Manganese can be used as a substitute for part of the nickel as an austenite stabilizer, although it does not significantly alter the corrosion resistance of high-chromium steels. Molybdenum strengthens the passive film and improves resistance to pitting in seawater. Other elements such as copper, aluminum and silicon also increase corrosion resistance.

Nickel has a relatively high corrosion resistance and is particularly useful for handling caustic alkalis. Nickel resists stress corrosion cracking in chloride environments, but may be susceptible in caustic environments if highly stressed and if it contains impurities in solution. Inconel, 78/16/6 Ni-Cr-Fe, is resistant to many acids and has outstanding resistance to nitriding at high temperatures. Nimonic alloys, based on the 80/20 Ni-Cr basic composition, have particularly good combination of high strength and oxidation resistance at high temperatures. Monel alloys, based on the 70/30 Ni-Cu composition, have similar resistance to pure nickel with the additional advantage of being less expensive and able to handle sea and brackish waters at high fluid velocities. Monel alloys present an economic means of handling hydrofluoric acid and are also resistant to other nonoxidizing acids. Monel alloys are not, however, resistant to oxidizing media such as nitric acid, ferric chloride, sulfur dioxide and ammonia.

Pure copper is a noble metal and is, therefore, highly corrosion resistant and is especially compatible with most industrial, marine and urban atmospheres in addition to water and seawater. When copper is alloyed with zinc in concentrations more than 15 percent, dezincification may occur in some environments. Addition of about 1 percent tin can reduce this problem. Tin bronzes are resistant to a variety of atmospheres, waters and soils. Phosphorus is added to impart oxidation resistance. Aluminum bronzes, containing about 10 percent Al, are resistant to corrosion from chloride potash solutions, nonoxidizing mineral acids and many organic acids. Cupronickels are widely used in saltwater and have excellent resistance to biofouling and stress corrosion cracking.

Over half of the tin production is used as protective coatings of steels and other metals. In addition to its corrosion resistance, tin is nontoxic and provides a good base for organic coatings. This explains its wide use in coating the steel cans (tin cans) used for storage of food products and beverages. Tin is normally cathodic to iron, but the potential reverses in most sealed cans containing food products and the tin acts as a sacrificial coating, thus protecting the steel. Tin is also resistant to relatively pure water and dilute mineral acids in the absence of air. This makes it suitable for coating copper pipes and sheets in contact with distilled water and medicants. Tin is attacked by strong mineral acids and alkalis.

A large proportion of lead production goes into applications where corrosion resistance is important, especially those involving sulfuric acid. The corrosion resistance of lead is due to the protective sulfates, oxides and phosphates which form on its surface as a result of reaction with corrosive environments. Lead containing about 0.06 percent copper is usually specified for process equipment in contact with sulfuric, chromic, hydrofluoric and phosphoric acids. It is also used for neutral solutions, seawater and soils. Lead is attacked by acetic, nitric, hydrochloric and organic acids.

Aluminum is a reactive metal, but it develops an aluminum oxide film that protects it from corrosion in many environments. The film is quite stable in neutral and many acid solutions but is attacked by alkalis. The aluminum oxide film is also resistant to a variety of organic compounds including fatty acids. This oxide film forms in many environments, but it can be artificially produced by anodizing. Pure aluminum and nonheat-treatable alloys exhibit high resistance to general corrosion but, because of their dependence on the surface oxide film, are liable to suffer local attack under deposits and in crevices. Heat treatable alloys in the 2000 series and those in the 7000 series that contain copper exhibit lower resistance to general corrosion and are used in applications where corrosion resistance is secondary to strength.

Titanium exhibits excellent corrosion resistance because of its stable, protective, strongly adherent surface oxide film. Titanium is immune to all forms of corrosive attack in seawater and chloride salt solutions at ambient temperatures and to hot strong oxidizing solutions. It also has very high resistance to erosion corrosion in seawater. Titanium also resists attack by moist chlorine gas, but if moisture concentration in the gas falls below 0.5 percent, rapid attack can result. Hydrofluoric acid is also among the substances that attack titanium by destroying the protective oxide film. Addition of alloying elements can affect corrosion resistance if they alter the properties of the oxide film.

Tantalum is inert to practically all organic compounds at temperatures below 150°C (300°F). Exceptions are hydrofluoric acid and fuming sulfuric acid. Zirconium is resistant to mineral acids, molten alkalis, alkaline solutions and most organic and salt solutions. It has excellent oxidation resistance in air, steam, CO_2, SO_2 and O_2 at temperatures up to 400°C (750°F). Zirconium is attacked by corrosion in hydrofluoric acid, wet chlorine, aquaregia, ferric chloride and cupric chloride solutions. Tantalum and zirconium will seldom be economic, but they are the only available resistant materials for a few applications.

Metallic glasses (amorphous alloys) are produced by quenching from the liquid state and have undercooled liquid structures similar to those of ceramic glasses. The compositions of these materials are adjusted to be close to low-melting, stable eutectics that yield noncrystalline structures on rapid solidification. Some of the iron-based metallic glasses have corrosion resistances approaching those of tantalum or the noble metals. Typical compositions include 8–20% Cr, 13% P, 7% C, remainder Fe and 10% Cr, 5–20% Ni, 13% P, 7% C, remainder Fe. These materials passivate very easily and at 8% Cr they are superior to conventional stainless steels. Their pitting resistance is equal to or greater than that of the high nickel alloys, Hastelloy C, and titanium. These materials may be susceptible to hydrogen embrittlement under certain conditions.

Plastics do not corrode in the same way as metals as they do not have the free electrons which are necessary for electrochemical processes. Because of their corrosion resistance, plastics and composites have replaced metals in many applications. Examples from the automotive industry include fenders, hoods and other body components. However, there are several environmental effects that should be considered when selecting plastics and FRP. Several plastics absorb moisture which causes swelling and distortion in addition to degrading their strength and electrical resistance. Polymers can also be attacked by organic solvents. Table 8.4 gives the relative water absorption and chemical stability of some polymers. Generally, crystalline thermoplastics, such as fluorocarbons, teflon and nylon, have superior chemical stability than amorphous types like polycarbonate. Fluorocarbons, e.g. polytetrafluoroethylene (PTFE), are among the most chemically inert materials available to the engineer. They are inert to all industrial chemicals and resist the attack of boiling aqua regia, fuming nitric acids, hydrofluoric acid and most organic solvents. Other thermoplastics like polyketones and polyphenylene sulfide provide excellent chemical resistance, even at relatively elevated temperatures. Among thermosetting plastics, epoxies represent the best combination of corrosion resistance and mechanical properties. The general chemical resistance of polymers can be related to structural parameters like crystallinity, bond strength and type of bonding. Crystalline thermoplastics like PTFE and nylon often have superior chemical resistance compared to amorphous types like polycarbonate. There are several standard tests for measuring the chemical resistance of polymers and FRP. The immersion test, ASTM D 543, is used extensively as it measures the changes in weight, dimensions and mechanical properties that result from immersion in standard reagents. Table 8.6 gives the relative chemical resistance, expressed in terms of percent retention of tensile strength, of some plastics.

Most ceramic materials exhibit good resistance to chemicals, with the main exception of hydrofluoric acid. Glasses are among the most chemically stable materials and they have exceptionally good resistance to attack by water, aqueous solutions of most acids, alkalis and salts. However, their relative performance in various environments may vary considerably between different grades. For example, borosilicate and silica glasses show much higher resistance to boiling water and hot dilute acid solution than do soda lime and lead alkali glasses. Enamels, which are made of silicate and borosilicate glass with the addition of fluxes to promote adhesion, are highly resistant to corrosion and are widely used to protect steels and cast irons. The composition of the enamel should be adjusted to match its coefficient of thermal expansion with that of the underlying metal. Although most refractory ceramics have good resistance to chemical attack, a particular corrosive environment requires careful selection of the ceramic material used. A strong acid

Table 8.6 Retention of tensile strength in polymers after 24 h exposure to chemicals at 93°C (200°F)

	Polyphenylene sulfide	Nylon 6/6	Polycarbonate	Polysulfone
Hydrochloric acid (37%)	100	0	0	100
Sulfuric acid (30%)	100	0	100	100
Sodium hydroxide (30%)	100	89	7	100
Gasoline	100	80	99	100
Chloroform	87	57	0	0
Ethylene chloride	72	65	0	0
Phenol	100	0	0	0
Ethyl acetate	100	89	0	0

Source: Guide to Engineering Materials, Advanced Materials and Processes, 1988.

environment requires an acidic refractory, whereas alkalis call for a basic refractory material.

Occasionally, no material may offer an economical combination of corrosion resistance and other performance requirements. In such cases, a low-cost base material which satisfies the mechanical and physical requirements can be selected provided that it is adequately protected against corrosion. Protection can take the form of sacrificial coatings, passivation, corrosion inhibitors, barrier coatings or cathodic protection. The selection of materials for protective coatings will be discussed in more detail in Section 8.9.

8.7 SELECTION OF MATERIALS FOR TEMPERATURE RESISTANCE

It was shown in Section 7.9 that creep is a major factor which limits the service life of parts and structures at elevated temperatures. Experience shows that many of the methods used to improve the low-temperature strength of metallic materials become ineffective as the operating temperature approaches $0.5T_m$ (T_m is the melting temperature expressed in Kelvin). This is because atomic mobility becomes sufficient to cause softening of cold-worked structures and coarsening of unstable precipitates. At these high temperatures, the differences in creep resistance from one material to another depend on the stability of the structure and the hardening mechanism. The most important method of improving creep strength is to incorporate a fine dispersion of stable second-phase particles within the grains. These particles can be introduced by dispersion, as in the case of thoria particles in nickel (TD nickel), or by precipitation, as in the case of precipitation-hardened nickel alloys. To minimize particle coarsening, it is the practice to make the chemical composition of the precipitates as complex as possible and to reduce the thermodynamic driving force for coarsening by reducing the interfacial energy between the precipitates and the matrix. Precipitates at the grain boundaries are important in controlling creep rupture ductility as they control grain boundary sliding which causes premature failure. Care should be taken, however, to avoid continuous films of brittle phases at grain boundaries as they reduce impact resistance.

The mechanical strength of plastics at high temperatures is usually compared on the basis of deflection temperature under load (DTUL), also known as heat deflection temperature. DTUL is defined as the temperature at which a specimen deflects 0.010 in under a load of 66 or 264 lb/in^2. Generally, thermosets have higher temperature resistance than thermoplastics. However, adding glass and carbon fibers, as well as mineral and ceramic reinforcements, can significantly improve DTUL of crystalline thermoplastics, such as nylon, thermoplastic polyesters, PPS, and fluoroplastics. For example, at 30 percent glass-fiber the DTUL of nylon 6/6 at 264 lb/in^2 increases from about 71°C to 249°C (160 to 480°F).

While several plastics can withstand short excursions to high temperatures, up to 500°C (930°F), continuous exposure can result in dramatic drop in mechanical properties and extreme thermal degradation. This has led to the introduction of tests where properties are measured after long exposure times to high temperatures. The most widely used of such tests are the Underwriters Laboratories (UL) temperature index and the ASTM D3045 heat-ageing tests.

Because operating temperature is the single most important factor that affects the selection of materials for elevated-temperature service, it is normal practice to classify temperature-resistant materials according to the temperature range in which they are expected to be used.

Room temperature to 150°C (300°F)

Most engineering metals and alloys, with the exception of lead, can be used in this temperature range. Several unreinforced thermoplastics are suitable for continuous service at temperatures above 100°C (212°F). In addition, fluoroplastics, polycarbonates, polymides polysulfone, polyphenylenesulfides and the newly developed materials like PEEK and PPS can be used at temperatures up to 200°C. Several FRP, e.g. nylon 6/6–glass fiber, can also serve in this temperature range.

150 to 400°C (300 to 750°F)

Plain carbon or manganese-carbon steels provide adequate properties in this temperature range, although it may be necessary to use low-alloy steels if very long service, more than 20 years, is required. High-grade cast irons can be used at temperatures up to 250°C for engine casings. For short exposure times, magnesium alloys are used at temperatures up to 200°C (400°F) and the addition of thorium improves creep resistance. Aluminum alloys can be used at temperatures up to about 250°C (480°F) although some powder metallurgy alloys have been used for short intervals at about 480°C (900°F). High-temperature plastics can be used at temperatures up to 200°C (400°F) and will stand temperatures up to about 300°C (500°F) for short periods. These include polysulfones, polyphenylenesulfides, polyethersulfone, and fluoroplastics. Thermoset polyemides–graphite composites can serve in the range of 260–290°C (500–550°F). New experimental plastics, like polypara-phenylene benzobisthiazole, are expected to withstand temperatures up to about 370°C (700°F) for long periods.

400 to 600°C (750 to 1100°F)

Low-alloy steels and titanium alloys are the main materials used in this temperature range. Low-alloy steels are relatively inexpensive and are used if there are no restrictions on weight. The main alloying elements that are usually added to these steels include molybdenum, chromium and vanadium. Molybdenum is a carbide stabilizer and thus increases creep resistance even when added in small amounts (0.1 to 0.5 percent). Chromium, in small amounts (about 0.5 percent), is a carbide former and stabilizer. Vanadium is added to provide additional resistance to tempering. An example of such steels is the 0.2C-1Cr-1Mo-0.25V steel which is used for intermediate- and high-pressure steam turbine rotors. In applications at temperatures approaching 600°C (1100°F), oxidation resistance becomes an important factor in determining the performance of materials. In such cases, at least 8 percent chromium needs to be added to steels. Several steels are available with chromium contents in the range of 5 to 12 percent. These steels usually also contain molybdenum to improve their creep resistance.

   Titanium alloys of alpha-phase structure exhibit better creep resistance than those of beta-phase structure. An example of a near-alpha alloy is 5.5Al-3.5Sn-3Zr-1Nb-0.25Mo-0.25Si which can be used at temperatures up to about 600°C (1110°F). However, the alpha-beta 6Al-4V alloy is most widely used for general purposes and is limited to a maximum operating temperature of about 450°C (840°F) because of the beta phase.

600 to 1000°C (1100 to 1830°F)

The most widely used materials for this temperature range can be divided into the following groups: (a) stainless steels; (b) Fe-Ni-base superalloys; (c) Ni-base superalloys; and (d) Co-base superalloys. Oxidation and hot corrosion resistance become increasingly important with increasing operating temperature. The level of oxidation resistance in this temperature range is a function of chromium content. Aluminum can also contribute to oxidation resistance, especially at higher temperatures. Chromium is also important for hot corrosion resistance. Chromium content in excess of 20 percent appears to be required for maximum resistance. When the oxidation and hot corrosion resistance of a given alloy is not adequate, protective coatings may be applied. Diffusion coatings, CoAl or NiAl, are commonly used for protection. FeCrAl, FeCrAlY, CoNiAl, or CoNiAlY overlay coatings can also be used and they do not require diffusion for their formation.

The ferritic stainless steels of the 400 series are less expensive than the austenitic grades of the 200 and 300 series, see Section 2.4. The ferritic grades are usually used at temperatures up to 650°C (1200°F) in applications involving low stresses. Austenitic stainless steels of the 300 series can be used at temperatures up to 750°C (1380°F). Type 316 stainless steel with 2.5 percent Mo is widely used and has the highest stress-rupture strength of all the 300 series alloys. The more highly alloyed compositions 19-9 DL and 19-9 DX contain 1.25 to 1.5 Mo, 0.3 to 0.55 Ti, and 1.2 to 1.25 W have superior stress-rupture strengths than the 300 series alloys. These alloys can be used at temperatures up to 815°C (1500°F). Also the 202 and 216 Cr-Ni-Mn alloys have higher stress-rupture capabilities than the 300 series alloys.

The Fe-Ni-base superalloys consist mainly of FCC matrix strengthened by intermetallic compound precipitates. A common precipitate is gamma prime as in the case of A-286 and Incoloy 901. Other precipitates are also used, for example gamma double as in the case of Inconel 718, and carbides, nitrides and carbonitrides as in the case of CRM series. Most of the Fe-Ni alloys are used in the wrought condition although the CRM series was developed primarily for casting applications. Table A.12 gives the composition of representative samples of these alloys.

Ni-base superalloys (see Table A.12) consist also of FCC matrix strengthened by intermetallic compound precipitates. Gamma prime is used to strengthen alloys like Waspaloy and Udimet 700. Oxide dispersions are also used for strengthening as in the case of IN MA-754 and IN MA-6000E. Other Ni-base superalloys are essentially solid-solution strengthened in addition to some carbide precipitation, as in the case of Hastelloy X. Ni-base superalloys are used in wrought and cast forms. Special manufacturing processes, like powder metallurgy and directional solidification, have also been applied to process these alloys.

Co-base superalloys (Table A.12) are strengthened mainly by a combination of carbides and solid-solution hardeners. In terms of strength, Co-base alloys can only compete with Ni-base superalloys at temperatures above 980°C (1800°F). Co-base superalloys are used in the wrought form, e.g. Haynes 25, and in the cast form, e.g. X-40.

1000°C (1830°F) and above

The refractory metals, Mo, Nb, Ta and W and their alloys can be used for stressed applications at temperatures above 1000°C (1830°F). Table A.14 lists the composition of some commercial refractory-metal alloys. Mo-30W molybdenum alloy has a melting point of 2830°C (5125°F) and excellent resistance to liquid-metal attack. Niobium can be used in contact with liquid lithium and sodium-potassium alloys at high temperatures, even above 800°C (1470°F). Addition of 1%Zr to niobium increases its resistance to embrittlement due to oxygen absorption. Tantalum can be used for structural applications at temperatures in the range of 1370 to 1980°C (2500 to 3600°F) but it requires protection against oxidation. Tantalum is also used for heat shields and heating elements in vacuum furnaces. Tungsten has the highest melting point of all materials, which makes it the obvious candidate for structural applications at very high temperatures. Molybdenum is added to tungsten to improve its machinability and rhenium is added to improve resistance to cold fracture in lamp filaments. Surface protection is an important obstacle to widespread use of refractory metals in high-temperature oxidizing environments. Various aluminide and silicide coatings are available commercially but they all have a maximum temperature limit of about 1650°C (3000°F).

Ceramics can withstand extremely high temperatures and are being increasingly used for structural applications. Creep resistance, thermal conductivity, thermal expansion and thermal shock resistance are the major factors that determine the suitability of a ceramic material for high-temperature applications. Creep resistance of many ceramics is affected by intergranular phases. Because crystalline phases are generally more creep resistant than amorphous ones, it is the usual practice to reduce the amorphous intergranular phases as a means of improving creep resistance. Doping can also be used to improve the strength of grain-boundary phases, as in the case of doping Si_3N_4 with Y_2O_3 and ZrO_2. Silicon-based ceramics have a lower thermal expansion coefficient, which helps improve their thermal shock resistance. However, this may not be an advantage if the ceramic is used as a coating on metals where a large difference in expansion may present difficulties. Thermal shock resistance is a function of thermal conductivity, coefficient of thermal expansion, tensile strength and modulus of elasticity. For structural ceramics, thermal shock resistance is dependent on both material type and processing method. For example, silicon nitride has better thermal shock resistance when hot-pressed than when reaction-sintered. Generally, silicon carbide and tungsten carbide have better thermal shock resistance than zirconium oxide and aluminum oxide. Silicon nitride has good thermal shock resistance and good oxidation resistance which makes it a feasible candidate for service temperatures of about 1200°C (2192°F) in gas turbines.

8.8   SELECTION OF MATERIALS FOR WEAR RESISTANCE

As discussed in Section 7.6, wear is a surface phenomenon which involves a progressive loss of material as a result of contact with other parts, as in the case of sliding and rolling contact, or as a result of action of the environment, as in the case fluids containing foreign particles. In practice, wear failure may be caused by the combined effect of different types of wear. The main factors which influence the wear behavior of a material can be grouped as:

1. Metallurgical variables, including hardness, toughness, chemical composition and microstructure.
2. Service variables, including contacting materials, contact pressure, sliding speed, operating temperature, surface finish, lubrication and corrosion.

Although the performance of a material under wear conditions is generally affected by its mechanical properties, wear resistance cannot always be related to one property. In general, wear resistance does not increase directly with tensile strength or hardness although if other factors are relatively constant, hardness values provide an approximate guide to relative wear behavior among different materials. This is particularly true for applications involving metal-to-metal sliding. In such cases, increasing the hardness increases wear resistance as a result of decreasing penetration, scratching and deformation. Increasing toughness also increases wear resistance by making it more difficult to tear off small particles of deformed metal.

Because wear is a surface phenomenon, surface treatments and coatings play an important role in combating it. Improving the wear resistance of the surface avoids having to make the entire part of a wear resistant material, which may not provide all the other functional requirements or may be more expensive. Surface treatments include surface alloying, as in the case of carburizing which increases the hardness of the surface by increasing its carbon content; and surface heat treatment, as in the case of induction heating which allows hardening of the surface without affecting the bulk of the material. Surface coatings consist of wear-resistant materials which are applied to the surface by vapor deposition, sputtering, welding, spraying, ion plating, flame plating or electroplating. The materials used for wear-resistant coatings will be discussed in Section 8.9. In spite of the widespread use of surface treatments and surface coatings to combat wear, these solutions are not without problems. Not all materials or parts can be surface treated and surface coatings can fail by spalling. In many applications wear problems are solved, wholly or in part, by the proper selection of materials, as will be discussed in the following paragraphs.

Wear resistance of steels

Mild steels, although among the cheapest and most widely used materials, have poor wear resistance and can suffer severe surface damage during dry sliding. This can be avoided by selecting compatible mating materials, such as Babbitt alloys, and providing adequate lubrication. Increasing the carbon content of the steel improves the wear resistance but increases the cost. Surface-hardenable carbon or low-alloy steels are another step higher in wear resistance. Components made from these steels can be surface hardened by carburizing, cyaniding or carbonitriding to achieve better wear resistance at a still higher cost. An even higher wear resistance can be achieved either by nitriding medium-carbon chromium or chromium-aluminum steels or surface hardening high-carbon high-chromium steels. Precipitation hardening stainless steels can be used in applications involving wear, elevated temperature and corrosion.

Austenitic manganese steels are selected for a wide variety of applications where good abrasion resistance is important. The original austenitic manganese steel, Hadfield steel, contained 1.2 percent C and 12 percent Mn, however, several compositions are now available as covered by ASTM A128. These steels have carbon contents between 0.7 and 1.45 percent and manganese contents between 11 and 14 percent, with or without other elements such as chromium, molybdenum, nickel, vanadium and titanium. Compared with other abrasion-resistant ferrous alloys, austenitic manganese steels have superior toughness at moderate cost. They have excellent resistance to metal-to-metal wear, as in sheave wheels, rails and castings for railway trackwork. Manganese steels are also valuable in conveyors and chains subjected to abrasion and used for carrying heavy loads.

Wear resistance of cast irons

As-cast gray cast iron has adequate wear resistance for applications such as slideways of machine tools and similar sliding members. Better wear resistance is achieved with white pearlitic and martensitic irons which are used in chilled iron rolls and grinding balls. Alloyed white irons have even better wear resistance but are more expensive.

Nonferrous alloys for wear applications

Aluminum bronzes range from the soft and ductile alpha alloys, which are used for press guides and wear plates, to the very hard and brittle proprietary die alloys, which are used for tube bending dies and drawing die inserts. The softest alloys contain about 7 percent Al with some additions of Fe and Sn. Increasing the aluminum content increases the hardness. Aluminum bronzes are not self-lubricating and should only be used where adequate lubrication can be maintained. These alloys are recommended for applications involving high loads and moderate to low speeds. Increasing the hardness increases abrasion resistance but lowers conformability and embeddability of these alloys when used as sleeves for sliding bearings. Beryllium copper alloys are among the hardest and strongest of all copper alloys. Properly lubricated, they have better wear resistance than other copper alloys and many ferrous alloys. An alloy containing 98 percent Cu, 1.9 percent Be, 0.2 percent Co is usually specified for wear applications and has better load-carrying capacity than all other copper-base alloys. In addition, beryllium coppers exhibit excellent corrosion resistance in industrial and marine atmospheres. Wear properties of beryllium copper can be increased by oxidizing the surface of the alloy, by placing graphite in the surface and by using cast parts rather than machining them from wrought alloys.

Wrought cobalt-base wear-resistant alloys have excellent resistance to most types of wear in addition to good resistance to impact and thermal shock, good resistance to heat and oxidation, good resistance to corrosion and high hot hardness. The primary Co-base alloys for severe wear applications are:

1. Stellite 6B, contains 0.9–1.4% C, 28–32% Cr, 3% Ni, 1.5% Mo, 3.5–5.5% W, 3% Fe, 2% Mn, 2% Si, rem Co.
2. Stellite 6K, contains 1.4–2.2% C, 28–32% Cr, 3% Ni, 1.5% Mo, 3.5–5.5% W, 3% Fe, 2% Mn, 2% Si, rem Co.
3. Haynes 25, contains 0.05–0.10% C, 19–21% Cr, 9–11% Ni, 14–16% W, 3% Fe, 1–2% Mn, 1% Si, rem Co.

Stellite has better resistance to abrasive wear, while Haynes 25 has better resistance to adhesive wear. Wrought Co-base alloys are nearly identical in chemical composition with their hard-facing alloy counterparts but with small differences in boron, silicon or manganese levels. Another difference is the microstructure, which depends on the method of fabrication. Co-base hard-facing alloys will be discussed in Section 8.9.

Wear resistance of plastics

Wear-resistant, self-lubricating plastics are favorably competing with metals in many applications including bearings, cams and gears. In addition to ease of manufacture, these plastics have better lubricating properties and need less maintenance. Wear-resistant plastics are formulated with internal lubricating agents and are available in both unrein-forced and reinforced versions. A combination of lubricating additives is usually employed to achieve optimum wear resistance. For example, silicone and PTFE are usually added to thermoplastics to improve their performance at high speeds and pressures. Carbon and aramid fibers, which are usually added as mechanical reinforcement, are also known to improve wear resistance. Table 8.7 lists some commonly used wear-resistant plastics in order of decreasing resistance to wear when sliding against steel.

In spite of the advantages of plastics as wear-resistant materials, the following limitations should be kept in mind when selecting them for some applications. The first limitation arises when plastics rub against plastics. In such cases, wear is much more severe

Table 8.7 Wear properties of some lubricated plastics on steel

Plastic material	Reinforcing fibers	Wear factor[a]	Coefficient of friction[b]
Nylon 6/6-18% PTFE-2% Silicone		6	0.08
Nylon 6/6-13%PTFE-2%Silicone	30% Carbon	6	0.11
Polyester-13%PTFE-2%Silicone	30% Glass	12	0.12
Acetal-20% PTFE		13	0.13
Acetal-2%Silicone		27	0.12
Polyimide-10%PTFE	15% Carbon	28	0.12
Polypropylene-20%PTFE		33	0.11
Polyurethane-15%PTFE	30% Glass	35	0.25
Polystyrene-2%Silicone		37	0.08

[a]10^{10} in 3-min/ft lb h.
[b]Dynamic at 40 lb/in^2, 50 ft/min.

than in plastic-metal systems. The severity of wear can be reduced by adding PTFE or other internal lubricants and by similarly reinforcing the mating surfaces. Another limitation arises due to the sensitivity of wear resistance to seemingly small variations in temperature. For example, the wear rate of 15 percent PTFE, 30 percent glass-fiber nylon 6/6 at 200°C (c. 400°F) is about 40 times its wear rate at room temperature. A third limitation is the sensitivity of plastics to the surface roughness of the mating metallic surface. Finishes that are too rough or too smooth can result in excessive wear. Minimum wear of lubricated plastics is usually obtained with metallic surface roughness in the range 12 to 16 μm. Similarly the type of metal can strongly affect the results. For example, using an aluminum alloy instead of steel can dramatically increase the wear rate of plastics.

Wear resistance of ceramics

Ceramics can be used in a variety of applications where wear resistance is required. The wear behavior of ceramics is determined by the nature of the mating surfaces and the presence of surface films. In general, as grain size of the ceramic material increases, wear increases. Porosity has a negative effect on wear especially if present at grain boundaries. When metallic surfaces rub against ceramics, large differences in mechanical properties can cause plowing of the softer material, usually the metal. In addition, metals which form stable oxides adhere to the oxygen ions on the ceramic surface. Less active metals which do not form stable oxides produce lower friction and wear. The presence of surface films, such as water and oils, can affect adhesion and wear. For example, wear of partially stabilized zirconia increases in aqueous environments but decreases in fatty acids such as stearic acid. For engines and similar applications, silicon carbide against lubricated steel has lower friction and less scuffing than chilled cast iron which makes it suitable for engine valve train components and bearings.

8.9   SELECTION OF MATERIALS FOR PROTECTIVE COATINGS

Most engineering materials, when fabricated by the common production techniques, need one or more finishing operations before they are ready for service. For example, coatings can be applied to most metallic materials which do not possess inherently attractive colors and thus are lacking in sales appeal. In such cases, the effectiveness of a coating material can be measured by its color, brightness, reflectivity and opacity. Color retention is also important, as it indicates the length of time during which the coating will remain attractive under service conditions. Coatings can also be applied to provide:

1. Protection against corrosion or oxidation.
2. Protection against abrasion and wear.
3. Electrical and thermal conductivity or insulation.

Regardless of its function, a coating must adhere to the base material. Both surface texture and cleanliness can affect the degree of adhesion.

Protection against corrosion can be achieved in two ways:

1. Isolation of the surface from the environment.
2. Electrochemical action.

Isolation of the surface is usually performed by nonmetallic coatings and in such cases the thickness, soundness and strength of the coating will control its effectiveness as an isolator. Nonmetallic coatings can either be inorganic, as in the case of vitreous enamels, or organic, as in the case of varnishes and lacquers. Electrochemical action is achieved with metallic coatings. The coating metal can be more noble than the base metal and thus protects it, as in the case of tin coatings on steel. However, if pores or cracks are present in the coating, more severe attack could result than if the base metal had no coating. When the coating is anodic with respect to the base metal, the coating dissolves anodically while the base metal, which is the cathode in the galvanic cell, will not be attacked. Examples of such coatings include aluminum, cadmium and zinc which are anodic with respect to iron. The two ways of isolation and electrochemical action can be combined by first applying an anodic coating to the base metal followed by a nonmetallic finish. Metallic coatings can be applied by several methods including:

1. Electrodeposition as in the case of chromium and nickel plating of steel and gold plating of copper.
2. Hot dipping as in the case of coating steel with tin or zinc.
3. Cementation, where the metal to be coated is packed in a compound containing the coating material. Examples include sheradizing (zinc coatings), calorizing (aluminum coatings) and chromizing (chrome coating).
4. Cladding, where a layer of the protective material is diffusion-bonded to the base metal. An example is aluminum cladding on aluminum alloys or mild steel by hot or cold rolling.

5. Metal spraying which is achieved by impinging molten metal particles against the base metal surface. Examples include spraying zinc or aluminum on steel.
6. Conversion coatings differ from other coatings in that they are an integral part of the base metal and can be divided into phosphate, chromate and anodized coatings. Phosphate coatings on steels and zinc are mainly used as a base for painting as they do not provide much protection by themselves. Chromate coatings are more protective than phosphate coatings and are used on aluminum, magnesium or zinc parts for corrosion protection. Anodizing involves forming a thick protective oxide layer on the base metal and is mainly used with aluminum and magnesium parts.

   Organic coatings depend mainly on their chemical inertness and impermeability in providing protection against corrosion. An organic coating is made up of two principal components: a vehicle and a pigment. The vehicle contains the film forming ingredients that dry to form the solid film. It also acts as a carrier for the pigment. Vehicles can be either oil or resin, but oils have limited industrial uses. Nearly all polymers can be used as film formers and frequently two or more kinds are combined to give the required properties. The properties of polymers as coatings are usually similar to those of the bulk materials. Pigments, which may or may not be present, give the required color, opacity and flow characteristics, and can also contribute to the protection against corrosion of the base metal and against the destructive action of ultraviolet light on the polymeric vehicle. Varnishes consist of a drying oil, a resin and a solvent thinner. On application, the thinner evaporates and the oil/resin mixture oxidizes and polymerizes to form a clear dry film. Enamels consist of dispersions of pigment in a varnish or a resin vehicle. Varnishes and enamels are hard and tough and resist attack by most chemical agents. They are the most widely used organic coatings in industrial applications. Lacquers are noted primarily for their fast drying as they depend only on solvent evaporation. At least two coats of lacquer are required to give the protection afforded by one coat of varnish or enamel. Table 8.8 gives the relative properties of some commonly used organic coatings.
   Vitreous, or porcelain, enamels are inorganic coatings applied primarily to protect metal surfaces against corrosion. The composition should be adjusted so that the coefficient of thermal expansion closely matches that of the base metal. Two coats are usually needed. Ground-coats contain oxides that promote adhesion to the metal base, cobalt or nickel oxides for steel base and lead oxide for cast-iron base, and cover-coats improve the appearance and properties of the coating. Porcelain enamels can be applied either as a suspension of the finely milled material, frit, in water or by electrostatic spraying. After application, firing is carried out at temperatures which depend on the composition of the frit. For example, household refrigerators are fired at 790 to 805°C (1450 to 1480°F) for 2.5 to 4 min. The composition of porcelain enamels varies widely depending on the metal base and the application. Table 8.9 gives typical compositions of some porcelain enamels for steel and cast iron.
   Hard facing coatings are normally used for protection against wear. These coats may be applied to new parts made of soft materials to improve their resistance to wear, or to worn parts to restore them to serviceable condition. The selection of hard facing alloys for a given application is guided primarily by wear and cost considerations. However, other factors, such as impact resistance, corrosion and oxidation resistance, and thermal requirements should also be considered. In general, the impact resistance of hard facing alloys decreases as the carbide content increases. As a result, a compromise has to be made in applications where a combination of impact and abrasion resistance is required. Most

Table 8.8 Rating of organic coatings

	Cost	Abrasion resistance	Flexibility	Adhesion	Resistance to atmosphere (salt spray)	Exterior durability	Color retention	Resistance to chemicals (general)	Maximum service temperature rating
Alkyd	3	2	3	3	1	3	1	1	1
Amine-alkyd	3	3	2	3	1	3	1	1	2
Acrylic	2	2	3	2	3	3	3	1	1
Cellulose (butyrate)	1	2	3	2	3	2	3	1	2
Epoxy	1	3	3	3	3	1	1	3	2
Epoxy ester	2	3	1	3	3	2	1	1	2
Fluorocarbon	0.5	1	1	2	3	3	1	3	2
Phenolic	2	3	1	3	3	3	0	2	2
Polyamide	2	3	1	2	1	0	2	1	2
Plastisol	3	3	3	2	3	2	1	3	1
Polyester (oil free)	2	2	2	3	3	2	2	1	1
Polyvinyl fluoride (PVF)	0.5	3	3	2	3	3	2	3	1
Polyvinylidene fluoride (PVF2)	0.5	3	3	2	3	3	2	3	1
Silicone	1	2	1	1	3	3	3	1	3
Silicone alkyd	1	2	1	2	2	3	2	2	3
Silicone polyester	1	2	2	2	3	3	2	2	3
Silicone acrylic	1	2	1	2	2	3	3	2	3
Vinyl	2	2	3	1	3	3	2	1	1
Vinyl alkyd	2	2	2	2	2	1	2	1	1
Polyvinyl chloride (PVC)	1	3	3	3	3	2	1	3	1
Neoprene (rubber)	3	3	3	2	3	3	1	1	1
Urethane	0.5	3	3	3	3	3	1	1	2

Properties: 3 = Excellent, 2 = Very good, 1 = Fair, 0 = Poor.
Cost: 3 = Cheapest, 2 = Moderate price, 1 = Expensive, 0.5 = Very expensive.

Source: Farag, M. M.: *Materials and Process Selection in Engineering*, Applied Science Publishers, London 1979.

Table 8.9 Acid-resistant porcelain enamels for steel and cast iron

Constituent	Enamel for steel wt%		Enamel for cast iron wt%	
	ground coat	cover coat	ground coat	cover coat
SiO_2	56.44	41.55	77.7	37.0
B_2O_3	14.90	12.85	6.8	4.9
Na_2O	16.59	7.18	4.3	16.8
K_2O	0.51	7.96	–	1.7
Li_2O	0.72	0.59	–	–
CaO	3.06	–	–	2.0
ZnO	–	1.13	–	5.9
Al_2O_3	0.27	–	7.2	1.9
TiO_2	3.10	21.30	–	7.9
CuO	0.39	–	–	–
MnO_2	1.12	–	–	–
NiO	0.03	–	–	–
Co_3O_4	1.24	–	–	–
P_2O_5	–	3.03	–	–
F_2	1.63	4.41	–	–
PbO	–	–	4.0	8.8
Sb_2O_3	–	–	–	13.1

hard facing alloys are marketed as proprietary materials and are classified as low-alloy steels (group 1), high-alloy ferrous materials (groups 2 and 3), nickel-base and cobalt-base alloys (group 4), and carbides (group 5). Generally, both wear resistance and cost increase as the group number increases.

The alloys in group 1 contain up to 12 percent $Cr + Mo + Mn$ and have the greatest shock resistance of all hard facing alloys, except austenitic manganese steels. They are less expensive than other hard facing alloys and are extensively used where machinability is necessary and only moderate improvement over the wear properties of the base metal is required. Alloys in 2A and 2B contain up to 25 percent $Cr + Mo$ and are more wear resistant, less shock resistant, and more expensive than group 1 alloys. Alloys 2C and 2D contain up to 37 percent $Mn + Ni + Cr$ and are highly shock resistant, but have limited wear resistance unless subjected to work hardening. Group 3 alloys contain up to 50 percent $Cr + Mo + Co + Ni$. Their structure contains massive hypereutectic alloy carbides that improves wear resistance and gives them some degree of corrosion and heat resistance. Group 3 alloys are more expensive than alloys of group 1 and 2.

The nonferrous Ni and Co-base alloys of group 4A contain 50–100 percent $Co + Cr + W$ and are the most versatile hard facing alloys. They resist heat, abrasion, corrosion, impact, galling, oxidation, thermal shock, erosion and metal-to-metal wear. Some of these alloys retain useful hardness up to 825°C (1500°F) and resist oxidation up to 1100°C (2000°F). Alloys of group 4B and 4C contain 50–100 percent $Ni + Cr + Co + B$ and are the most effective for service involving both corrosion and wear. They retain useful hardness up to about 650°C (1200°F) and resist oxidation up to 875°C (1600°F). Group 5 materials provide maximum abrasion resistance under service conditions involving low or moderate impact. They are made of 75–96 percent carbides cemented by a metal matrix. Either WC or WC + TiC + TaC are used as the carbide phase, while Fe, Ni or Co-base alloys are used as the matrix material. Hard facing materials are usually available as bare cast or tubular rod, covered solid or tubular electrodes, solid wire and powder. Hard facing is usually applied by welding, plasma spraying or flame plating. Welding processes are usually used for applications requiring dense, relatively thick coatings with high bond strengths between the hard facing and the base metal. Thermal spraying processes, on the other hand, are preferred for applications requiring thin, hard coatings applied with minimal thermal distortion of the base metal.

8.10   REVIEW QUESTIONS AND PROBLEMS

8.1   Arrange the following materials in the order of increasing (a) density; (b) tensile strength; and (c) specific tensile strength: AISI 1045, AISI 4340, 316 stainless steel, 6061 T6 aluminum alloy, Ti-6 Al-4 V alloy, AZ31B H24 magnesium alloy, epoxy-70% glass fibers, and epoxy-62% aramid fibers.

8.2   Distinguish between the stiffness of a component and the modulus of elasticity of the material used in making that component.

8.3   How does ductility of the material affect the performance of a component in service?

8.4   How does surface hardness affect the performance of a component in service?

8.5   Why are stainless steels corrosion resistant? Explain the phenomena of passivation and sensitization.

8.6   Why are austenitic stainless steels more widely used than other types?

8.7   Compare the uses of titanium and aluminum alloys as supersonic aircraft wing-tip materials.

8.8   What are the main requirements for materials used for gas turbine blades?

8.9   What are the main requirements for materials used for the sleeves of lubricated journal bearings?

8.10   What are the main requirements for materials used for making motor car bodies?

8.11   What would be the advantages and disadvantages of substituting fiber reinforced plastics for steel in making motor car bumpers?

8.12   What are the main material requirements for knife blades? What are the materials that are normally used for the manufacture of such blades? Can ceramics be used in this application?

8.13   What are the differences between galvanizing and tinning of steel parts? Compare the merits of using each of these methods for: (a) food cans; (b) outdoor fencing.

8.14   What are the differences between organic coatings and vitreous enamels? Give examples of the uses of each type of coating.

BIBLIOGRAPHY AND FURTHER READING

Billmeyer, F. W., *Textbook of Polymer Science*, John Wiley & Sons, Sussex, 1984.
Boyer, H. E. and Gall, T. L., *Metals Handbook, Desk Edition*, ASM, Ohio, 1985.
Bracke, P. *et al.*, *Inorganic Fibers and Composite Materials*, Pergamon Press, London, 1984.
Crane, F. A. A. and Charles, *Selection and Use of Engineering Materials*, Butterworths, London, 1984.
Crawford, R. J., *Plastics Engineering*, Pergamon Press, London, 1987.
Farag, M. M., *Materials and Process Selection in Engineering*, Applied Science, London, 1979.
Flinn, R. A. and Trojan, P. K., *Engineering Materials and their Applications*, 2nd ed., Houghton Mifflin Co., Boston, 1981.
Fontana, M. G., *Corrosion Engineering*, 3rd ed., McGraw-Hill, New York, 1986.
Hanley, D. P., *Selection of Engineering Materials*, Van Nostrand Reinhold Co., New York, 1980.
Harris, B., *Engineering Composite Materials*, Institute of Metals, London, 1986.
Keyser, C. A., *Materials Science in Engineering*, 4th ed., Charles Merrill, Columbus, 1986.

Khobaib, M. and Krutenant, R. C., *High-Temperature Coatings*, The Metals Soc., New York 1986.
Metals Data Sourcebook, Institute of Metals, London, 1987.
Sims, C. T., Stoloff, N. and Hagel, W. C., *Superalloys*, 2nd ed., John Wiley & Sons, New York, 1987.
Thornton, P. A. and Colangelo, V. J., *Fundamentals of Engineering Materials*, Prentice Hall, New Jersey, 1985.
Yam, M. G. *et al.*, *Mechanical Behavior of Materials V*, Pergamon Press, London, 1988.

PART II

DESIGN AND MANUFACTURE OF ENGINEERING COMPONENTS

The Successful Designer

The designer bent across his board,
Wonderful things in his head were stored,
And he said as he rubbed his throbbing bean,
'How can I make this hard to machine?'

'If this part here were only straight,
I'm sure the thing would work first rate,
But 't would be so easy to turn and bore,
It would never make the machinests sore.

'I'd better put in a right angle there,
Then watch those babies tear their hair,
Now I'll put the holes that hold the cap
'Way down here where they're hard to tap.

'Now this piece won't work, I'll bet a buck,
For it can't be held in a shoe or chuck;
It can't be drilled or it can't be ground,
In fact the design is exceedingly sound.'

He looked again and cried, 'At last—
Success is mine, it can't even be cast!'

Author unknown

Chapter 9

Elements of Engineering Design

9.1 INTRODUCTION

Engineering design can be defined as the creation of a product which satisfies a certain need. A good design should result in a product that performs its function efficiently and economically within the imposed constraints. Cost, reliability, safety, level of performance, legal requirements, sociological considerations, pollution and energy consumption are among the major constraints.

Although the designer is not expected to be an expert on all the different factors that influence the design, he, or she, should be aware of their importance and should be able to communicate effectively with various specialists in the different departments of the company or organization. The different departments usually include customer service, marketing and sales; legal and patents; safety, codes and regulations; research and development (R&D); and, most importantly, manufacturing and fabrication. The designer is often required to make compromises to satisfy the conflicting requirements of the different departments.

Design work in engineering is usually performed on three different levels:

1. Development of existing products or designs by introducing minor modifications in size, shape or materials to overcome difficulties in production or performance. This type of work represents a large proportion of the design effort in industry.
2. Adaptation of an existing product or design to operate in a new environment or to perform a different function. In some cases, the new design may be widely different from the starting one.
3. Creation of a totally new design that has no precedent. This type of work is most demanding in experience and creativity of the designer and is not performed as often as the other types of design work. It often requires the solution of problems which have not been encountered before and could require the addition of new resources to the organization. The introduction of a totally new product usually involves considerable innovation and a large amount of experimental and prototype work.

9.2 FACTORS INFLUENCING DESIGN

Design is a multifaceted activity which has to take into consideration a large number of diverse factors. These factors and their interrelationship depend on the nature of the design problem but can be generally grouped in the following categories:

1. Factors related to product specifications, such as capacity, size, weight, expected service life, safety, reliability, maintenance, human factors, ease of operation, ease of repair, frequency of failure, initial cost, operating costs, styling, noise level, pollution, intended service environment and possibility of use after retirement.
2. Factors related to design specifications, such as design codes, patents, complexity, number of parts in the system, operating loads, flexibility, lubrication, thermal considerations, electrical considerations and expected life.
3. Material-related factors, such as strength, ductility, toughness, stiffness, density, corrosion resistance, wear resistance, friction coefficient, cost, available stock size, availability, delivery time, melting point, thermal and electrical conductivity, processability and possibility of recycling.
4. Manufacturing related factors such as available fabrication process, accuracy, surface finish, shape, size, required quantity, delivery time, cost and required quality.

This is not an exhaustive list and not all of the above factors are applicable or are of equal importance for all design situations. Discussions in Chapters 9 to 14 are mainly concerned with the effect of the above factors on design.

9.3 MAJOR PHASES OF DESIGN

Engineering design is usually an iterative process which involves a series of decision-making steps where each decision establishes the framework for the next one. There is no single universally recognized sequence of steps that leads to a workable design, as these depend on the nature of the problem being solved as well as the size and structure of the organization. Generally, however, a design usually passes through most of the phases which are shown in Fig. 9.1 and described in the following paragraphs:

1. Identification of the problem and evaluating the need in order to define the objective of the design represent the first phase of design in most cases. The major constraints such as cost, safety and level of performance and the overall specifications are also defined at this stage. Effective communication with other departments in the organization such as marketing, legal, R&D and manufacturing is essential at this stage. Unavailable information is identified at this stage and strategy for obtaining it is outlined. This may call for a search through company files, research department, published literature or patent office files. If vital information is still unavailable, a research project may have to be initiated either in-house or in a research institution.
2. Functional requirements and operational limitations are directly related to the required characteristics of the product and are specified as a result of the activities of phase 1 above. It is not always possible to assign quantitative values to these product characteristics. Nevertheless, in one way or another, they may be related as precisely as possible to measurable quantities which will allow future evaluation of the product performance. This is sometimes called the conceptual design stage.
3. System definition, concept formulation and preliminary layout are usually completed, in this order, before evaluating the operating loads and determining the form of the different components or structural members. Allowances must be made for uncertainties of loading and approximations in calculations. The consequences of component failure must also be considered at this stage. Whether the component can be easily and cheaply replaced or whether large costs will be incurred, will significantly influence the

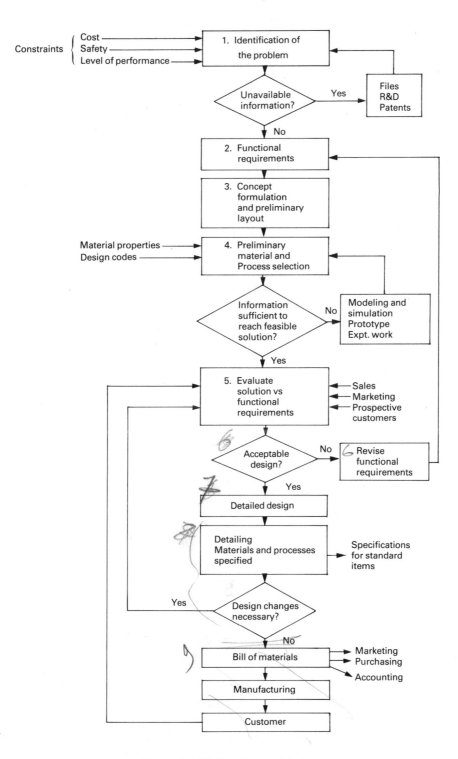

Figure 9.1 Major phases of design.

design. Design criteria like unlimited-life designs, safe-life designs, fail-safe designs and damage-tolerant designs will be discussed in Chapter 11.

4. Consulting design codes and collecting information on material properties will allow the designer to perform preliminary materials selection, preliminary design calculations and rough estimation of manufacturing requirements. Preliminary design begins by expanding the conceptual design into a detailed structure of subsystems and sub-subsystems. Some of these elements are familiar to the designer and can be easily designed, e.g. machine frames and standard power supplies. Other elements may be new and their design may require careful analysis. If the available information is not sufficient or is unreliable, models, prototypes, or experimental work may be necessary before arriving at an acceptable design for critical components. A factor of safety is usually employed to account for such uncertainties. In many cases several solutions to the design problem can be proposed at this stage.

5. The evaluation phase involves a comparison of the expected performance of the design with the performance requirements established in phase 2 above. Evaluation of the different solutions and selection of the optimum alternative can be performed using decision making techniques, modeling techniques, experimental work and/or prototypes. Decision making techniques will be described in Chapter 10. Computers are playing an increasingly important role in solving design problems, as will be discussed in Chapter 14. Decision making at this stage often involves compromises between what could be required in an ideal situation and what can be achieved in reality. Several optimization techniques have been developed to help in the selection of the best values of important design parameters, as will be discussed in Chapter 10. Having arrived at a compromise solution, it is often necessary to revise the design and to make more precise design calculations as well as to specify materials in more detail. Iteration steps of this type are easier to perform if computer techniques are used in design and materials selection.

6. In some cases, it is not possible to arrive at a design that fulfills all the requirements and complies with all the limitations established in phase 2 above. This means that these requirements and limitations have to be reconsidered and phases 3 to 5 repeated until an acceptable design is arrived at.

7. Having arrived at a final design, the project then enters the detailed design stage where it is converted into a detailed and finished form suitable for use in manufacturing. The preliminary design layout, any available detail drawings, models and prototypes and access to the developer of the preliminary design usually form the basis of the detailed design. As the detailed design is the final stage, it should be so complete as to ensure that the different components will perform satisfactorily and to allow their fabrication and assembly without further instructions. The increasing use of computers in industry has revolutionized this design phase. All the laborious calculations of strength, stiffness, size, weight, compatibility of assembled parts, etc., can now be performed faster and more accurately with the aid of computers, as will be discussed in Chapter 14.

8. The next step in the detailed design phase is detailing, which involves the creation of detailed drawings for every part. All the information that is necessary unambiguously to define the part should be recorded in the detail drawing. The drawing should include the different views and cross-sections that are necessary to define the shape as well as dimensions, tolerances and surface finish. The material of the part should also be selected and specified by reference to standard codes. The temper condition of the

stock material, the necessary heat treatment and the expected hardness may also be specified for quality control purposes. Standard parts and purchased items like bolts, nuts, washers, ball bearings, electric motors and control panels need only to be specified but not detailed. The vendors will supply the drawings if needed. In the course of detailing it may become clear that the performance of some parts of the layout could be improved if they are changed. In this case, communication should take place between the detail designer and the preliminary designer to agree on the proposed changes. The task of design modification as well as the cost and time needed to prepare detail drawings can be greatly reduced by using the automated drafting facility of computer aided design (CAD).

9. An important part of the detail design phase is the preparation of the bill of materials, sometimes called parts list. The bill of materials is a hierarchical listing of everything that goes into the final product including fasteners and purchased parts. Using computers and word processing makes the task of preparing and updating the bill of materials much easier. The bill of materials is used by a variety of departments including purchasing, marketing and accounting. When the detailed design is released for manufacturing, a working bill of materials should go with it. The manufacturing plan, production planning and assembly will all be based on the bill of materials. An appropriate version of the bill of materials is also shipped with the finished product for guidance in operating and maintenance. Close interaction between design, manufacturing and materials engineers is important at this stage. CAD techniques have made it possible to communicate such designs to the manufacturing department and allow direct link to computer aided manufacturing (CAM) facilities, as will be discussed in Chapter 16.

10. It must always be kept in mind that the purpose of the design is to satisfy the needs of a client. Communication with other departments of the organization, such as marketing and sales, or even with prospective customers, should take place during the different stages of design to ensure that the different points of view are considered and that important decisions are acceptable to interested parties. The communication is usually by oral presentation as well as by written design report. Detailed engineering drawings, computer programs and working models are sometimes part of the communication process.

11. The relationship between the designer and the product does not usually end at the manufacturing or even delivery stages. The manufacturing engineer may ask the detail designer for a change in some parts to make fabrication easier or cheaper. Finally, when the product gets into use, the reaction of the consumer and the performance of the product in service are of concern to the designer as the feedback represents an important source of information for future design modifications.

9.4   DESIGN CODES AND STANDARDS

A design code is a set of specifications for the analysis, design, manufacture and construction of a structure or a component. Codes of practice are set by professional groups and government bodies in order to achieve a specified degree of safety, efficiency, performance or quality as well as a common standard of good design practice. Codes serve to disseminate proved data and research results to the average designer who is not

Table 9.1 Sample of specification and code writing groups

Aluminum Association (AA), USA .
American Gear Manufacturing Association (AGMA)
American Iron and Steel Institute (AISI)
American National Standards Institute (ANSI)
American Petroleum Institute (API)
American Society of Mechanical Engineers (ASME)
American Society for Metals (ASM)
American Society for Testing and Materials (ASTM)
American Welding Society (AWS)
British Standards (BS)
Deutsches Institut für Normung (DIN), West Germany
International Standards Organization (ISO)
Japan Industrial Standards (JIS)
National Bureau of Standards (NBS), USA
National Physical Labs (NPL), UK
Society of Automotive Engineers (SAE), USA

expected to have the expertise to appreciate and examine critically all the specialized information associated with the part being designed.

A standard specification is a published document that describes the characteristics of a part, material, or process which is acceptable for a wide range of applications. Specifications could also cover complex units manufactured from standard materials and components. When it is cited by the purchaser and accepted by a supplier, it becomes part of the purchase agreement. The widespread use of standards has benefited companies by reducing the number of products, materials or components that need to be manufactured or held in stock. Specification of dimensions, shapes and sizes also helps in achieving interchangeability of components. The American Society for Testing and Materials (ASTM) issues very comprehensive and widely used standard specifications. Other engineering societies, associations and institutes whose members make, specify or purchase materials or products publish standard specifications to cover their areas of interest. Some of the important organizations and societies which publish standard specifications or design codes are listed in Table 9.1.

As a specification is a document that can be used to control procurement, it should contain both technical and commercial requirements. Specifications normally cover the following information:

1. Product classification, scope of application, size range, condition and processing details that could help either the supplier or the user.
2. Allowable ranges of chemical composition.
3. All physical and mechanical properties necessary to characterize the product are given in addition to the test methods used to determine these properties.
4. If applicable, other requirements such as special tolerances, surface preparation, loading instructions, packaging, etc., are included in the specifications.

Specifications differ greatly in quality and completeness and are periodically updated. The designer should always refer to the latest edition of the specifications he is using. Care should also be exercised when comparing different materials or products which are based on different specifications. When two different materials or product specifications are considered equivalent, this implies that they will be interchangeable in the given application and will perform in a known and similar manner. Generally, however, standards are not exchangeable on the international level and although some important work has been done by the International Standards Organization (ISO), their standards are not yet generally accepted. The development of international standards has been hampered by the variety of measurement systems used in different countries. With the multiplicity of standards, an extra burden is placed on designers and production engineers in a company which depends heavily on exports to other countries.

9.5 PROBABILISTIC DESIGN

Traditionally, design engineers have followed the deterministic approach in making their designs. Uncertainties in the applied load, variability in material properties and inaccuracies in manufacturing are accounted for by the factor of safety or derating factors, as will be discussed in Section 9.6. In the design of critical components, where reliability needs to be estimated, there is a growing trend to use a probabilistic approach in design. The statistical principles and the definition of statistical parameters that are needed for

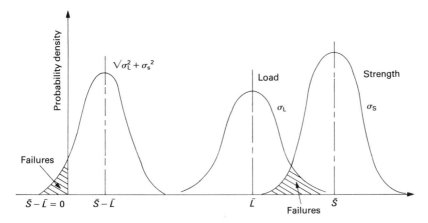

Figure 9.2 Effect of variations in load and strength on the failure of components.

probabilistic design are given in Appendix C. In this case, loads acting on the component and the strength of the component are assumed to vary according to a certain distribution, as illustrated in Fig. 9.2. Failure will occur when the load exceeds the strength, i.e. the area of overlap of the two distributions in Fig. 9.2. The figure shows that even though the average load ($\bar{L}$) is lower than the average strength ($\bar{S}$), the weakest components will fail if subjected to the highest loads. The probability of failure in this case can be estimated from the distribution curve which results from the subtraction of the load and the strength distributions according to (C9) and (C10), Appendix C, as shown in Fig. 9.2. The negative part of the curve ($\bar{S} - \bar{L}$) represents the possible failures. In order to control the magnitude of failures it is necessary that the mean of the ($\bar{S} - \bar{L}$) curve is an appropriate number of standard deviations above zero. Hence:

$$\bar{S} - \bar{L} = a\sqrt{(\sigma^2_{\text{L}} + \sigma^2_{\text{S}})} \tag{9.1}$$

where a is a number which defines the proportion of acceptable failures.

If the strength and load distribution curves are assumed to be normal, it follows that their difference will also be normally distributed. Fig. C.6, Appendix C, shows that the area bounded by $\mu \pm 2\sigma$ is 95.46 percent of the total area. According to (9.1), if a is taken as 2, the negative area under the $S - L$ curve will be $(1 - 0.9546)/2 = 0.0227$, i.e. the probability of failure is 2.27 percent. If a is taken as 3 the probability of failure is then 0.135 percent. It should be noted that in the above discussion the strength of the component is taken as the load that will cause it to fail. This is affected by the strength of the material but not equal to it. Components that are made of the same material may fail under different loads if they have different surface finishes or serve under different environments.

The above discussion shows that the leading end of the load distribution curve and the trailing end of the strength distribution curve play an important role in determining the value of a. In conventional applications of statistics, a distribution is considered a good fit if it agrees with the bulk of the results with little concern to the extreme values at the tails of the distribution. However, for factor of safety and reliability calculations, inaccuracies in this region can lead to significant errors. Added difficulties arise from the sensitivity of extreme values to changes in quality control and manufacturing techniques and form the tendency to find peaks at the extreme ranges of some real-life operating systems.

9.6   FACTOR OF SAFETY AND DERATING METHODS

The term factor of safety is applied to the factor used in designing a component to ensure that it will satisfactorily perform its intended function. The factor of safety is normally used to divide into the strength of the material to obtain the allowable stress. The definition of strength depends on the type of material and loading conditions. For example, under static loading the strength is defined as the ultimate tensile strength in the case of brittle materials and as the yield strength in the case of ductile materials. Under cyclic loading the endurance limit is used to represent the strength of the material. The main parameters that affect the value of the factor of safety, which is always greater than one, can be grouped into two main categories:

1. Uncertainties associated with material properties and manufacturing processes. These uncertainties result in the observed statistical variations in strength values.
2. Uncertainties in loading and service conditions.

Traditional ductile materials which are produced in large quantities generally show less property variations than brittle materials and advanced composites that are produced by small batch processes. In the latter materials small variations in fiber orientation or volume fraction can have considerable effect on properties, as discussed in Chapter 6. Manufacturing processes can also add to the variations in component behavior. For example, parts manufactured by processes such as casting, forging and cold forming are known to have variations in properties from point to point. Dimensional and geometrical variations resulting from manufacturing process tolerances can also affect the load-carrying capacity of components. Improved quality control techniques should result in more uniform material properties and more consistent component behavior in service and therefore lower values for the factor of safety. The nominal strength, S, and allowable strength, S_a values are related by the formula:

$$n_s = S/S_a \qquad (9.2)$$

where ns is the factor of safety.

In very simple components, S_a in the above equation can be viewed as the minimum allowable strength of the material. However, there is some danger involved in this use especially in the cases where the load-carrying capacity of a component is not directly related to the strength of the material used in making it. Examples include long compression members, which could fail as a result of buckling, and components of complex shapes, which could fail as a result of stress concentration. The earlier point was illustrated in the example of Section 8.2, where buckling was shown to be the expected cause of failure in stronger materials and in those with a lower elastic constant. Under these conditions it is better to consider S_a as the load carrying capacity which is a function of both material properties and geometrical characteristics of the component.

In assessing the uncertainties in loading, two types of service conditions have to be considered:

1. Normal working conditions which the component has to endure during its intended service life.
2. Limit working conditions, such as overloading, which the component is only intended to endure on exceptional occasions and which if repeated frequently could cause premature failure of the component.

In a mechanically loaded component, the stress levels corresponding to both normal and limit working conditions can be determined from a duty cycle. The normal duty cycle for an airframe, for example, includes towing and ground handling, engine run, take-off, climb, normal gust loadings at different altitudes, kinetic and solar heating, descent and normal landing. Limit conditions can be encountered in abnormally high gust loadings or emergency landings. Analyses of the different loading conditions in the duty cycle lead to determination of the maximum load that will act on the component. This maximum load can be used to determine the maximum stress, or damaging stress, which if exceeded would render the component unfit for service before the end of its normal expected life. The factor of safety, n_1, in this case can be taken as:

$$n_1 = L/L_a \tag{9.3}$$

where L is the maximum load and L_a is allowable load.

The total or overall factor of safety, n, which combines the uncertainties in the load-carrying capacity of the component and the external loading conditions can be calculated as:

$$n = n_s * n_1 \tag{9.4}$$

If more than one load is acting on the component the uncertainties associated with each of them can be introduced in the above equation which then becomes:

$$n = n_s * n_1 * n_2 * n_3 \tag{9.5}$$

where n_1, n_2 and n_3 are the factors of safety given to first, second and third loads respectively.

Factors of safety ranging from 1.1 to 20 have been used but common values range from 1.5 to 10.

In some applications a designer is required to follow certain established codes when making his design, e.g. pressure vessels, piping systems, etc. Under these conditions the factors of safety used by the writers of the codes may not be specifically stated but an allowable working stress is given instead.

Derating factors are numbers less than unity and are used to reduce material strength values to take into account manufacturing imperfections and the expected severity of service conditions. When a component is subjected to fatigue loading, for example, several derating factors can be used to account for imperfections in surface finish, d_1; size of the component in relation to sample size, d_2; stress concentration, d_3; operating temperature, d_4; possibility of impact loading, d_5; presence of corrosive environment, d_6; desired reliability of the component, d_7; etc., as will be discussed in Chapter 11. In this case, the working stress, S_w, is related to the endurance limit of the material S by the formula:

$$S_w = S \, d_1 \, d_2 \, d_3 \, d_4 \, d_5 \, d_6 \, d_7 \qquad (9.6)$$

Factors of safety and derating factors can be combined as given in the following relationship:

$$S_w = S \, d_1 \, d_2 \, d_3 / n_1 \, n_2 \, n_3 \qquad (9.7)$$

where $d_1 \, d_2 \, d_3$ are derating factors and $n_1 \, n_2 \, n_3$ are factors of safety.

Statistical consideration of factor of safety

In real life, the actual strength of the material in a component could vary from one point to another and from one component to another due to variations in manufacturing conditions. In addition, it is usually difficult precisely to predict the external loads acting on the component under actual service conditions. To account for these variations and uncertainties both the load-carrying capacity S and the externally applied load L have to be expressed in statistical terms. As both S and L depend upon many independent factors, it would be reasonable to assume that they can be described by normal distribution curves. Consider that the load carrying capacity of the population of components has an average of $\bar{S}$ and a standard deviation s_s while the externally applied load has an average of $\bar{L}$ and a standard deviation s_1. The relationship between the two distribution curves is important in determining the factor of safety and reliability of a given design. Figure 9.2 shows that failure takes place in all the components that fall in the area of overlap of the two curves, i.e. when the load-carrying capacity is less than the external load. This is described by the negative part of the $(\bar{S} - \bar{L})$ curve. Transforming the distribution $(\bar{S} - \bar{L})$ to the standard normal deviate according to (C.16), Appendix C, $z = (x - \mu)/\sigma$, the following equation is obtained:

$$z = \{(S - L) - (\bar{S} - \bar{L})\} / \sqrt{\{(s_s)^2 + (s_1)^2\}} \qquad (9.8)$$

From Fig. 9.2, the value of z at which failure occurs is:

$$z = 0 - (\bar{S} - \bar{L}) / \sqrt{\{(s_s)^2 + (s_1)^2\}}$$

$$z = -(\bar{S} - \bar{L}) / \sqrt{\{(s_s)^2 + (s_1)^2\}} \qquad (9.10)$$

For a given reliability, or allowable probability of failure, the value of z can be determined from the cumulative distribution function for the standard normal distribution (SND), Table C.3, Appendix C. Table 9.2 gives some selected values of z that will result in different values of probabilities of failure.

Knowing s_s, s_1 and the expected $\bar{L}$, the value of $\bar{S}$ can be determined for a given reliability level. As defined earlier in the section, the factor of safety in the present case is simply $\bar{L}/\bar{S}$.

Table 9.2 Values of z to give various levels of reliability and probability of failure

Reliability	Probability of failure	z
0.9	10^{-1}	-1.28
0.99	10^{-2}	-2.33
0.999	10^{-3}	-3.09
0.9999	10^{-4}	-3.72
0.99999	10^{-5}	-4.26
0.999999	10^{-6}	-4.75

To illustrate the use of the above concepts in design, consider the case of a structural element which is made of epoxy-70 percent glass fibers with an average tensile strength of 2100 MPa. The element is subjected to a static tensile stress of an average value of 1600 MPa. If the variations in material quality and load cause the strength and stress to vary according to normal distributions with standard deviations of $\sigma_1 = 400$ and $\sigma_2 = 300$ respectively, what is the probability of failure of the structural element? From Fig. 9.2, $\bar{S} - \bar{L} = 2100 - 1600 = 500$ MPa, $\sigma_3 =$ standard deviation of the curve $(\bar{S} - \bar{L})$ $= \sqrt{\{(400)^2 + (300)^2\}} = 500$

From (9.10), $z = -500/500 = -1$.

From Table C3, Appendix C, the area a is 0.1587. This represents the probability of failure of the structural element and is too high for practical applications. One solution to reduce the probability of failure is to impose better quality measures on the production of the material and thus reduce the standard deviation of the strength. Another solution is to increase the cross-sectional area of the element in order to reduce the stress. For example, if the standard deviation of the strength is reduced to $\sigma_1 = 200$, $\sigma_3 = \sqrt{\{(200)^2 + (300)^2\}} = 360$, $z = -500/360 = -1.4$, which gives a more acceptable probability of failure value of 0.08. Alternatively, if the average stress is reduced to 1400 MPa, $\bar{S} - \bar{L} = 700$ MPa, $z = -700/500 = -1.4$, with a similar probability of failure as the first solution.

As the above discussion shows, statistical analysis allows the generation of data on the probability of failure and reliability, which is not possible when a deterministic safety factor is used. However, difficulties arise because the accuracy of the calculations depends on the shape of the tail ends of the load and strength distributions which cannot usually be accurately determined. In this case, the theory must be applied with care and with awareness of its limitations.

9.7   MODELING AND SIMULATION IN DESIGN

Modeling is the representation of a system or part of a system in physical or mathematical form that aids in the analysis of the behavior of the system. Models can be either dynamic or static depending on whether time-dependent variables are considered or not. Models can also be deterministic or probabilistic depending on the certainty of the events they describe. Models can generally be classified as:

1. Iconic models.
2. Analog models.
3. Symbolic models.

Iconic models are representations that look like the reality and may be two-dimensional or three-dimensional. Iconic models can be miniatures, as in road maps and model trains; enlargements, as in three-dimensional models of polymer molecules and photographs of material microstructures; or duplicates made to the same scale, as in the case of engineering working drawings and mockups of a full-size automobile. Such models are useful in visualizing parts, visually checking relationships between different components in the system and in getting customer reaction to new designs. However, iconic models have limited use in optimizing the performance of the product under development.

Analog models are systems that follow the same principles and behave like the real system, but may not look like it. Such models can be used to determine numerical results and to check interactions quantitatively. For example, an electrical circuit made up of resistances connected in series and in parallel can be used to study heat transfer and fluid flow in complex systems. Another example is stress–strain diagrams which can be considered as analog models for material behavior.

Symbolic models are abbreviated abstractions of the relevant quantifiable parts of the system. A common example is the mathematical equation which describes the relationship between the input and output parameters of the system. Symbolic models offer maximum generality in solving problems and yield quantitative results. Symbolic models can be based on established laws of nature, theoretical models, or based on best fits to experimental data, as in empirical models.

An important step in the development of a model is validation to determine the accuracy with which it represents the system and the range of its applicability. This can be done by comparing the theoretical results with the actual performance of the system. For example, the rule of mixtures given by (6.10) is a symbolic model which describes the behavior of materials reinforced with continuous fibers. The limit of applicability of (6.10) is that the volume fraction of the fibers should be greater than the critical volume fraction.

Simulation is the manipulation of the model and subjecting it to various inputs or working conditions to demonstrate the performance of the system in different environments. Simulation may involve subjecting physical models to the actual environment, or it may involve subjecting symbolic models to mathematical disturbance functions that represent the expected service conditions. Physical scale models are usually less than full size and are normally tested in a pilot plant which reproduces the important aspects of the process. It is important in such cases to establish the conditions under which similitude prevails between the model and the full-scale product. The different types of similitude which relate certain aspects of the model to the full scale product include:

1. Geometric similarity, which relates the shapes and dimensions.
2. Static similarity, which relates stresses and deflections.
3. Kinematic similarity, which relates time and distance or velocity and position.
4. Dynamic similarity, which relates forces and velocity.
5. Thermal similarity, which relates temperature profiles.
6. Chemical similarity, which relates chemical reactions at different locations.

Numerical methods, which easily lend themselves to iterative procedures, have established computer modeling and simulation techniques among the most powerful tools in problem solving and decision making. Computer-generated geometric models play a vital role in computer aided design (CAD) and computer aided manufacture (CAM) and will be discussed in Section 14.3. Another important computer-based method of modeling is finite element analysis which will be discussed in Section 14.5.

9.8   GENERAL CONSIDERATIONS IN MECHANICAL DESIGN

The design of mechanical elements and systems is influenced by a variety of factors which can be classified into the following broad categories:

1. Functional requirements.
2. Manufacturing considerations.
3. Safety and environmental requirements.
4. Marketing and aesthetic parameters.
5. Economic factors.

Functional requirements represent the minimum level of performance that any acceptable design must have. With increasing competition between different manufacturers and with more emphasis on product liability, designs are required to meet the additional requirements of reliability, safety, marketability and cost. The different aspects of reliability and safety will be discussed in Chapter 13. Manufacturing considerations affect the feasibility of making the product at a competitive price and will be discussed in Chapter 12, while economic considerations will be discussed in Part III.

Service life represents an important design parameter as it affects both reliability and economics of the product. Service life of a component can be estimated according to safe-life or fail-safe criteria. The safe-life criterion can be applied to components in which undetected crack or other defects could lead to catastrophic structural failure, and a life limitation must therefore be imposed on their use. The fail-safe criterion can be applied to structures in which there is sufficient tolerance of a failure to permit continuous service until discovered by routine inspection procedure, or by obvious functional deficiencies. The majority of engineering components can be designed according to the fail-safe criterion. Even a critical component can be designed according to the fail-safe criterion if failure is detectable by the maintenance program, which must define both the timing and the methods of inspection to be applied. Redistribution of the load into sufficiently robust adjacent components if failure occurs is an added safety precaution. If the use of a safe-life component is unavoidable, its safe service life must be estimated by testing, and its replacement life calculated by applying an appropriate factor of safety.

9.9   REVIEW QUESTIONS AND PROBLEMS

9.1   A manufacturer of sports equipment is considering the possibility of using fiber reinforced plastics in making racing bicycle frames. It is expected that fatigue failures of the joints could be a problem in this case. Describe a design and testing program that can solve this problem.

9.2   Two batches of steel components were heat treated in two different shops. The RC hardness results were as follows:
   Shop 1: 48, 51, 52, 49, 50, 50, 47, 50, 51 and 47.
   Shop 2: 50, 49, 47, 48, 50, 48, 49, 52, 51 and 48.
   Did treating the steel in different shops make a significant difference? [Answer: no]

9.3   Distinguish between the factor of safety and the derating factor. What are the main factors that affect the value of the factor of safety?

9.4   Why should caution be exercised when substituting a new material for an established one?

9.5   A structural member is made of steel of mean yield strength of 200 MPa (28.6 ksi) and a standard deviation on strength of 30 MPa (4.3 ksi). The applied stress has a mean value of 150 MPa (14.3 ksi) and a standard deviation of 5 MPa (700 lb/in^2).

(a) what is the probability of failure? [Answer: 4.95%]
(b) what factor of safety is required if the allowable failure rate is one percent? [Answer: 1.55]

BIBLIOGRAPHY AND FURTHER READING

Brichta, A. M. and Sharp, E. M., *From Project to Production*, Pergamon Press, London, 1970.
Dieter, G. E., *Engineering Design, a Materials and Processing Approach*, McGraw-Hill, New York, 1983.
Edel, D. H., ed., *Introduction to Creative Design*, Prentice Hall, New Jersey, 1967.
Hubka, V., *Principles of Engineering Design*, Butterworth Scientific, London, 1982.
Polak, P., *A Background to Engineering Design*, The Macmillan Press, London, 1976.
Ray, M. S., *Elements of Engineering Design*, Prentice Hall, New Jersey, 1985.
Shigley, J. E. and Mitchell, L. D., *Mechanical Engineering Design*, 4th ed., McGraw-Hill, New York, 1983.
Woodson, T. T., *Introduction to Engineering Design*, McGraw-Hill, New York, 1966.

Chapter 10

Decision Making

10.1 INTRODUCTION

Many engineering problems, especially in design and materials selection, have several possible solutions which create the need for making decisions or choosing between alternatives. In many cases a decision has to be made in the face of poorly defined problems or inadequate information. In such cases a problem statement can be used to define the problem as clearly as possible. The essential elements of the problem statement can be generally written as:

1. Definition of the need.
2. Determination of the required level of performance, allowable lead time and expected cost level.
3. Identification of goals, or objectives, to determine which ones to be achieved, rather than how to achieve them. These goals are sometimes considered as design specifications.
4. Identification of imposed constraints, limitations and possible trade-offs. Constraints could be legal or social, as in the case of pollution and noise levels, or could be imposed by the client to suit his own use.
5. Setting criteria for evaluating the possible solutions and screening out unsatisfactory alternatives. Such criteria are easier to apply when they provide a quantitative means of evaluation.

Decision making can be classified according to the certainty of the available information as:

1. Decision under certainty, where each action results in a known and well-defined outcome, i.e. probability = 1. In some applications uncertainty is suppressed to take advantage of the simplicity of arriving at these decisions; examples include assigning single-value estimates for strength, operating cost, service life, etc.
2. Decision under risk, where several parameters and the probabilities of their occurrence are considered. In some decision problems the probability values may be objectively known from service records, experimental work or previous experience.
3. Decision under uncertainty, where several parameters can influence the decision but probability of their occurrence is not known. Several principles have been introduced to solve decision problems under such conditions, as will be discussed later.
4. Decision under conflict, where the goals of decision making can only be described in terms of multiple objectives which can be in conflict.

The use of the above types of decision making in design will be illustrated using the decision matrix and decision tree models, Sections 10.2 and 10.3 respectively. Other widely used quantitative methods of decision making will be discussed in the different sections of this chapter.

10.2 DECISION MATRIX

The decision matrix model can be used to select one solution out of the mutually exclusive possible alternatives. The different strategies which can be controlled by the decision maker are usually taken as the rows of the decision matrix, while the objectives, or criteria for evaluating the different solutions, are taken as the columns. The matrix is made up of the different outcomes of each strategy-versus-objective combination. The decision can be made more easily if the various outcomes are first normalized to have a common base and then multiplied by a weighting factor to arrive at the value of each outcome. The summation of the different values of outcomes for a given strategy is called the overall satisfaction and represents its effectiveness in satisfying the different design criteria. This procedure is illustrated in the following example.

Consider a structural element of rectangular cross-section in the form of a simply-supported beam which is one meter long and 100 mm wide. The beam is subjected to a concentrated load of 20 kN which acts in the middle. The maximum deflection under maximum load should not exceed 10 mm. The objectives, or selection criteria, in this case are the weight, cost and fatigue strength of the beam. The relative importance, or weighting factors, of these criteria are: 0.2 for the weight, 0.5 for the cost and 0.3 for the fatigue strength. It is required to compare the different strategies of making the beam out of AISI 1020 steel, 6061 T6 aluminum or epoxy-70 percent glass fibers. The relevant properties of these materials are given in Table 10.1.

The first step in this case is to calculate the thickness of the beam which will give a maximum deflection of 10 mm. For a simply supported beam with a concentrated load in the middle, the maximum deflection is given by:

$$d = \frac{L\, l^3}{4Ewt^3} \tag{10.1}$$

where d = maximum deflection = 10 mm
     L = load = 20 kN
     l = length of beam = 1000 mm
     w = width of beam = 100 mm
     E = Young's modulus of the beam material
     t = thickness of the beam

Using the calculated thickness and the specific gravity, the mass of the beam is calculated for the different materials. The results are given in the decision matrix of Table 10.2.

The outcomes in Table 10.2 are difficult to compare as they are of different units and of different preference, i.e. it is preferable to have lower mass and lower cost but it is preferable to have higher strength. The outcomes are normalized by giving the best value in a criterion column 100 percent, e.g. minimum weight, minimum cost, or maximum strength, and relating the other values in proportion. Table 10.3 gives the normalized outcomes.

Table 10.1 Comparison of materials for simply-supported beam

	Young's modulus (GN/m^2)	Fatigue strength (MN/m^2)	Specific gravity	Cost $(\$/kg)$
Steel AISI 1020	207	180	7.86	1.05
Aluminum 6061 T6	68.9	63	2.70	6
Epoxy-70% glass	62.3	85	2.11	8.3

Table 10.2 Decision matrix for a structural member

Criteria / Strategy	Mass (kg)	Cost (\$)	Fatigue strength (MPa)
Steel AISI 1020	28.6	30	180
Aluminum 6061 T6	14.2	85	63
Epoxy-70% glass	11.45	95	85

Table 10.3 Normalized outcomes for a structural member

Criteria / Strategy	Normalized mass	Normalized cost	Normalized fatigue strength
Steel AISI 1020	40	100	100
Aluminum 6061 T6	80.6	35.3	35
Epoxy-70% glass	100	31.6	47.2

Table 10.4 Values of outcomes and overall satisfaction (1)

Criteria / Strategy	Normalized mass *0.2	Normalized cost *0.5	Normalized strength *0.3	Overall satisfaction
Steel AISI 1020	8	50	30	88
Aluminum 6061 T6	16.12	17.65	10.50	44.27
Epoxy-70% glass	20	15.8	14.16	49.96

The values of the outcomes are then calculated by multiplying the normalized outcomes by the weighting factors, as shown in Table 10.4. The values of outcomes for a given strategy are then summed for each strategy to give the overall satisfaction, as shown in Table 10.4. The results show that, on the basis of the allocated weighting factors, steel gives the highest overall satisfaction.

If the weighting factors are changed such that the mass of the structure is more important than the cost, e.g. weighting factors of 0.5 for weight, 0.2 for cost, and 0.3 for strength, epoxy + 70 percent glass fibers gives slightly better overall satisfaction than steel, as shown in Table 10.5.

The decisions in the above example can be considered to have been taken under certainty since the values involved were assumed to occur with probability = 1. If, however, the applied load or the fatigue resistance of the different structures are assumed

Table 10.5 Values of outcomes and overall satisfaction (2)

Criteria Strategy	Normalized mass *0.5	Normalized cost *0.2	Normalized strength *0.3	Overall satisfaction
Steel AISI 1020	20	20	30	70
Aluminum 6061 T6	40.3	7.06	10.50	56
Epoxy-70% glass	50	6.32	13.16	70.48

to vary according to a certain probability function, the decisions will then be taken under risk. Consider the load in the above example to vary as follows:

Concentrated load (kN)	16	18	20	22	24
Probability of occurrence	0.1	0.2	0.4	0.25	0.05

The design load can be taken as:

$$16(0.1) + 18(0.2) + 20(0.4) + 22(0.25) + 24(0.05) = 19.9 \, kN$$

The load of 19.9 kN can then be used for static design calculation as before.

Similar treatment can be applied to any of the variables involved in the design if probability values can be assigned to the occurrence of each state of the variable.

10.3   DECISION TREES

The decision matrix model discussed above is useful when the decision situations happen at the same time. There are cases, however, where decisions must be made in sequential fashion. Such sequential decisions are best represented by a decision tree model, similar to that shown in Fig. 10.1. A decision point in the decision tree is usually indicated by a triangle, and a branch emanating from a decision point represents an alternative that can be chosen at this point. A fork, or node, in the tree is represented by a circle and indicates a chance event which is outside the control of the decision maker. A branch emanating from a chance event represents the probabilistic outcome for a given alternative. Such branches are mutually exclusive and collectively exhaustive, i.e. the sum of their probabilities is 1.0. The value associated with a particular outcome (branch) is usually written within a square at the end of the branch.

As an illustration of the use of decision trees, consider the case of the structural element discussed in Section 10.2. The analysis showed that AISI 1020 steel is the most suitable material when the cost is important. The next step is to select the least expensive method of fabrication. Figure 10.1 shows the sequence of decisions associated with the three available methods of fabrication as well as the total cost of fabrication in each case. Selection of riveting as a method of fabrication is relatively simple as it involves no further decisions. On the other hand if welding is selected, another decision should be made to select the type of welding. If arc welding is selected a third decision should then be made to select the type of shield. The option of casting the part involves uncertainty as to whether a suitable pattern is available, probability of 0.3, or whether a new one has to be made,

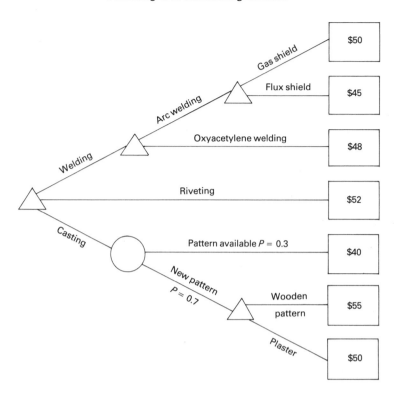

Figure 10.1 Decision tree for the manufacture of a structural element.

probability of 0.7. If a pattern is not available a decision should be made to decide on the material of the new pattern. Figure 10.1 shows how the cost of casting is calculated under these conditions. Starting at the end of the casting branches, the plaster pattern is selected as it is less expensive than the wooden pattern. The probable cost of fabrication by casting can then be calculated as:

$$\text{Cost of casting} = 0.3\,(40) + 0.7\,(50) = \$47$$

Comparison of the different methods of fabrication shows that the least expensive method of fabricating the structural element is by flux-shielded arc welding.

10.4   PLANNING AND SCHEDULING MODELS

As pointed out in Section 1.5, planning consists of identifying the major activities in a project and ordering them in the sequence in which they should be performed, while scheduling consists of putting the plan into a time frame. Normally, scheduling and planning of activities should be done together, which means establishing the timing and interdependence of the various activities during the planning stge. Several planning and scheduling models have been developed to help in organizing and coordinating the large number of activities involved in complex engineering projects. The Gantt chart, the critical path method (CPM) and the program evaluation and review technique (PERT) are among the widely used planning and scheduling models. The use of these models will be illustrated in the following simple example.

Consider the activities involved in installing a plastics injection molding machine and getting it ready for production. The major tasks are:

I—Preparation of site;
II—Installation of the machine;
III—Preparation of the machine for production.

The above major tasks can be divided into the simple activities shown in Table 10.6. The sequence in which the activities should be performed and the time required to complete each activity are also included.

The bar, or Gantt, chart is the simplest scheduling tool. In this method, each activity is listed together with its starting and finishing date, as shown in Fig. 10.2 for the injection molding machine example. In this form, the Gantt chart does not show how one activity is related to other activities or the latest time that each activity can begin in order that its completion time does not interfere with the beginning of other activities. This limits its use in complex projects where analysis of schedule is needed to ensure completion in the optimum time. Such an optimum schedule for a project can be determined by using network planning techniques which have been developed to meet this need. The critical path method (CPM) and the program evaluation and review technique (PERT) are computer-based scheduling systems which employ network analysis techniques.

The basic tool of CPM is a network diagram similar to that shown in Fig. 10.3. The main elements of the diagram are events, represented by circles, linked by activities, represented by arrows. An event is a point of accomplishment and/or decision while an activity is part of the project that requires a certain length of time to complete. The length of the arrow representing the activity is irrelevant, but its direction shows the sequence of events. Sometimes a dummy or phantom activity is introduced for clarity of logic and to indicate interdependencies although it represents no real physical activity. A dummy activity is represented by a dashed arrow and requires zero time. The network of Fig. 10.3 represents the activities of installing and preparing for operation of the injection molding machine, shown in Table 10.6. The path that takes the longest time is the critical path. If any activity

Table 10.6 Installing and preparing for operation of an injection molding machine

Major task	Activity	Description	Immediate predecessor	Time (h)
I	a	Excavate foundations	–	5
	b	Pour concrete in foundation	a	2
	c	Unpack machine parts	–	3
II	d	Place machine body on foundations	b, c	2
	e	Level machine body	d	1
	f	Assemble rest of machine parts	c, e	3
	g	Connect electrical wiring	f	1
	h	Connect cooling water and drainage	f	2
III	i	Install injection molding die	g, h	3
	j	Calibrate temperature controller	i	2
	k	Place plastic pellets in hopper	f	1
	l	Adjust plastic metering device	k	1
	m	Perform experimental runs	j, l	2

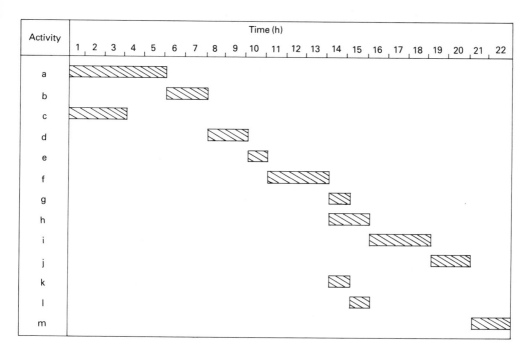

Figure 10.2 Bar chart describing the activities of installing and preparing an injection molding machine for operation. See Table 10.5 for description of activities.

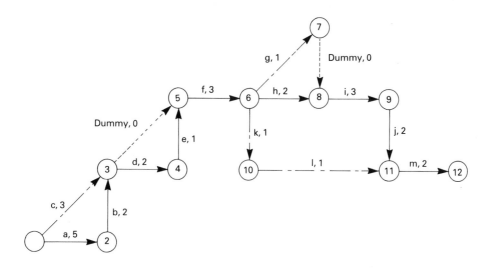

Figure 10.3 Network diagram describing the activities of installing and preparing an injection molding machine for operation. See Table 10.5 for description of activities.

on this path is delayed, the whole project will be affected, whereas delays on other paths need not prolong the project time. The critical path can be determined by inspection for a relatively simple network like the one in Fig. 10.3. For more complex systems, however, the following parameters are defined and used to determine the critical path:

ES = Earliest start time for an activity when all preceding activities are completed as rapidly as possible.

LS = Latest start time for an activity to be initiated without delaying the minimum completion time of the project.

EF = Earliest finishing time for an activity.
   = ES + duration of activity.

LF = Latest finishing time for an activity.
   = LS + duration of activity.

TF = Total float which is the slack between earliest and latest start
   = LS − ES.

The critical path is defined by the activities with zero total float, TF = 0. As an illustration, the above parameters will be used to analyze the network shown in Fig. 10.3 and the results are shown in Table 10.7. The ES times are estimated by starting at the first event and moving forward through the network by adding the duration of each activity to the ES of the preceding activity. The LS times are estimated using a reverse procedure by starting with the last event and moving backwards by subtracting the duration of each activity from the limiting LS at each event. When all the values of ES and LS have been estimated, EF, LF and TF can then be calculated by simple addition and subtraction as shown in Table 10.7. Activities which fall on the critical path, with zero total float, are then marked.

In the above discussion of CPM the time was taken as the most likely estimate which can be based on experience with similar or related projects. Not all projects, however, can be related to previous work, and time estimates in these cases cannot be made accurately. PERT uses the same ideas as CPM but it employs a probabilistic estimate of time for

Table 10.7 Determining the critical path for the installation and preparation for operation of the injection molding machine

Major task	Activity	Immediate predecessor	Time (h)	ES	LS	EF	LF	TF
	a	–	5	0	0	5	5	0
I	b	a	2	5	5	7	7	0
	c	–	3	0	4	3	7	4
	d	b, c	2	7	7	9	9	0
	e	d	1	9	9	10	10	0
II	f	e	3	10	10	13	13	0
	g	f	1	13	14	14	15	1
	h	f	2	13	13	15	15	0
	i	g, h	3	15	15	18	18	0
	j	i	2	18	18	20	20	0
III	k	f	1	13	18	14	19	5
	l	k	1	14	19	15	20	5
	m	j, l	2	20	20	22	22	0

completion of an activity. The time estimates are assumed to follow a beta frequency distribution that gives the expected time, t_e, as:

$$t_e = (t_o + 4 * t_n + t_p)/6 \tag{10.2}$$

where t_o = time duration under the most favorable conditions (optimistic time)
  t_n = time duration under normal conditions and
  t_p = time under the least favorable conditions (pessimistic time)

In PERT the expected time is computed for each activity and used to determine the critical path as in the case of CPM.

10.5   OPTIMIZATION METHODS

Optimization is generally considered as the process of maximization of a desirable quantity, like profit or component life, or minimization of an undesirable quantity, like expenses or material losses. In many cases, engineering problems do not have one optimum solution and the answer may depend on the model used for analysis. For example, one analyst may choose to improve the financial standing of an organization by maximizing profit, while another may prefer to minimize cost. In such cases, the optimum solution may be defined as the best relative to the model. The optimum solution will be the best for the problem only if the specified criterion is a true representation of the goals of the entire organization in which the problem exists. In practice, however, there is often more than one conflicting goal. For example, in making an optimum design for a tennis racket, the designer must include the conflicting goals of low cost, light weight, high stiffness, high toughness and playability. If the design criterion represents some, but not all, of the conflicting goals, the result is a suboptimal solution. Naturally, the analyst or designer would try to minimize the consequences of suboptimality. One possibility is to combine the major conflicting goals in one function by assigning appropriate weights to each goal and then adding them together. This approach is similar to the decision matrix method discussed in Section 10.2. Another possibility is to select one predominant goal as the objective function and to reduce the other goals to the status of constraints or specifications. Such specifications can be considered as target values which are subject to trade-offs when the final decision is made.

The methods used for optimizing engineering problems can be classified as:

1. Optimization by evolution, which involves improving existing designs or products by introducing different modifications. Survival of a given alternative solution will depend on the natural selection of user acceptance.
2. Optimization by intuition, which involves making decisions without being able to formulate a justification. Even with the present knowledge and analytical tools, intuition continues to play an important role in solving many engineering problems.
3. Optimization by analytical methods that deal with the properties of maxima and minima and how to find maxima and minima numerically. These methods can be grouped as differential calculus, search methods, analytical-graphical methods, linear programming, integer programming, dynamic (multistage) programming, geometric programming and nonlinear programming. Some of these methods will be discussed briefly here. The textbooks listed in the bibliography should be consulted for more detail.

In the analytical methods of optimization, it is assumed that all the relevant variables, parameters and constraints are quantifiable. An objective function that defines the problem in terms of the independent variables is first formulated. Typical objective functions could be cost, weight, load-carrying capacity, reliability, or energy consumption. Generally the objective function is subject to constraints which arise from physical laws or from compatibility conditions on the individual variables. Thus, if X_j, where $j = 1, 2, \ldots$ n, represents the n decision variables of the problem under study, and if the system is subject to m constraints, the general mathematical model can be written in the form:

optimize $Z = f(X_1, X_2, . X_j, . ., X_n)$ objective function, subject to

$$\left. \begin{array}{l} G_i(X_1, X_2, \ldots, X_n) < L_i, \, i = 1, 2, \ldots, m \\ X_1, X_2, \ldots, X_n > 0 \end{array} \right\} \text{constraints}$$

The constraints $X_1, \ldots, X_n > 0$ are called the nonnegativity restrictions. They restrict the variables to zero or positive values, which is the case in most real-life situations.

To illustrate definitions, consider the case of a pipeline carrying hot fluid. If the pipeline is not sufficiently insulated, heat will be lost, as shown in Fig. 10.4. However, if too much insulation is used then the cost of insulation may exceed the savings in the cost of heat lost during the expected life of the insulator. The optimum thickness can be determined as the point of minimum total cost curve (cost of heat saved plus cost of insulation) as shown in Fig. 10.4. The figure also shows the constraints which may be imposed on the possible solution. The first constraint is decided by the minimum thickness of commercially available insulator while the second constraint is imposed by the available space for insulation. In the figure, the optimum thickness falls between the constraints and if selected it would represent the optimum solution. However, if the energy costs increase, the optimum insulation thickness would then increase accordingly. If the available space is

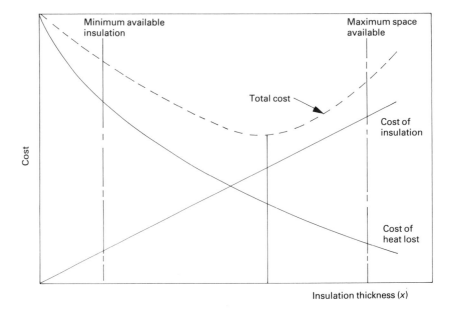

Figure 10.4 Optimization of insulation thickness in terms of total cost.

less than the most economic thickness, the optimum design cannot be achieved. In this case, using the maximum allowable insulation thickness yields an optimal design, which is the best of all feasible designs.

10.6   OPTIMIZATION BY DIFFERENTIAL CALCULUS

The above problem can be represented analytically by the following relationship:

$$C = A\frac{1}{X} + BX \qquad (10.3)$$

where C is the total cost.

      X is the insulation thickness which should not be less than the minimum available thickness or more than the maximum allowable space.

      A is the cost of energy loss.

      B is the cost of insulation.

A simple method of optimizing (10.3) is differential calculus where the derivative is zero at the maximum or minimum points. In this case,

$$\frac{dC}{dX} = -A\frac{1}{X^2} + B = 0 \qquad (10.4)$$

$$X = \sqrt{\frac{A}{B}} \qquad (10.5)$$

If the value of X given by (10.5) is within the constraints shown in Fig. 10.4, it can be used and the design is optimum. If X falls outside the constraints, only the optimal design is possible. In the latter case, the insulation thickness is taken as the allowable limit nearest to the optimum value.

10.7   SEARCH METHODS OF OPTIMIZATION

Frequently, engineering problems cannot be expressed by simple analytical functions. In such cases, an efficient search method has to be devised to identify the set of conditions that will yield the optimum outcome or solution. When the available data are free from error, the search is called deterministic. Frequently, however, the available data are subject to experimental or human errors and the search is called stochastic. Several search techniques have been devised to solve problems which involve one variable whose behavior is unimodal, i.e. single optimum value or peak.

In the uniform search method, the allowable range of values of the variable is divided by equally spaced trial points and the outcome of the system is evaluated at each trial point. The optimum outcome is expected to fall on either side of the trial point with the highest outcome, as shown in Fig. 10.5a. The interval of uncertainty, i.e. the interval of values of the variable which contain the optimum, depends on the spacing of the trial points.

In sequential search, a pair of trial points, a and b in Fig. 10.5b, near the center of the allowable range of values of the variable is evaluated to determine the direction of change

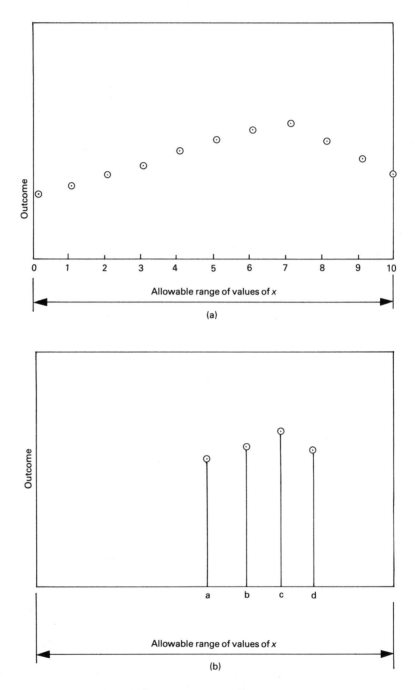

Figure 10.5 Search methods of optimization. (a) Uniform search method. (b) Sequential search method. (c) The golden section search method.

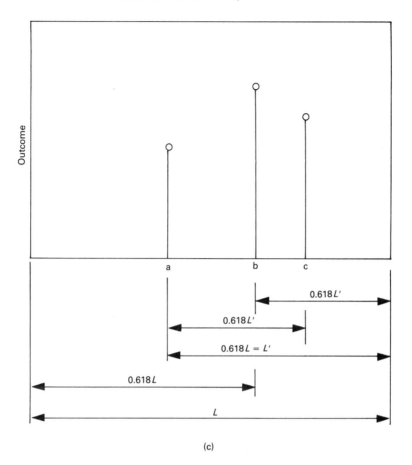

(c)

Figure 10.5 – *contd.*

and to allow further evaluations to be made in sequence to arrive at the peak. In Fig. 10.5b, as the outcome at point b is higher, point c is evaluated followed by point d. As the outcome at point d is lower, the optimum value of the outcome falls on either side of point c.

The golden section search method places the first two trial points at $0.618L$ from either end of the range, where L is the total range. Based on the value of the outcome of the two points a and b of Fig. 10.5c, the interval between one end and the lower point is eliminated. With the new reduced range L', which equals $0.618L$, additional trial points, located at $0.618L'$ from either end, are evaluated. In this case only one new point c is needed, since the point from the previous trial, b in this case, lies at the correct location for the new length L'. As the procedure continues, the range successively decreases to 0.618 of the preceding range, with each new range needing one extra trial point. The process can be repeated until the maximum is located to within as small an interval as desired.

When the objective function is affected by two variables, X_1 and X_2, the values of outcome fall on a three-dimensional surface which can be projected on the X_1–X_2 plane with the different levels represented by contours as in a map. In the lattice search method, a grid lattice is placed over the X_1–X_2 plane to form trial points as in the case of the uniform

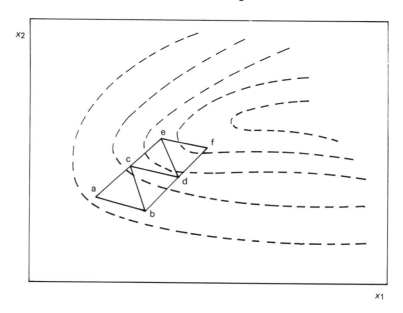

Figure 10.6 Sequential simplex search method for optimization of two-variable objective functions.

single-variable search. A coarse grid can be used initially to narrow the field of search followed and a finer grid for better location of the maximum.

Another technique of optimizing such objective functions is the sequential simplex search method. For an objective function with two variables, the simplex is an equilateral triangle. The optimization method is illustrated in Fig. 10.6. Beginning at point a, two additional objective function values are calculated at points b and c such that a, b, c, form an equilateral triangle, the simplex. The basic simplex technique rejects the least point, considers it to be a in this case, and moves away from it in a direction that creates a new simplex, comprised of points b, c, and d. By continually constructing new simplexes, the path eventually leads to an optimum value. This technique can be used to optimize objective functions with more than two variables. For example, three variables can be optimized be defining a tetrahedron simplex. The simplex search stalls when two equally nonoptimum values cause the search to oscillate. Therefore, two additional rules are applied: (a) a return to the point that has just been left is invalid, and movement goes to the next highest rejected value; (b) the search is terminated after a specified number of iterations. This reduces oscillation about the maximum. The reader should consult the textbooks listed in the bibliography for more information on the optimization of multivariable problems.

10.8   LINEAR PROGRAMMING

Linear programming is widely used in solving optimization problems that involve linear objective functions subject to linear constraints. Such problems exist in many engineering and nonengineering applications including industry, military, agriculture, economics, transportation, health systems and social sciences. Several models have been developed for the solution of specific problems, for example transportation model deals with transportation or distribution problems in the area of shipping commodities between sources and destinations at minimum transportation costs. The basic concept of linear programming will be illustrated by solving a two-variable problem using a graphical method.

Consider the case where a foundry produces castings from two main metals A and B. The maximum availability of metal A is 8 ton/day and that of B is 10 ton/day. Castings are produced in two types of alloys: A-rich alloy, containing 2/3 A + 1/3 B, and B-rich alloy, containing 1/3 A + 2/3 B. For more balanced use of melting facilities, it is decided that the daily production of the A-rich alloy castings cannot exceed that of B-rich alloy castings by more than 5 ton. Market survey shows that the maximum demand for A-rich alloy castings is limited to a maximum of 8 ton/day. The sale price for A-rich alloy castings is $300 and that for B-rich alloy castings is $200. How much of A-rich and B-rich castings should the foundry produce to maximize its gross income?

The first step is to identify the variables, which in this case can be defined as:

$$X_A = \text{tons produced daily of A-rich alloy castings}$$
$$X_B = \text{tons produced daily of B-rich alloy castings}$$

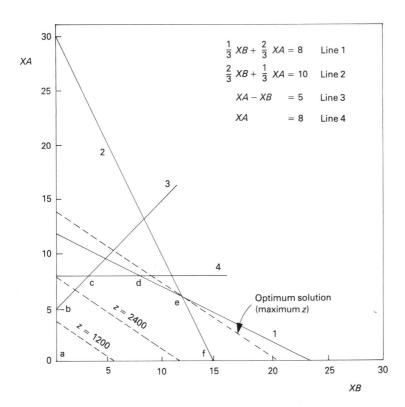

Figure 10.7 Using linear programming to maximize the gross income of a foundry.

The second step is to determine the objective function, which is to maximize the gross income, z, in this case.

$$z = 300\,X_A + 200\,X_B$$

The third step is to determine the constraints, which in this case are: usage of metals $<$ or $=$ maximum metal availability. This leads to the following relationships:

$$1/3\,X_B + 2/3\,X_A < 8 \text{ for metal A, ignoring losses in melting}$$
$$2/3\,X_B + 1/3\,X_A < 10 \text{ for metal B, ignoring losses in melting}$$

The second restriction on demand can be written as:

$$X_A - X_B < 5 \quad \text{excess of A-rich alloy over B-rich alloy}$$
$$X_A < 8 \qquad\quad \text{maximum demand for A-rich alloy}$$

The next step is to plot the solution space, which satisfies all the constraints simultaneously. Fig. 10.7 shows the required solution space. The space enclosed by the constraints is determined by replacing ($<$) by ($=$) for each constraint, thus obtaining a straight-line equation, which is plotted in Fig. 10.7. The resulting solution space is shown by the area abcdef in Fig. 10.7. All points within or on the boundary of the solution space abcdef satisfy all the constraints and thus represent a possible solution. The optimum solution can be determined by observing the direction in which the value of the objective function, z, increases. This is illustrated in Fig. 10.7 for arbitrary values of $z = 1200$ and 2400. To find the optimum solution, i.e. maximum possible z, the line which represents z is moved to the point where any further increase in z would violate one or more of the constraints. The figure shows that this condition is represented by point e. Since point e is the intersection of lines 1 and 2, the values of X_A and X_B are determined by solving the following two equations simultaneously:

$$1/3\,X_B + 2/3\,X_A = 8$$
$$2/3\,X_B + 1/3\,X_A = 10$$

The two equations yield $X_A = 6$ and $X_B = 12$, which represents the optimum condition. Under this condition, the maximum gross income is:

$$z = 300*6 + 200*12 = 4200.$$

10.9 SENSITIVITY ANALYSIS

Sensitivity analysis is usually performed in order to determine how the optimum solution is affected by small changes in the original model. This gives the model a dynamic characteristic that allows the analyst to check the effect of possible changes in variables or constraints on the objective function. Sensitivity analysis will also allow the analyst to determine the effect of errors or uncertainty in some of the estimated values on the final answer. In the above example, it may be of interest to know how the optimum gross income is affected by relative change in demand for castings and/or availability of metals.

In performing sensitivity analysis, the different constraints can be classified as hard, or binding, and soft, or nonbinding. A hard constraint passes through the point of optimum solution, as in the case of constraint 1 and 2 in the above foundry example. Changes in the hard constraints will directly affect the solution. On the other hand, some changes in the soft constraints may not affect the solution as they do not pass through the optimum point. In the foundry example, Fig. 10.7 shows that constraints 3 and 4 are soft and the demand for A-rich alloy, constraint 4, can decrease from 8 to about 6 before the gross income is affected. The figure also shows that the solution is insensitive to relatively large changes in constraint 3, which is related to the loading of the melting facilities.

10.10   GEOMETRIC PROGRAMMING

Geometric programming is a method of optimization of nonlinear objective functions, particularly when they are in the form of polynomial terms. When all the terms are positive, the polynomial is called posynomial. The objective function can be written as:

$$z = \sum_{j=1}^{N} U_j \tag{10.6}$$

where

$$U_j = c_j \prod_{i=1}^{n} (X_i)^{aij} \, j = 1, 2, \ldots, N. \tag{10.7}$$

It is assumed that all $c_j > 0$. The exponents aij are unrestricted in sign.

To illustrate the use of geometric programming method, consider the insulated pipe discussed earlier. Instead of representing the total cost, C, by (10.3), a better representation would be in the form:

$$C = \left(A \frac{1}{X^4} \right) + (BX) \tag{10.8}$$

The objective function represented by (10.8) can be written as:

$$C = U_1 + U_2 = A X^{a_1} + B X^{a_2} \tag{10.9}$$

where $a_1 = -4$ and $a_2 = 1$.

The number of terms T in this case is 2 and the number of variables N is 1. A degree of difficulty in this case is defined as $\{T - (N + 1)\} = 0$.

The function g is then written as:

$$g = \left(\frac{U_1}{w_1} \right)^{w_1} \left(\frac{U_2}{w_2} \right)^{w_2} = \left(\frac{A X^{a_1}}{w_1} \right)^{w_1} \left(\frac{B X^{a_2}}{w_2} \right)^{w_2} \tag{10.10}$$

where w_1 and w_2 are weighting functions and

$$w_1 + w_2 = 1 \tag{10.11}$$

The values of w_1 and w_2 can be changed in order to optimize the value of the objective function, z'. Under optimal conditions,

$$z' = U'_1 + U'_2 \tag{10.12}$$

and

$$w_1 = \frac{U'_1}{U'_1 + U'_2} \quad \text{and} \quad w_2 = \frac{U'_2}{U'_1 + U'_2} \tag{10.13}$$

This shows that w_1 and w_2 are >0, and represent the relative contribution of each of the posynomial terms to the optimal value of the objective function. The necessary condition for the optimal value can be written as:

$$a_1 w_1 + a_2 w_2 = 0 \tag{10.14}$$

In the case of the insulated pipe, $-4 w_1 + w_2 = 0$, and from (10.11), the values of w_1 and w_2 can be calculated as: $w_1 = 0.2$ and $w_2 = 0.8$. These values can be used to determine U'_1, U'_2, and z'. Taking the values of A and B of (10.8) as 10 000 and 100 respectively, the optimum value of C is given from (10.10) as:

$$C' = g' = \left(\frac{10\,000}{0.2\,X^4}\right)^{0.2} \left(\frac{100\,X}{0.8}\right)^{0.8} = 414$$

From (10.13) and the value of C', x can be calculated as 3.3 in units of thickness, i.e. mm or in. depending on the original units of (10.8).

The above solution can be subjected to sensitivity analysis, as discussed earlier to study the effects of variations in insulation and energy costs on the optimal cost and insulation thickness.

The above example had a zero degree of difficulty and could, therefore, yield a unique solution. The problem becomes more complex when $T > N + 1$. An approach to the solution is to use the technique of condensation, where two of the terms are combined so as to reduce the problem to one of 0 degree of difficulty. In this case it is better to combine the terms with similar exponents. Other techniques that can be used when the degree of difficulty is not 0 include partial invariance and dual geometric programming. The text by Wilde, D. J. on 'Globally optimum design', Wiley-Interscience, N.Y., 1987, gives an excellent account of such techniques.

10.11   REVIEW QUESTIONS AND PROBLEMS

10.1   Find the maximum value of the function:

$$M = 15\frac{1}{X} + 25\,X$$

[Answer: 38.74]

10.2   Find the maximum value of the function:

$$y = 15\,x - x^2$$

Use the golden section search method. Carry out the search until the difference between the two largest calculated values of y is less than 0.05. [Answer: $y = 56.25$ at $x = 7.476$]

10.3   A forging shop is producing two products A and B. One unit of product A requires 3 min for hammering, 2 min for trimming, and 5 min for cleaning. The profit for each unit is $1.25. In the case of product B, one unit requires 3 min for hammering, 7 min for trimming, and 6 min for cleaning. The profit from each unit of product B is $1.55. The total actual daily capacity of the hammer is 240 min; that of the trimming press is 200 min; and that of the cleaning shop is 300 min. What is the optimum mix between the two products to maximize the total profit? [Answer: 39 units of A and 17 units of B]

BIBLIOGRAPHY AND FURTHER READING

Beightler, C. S. *et al.*, *Foundations of Optimization*, 2nd ed., Prentice Hall, New Jersey, 1979.

Brichta, A. M. and Sharp, E. M., *From Project to Production*, Pergamon Press, London, 1970.

Buffa, E. S., *Modern Production Planning*, John Wiley & Sons, New York, 1973.

Dieter, G. E., *Engineering Design, a Materials and Processing Approach*, McGraw-Hill, New York, 1983.

Gajda, W. J. and Biles, W. E., *Engineering Modeling and Computation*, Houghton Mifflin, Boston, 1978.

Johnson, R. C., *Optimum Design of Mechanical Elements*, 2nd ed., John Wiley & Sons, New York, 1980.

Siddall, J. N., *Analytical Decision Making in Engineering Design*, Prentice Hall, New Jersey, 1979.

Vanderplaats, G. N., *Numerical Optimization Techniques for Engineering Design*, McGraw-Hill, New York, 1984.

Woodson, T. T., *Introduction to Engineering Design*, McGraw-Hill, New York, 1966.

Chapter 11

Effect of Material Properties on Design

11.1   FACTORS AFFECTING THE BEHAVIOR OF MATERIALS IN COMPONENTS

As it became clear from the discussions of Chapter 9, a successful design should take into account the function, material properties and manufacturing processes, as shown in Fig. 11.1. The figure also shows that there are other secondary relationships between material properties and manufacturing processes, between function and manufacturing processes, and between function and material properties. The relationship between design and material properties will be discussed in this chapter and the relationship between design and manufacturing processes will be discussed in Chapter 12.

The relationship between design and material properties is complex because the behavior of the material in the finished product can be quite different from that of the stock material used in making it. This point is illustrated in Fig. 11.2, which shows the direct influence of stock material properties, production method and component geometry and external forces on the behavior of materials in the finished component. The figure also shows that secondary relationships exist between geometry and production method, between stock material and production method, and stock material and component geometry. The effect of stock material properties, component geometry and applied forces on the behavior of materials will be discussed in this chapter. The effect of production method on material behavior was discussed in Part I and the effect of production method on component geometry will be discussed in Chapter 12.

11.2   STATISTICAL VARIATION OF MATERIAL PROPERTIES

In making design calculations, material properties are often presumed to be homogeneous and isotropic. In practice, however, material properties are seldom homogeneous or isotropic as they are sensitive to variations in parameters such as composition, heat treatment and processing conditions. Surface roughness, internal stresses, sharp corners and other stress raisers can also influence material behavior. In addition, material properties are also sensitive to environmental variables such as time, temperature, humidity and ambient chemicals.

The problem of variability of material properties can be solved by choosing a factor of safety which is simply a multiplier of the actual measured values of a property, as discussed in Section 9.6. The factor of safety accommodates unknown influences that affect the life

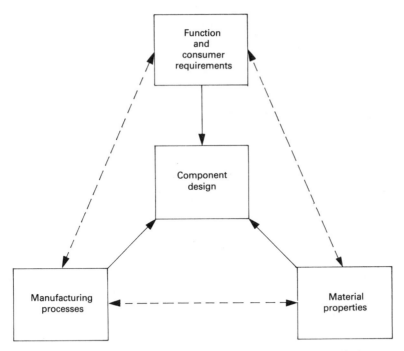

Figure 11.1 Factors that should be considered in component design.

Source: Farag, M. M., *Materials and Process Selection in Engineering*, Applied Science Publishers, London, 1979.

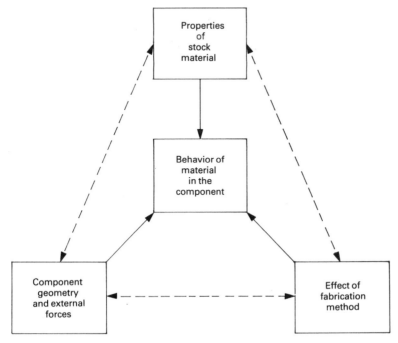

Figure 11.2 Factors that should be considered in anticipating the behavior of material in the component.

Source: Farag, M. M., *Materials and Process Selection in Engineering*, Applied Science Publishers, London, 1979.

of the component in service, but that are not part of the experimental testing procedure. The use of a factor of safety becomes unsuitable when estimating the reliability of components in service under random loading conditions.

Material properties can be statistically described by the mean value, standard deviation and coefficient of variation, which were discussed in Appendix C. The coefficient of variation, which is a dimensionless quantity, is useful in assessing the relative variability of data from different sources. With large samples, material properties usually follow a normal distribution. Smaller values of standard deviation and coefficient of variation indicate more homogeneous material.

From a statistical point of view, materials testing usually produces limited observations only. Although not all material properties are normally distributed, a normal distribution can be assumed as a first approximation which usually results in a conservative design. From the normal distribution, confidence limits can be determined and assigned to the calculated mean values. It was shown in Fig. C.6, Appendix C, that the probabilities of obtaining values at plus or minus one, two, or three standard deviations from the mean value are 68.26, 95.46 and 99.73 percent respectively. Therefore the probability that a value would occur outside the range represented by plus or minus two standard deviations from the mean is 4.54 percent, or the confidence level that the value will be found within the limits is 95.46 percent.

Assuming that the experimental data were obtained from a reasonably large number of samples, more than 100, and that they follow a normal distribution, it is possible to estimate statistical data from nonstatistical sources that give only ranges or tolerance limits. In this case the standard deviation S is approximately given by:

$$S = (\text{max. value of property} - \text{min. value})/6 \qquad (11.1)$$

This procedure is based on the assumption that the given limits are bounded between plus and minus three standard deviations. For example, if the range of strength of an alloy is given as 800 to 1200 MPa, the mean value can be taken as 1000 MPa and the standard deviation S can be estimated as:

$$S = (1200 - 800)/6 = 66.67 \text{ MPa}$$

The coefficient of variation v is then:

$$v = 66.67/1000 = 0.0667.$$

If the results are obtained from a sample of about 25 tests, it may be better to divide by 4 in (11.1) instead of 6, and with a sample of about 5 to divide by 2.

In the absence of test data the following values of coefficient of variation can be taken as typical for metallic materials:

$v = 0.05$ for ultimate tensile strength.
$v = 0.07$ for yield strength.
$v = 0.08$ for endurance limit for steel.
$v = 0.07$ for fracture toughness.

11.3   STRESS CONCENTRATION

In almost all cases, engineering components and machine elements have to incorporate design features which introduce changes in their cross-section. For example, shafts must have shoulders to take thrust loads at the bearings and must have keyways or splines to transmit torques to or from pulleys and gears mounted on them. Other features which introduce changes in cross-section include oil holes, fillets, undercuts, bolt heads, screw threads and gear teeth. These changes cause localized stress concentrations which are higher than those based upon the nominal cross-section of the part. The severity of the stress concentration depends on the geometry of the discontinuity and the nature of the material. A geometric, or theoretical, stress concentration factor, K_t, is usually used to relate the maximum stress, S_{max}, at the discontinuity to the nominal stress, S_{av}, according to the relationship:

$$K_t = \frac{S_{max}}{S_{av}} \qquad (11.2)$$

The value of K_t depends only on the geometry of the part and for the simple case of an elliptical hole in an infinitely large plate, it is given by:

$$K_t = 1 + \frac{2b}{a} \qquad (11.3)$$

where $2b$ is the dimension of the hole perpendicular to the stress direction and $2a$ is the dimension of the hole parallel to the stress direction.

In the case of a circular hole in an infinite plate, a is equal to b and $K_t = 3$. The value of K_t for other geometries can be determined from stress concentration charts, such as those given by Peterson and Shigley (see bibliography). Other methods of estimating K_t for a certain geometry include photoelasticity, brittle coatings and finite element techniques. The latter method will be discussed in Section 14.5. Table 11.1 gives some typical values of K_t.

Experience shows that, under static loading, K_t gives an upper limit to the stress concentration value and applies only to brittle and notch sensitive materials. With more ductile materials, local yielding in the very small area of maximum stress causes consider-able relief in the stress concentration. Consequently, for ductile materials under static loading it is not usually necessary to consider the stress concentration factor. However, due consideration should be given to the stress concentration when designing with high-strength, low-ductility, case-hardened and/or heavily cold-worked materials.

Stress concentration should also be considered in designing components that are subject to fatigue loading. Under such conditions, a fatigue stress-concentration factor, or fatigue-strength reduction factor, K_f, is usually defined as:

$$K_f = \frac{\text{endurance limit of notch-free part}}{\text{endurance limit of notched part}} \qquad (11.4)$$

The relationship between K_f and K_t was discussed in Section 8.4 and a notch sensitivity factor, q, was defined in (8.8). The value of q was shown to vary between 1 and zero. When $K_f = K_t$, the value of q is 1 and the material is fully sensitive to notches. On the other hand when the material is not at all sensitive to notches, $K_f = 1$ and $q = 0$. In making a design, K_t is usually determined from the geometry of the part. Then, when the material is selected, q can be specified, and (8.8) is solved for K_f. Generally, the value of q approaches unity as the material strength increases, e.g. UTS more than 1400 MPa (200 ksi) for steels, and as the fillet radii increase, e.g. more than about 1 mm (0.04 in). Whenever in doubt, the designer can take $K_f = K_t$ and err on the safe side.

Table 11.1 Approximate values of stress concentration factor (K_t)

Component shape	Value of critical parameter	K_t
Round shaft with transverse hole		
Bending	$d/D = 0.025$	2.65
	$= 0.05$	2.50
	$= 0.10$	2.25
	$= 0.20$	2.00
Torsion	$d/D = 0.025$	3.7
	$= 0.05$	3.6
	$= 0.10$	3.3
	$= 0.20$	3.0
Round shaft with shoulder		
Tension	$d/D = 1.5, r/d = 0.05$	2.4
	$r/d = 0.10$	1.9
	$r/d = 0.20$	1.55
	$d/D = 1.1, r/d = 0.05$	1.9
	$= 0.10$	1.6
	$= 0.20$	1.35
Bending	$d/D = 1.5, r/d = 0.05$	2.05
	$r/d = 0.10$	1.7
	$r/d = 0.20$	1.4
	$d/D = 1.1, r/d = 0.05$	1.9
	$r/d = 0.10$	1.6
	$r/d = 0.20$	1.35
Torsion	$d/D = 1.5, r/d = 0.05$	1.7
	$r/d = 0.10$	1.45
	$r/d = 0.20$	1.25
	$d/D = 1.1, r/d = 0.05$	1.25
	$r/d = 0.10$	1.15
	$r/d = 0.20$	1.1
Grooved round bar		
Tension	$d/D = 1.1, r/d = 0.05$	2.35
	$r/d = 0.10$	2.0
	$r/d = 0.20$	1.6
Bending	$d/D = 1.1, r/d = 0.05$	2.35
	$r/d = 0.10$	1.9
	$r/d = 0.20$	1.5
Torsion	$d/D = 1.1, r/d = 0.05$	1.65
	$r/d = 0.10$	1.4
	$r/d = 0.20$	1.25

The above discussion shows that stress concentration can be a source of failure in many cases, especially when designing with high-strength materials and under fatigue loading. In such cases, the following design guidelines should be observed if the deleterious effects of stress concentration are to be kept to a minimum:

1. Abrupt changes in cross-section should be avoided. If they are necessary, generous fillet radii or stress relieving grooves should be provided (Fig. 11.3a).
2. Slots and grooves should be provided with generous run-out radii and with fillet radii in all corners (Fig. 11.3b).

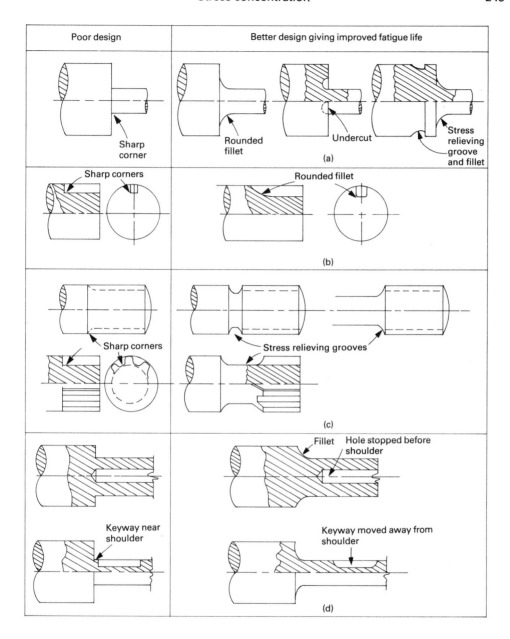

Figure 11.3 Design guidelines for shafts subjected to fatigue loading.

3. Stress relieving grooves or undercuts should be provided at the end of threads and splines (Fig. 11.3c).
4. Sharp internal corners and external edges should be avoided.
5. Oil holes and similar features should be chamfered and the bore should be smooth.
6. Weakening features like bolt and oil holes, identification marks and part numbers should not be located in highly stressed areas.
7. Weakening features should be staggered to avoid the addition of their stress concentration effects (Fig. 11.3d).

11.4 DESIGNING FOR STATIC STRENGTH

The design and materials selection of a component or structure can be based on static strength, stiffness, stability, fatigue strength, or creep, depending on the service conditions and the intended function. Designs based on the static strength are usually aimed at avoiding yielding of the component in the case of soft, ductile materials and at avoiding fracture in the case of strong, low-toughness materials. Designs based on soft, ductile materials will be discussed in this section, while those based on strong, low-toughness materials will be discussed in Section 11.5. In addition to being strong enough to resist the expected service loads, there may also be the added requirement of stiffness to ensure that deflections do not exceed certain limits. Stiffness is important in machine elements to avoid misalignment and to maintain dimensional accuracy. Elastic instability becomes an important design criterion in the case of columns, struts and thin-wall cylinders subjected to compressive axial loading where failure can take place by buckling. Fatigue is a major concern when designing components that are expected to carry fluctuating loads, as will be discussed in Section 11.6, and creep is usually considered when designing for high-temperature applications, as will be discussed in Section 11.7.

Designing for simple axial loading

Components and structures made from ductile materials are usually designed so that no yield will take place under the expected static loading conditions. When the component is subjected to uniaxial stress, yielding will take place when the local stress reaches the yield strength of the material. The critical cross-sectional area, A, of such component can be estimated as:

$$A = \frac{K_t n L}{YS} \qquad (11.5)$$

where K_t = stress concentration factor, as determined in Section 11.3
L = applied load
n = factor of safety, as determined in Section 9.7
YS = yield strength of the material

In some cases the stiffness of the component, rather than its strength, is the limiting factor. In such cases, limits are set on the extension in the component. This point was illustrated in the example of Section 2.8.

Designing for torsional loading

The critical cross-sectional area of a circular shaft subjected to torsional loading can be determined from the relationship:

$$\frac{2I_p}{d} = \frac{K_t n T}{\tau_{max}} \qquad (11.6)$$

where I_p = polar moment of inertia of the cross-section

$$I_p = \frac{\pi d^4}{32} \text{ for a solid circular shaft}$$

$$I_p = \frac{\pi(d_o^4 - d_i^4)}{32} \text{ for a hollow circular shaft of inner diameter } d_i \text{ and outer diameter } d_o$$

d = shaft diameter at the critical cross-section
τ_{max} = shear strength of the material
T = transmitted torque

While (11.6) gives a single value for the diameter of a solid shaft, a large combination of inner and outer diameters can satisfy the relationship in the case of a hollow shaft. Under such conditions, one of the diameters or the required thickness has to be specified in order to calculate the other dimension. The ASTME code of recommended practice for transmission shafting gives an allowable value of shear stress of 0.3 of the yield or 0.18 of the ultimate tensile strength, whichever is smaller. With shafts containing keyways, ASTM recommends a reduction of 25 percent of the allowable shear strength to compensate for stress concentration and reduction in cross-sectional area.

The torsional rigidity of a component is usually measured by the angle of twist, θ, per unit length. For a circular shaft, θ is given in radians by:

$$\theta = \frac{T}{GI_p} \tag{11.7}$$

where G = modulus of elasticity in shear

$$= \frac{E}{2(1+v)} \tag{11.8}$$

where v = Poisson's ratio.

The usual practice is to limit the angular deflection in shafts to about 1 degree, i.e. $\pi/180$ radians, in a length of 20 times the diameter.

Designing for bending

When a relatively long beam is subjected to bending, the bending moment, the maximum allowable stress and dimensions of the cross-section are related by the equation:

$$Z = \frac{nM}{YS} \tag{11.9}$$

where M = bending moment
Z = section modulus = I/c
I = moment of inertia of the cross-section with respect to the neutral axis normal to the direction of the load
c = distance from center of gravity of the cross-section to the outermost fiber.

Figure 8.3 gives the formulas for calculating the value of I for some commonly used cross-sections.

When an initially straight beam is loaded, it becomes curved as a result of its deflection. As the deflection at a given point increases, the radius of curvature at this point decreases. The radius of curvature, r, at any point on the curve is given by the relationship:

$$r = \frac{EI}{M} \tag{11.10}$$

Equation 11.10 shows that the stiffness of a beam under bending is proportional to the elastic constant of the material, E, and the moment of inertia of the cross-section, I. Selecting materials with higher elastic constant and efficient disposition of material in the cross-section is essential in designing of beams for stiffness. The effect of the elastic constant of the material on design was analyzed in the example of Section 10.2. Placing as much as possible of the material as far as possible from the neutral axis of bending, is generally an effective means of increasing I for a given area of cross-section. This point was discussed in Section 8.3.

Designing for combined loading

When a component or a structural member is subjected to combined loading, the behavior under the resulting multiaxial stresses cannot be directly predicted and a criterion for yielding must first be established. There are two commonly used yield criteria, namely maximum shear stress and maximum distortion energy.

The maximum shear stress criterion, also called the Tresca yield criterion, assumes that yield will take place when the value of the maximum shear stress in a component reaches the value of shear stress in a tensile-test specimen as it starts to yield. According to this criterion, yield will take place in a component subjected to multiaxial loading when:

$$YS = 2\tau_{max} = \sigma_1 - \sigma_3 \tag{11.11}$$

where YS = yield strength of the material
τ_{max} = maximum shear stress in the component
σ_1 = algebraically greatest principal normal stress
σ_3 = algebraically smallest principal normal stress

The maximum shear stress criterion is in good agreement with experimental results, being slightly on the safe side, and is widely used by designers for ductile materials.

The maximum distortion energy criterion, also called von Mises yield criterion, is based on the determination of the energy associated with changes in shape of the material and gives better fit with experimental results. According to this criterion, yield takes place in a component when the value of the distortion energy per unit volume equals the distortion energy per unit volume required to cause yield in a tensile-test specimen. This criterion can be written as:

$$YS = \frac{1}{\sqrt{2}} [(\sigma_1 - \sigma_2)^2 + (\sigma_2 - \sigma_3)^2 + (\sigma_3 - \sigma_1)^2]^{1/2} \tag{11.12}$$

where σ_1, σ_2, and σ_3 are the three principal normal stresses acting on the component.

The yield criteria discussed above give identical results for conditions of uniaxial stress and balanced biaxial stress, σ_1 equals σ_2. The greatest diversion between the two criteria occurs for a state of pure shear, $\sigma_1 = -\sigma_2$. In this case the maximum shear stress criterion predicts a yield stress which is 15 percent lower than the value given by the maximum distortion energy criterion.

As an illustration of the use of the above yield criteria, consider a solid round shaft subjected to combined bending and torsional loading. The bending and torsional stresses are given by:

$$\sigma_B = \frac{32M}{\pi d^3} \qquad\qquad \tau_s = \frac{16T}{\pi d^3}$$

where σ_B = bending stress
τ_s = torsional stress
d = shaft diameter
M = bending moment at critical section
T = torsional moment at critical section

Using Mohr's circle it is found that the maximum shear stress, τ_{max}, is

$$\tau_{max} = \left[\left(\frac{\sigma_B}{2} \right)^2 + \tau_s^2 \right]^{1/2}$$

$$= \frac{16}{\pi d^3} [M^2 + T^2]^{1/2} \qquad\qquad (11.13)$$

The maximum shear stress criterion shows that $2\tau_{max}$ should not exceed YS and with a factor of safety n, (11.11) can be written as:

$$\frac{YS}{2n} = \frac{16}{\pi d^3} [M^2 + T^2]^{1/2}$$

$$d = \left[\left(\frac{32n}{\pi YS} \right) (M^2 + T^2)^{1/2} \right]^{1/3} \qquad\qquad (11.14)$$

The maximum distortion energy criterion can be similarly applied to give:

$$d = \left[\left(\frac{32n}{\pi YS} \right) \left(M^2 + \frac{3T^2}{4} \right)^{1/2} \right]^{1/3} \qquad\qquad (11.15)$$

The value of d calculated according to (11.14) is larger than that calculated according to (11.15).

Designing of columns

Columns and struts, which are long slender parts, are subject to failure by elastic instability, or buckling, if the applied axial compressive load exceeds a certain critical value, P_{cr}. The Euler column formula is usually used to calculate the value of P_{cr}, which is a function of the material, geometry of the column and restraint at the ends. For the fundamental case of a pin-ended column, i.e. ends are free to rotate around frictionless pins, P_{cr} is given as:

$$P_{cr} = \frac{\pi^2 EI}{L^2} \qquad (11.16)$$

where I = the least moment of inertia of the cross-sectional area of the column
       L = length of the column

The above equation can be modified to allow for end conditions other than the pinned ends. The value of P_{cr} for a column with both ends fixed, i.e. built-in as part of the structure, is four times the value given by (11.16). On the other hand, the critical load for a free-standing column, i.e. one end is fixed and the other free as in a cantilever, is only one-quarter of the value given by (11.16).

The Euler column formula given above shows that the critical load for a given column is only a function of E and I and is independent of the compressive strength of the material. This means that resistance of a column of a given cross-sectional area to buckling can be increased by distributing the material as far as possible from the principal axes of the cross-section. Hence, tubular sections are preferable to solid sections. Reducing the wall thickness of such sections and increasing the transverse dimensions increases the stability of the column. However, there is a lower limit for the wall thickness below which the wall itself becomes unstable and causes local buckling.

Experience shows that the values of P_{cr} calculated according to (11.16) are higher than the buckling loads observed in practice. The discrepancy is usually attributed to manufacturing imperfections, such as lack of straightness of the column and lack of alignment between the direction of the compressive load and the axis of the column. This discrepancy can be accounted for by using an appropriate imperfection parameter or a factor of safety. For normal structural work a factor of safety of 2.5 is usually used. As the extent of the above imperfections is expected to increase with increasing slenderness of the column, it is suggested that the factor of safety be increased accordingly. A factor of safety of 3.5 is recommended for columns with $[L(\sqrt{A/I})] > 100$, where A is cross-sectional area.

Equation 11.16 shows that the value of P_{cr} increases rapidly as the length of the column, L, decreases. For a short enough column, P_{cr} becomes equal to the load required for yielding or crushing of the material in simple compression. Such a case represents the limit of applicability of the Euler formula as failure takes place by yielding or fracture rather than elastic instability. Such short columns are designed according to the procedure described for simple axial loading. This design procedure was illustrated in the example of Section 8.2.

11.5 DESIGNING WITH HIGH-STRENGTH LOW-TOUGHNESS MATERIALS

High-strength materials are being increasingly used in designing critical components to save weight or to meet difficult service conditions. Unfortunately these materials tend to be less tolerant of defects than the traditional lower-strength, tougher materials. While a cracklike defect can safely exist in a part made of low-strength ductile material, it can cause catastrophic failure if the same part is made of a high-strength low-toughness material. This has led to more demand for accurate calculation of acceptable defect levels and to

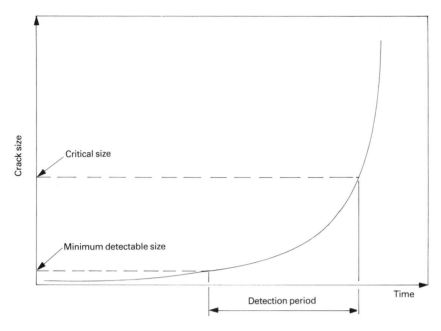

Figure 11.4 Principles involved in fail-safe designs.

increased use of nondestructive testing (NDT) in manufacture. These defects can be the result of:

1. Initial flaws in the material.
2. Production deficiencies, e.g. welding defects.
3. Service conditions, e.g. fatigue cracks or stress corrosion cracks.

   Fail-safety requires a structure to be sufficiently damage tolerant to allow defects to be detected before they develop to a dangerous size. This means that inspection has to be conducted before the structure is put into service to ensure that none of the existing defects exceed the critical size. In addition, the structure has to be inspected periodically during its service life to ensure that none of the subcritical defects grow to a dangerous size, as illustrated in Fig. 11.4. The figure shows that it is not strictly necessary to select a material with a low crack propagation rate. In principle the structure can be made fail-safe when cracks propagate fast if the inspection interval is short enough. However, short inspection periods are not always possible or cost effective. A better alternative is to use a more sensitive inspection method to reduce the minimum detectable defect size.

   In designing with high-strength low-toughness materials the interaction between fracture toughness of the material, the allowable crack size and the design stress should be considered. An analogy can be drawn between these parameters, the yield strength and the nominal stress which are considered in designing with ductile unflawed part. In the latter case, as the load increases the nominal stress increases until it reaches the yield stress and plastic deformation occurs. In the case of high-strength low-toughness material, as the design stress increases (or as the size of the flaw increases) the stress concentration at the edge of the crack, stress intensity K_I, increases until it reaches K_{Ic} and fracture occurs. Thus the value of K_I in a structure should always be kept below K_{Ic} value in the same manner that the nominal design stress is kept below the yield strength. It was shown in

Section 7.3 that the condition for failure under plane-strain conditions, where a crack of length $2a$ exists in a thick, infinitely large plate, is given by:

$$K_I = K_{Ic} = \sigma\sqrt{\pi a} \qquad (11.17)$$

In this case, K_{Ic} is a material property and is controlled by material selection, σ is a design parameter and is controlled by the applied load and shape of the part, and a is a quality control parameter which is controlled by the manufacturing method and NDT technique used. Equation 11.17 shows that the critical crack size in a structure is a function of the stress level and is not a single value for a particular material. The above relationship may be used in several ways to design against failure. For example, selecting a material to resist other service requirements automatically fixes K_{Ic}. In addition, if the minimum crack size that can be detected by the available NDT method is known, (11.17) is used to calculate the allowable design stress which must be less than: $K_{Ic}/\sqrt{\pi a}$. Alternatively, if the space and weight limitations necessitate a given material and operating stress, the maximum allowable crack size can be calculated to check whether it can be detected using routine inspection methods.

Equation 11.17 can be rewritten in a more general form to take into account different plate and crack shapes, see Section 7.3. Thus:

$$K_I = K_{Ic} = Y\sigma\sqrt{\pi a} \qquad (11.18)$$

where Y is a dimensionless shape factor which is a function of crack geometry. The value of Y can be estimated experimentally, analytically, or numerically. Figure 11.5 and Table 11.2 give solutions to some common types of flaws that occur in structural members. More complex cases can often be solved by approximation to one of these models or by reference to the literature, e.g. crack intensity handbooks and ASTM STP 380 and 410.

Figure 11.6 gives a flow chart of the steps which can be followed in designing fracture-resistant structures. As an example, consider the design of a pressure vessel of the following specifications:

Internal pressure, $p = 35$ MN/m^2 (5076 lb/in^2)
Internal diameter, $D = 800$ mm (31.5 in)
The pressure vessel will be manufactured by welding of sheets and the welded joints will then be inspected using an NDT technique which is capable of detecting surface cracks of sizes greater than 15 mm ($c.$ 0.59 in).

As a preliminary step consider the use of AISI 4340 in the manufacture of the pressure vessel. When tempered at 260°C (500°F) this steel has a yield strength 1640 MPa (238 ksi) and K. 50.0 MPa m$^{1/2}$ (45.8 ksi in $^{1/2}$).

Treating the pressure vessel as a thin-wall cylinder, the wall thickness, t, can be calculated as:

$$t = \frac{pD}{2\sigma w} \qquad (11.19)$$

where σw is the working stress.

Taking a factor of safety of 2, the working stress is 820 MPa (119 ksi) and the wall thickness is 17.1 mm (0.67 in). The critical surface crack size can then be calculated from (11.18) and Fig. 11.5 as:

$$K_{Ic} = \frac{1.1}{\sqrt{Q}}\sigma\sqrt{\pi a} \qquad (11.20)$$

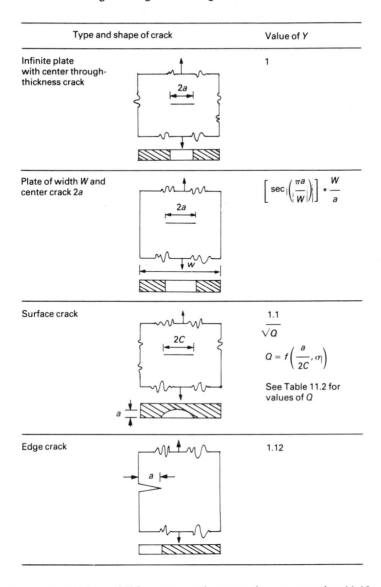

Type and shape of crack		Value of Y
Infinite plate with center through-thickness crack		1
Plate of width W and center crack $2a$		$\left[\sec\left(\dfrac{\pi a}{W}\right)\right]^* \dfrac{W}{a}$
Surface crack		$\dfrac{1.1}{\sqrt{Q}}$ $Q = f\left(\dfrac{a}{2C}, \sigma\right)$ See Table 11.2 for values of Q
Edge crack		1.12

Figure 11.5 Values of Y for some crack geometries, see equation 11.18.

Taking the conservative case of a semicircular crack, i.e. $a/2c = 0.5$, the value of Q can be estimated from Table 11.2 for $\sigma/YS = 0.5$ as $Q = 2.35$. From Equation 11.20, the critical crack length is found to be 4.6 mm (0.18 in). This length is too small to be detected by the available NDT technique, which makes the selected steel unsuitable.

Taking the same steel, AISI 4340, but tempering at 425°C (800°F) gives a yield strength of 1420 MPa (206 ksi) and K_{Ic} of 87.4 MPa m$^{1/2}$ (80 ksi in$^{1/2}$). Following the same procedure as before, $t = 19.7$ mm (0.78 in) and critical crack length is 18.74 mm (0.74 in) which can be detected by the available NDT technique and a suitable monitoring technique can be arranged. This means that the steel in this tempered condition is acceptable for making the pressure vessel from the fracture mechanics point of view.

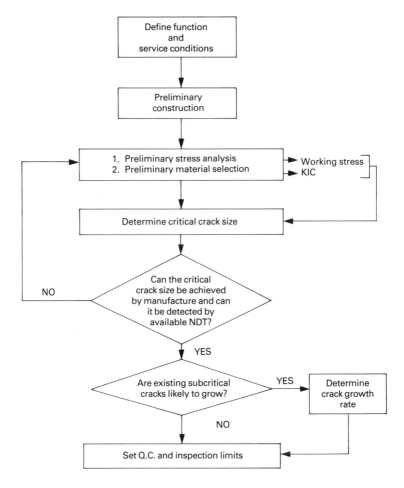

Figure 11.6 Flow chart giving the steps which can be followed in designing fracture-resistant structures.

Table 11.2 Values of flaw shape parameter Q for surface and elliptically-shaped internal cracks (see Fig. 11.5)

$\dfrac{a}{2C}$	$\dfrac{\sigma}{YS} = 0$	$\dfrac{\sigma}{YS} = 0.4$	$\dfrac{\sigma}{YS} = 0.6$	$\dfrac{\sigma}{YS} = 0.8$	$\dfrac{\sigma}{YS} = 1.0$
0	1.0	0.95	0.90	0.86	0.77
0.1	1.1	1.06	1.02	0.97	0.88
0.2	1.32	1.27	1.23	1.18	1.10
0.3	1.62	1.58	1.55	1.50	1.40
0.4	2.00	1.95	1.90	1.85	1.80
0.5	2.42	2.37	2.32	2.27	2.20

Other factors like availability of material, weldability, weight of vessel and cost have to be taken into consideration before this steel is finally selected.

Another design approach which utilizes the fracture mechanics approach is the leak-before-burst concept which can be used in designing pressure vessels and similar structures. This approach is based on the concept that if a vessel containing pressurized gas or liquid contains a growing crack, the toughness should be sufficiently high to tolerate a defect size which will allow the contents to leak out before it grows catastrophically. For leakage to occur, the crack must grow through the vessel wall thickness, t. This means that the crack length $2a$ is about $2t$. From (11.17) it can be seen that this condition is satisfied when $(K_{Ic}/\sigma)^2$ is larger than πt.

11.6   DESIGNING AGAINST FATIGUE

The fatigue behavior of materials is usually described by means of the S–N diagram which gives the number of cycles to failure, N, as a function of the maximum applied alternating stress, S_a, as shown in Fig. 7.10. In the majority of cases, the reported fatigue strengths or endurance limits of materials are based on tests of carefully prepared small samples under laboratory conditions. Such values cannot be directly used for design purposes because the behavior of a component or structure under fatigue loading does not only depend on the fatigue or endurance limit of the material used in making it, but also on several other factors including: size and shape of the component or structure, type of loading and state of stress, stress concentration, surface finish, operating temperature, service environment and method of fabrication. The effect of some of these parameters on the fatigue strengths of some steels is shown in Table 11.3. The influence of the above factors on fatigue behavior of a component can be accounted for by modifying the endurance limit of the material using a number of factors, as discussed in the following paragraphs.

Table 11.3 Effect of surface condition and environment on the fatigue strength of steels

UTS	Fatigue strength as % of maximum endurance limit						
	Mirror polish	Polished	Machined	0.1 mm notch	Hot-worked surface	Under fresh water	Under salt water
280 MPa (40 ksi)	100	95	93	87	82	72	52
560 MPa (80 ksi)	100	92	88	77	63	53	36
840 MPa (122 ksi)	100	90	84	66	47	37	25
1120 MPa (162 ksi)	100	88	78	55	37	25	17
1400 MPa (203 ksi)	100	88	72	44	30	19	14
1540 MPa (223 ksi)	100	88	69	39	30	19	12

Endurance-limit modifying factors

A variety of modifying factors, or derating factors as discussed in Section 9.6, are usually used to account for the main parameters that affect the behavior of components or structures in service. The numerical value of each of the modifying factors is less than unity and each one is intended to account for a single effect. This approach is expressed as follows:

$$S_e = k_a k_b k_c k_d k_e k_f k_g k_h S'_e \qquad (11.21)$$

where S_e = endurance limit of the component or structure
$\quad S'_e$ = endurance limit of the material as determined by laboratory fatigue test
$\quad k_a$ = surface finish factor
$\quad k_b$ = size factor
$\quad k_c$ = reliability factor
$\quad k_d$ = operating temperature factor
$\quad k_e$ = loading factor
$\quad k_f$ = stress concentration factor
$\quad k_g$ = service environment factor
$\quad k_h$ = manufacturing processes factor

Equation 11.21 can be used to predict the behavior of a component or a structure under fatigue conditions provided that the values of the different modifying factors are known. The effect of some of the above factors on fatigue behavior is known and can be estimated accurately, others are more difficult to quantify, as the following discussion shows.

Surface finish factor, k_a, is introduced to account for the fact that most machine elements and structures are not manufactured with the same high-quality finish that is normally given to laboratory fatigue test specimens. The value of k_a can vary between unity and 0.2 depending on the surface finish and the strength of the material. As shown in Tables 11.3 and 11.4, stronger materials are more sensitive to surface roughness variations.

Size factor, k_b, accounts for the fact that large engineering parts have lower fatigue strengths than smaller test specimens. In general, the larger the volume of the material under stress, the greater is the probability of finding metallurgical flaws that could cause fatigue crack initiation. Although there is no quantitative agreement on the precise effect of size, the following values can be taken as rough guidelines:

k_b = 1.0 for component diameters less than 10 mm (0.4 in).
k_b = 0.9 for diameters in the range 10 to 50 mm (0.4 to 2.0 in).
$k_b = 1 - [(D - 0.03)/15]$, where D is diameter expressed in inches, for sizes 50 to 225 mm (2 to 9 in).

Reliability factor, k_c, accounts for the random variations in fatigue strength. The published data on endurance limit usually represent average values representing 50 percent survival in fatigue tests. Since most designs require higher reliability, the published values of endurance limit must be reduced by the reliability factor, k_c. The following values can be taken as guidelines:

k_c = 0.900 for 90 percent reliability.
k_c = 0.814 for 99 percent reliability.
k_c = 0.752 for 99.9 percent reliability.

Table 11.4 Effect of surface finish and UTS on surface finish factor (k_a) for steels

UTS	Forged	Hot rolled	Machined or cold drawn	Ground	Polished
	$R_a = 500–125$	$R_a = 250–63$	$R_a = 125–32$	$R_a = 63–4$	$R_a < 16$
420 MPa (60 ksi)	0.54	0.70	0.84	0.90	1.00
700 MPa (100 ksi)	0.40	0.55	0.74	0.90	1.00
1000 MPa (143 ksi)	0.32	0.45	0.68	0.90	1.00
1400 MPa (200 ksi)	0.25	0.36	0.64	0.90	1.00
1700 MPa (243 ksi)	0.20	0.30	0.60	0.90	1.00

Operating temperature factor, k_d, accounts for the difference between the test temperature, which is normally room temperature, and the operating temperature of the component or structure. For carbon and alloy steels, the fatigue strength is not greatly affected by operating temperature in the range −45 to 450°C (−50 to 840°F) and, therefore, k_d can be taken as 1.0 in this temperature range. At higher operating temperatures k_d can be calculated according to the following relationships:

$$k_d = 1 - 5800 \ (T - 450) \text{ for } T \text{ between 450 and 550°C or}$$
$$k_d = 1 - 3200 \ (T - 840) \text{ for } T \text{ between 840 and 1020°F}$$

Loading factor, k_e, can be used to account for the differences in loading between laboratory tests and service. Transient overloads, vibrations, shocks and changes in load spectrum which may be encountered during service can greatly affect the fatigue life of a component or structure. Experience shows that repeated overstressing, i.e. stressing above the fatigue limit, can reduce the fatigue life. Under such conditions k_e should be given a value less than unity. The type of loading also affects the fatigue life. Most published fatigue data are based on reversed bending test. Other types of loading, e.g. axial or torsional, generate different stress distributions in the material which could affect the fatigue results. The factor k_e can be used as a correction factor to allow the use of reversed bending data in a different loading mode. Thus:

$k_e = 1$ for applications involving bending.
$k_e = 0.9$ for axial loading.
$k_e = 0.58$ for torsional loading.

Stress concentration factor, k_f, accounts for the stress concentrations which may arise due to changes in cross-section or similar design features, as was discussed in Section 11.4. Experience shows that low-strength, ductile steels are less sensitive to notches than high-strength steels.

Service environment factor, k_g, accounts for the reduced fatigue strength due to the action of hostile environment. The sensitivity of the fatigue strength of steels to corrosive environments is also affected by their strength, as shown in Table 11.5.

Manufacturing process factor, k_h, accounts for the influence of fabrication parameters like heat treatment, cold working, residual stresses and protective coatings on the fatigue strength of the material. Although the factor k_h is difficult to quantify, it is included here as

Table 11.5 Effect of environment and UTS on the service environment factor (k_g) for steels

UTS	k_g under fresh water	k_g under salt water
280 MPa (40 ksi)	0.72	0.52
560 Mpa (80 ksi)	0.52	0.36
840 MPa (122 ksi)	0.37	0.25
1120 MPa (162 ksi)	0.25	0.17
1400 MPa (203 ksi)	0.19	0.14
1540 MPa (223 ksi)	0.19	0.12

a reminder that the above parameters should be taken into account. The following example illustrates the use of the above parameters in design.

Calculate the diameter of the two rear-axle shafts for a truck of gross mass of 6000 kg when fully loaded. Assume that 2/3 of the mass of the truck is supported by the rear axles and the construction is such that the axles can be treated as cantilever beams each of one metre length with the load acting on its end. The shaft material is heat treated 4340 steel of tensile strength 952 MPa. The shaft construction requires a change in diameter to allow the fitting of bearings. From Table 7.1, the endurance ratio of the steel is 0.56 and the endurance limit is 532 MPa. Assuming that the shaft will be finished by machining, the surface finish factor k_a can be taken as 0.68, Table 11.4. As the shaft diameter is expected to be in the range 50–225 mm, the size factor k_b can be taken as 0.8, assuming 3 in diameter. The reliability of the shaft should be high and the reliability factor is taken as $k_c = 0.752$. The loading factor can be taken as $k_e = 1$, as the shaft is loaded in bending. Stress concentration due to change in diameter can be reduced by taking a relatively large fillet radius at the change in diameter, in this case $k_f = 0.7$. From (11.21), the endurance limit of the shaft S_e is:

$$S_e = 532 \star 0.68 \star 0.8 \star 0.752 \star 1.0 \star 0.7 = 152.3 \text{ MPa}.$$

The load acting on each shaft $= 6000 \star 2/3 \star 1/2 \star 9.8 = 19\,600 \text{ N}$. Bending moment $M = 19\,600 \star 1 = 19\,600 \text{ N m}$.

$$\text{Stress acting on the surface of the shaft} = S = \frac{M}{Z} = \frac{32M}{\pi D^3}$$

$$D^3 = 32 \star 19\,600 \star 1000/\pi \star 152.3 = 1\,310\,860$$

$$D = 109.5 \text{ mm}.$$

Effect of mean stress

In the majority of cases, the fatigue behavior is determined using the rotating bending test which applies alternating tension-compression and a stress ratio $R = -1$, as shown in Fig. 7.9. In practice, however, conditions are often met where a static mean stress, S_m, is also present, as shown in Fig. 7.9. Several methods are available for describing the fatigue behavior of materials under such conditions, as shown in Fig. 11.7. The point S_e represents the fatigue or endurance limit under a stress ratio $R = -1$, while the points UTS and YS represent the ultimate tensile strength and yield strength under static loading, i.e. $S_a = 0$. In general, experimental results for ductile materials fall between the Goodman line and

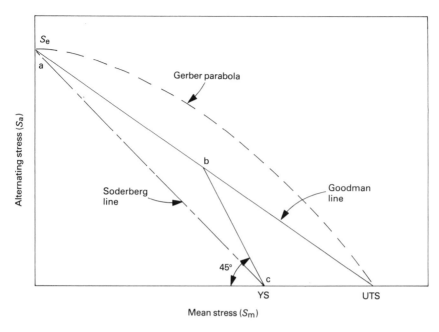

Figure 11.7 Fatigue behavior under combined static and alternating stresses.

the Gerber parabola, but because of scatter in the results, the Goodman line is usually preferred for design. The Goodman line can be represented by the relationship:

$$\frac{S_m}{UTS} + \frac{S_a}{S_e} = 1 \tag{11.22}$$

The Soderberg line is more conservative and uses the yield strength as the limiting mean stress instead of the ultimate tensile strength, as shown in Fig. 11.7. The Soderberg line can be represented by (11.22) by substituting YS for UTS.

A criterion which follows the line abc of Fig. 11.7 is not as conservative as the Soderberg criterion but avoids gross yielding at high mean stresses. Along the line bc, which is drawn from the yield stress at 45°, the sum of the mean and alternating stresses equals the yield stress. Operating below this line avoids gross yielding of smooth components under combined alternating and static stresses. As shown in Section 11.4, when the component contains a stress concentration, the maximum allowable static stress is reduced by an amount proportional to the stress-concentration factor, K_t, and the fatigue strength is reduced by an amount proportional to the fatigue stress-concentration factor, K_f. Under such conditions, the Goodman line is moved to the position dee′, as shown in Fig. 11.8. In the region de the local stresses at the stress concentration remain below the yield strength of the material and the behavior can be predicted by substituting $(K_t \star S_m)$ for S_m and $(K_f \star S_a)$ for S_a in (11.22). As the mean stress increases to point e, the material at the stress concentration begins to yield. Point e is graphically determined by drawing a 45° line from the point YS/K_t which satisfies the condition:

$$(K_t \star S_m) + (K_f \star S_a) = YS.$$

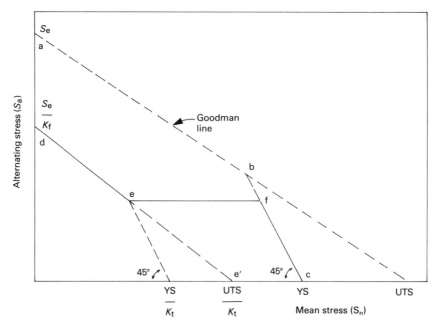

Figure 11.8 Effect of stress concentration on the design stresses under combined static and fatigue loading.

Beyond point e, and assuming that the yielded area is a small fraction of the total cross-section, further increase in the mean stress does not cause a corresponding change in the maximum of the local stress cycle as it cannot exceed the yield stress. This is indicated by the horizontal line ef, Fig. 11.8. Further increase in the mean stress increases the yielded area until gross yielding takes place at point f which satisfies the condition:

$$S_a + S_m = YS.$$

The high mean stresses needed to reach point f in a notched component are usually outside the scope of normal design.

Factor of safety

In order to use Fig. 11.8 for design purposes, a factor of safety must be applied. It is generally preferable to employ separate factors of safety for the static and alternating strengths of the material, n_m and n_a respectively. The factors of safety, n_m and n_a, and the stress-concentration factors, K_t and K_f, can be incorporated in (11.22) which will allow it to be used for design, provided that the mean stresses are less than those satisfying the condition for point e. Thus:

$$\frac{n_m K_t S_m}{UTS} + \frac{n_a K_f S_a}{S_e} = 1 \tag{11.23}$$

Combined loading

In many cases, a component or a structure is subjected to complex fatigue loading that induces three-dimensional stresses. An example is that of a rotating shaft transmitting a torque and subjected to bending load where the principal stresses do not maintain the same orientation relative to a point on the surface. Experimental fatigue tests with different combinations of bending and torsion show that the distortion-energy theory provides the best agreement for ductile materials. On the other hand, failure of brittle materials is described better by the maximum principal stress theory. A prediction of the fatigue performance of a combined stress system can be made by separately applying the failure criterion to the static mean stresses and to the alternating stresses. The mean and alternating von Mises stresses can then be defined as:

$$S'_m = \sqrt{(S^2_{xm} - S_{xm}S_{ym} + S^2_{ym} + 3\tau^2_{xym})}$$
$$S'_a = \sqrt{(S^2_{xa} - S_{xa}S_{ya} + S^2_{ya} + 3\tau^2_{xya})} \qquad (11.24)$$

Equation 11.24 can be used to combine any set of shear, normal, and bending stresses into two stress components which can be used in (11.23) in the same way as the simple $S_m - S_a$ stress system.

Cumulative fatigue damage

Engineering components and structures are often subjected to different fatigue stresses in service. Estimation of the fatigue life under variable loading conditions is normally based on the concept of cumulative fatigue damage which assumes that successive stress cycles cause a progressive deterioration in the component. The Palmgren–Miner rule, also called Miner's rule, proposes that if cyclic stressing occurs at a series of stress levels $S_1, S_2, S_3, \ldots S_i$, each of which would correspond to a failure life of $N_1, N_2, N_3, \ldots N_i$ if applied singly, then the fraction of total life used at each stress level is the actual number of cycles applied at this level $n_1, n_2, n_3, \ldots n_i$ divided by the corresponding life. The part is expected to fail when the cumulative damage satisfies the relationship:

$$\frac{n_1}{N_1} + \frac{n_2}{N_2} + \frac{n_3}{N_3} + \ldots + \frac{n_i}{N_i} = C \qquad (11.25)$$

The constant C can be determined experimentally and is usually found to be in the range 0.7 to 2.2. When such experimental information is lacking, C can be taken as unity.

The Palmgren–Miner rule does not take into account the sequence of loading nor the effect of mean stress and should only be taken as a rough guide to design.

Other fatigue-design criteria

Components made of steel or other materials that have well-defined endurance limits can be designed for an indefinite fatigue life provided that working stresses do not exceed this critical value. According to the above discussion, the endurance limit has to be reduced to account for adverse service environment as well as material, manufacturing and design inaccuracies. In some cases, components and structures whose design is based on the resulting working fatigue strength would be too heavy or too bulky for the intended application. In such cases, other design criteria like safe-life, fail-safe or damage-tolerance may be employed.

Safe-life, or finite-life, design is based on the assumption that the component or structure is free from flaws but the stress level in certain areas is higher than the endurance limit of the material. This means that fatigue-crack initiation is inevitable and the life of the component is estimated on the basis of the number of stress cycles which are necessary to initiate such a crack. Fail-safe design is based on the philosophy that cracks that form in service will be detected and repaired before they can lead to failure. Materials with high fracture toughness, crack stopping features and a reliable NDT program should be employed when the fail-safe criterion is adopted. Damage-tolerant design is an extension of the fail-safe criterion and assumes that flaws exist in engineering components and structures before they are put in service. Fracture mechanics techniques are used to determine whether such cracks will grow large enough to cause failure before they are detected during a periodic inspection. Fracture mechanics techniques are discussed in Sections 7.3 and 11.5.

Designing for finite life

Designing for a finite life can be based on a safe-life criterion and allows stresses to exceed the endurance limit, which means that components would fail after a certain number of loading cycles. Such stresses are usually present at local discontinuities and other areas of stress concentration, as discussed in Section 11.4. Finite-life designs are affected by low-cycle fatigue behavior of the material, which is a function of both its strength and ductility. Low-cycle fatigue properties are conveniently described by the strain-life curves which give the relation between the total strain amplitude, $\Delta\varepsilon$, and the number of strain reversals to failure, $2N$. The relationship between the strain amplitude and fatigue life can be expressed as:

$$\frac{\Delta\varepsilon}{2} = \frac{\sigma'f}{E}(2N)^b + \varepsilon'_f(2N)^c \tag{11.26}$$

Where σ'_f = fatigue strength coefficient, which is the true stress corresponding to fracture in one stress reversal fatigue strength coefficient and can be approximately taken as true stress at fracture in a tension test

E = elastic modulus

b = fatigue strength exponent and is the power to which the life $2N$ must be raised to be proportional to the true stress amplitude. The value of b ranges between -0.06 and -0.14 depending on the material

e = fatigue ductility exponent and is the power to which the life $2N$ must be raised to be proportional to the true plastic strain amplitude. The value of c ranges between -0.4 and -0.9

ε'_f = fatigue ductility coefficient, which is the true strain corresponding to fracture in one reversal and is a fraction, 0.35 to 1.0, of the true fracture strain measured in a static tension test.

Equation 11.26 is known as the Manson–Coffin relationship and has been simplified by Manson (*Exp. Mech.* **5**, (7) p. 193, 1965), to give:

$$\Delta\varepsilon = 3.5\frac{S_u}{EN^{0.12}} + \left(\frac{\varepsilon_f}{N}\right)^{0.6} \tag{11.27}$$

where S_u = ultimate tensile strength

ε_f = true strain at fracture in tension

Equation 11.27 simplifies the estimation of fatigue life as the three factors S_u, ε_f, and E can be obtained from the simple tension test. This equation can be used as a first approximation for the strain–life curve for unnotched components under fully reversed fatigue loading. The main difficulty with (11.26) and (11.27) is how to determine the value of $\Delta\varepsilon$, especially in the case of notches or similar discontinuities. Finite element analysis is a useful tool in this case.

11.7 DESIGNING UNDER HIGH-TEMPERATURE CONDITIONS

From discussions in Sections 7.9 and 8.7 it becomes clear that the service temperature has a considerable influence on the strength of materials and, consequently, on the working stress used in design. Depending on the temperature range, the design can be based on:

1. Short-time properties of the material for moderate temperatures.
2. Both short-time and creep properties for intermediate temperature range.
3. Creep properties of the materials for high temperatures.

For example, in the case of carbon steels 300°C (575°F) and 400°C (750°F) can be taken as the upper and lower temperature limits of ranges 1 and 3 above. Adding alloying elements to the steel generally increases the limiting temperatures of the three ranges. For example, short-time properties of 18-8 stainless steel can be used at temperatures up to 425°C (800°F).

In addition to creep, the other factors which must be taken into consideration when designing for elevated temperatures include:

1. Metallurgical and microstructural changes which occur in the material due to long-time exposure to elevated temperature.
2. Influence of method of fabrication, especially welding, on creep behavior.
3. Oxidation and hot corrosion which may take place during service and shut-down periods.

For design purposes, creep properties are usually presented on plots which yield reasonable straight lines. Common methods of presentation include log–log plots of stress vs. steady state creep rate and stress vs. time to produce different amounts of total strain (instantaneous strain plus creep strain), as shown schematically in Fig. 11.9. A change in microstructure of the material is usually accompanied by a change in creep properties, and consequently a change in the slope of the line.

Generally, designing under high-temperature conditions is carried out according to well-established codes. Most countries have one or more design codes which cover plants and structures operating at high temperatures. Examples of such codes include ASME Boiler and Pressure Vessel Code: section I (power boilers) and section VIII (pressure vessels); also BS 806:1975 (piping for land boilers) and BS 5500:1976 (unfired fusion-welded pressure vessels). The common feature of such codes is that calculations and stress analysis are kept to a minimum. Design of local areas such as branch connections and supports is provided by simple formulas and by reference to charts.

For moderate temperatures, range 1, both the ASTM Pressure Vessel Code and Boiler Code specify the allowable stresses as the lowest obtained from:

1. 25 percent of the tensile strength at room temperature.

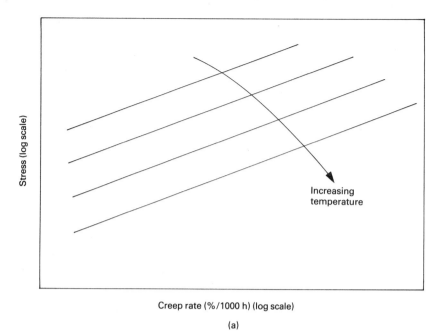

Creep rate (%/1000 h) (log scale)

(a)

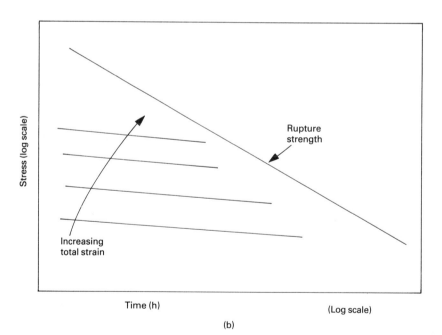

Time (h)   (Log scale)

(b)

Figure 11.9 Presentation of creep data for design purposes. (a) Variation of stress with steady-state creep rate at various temperatures. (b) Variation of stress with time to produce different amounts of total strain at a given temperature. The uppermost curve is the stress-rupture which occurs at different total strains to failure.

2. 25 percent of the tensile strength at temperature.
3. 62.5 percent of the yield strength (or 0.2 percent offset) at temperature.

For high temperatures, range 3, the Boiler Code specifies that the stress values are based on 60 percent of the stress to produce a creep rate of 1/100 percent per 1000 h. In addition, the stress values are also limited to 80 percent of the stress to produce rupture at the end of 100 000 h. Generally, service temperatures and pressures are lower than design values, wall thicknesses are often increased by a corrosion allowance and material properties are usually higher than those specified. All these factors result in increased factor of safety.

At intermediate temperatures, range 2, the code limits the stress to values obtained from a smooth curve joining the values for the low and high temperature ranges.

To illustrate the above discussion, the allowable design stress for 5 percent Cr – 0.5 percent Mo steel according to the ASME Boiler Code is shown in Fig. 11.10. The figure is based on data given by Clark (see appendix).

In many cases, creep data are incomplete and have to be supplemented or extended by interpolation or, more hazardously, extrapolation. This is particularly true of long-time

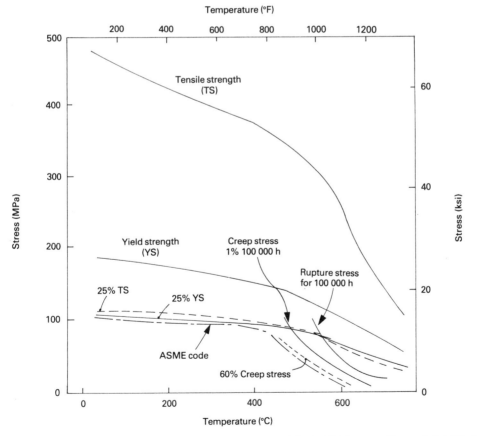

Figure 11.10 Determination of the design stress for steel 5%Cr, 0.5%Mo according to ASTM Boiler Code.

creep and stress-rupture data where the 100 000 hour (11.4 years) creep resistance of newly developed materials is required. Reliable extrapolation of creep and stress rupture curves to longer times can be made only when no structural changes occur in the region of extrapolation. Such changes can affect the creep resistance which would result in considerable errors in the extrapolated values. Since structural changes usually occur at shorter times for higher temperatures, the long-time behavior can be checked by testing the material at a higher temperature than the service temperature. For example, if no change in slope occurs in 1000 hours at 100°C (180°F) above the service temperature, extrapolation of the lower temperature results to 100 000 h may be possible. As an aid in extrapolating creep data, several time–temperature parameters have been developed for trading off temperature for time. The basic idea of these parameters is that they permit the prediction of long-time creep behavior from the results of shorter time tests at higher temperatures at the same stress. A widely used parameter for correlating the stress rupture data is the Larson–Miller parameter, LMP, which is given as:

$$\text{LMP} = T(C + \log t_r) \tag{11.28}$$

where T is the test temperature in Kelvin (°C + 273) or degrees Rankin (°F + 460)
t_r is time to rupture in hours, the log is to the base 10
C is the Larson–Miller constant, generally falls between 17 and 23, but is often taken to be 20.

Consider the following example as an illustration to the use of the Larson–Miller parameter. Assuming that a turbine blade made of Nimonic 105 alloy had a life of 10 000 hours at 150 MPa with a service temperature of 810°C, what is the expected life at the same stress but with a service temperature of 750°C? From (11.28):

$$\text{LMP} = (810 + 273)(20 + 4) = (750 + 273)(20 + \log t_r)$$

The expected life at the new service temperature is 255 6387.5 hours.

The Larson–Miller parameter can also be expressed in terms of time to give a specified strain, t_s, thus $\text{LMP} = T(C + \log t_s)$. Figure 11.11 gives a schematic representation of creep properties expressed according to the Larson–Miller parameter.

Although the Larson–Miller parameter is widely used in design at elevated temperatures, there are cases where the creep data do not fall on a single curve, even when C is adjusted for the best fit. In such cases, the Manson–Haferd parameter, with its additional flexibility, may give a better fit for the creep results. The Manson–Haferd parameter, MHP, is given as:

$$\text{MHP} = \frac{T - T_a}{\log t_r - \log t_a} \tag{11.29}$$

where T is the test temperature, °C or °F
t_r is rupture or creep time, hours
T_a and t_a are constants derived from test data.

The Manson–Haferd parameter can be used in a similar way to the Larson–Miller parameter.

When a part or a structure is subjected to combined stress during steady-state creep, the basic assumptions of the plasticity theory can be applied, provided no volume changes take place during creep. Accordingly, it can be assumed that: (a) hydrostatic stress does

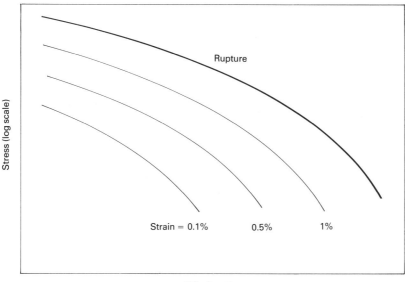

Figure 11.11 Presentation of creep data at different temperatures according to Larson–Miller parameter.

not affect creep; (b) axis of principal stress and strain-rate are coincident. Under these conditions, the effective strain rate can be correlated to effective stress using the same relationships used for uniaxial loading. The effective stress can be calculated using the von Mises or Tresca relationships.

The stress-rupture life of a part or a structure which is subjected to variable loading can be roughly estimated if the expected life at each stress level is known. Under such conditions, the life fraction rule assumes that rupture occurs when:

$$\frac{t_1}{t_{r1}} + \frac{t_2}{t_{r2}} + \frac{t_3}{t_{r3}} + \ldots = 1 \qquad (11.30)$$

where t_1, t_2, t_3 . . . are the times spent by the part under stress levels 1, 2, 3, . . . respectively.

t_{r1}, t_{r2}, t_{r3}, . . . are the rupture lives of the part under stress levels 1, 2, 3, . . . respectively.

Similar reasoning can also be applied to predict the life of a part or a structure when subjected to combined creep and fatigue loading. Cumulative fatigue damage laws, e.g. Palmgren–Miner law, can be combined with the life fraction rule, given in (11.30), to give a rough estimate of expected life under combined creep-fatigue loading. Thus:

$$\frac{t_1}{t_{r1}} + \frac{t_2}{t_{r2}} + \frac{t_3}{t_{r3}} + \ldots + \frac{n_1}{N_1} + \frac{n_2}{N_2} + \frac{n_3}{N_3} + \ldots = 1 \qquad (11.31)$$

where n_1, n_2, n_3, . . . are the number of cycles at stress levels 1, 2, 3, . . . respectively
N_1, N_2, N_3, . . . are fatigue lives at stress levels 1, 2, 3, respectively.

11.8   REVIEW QUESTIONS AND PROBLEMS

11.1 An aluminum 2014 T6 tube of 75 mm (3 in) outer diameter and 1 mm (0.04 in) thickness is subjected to internal pressure of 8.4 MPa (1200 lb/in^2). What is the factor of safety that was taken against failure by yielding when the pipe was designed? [Answer: 1.37]

11.2 Determine the dimensions of a cantilever beam of length 1 m (40 in) and rectangular section of depth-to-width ratio 2:1. The cantilever is expected not to deflect more than 50 mm (2 in) for every 1000 N (220 lb) increment of load at its tip. The material used in making the beam is steel AISI 4340 with a yield strength of 1420 MPa (206 ksi) and UTS 1800 MPa (257 ksi). What is the maximum permissible load? Assume a suitable factor of safety. [Answer: 2033 N]

11.3 If the minimum detectable crack in the above beam is 3 mm (0.118 in) what is the maximum permissible load? KIC of the beam material is 87.4 MPa m$^{1/2}$ (80 ksi in$^{1/2}$). Compare this deflection with that calculated in question 11.2. [Answer: 1824 N]

11.4 If the load of the cantilever beam in question 11.2 is fluctuating around a mean value of 1000 N (220 lb), what is the maximum alternating load that can be applied without causing fatigue failure? The endurance ratio for this steel is 0.56. Use a suitable factor of safety and modifying factors for the endurance limit. [Answer: 440 N]

11.5 If the cantilever beam in question 11.2 is made of 5% Cr, 0.5% Mo steel and is to serve at 300°C, what is the maximum allowable load that can be endured at least 100 000 hours? Use information in Fig. 11.10. [Answer: 193 N]

BIBLIOGRAPHY AND FURTHER READING

Bernasconi, G. and Piatti, G. Rd, *Creep of Engineering Materials*, Applied Science, London, 1979.
Black, P. H., *Machine Design*, 3rd ed., McGraw-Hill, New York, 1968.
Clark, C. L., *High-temperature Alloys*, Pitman, New York, 1953.
Cook, N. H., *Mechanics and Materials for Design*, McGraw-Hill, New York, 1985.
Crane, F. A. A. and Charles, J. A., *Selection and Use of Engineering Materials*, Butterworths, London, 1984.
Dieter, G. E., *Engineering Design: a Materials and Processing Approach*, McGraw-Hill, New York, 1983.
Farag, M. M., *Materials and Process Selection in Engineering*, Applied Science, London, 1979.
Hanley, D. P., *Selection of Engineering Materials*, Van Nostrand Reinhold, New York, 1980.
Maleev, V. L. and Hartman, J. B., *Machine Design*, 3rd ed., International Textbook, Scranton, Pennsylvania, 1960.
Parker, A. P., *The Mechanics of Fracture and Fatigue*, E. and F. N. Spon, London, 1981.
Peterson, R. E., *Stress-concentration Design Factors*, John Wiley & Sons, New York, 1974.
Shigley, J. E. and Mitchell, L. D., *Mechanical Engineering Design*, 4th ed., McGraw-Hill, New York, 1983.
Smith, C. O., *Introduction to Reliability in Design*, McGraw-Hill, 1976.
Timoshenko, S. and Young, D. H., *Elements of Strength of Materials*, 5th ed., D. Van Nostrand, New York, 1968.
Wilshire, B. and Owen, D. R. J., ed., *Engineering Approaches to High-temperature Design*, Pineridge Press, Swansea, 1983.

Chapter 12

Effect of Manufacturing Processes on Design

12.1 INTRODUCTION

Discussions in Chapter 1 have shown that design, materials selection and manufacturing are intimately related activities which cannot be performed in isolation of each other. Creative designs may never develop into marketable products unless they can be manufactured economically at the required level of performance. In many cases, design modifications are made to achieve production economy or to fit existing production facilities and environment. This is because every factory has its own facilities, skills, practices and customs that have to be appreciated and addressed by the designer if the most effective solution is to be found. Modifications of design may also be made in order to improve quality and performance, in which case the cost of production may increase. Recognition of the capabilities and limitations of the available processes would make it possible to specify the correct manufacturing instructions and to select the processing route that will yield the required product quality at the optimum cost. Assembly and inspection are normally considered as important manufacturing activities. Ease of assembly reduces the manufacturing cost, while ease of inspection could improve quality and reliability of the product. Products that are easy to assemble are usually easy to disassemble and this improves maintainability.

Discussions in this chapter are related to the relationship between the different manufacturing processes and the various aspects of component design.

12.2 DESIGN CONSIDERATIONS FOR CAST COMPONENTS

Casting covers a wide range of processes which can be used to shape almost any metallic material and some plastics in a variety of shapes, sizes, accuracy and surface finish. The number of castings can vary from very few to several thousands. In some cases, casting represents the obvious and only way of manufacturing, as in the case of components made of the different types of cast iron or cast alloys. In many other applications, however, decisions have to be made whether it is advantageous to cast a product or whether to use another method of manufacture. In such cases, the following factors should be considered:

1. Casting is particularly suited for parts which contain internal cavities that are inaccessible, too complex, or too large to be easily produced by machining.

2. It is advantageous to cast complex parts when required in large numbers, especially if they are to be made of aluminum or zinc alloys.
3. Casting techniques can be used to produce a part which is one of a kind in a variety of materials, especially when it is not feasible to make it by machining.
4. Precious metals are usually shaped by casting, since there is little or no loss of material.
5. Parts produced by casting are isotropic, which could be an important requirement in some applications.
6. Casting is not competitive when the parts can be produced by punching from sheet or by deep drawing.
7. Extrusion can be preferable to casting in some cases, especially in the case of lower-melting point nonferrous alloys.
8. Casting is not usually a viable solution when the material is not easily melted, as in the case of metals with very high melting points such as tungsten.

When casting is selected as the manufacturing process, it is important for the designer to observe the general rules which are related to solidification, material of the casting, position of the casting in the mold and required accuracy of the part. Designs which violate the rules of sound casting can make production impossible or possible only at higher expense and large rejection rates.

A general rule of solidification is that the shape of the casting should allow the solidification front to move uniformly from one end toward the feeding end, i.e. directional solidification. This can most easily be achieved when the casting has virtually uniform thickness in all sections. In most cases this is not possible. However, when section thickness must change, such change should be gradual, in order to avoid feeding and shrinkage problems. Sudden changes in section thickness give rise to stress concentration and possible hot tears in the casting. Figure 12.1 gives some guidelines to avoid these defects. Another problem which arises in solidification is caused by sharp corners, as these also give rise to stress concentration and should be replaced by larger radii. When two sections cross or join, the solidification process is interrupted and a hot spot results. Hot spots retard solidification and usually cause porosity and shrinkage cavities. Some solutions to this problem are given in Fig. 12.1. Large unsupported flat areas should also be avoided as they tend to warp during cooling.

The type and composition of the material play an important part in determining the shape, minimum section thickness and strength of the casting. Materials which have large solidification shrinkage and contain low-melting phases are susceptible to hot tears. Collapsible mold materials and a casting shape which allows shrinkage with least stresses in the casting are required in such cases. Another material variable is castability, which can be related to the minimum section thickness which can be easily achieved. Table 3.5 gives some guidelines to minimum section thickness which can be economically obtained for different casting processes. It should be noted that the shape and size of the casting as well as the casting process and foundry practice can affect the minimum section thickness. As cooling rates are directly related to section thickness, sections of different thickness can develop different structures and mechanical properties. In cast steel, the main effect of thicker section is increased grain size. In cast iron, however, the grain size as well as the graphite size and the amount of combined carbon are affected, which greatly influences the strength and hardness. Thinner sections result in faster cooling rates and higher strength and hardness in the different classes of gray cast iron.

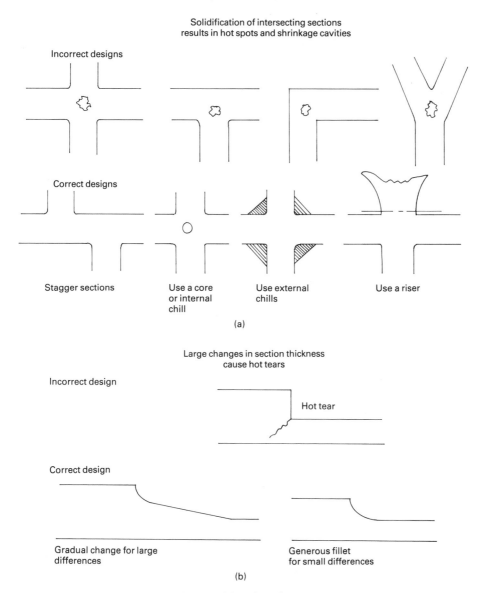

Figure 12.1 Design considerations for cast components.

The tolerances and surface finishes which can be achieved by standard commercial foundry practice are given in Tables 3.1 and 3.5.

12.3   DESIGN CONSIDERATIONS FOR MOLDED PLASTIC COMPONENTS

As shown in Chapter 4, compression, transfer and injection molding processes are the commonly used methods of molding plastic components. These processes involve the introduction of a fluid or semifluid material into a mold cavity and permitting it to solidify into the desired shape. Although section thickness, dimensional accuracy, or the incorporation of inserts would make it desirable to use one molding technique in preference to another, there are general design rules that should be observed in order to ensure the quality of the product. Experience shows that the mechanical, electrical and chemical properties of molded components are influenced by the flow of the molten plastic as it fills the mold cavity. Streamlined flow will avoid gas pockets in heavy-sectioned areas.

An important common feature in molding processes is draft, which is required for easy ejection of molded parts from the mold cavity. A taper of 1 to 4° is usual for polymers but tapers of less than 1° can be used for deep articles. Another common feature is uniformity of wall thickness. In general, molded parts should be designed to have uniform wall thickness. Nonuniformity of thickness in a molded piece tends to produce nonuniform cooling and unbalanced shrinkage, leading to internal stresses and warpage. If thickness variations are necessary, generous fillets should be used to allow gradual change in thickness. The effect of junctions and corners can also be reduced by using a radius instead, as shown in Fig. 12.2.a. The nominal wall thickness must obviously be such that the part is sufficiently strong to carry the expected service loads. However, it is better to adjust the shape of the part to cope with the applied load than to increase the wall thickness. Thick sections should be avoided as they retard the molding cycle and require more material. Ribs, beads, bosses, edge stiffeners and flanges should be used instead. However, shrinkage dimples (sink marks) may appear opposite ribbed surfaces if they are not proportioned correctly. Adopting the proportions shown in Fig. 12.2b may eliminate sink marks. Large, plain, flat surfaces should be avoided as they are prone to warping and lack rigidity. Such surfaces should be strengthened by ribbing or doming.

The presence of holes disturbs the flow of the material during molding and a weld line occurs on the side of the hole away from the direction of flow. This results in a potentially weak point and some form of strengthening, such as bosses, may be necessary, as shown in Fig. 12.2.c. Through holes are preferred to blind holes from a manufacturing standpoint. This is because core pins can often be supported in both halves of the mold in the case of through holes, but can only be supported from one end in the case of blind holes. Undercuts are undesirable features in molded parts as they cause difficulties in ejection from the mold. Examples of external and internal undercuts are shown in Fig. 12.2d. Some parts with minor undercuts may be flexible enough to be stripped from the mold without damage. Many thermoplastics can tolerate about 10 percent strain during ejection from the mold. Parts with external or internal threads can be made by molding and are usually removed from the mold by unscrewing. The mold is usually costly, since unscrewing devices may need to be incorporated. External threads may be produced without special devices if they are located on the parting plane of the mold. The thread may need secondary finishing in this case. Threads can also be formed by tapping, especially in the case of diameters less than 8 mm (5/16 in).

Inserts are used in molded parts to strengthen them or to meet special functional requirements. To anchor the insert to the plastic part, a mechanical or friction lock must be provided. Inserts are usually made of steel, brass or aluminum, but other materials can also be used. Inserts can cause difficulties in assembly, especially when clips or threads are used. When it is necessary to use inserts, they should be surrounded by enough material to avoid having the insert break loose in service. Inserts with sharp

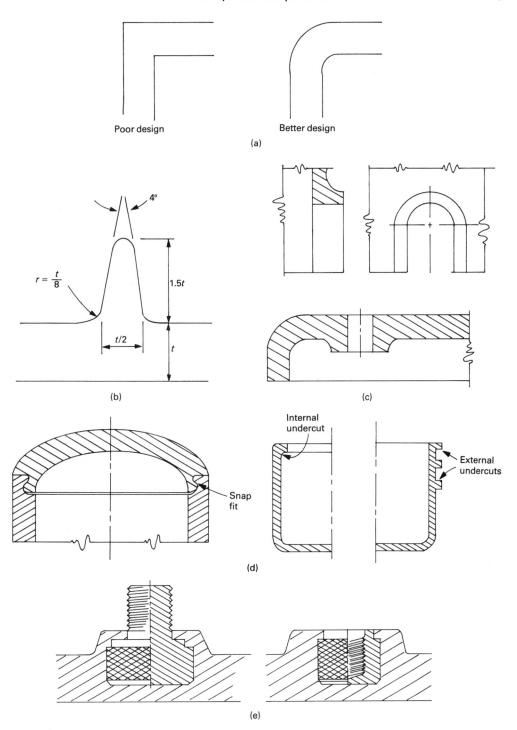

Figure 12.2 Some design features of plastic parts. (a) Using radii instead of sharp corners. (b) Recommended rib proportions to avoid sink marks. (c) Use of bosses to strengthen areas round holes and slots. (d) Examples of undercuts. (e) Examples of threaded inserts for molded plastic parts.

corners cause stress concentration in the surrounding plastic and should be avoided. Figure 12.2e shows two examples of threaded inserts.

Plastic parts can be given a wide variety of surface finishes including mirror-like finish, dull satin, wood grain, leather grain and other decorative textures. A highly smooth surface is usually required for surfaces that are to be painted or vacuum metallized. Decorative textures are often helpful in hiding any possible surface imperfections such as flow lines or sink marks. Raised letters on a molded part are easier and cheaper to produce than depressed letters because the lettering is machined into the mold cavity. The position of the parting plane of mold should be carefully considered as it is normally accompanied by an unsightly flash line.

Dimensional tolerances in molded plastic parts are affected by the type and constitution of the material, shrinkage of the material, heat and pressure variables in the molding process and the toolmaker's tolerances on the mold manufacture. Generally, shrinkage has two components: (a) mold shrinkage, which occurs upon solidification; and (b) after shrinkage, which occurs in some materials after 24 h. For example, a thermosetting plastic like melamine has a mold shrinkage of about 0.7 to 0.9 per cent and an after shrinkage of 0.6 to 0.8 percent. Thus a total shrinkage of about 1.3 to 1.7 percent should be considered. On the other hand, a thermoplastic like polyethylene may shrink as much as 5 percent and nylon as much as 4 percent. When the mold contains more than one cavity, to make multiple parts per mold closure, tight tolerances become more difficult. Thus, some relaxation of tolerances should be allowed if a multiple cavity mold is desired. Larger tolerances should also be allowed in filled polymers. In addition, the value of the tolerance depends on the size of the part and the direction in relation to the parting plane. Larger dimensions are normally accompanied by larger tolerances. For example, dimensions less than 25 mm (1 in) can be held within $\pm50\,\mu$m (±0.002 in). Larger dimensions are usually given tolerances of ±10 to $20\,\mu$m/cm (±0.001 to 0.002 in/in). Generally, dimensions at right angles to the parting plane should be given higher tolerance than dimensions parallel to it. As a rough guide, if a tolerance of $\pm50\,\mu$m is allowed in the direction parallel to the parting plane of the mold, a tolerance of ±0.2 mm should be allowed at right angles. Generous dimensional tolerances make economical production possible.

12.4   DESIGN CONSIDERATIONS FOR FORGED COMPONENTS

Forging processes represent an important means of producing relatively complex parts for high-performance applications. In many cases forging represents a serious competitor to casting, especially for solid parts that have no internal cavities. Forged parts have wrought structures which are usually stronger, more ductile, contain less segregation and are likely to have less internal defects than cast parts. This is because the extensive hot working which is usually involved in forging closes existing porosity, refines the grains and homogenizes the structure. On the other hand, cast parts are more isotropic than forged parts, which usually have directional properties. This directionality is due to the fiber structure which results from grain flow and elongation of second phases in the direction of deformation. Forged components are generally stronger and more ductile in the direction of fibers than across the fibers. This directionality can be exploited in some cases to enhance the mechanical performance of the forged part, as shown in Fig. 12.3.

When forging is selected as the manufacturing process, it is important for the designer

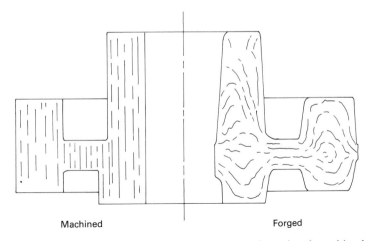

Machined   Forged

Figure 12.3 Schematic comparison of the grain flow in forged and machined parts.

to observe the general rules which are related to the flow of material in the die cavity. As with casting, it is better to maintain uniform thickness in all sections. Rapid changes in thickness should be avoided because these could result in laps and cracks in the forged metal as it flows in the die cavity. To prevent these defects, generous radii must be provided at the locations of large changes in thickness. Another similarity with casting is that vertical surfaces of a forging must be tapered to permit removal from the die cavity. A draft of 5 to 10° is usually provided.

It is better to locate the parting line near the central height of the part. This avoids deep impressions in either half of the die and easier filling of the die cavity. Inaccuracies in die-forging result from mismatch between the die halves, due to the lateral forces that occur during forging, and from incomplete die closure, which is usually introduced to avoid die-to-die contact. A design would be more economically produced by forging if dimensions across the parting line are given appropriate mismatch allowance and parallel dimensions are given a reasonable die closure allowance. Specifying close tolerances to these dimensions could require extensive machining which would be expensive. Allowance must also be made for surface scale and warpage in hot forged parts. The dimensional tolerances and surface roughness that are commercially achieved in forging processes are shown in Table 3.1.

12.5   DESIGN OF POWDER METALLURGY PARTS

Powder metallurgy (P/M) techniques can be used to produce a large number of small parts to the final shape in few steps, with little or no machining and at high rates. Many metallic alloys, ceramic materials and particulate reinforced composites can be processed by powder metallurgy techniques. Generally, parts prepared by the traditional P/M techniques, which involve mechanical pressing followed by sintering, contain from 4 to 10 volume percent porosity. The amount of porosity depends on part shape, type and size of powder, lubricant used, pressing pressure, sintering temperature and time, and finishing treatments. The distribution and volume fraction of porosity greatly affect the mechanical, chemical and physical properties of parts prepared by P/M techniques.

Using higher compaction pressures or employing techniques like P/M forging and hot isostatic pressing (HIP) can greatly decrease porosity and provide strength properties close to those of wrought materials. The HIP process is particularly suited to producing parts from high-temperature alloys that are difficult to forge and machine. An added advantage of P/M is versatility. Materials that can be combined in no other way can be produced by P/M. Examples include aluminum-graphite bearings, copper-graphite electrical brushes, cobalt-tungsten carbide cutting tools (cermets) and porous bearings and filters. P/M is also the only practical way of processing tungsten and other materials with very high melting points.

The final tolerances in mechanically pressed and sintered components are comparable to those achievable by machining on production machine tools. Tolerances in the axial direction, die-fill direction, are usually about +2 percent of the dimension. Closer tolerances in diameter are usually possible and can be less than +0.5 per cent of the diameter.

On a unit weight basis, powdered metals are considerably more expensive than bulk wrought or cast materials. However, the absence of scrap, elimination of machining, the fewer production steps and the higher rates of production often offset the higher material cost. Dies needed for mechanical pressing are also an expensive item in P/M techniques. Production volumes of less than 10 000 parts are usually not practical for mechanically pressed parts. When HIP is used to produce relatively large parts using materials that are difficult to form by other techniques, production runs as low as 20 parts could be economical.

Unlike forging or casting processes, mechanical compaction of powders is restricted to two dimensions. It is impractical to apply pressure to the sides of mechanical dies, so that the flow of powders during compaction is almost entirely axial. It is also necessary to be able to eject the compact. These limitations give rise to certain design rules which have been established by the Powder Metallurgy Parts Association and the Metal Powder Industries Federation. These rules can be summarized as follows:

1. The shape of the part must permit ejection from the die, Fig. 12.4a.
2. Parts with straight walls are preferred. No draft is required for ejection from lubricated dies.
3. Parts with undercuts or holes at right angles to the direction of pressing cannot be made, Fig. 12.4b.
4. Straight serrations can be made easily, but diamond knurls cannot, Fig. 12.4.c.
5. The shape of the part should be such that the powder is not required to flow into thin walls, narrow splines or sharp corners. Side walls should be thicker than 0.75 mm (0.030 in).
6. The shape of the part should permit the construction of strong tooling and punches should have no sharp or feather edges, Fig. 12.4.d.
7. The part should be designed with as few changes in diameter and section thickness as possible.
8. Since pressure is not transmitted uniformly through a deep bed of powder, the length/diameter ratio of a mechanically pressed part should not exceed about 2.5:1.
9. Take advantage of the fact that certain materials, such as cermets and porous components, can be produced by P/M which are impossible, impractical, or uneconomical to obtain by any other method.

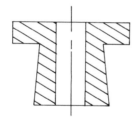

(a) Reverse taper should be avoided, use parallel sides and machine the required taper after sintering

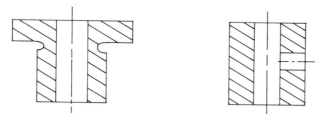

(b) Undercuts and holes at right angles to pressing direction should be avoided; if necessary such features are introduced by machining after sintering

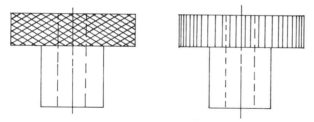

(c) Diamond knurls should be replaced by straight serrations

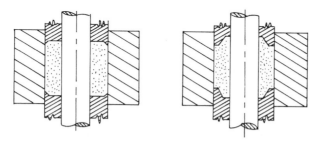

(d) Part shapes that require feather edges on punches and similar weakening features should be avoided

Figure 12.4 Design considerations for powder metallurgy components.

10. P/M parts may be bonded by assembling in the green condition and then sintering together to form a bonded assembly. Other joining methods are also possible and can be used to join P/M parts to castings or forgings.

12.6   DESIGN OF SHEET METAL PARTS

Parts made from sheet metal cover a wide variety of shapes, sizes and materials. Many examples are found in automotive, aircraft and consumer industries. Generally, sheet metal parts are produced by shearing, bending and/or drawing. The quality of the sheet material plays an important role in determining the quality of the finished product, as well as the life of tools and the economics of the process. Grain size of the sheet material is important and should be closely controlled. Steel of 0.035–0.040 mm (0.001–0.0016 in) grain size is generally acceptable for deep drawing applications. When formability is the main requirement in a sheet material, drawing quality low-carbon steels represent the most economic alternative. However, with the increasing demand for lighter energy-conserving products, other steel grades are increasingly used to make thinner and stronger components. Examples include control-rolled high-strength low-alloy (HSLA) steels and dual-phase steels.

The construction, operation and life of shearing tools as well as blanking and punching dies can be improved if the following rules are followed:

1. Diameters of punched holes should be more than the sheet thickness with a minimum of about 0.6 mm (0.025 in). Smaller holes result in extensive punch breakage and should be drilled.
2. The distance between holes, or between a hole and sheet edge, should be at least equal to sheet thickness.
3. The width of any slot should be at least 1.5 times the sheet thickness with a minimum of 3 mm (1/8 in).
4. Tolerances of about ±0.08 mm (±0.003 in) are usually achieved in blanking and punching. Closer tolerances can be achieved by shaving at an increased cost.
5. If possible the design should be changed to allow nesting of the parts to reduce scrap.
6. It may be less expensive to design the component from several simple parts than to make an intricate blanked part.
7. Blanks with sharp corners are more expensive to produce.

Metal sheets are usually anisotropic, which means that their strength and ductility vary when measured at different angles with respect to the rolling direction. This anisotropy is caused by the elongated inclusions, i.e. stringers, and by preferred orientation in the grains, i.e. texture. The most important factor which should be considered when designing parts that are to be made by bending is bendability. This is related to the ductility of the material and is expressed in terms of the smallest bend radius that does not crack the material. Bendability of a sheet is usually expressed in multiples of sheet thickness, such as $2T$, $3T$, $4T$, etc. A $2T$ material has greater bendability than a $3T$ material. Because of anisotropy, bendability of a sheet is usually greater when tested such that line of bend is at right angles to the rolling direction of the sheet. It is inadvisable to specify sharp bends in steel sheets parallel to the rolling direction for tempers above 1/4-hard. If two perpendicular bed axes are involved, the sheet should be oriented so they are at 45° to the rolling direction.

Another factor which should be considered when designing for bending is springback which is caused by the elastic recovery of the material when the bending forces are removed. One way of compensating for springback is to overbend the sheet. Another method is bottoming, which eliminates the elastic recovery by subjecting the bend area to high localized stresses. A tolerance of ± 0.8 mm ($\pm 1/32$ in) or more should be allowed in bent parts.

Drawing and stretching operations produce thin-walled hollow parts, such as seamless can bodies, household utensils and several components of the motor car body. One of the most widely used operations in this category is deep drawing, Fig. 3.8. This is used to make circular cups, noncircular symmetrical shapes and asymmetrical complex shapes. In many cases, complicated stress patterns arise as the metal tries to move in several directions simultaneously. Sharp changes over the punch or die edges can impose excessive strain on the sheet and lead to failure. The die aperture edge radius is related to the sheet thickness T and is usually kept in the range of $4T$ to $10T$ according to the severity of the draw. The punch nose radius is not so critical, but should not fall below $2T$. However, too large a radius can also cause difficulties with wrinkling of the unsupported sheet.

12.7   DESIGNS INVOLVING JOINING PROCESSES

Joining can be considered as a method of assembly, where parts made by other processes are joined to make more complex shapes or larger structures. In this respect, joining extends the capabilities of processes like casting, forging and sheet metal working and allows the manufacture of products like machine frames, steel structures, motor car bodies, beverage and food cans, storage tanks and piping systems. Some joints are temporary and can be dismantled easily, as in the case of bolted joints, while others provide permanent assembly of joined parts, as in the case of rivets and weldments. Normally, the major function of a joint is to transmit stresses from one part to another and in such a case the strength of the joint should be sufficient to carry the expected service loads. In some applications, tightness of the joint is also necessary to prevent leakage. Because joints represent areas of discontinuities in the assembly, they should be located in low stress regions, especially in dynamically loaded structures. Other design considerations which are applicable to the main joining processes will be discussed in the following paragraphs.

Welding

Welding has replaced riveting in many applications including steel structures, boilers, tanks and motor car chassis. This is because riveting is less versatile and always requires lap joints. Also, the holes and rivets subtract from strength and a riveted joint can only be about 85 percent as strong, whereas a welded joint can be as strong, as the parent metal. Welded joints are easier to inspect and can be made gas- and liquid-tight without the caulking which has to be done in riveted joints. On the negative side, however, is that structures produced by welding are monolithic and behave as one piece. This could adversely affect the fracture behavior of the structure. For example, a crack in one piece of a multipiece riveted structure may not be serious, as it will seldom progress beyond that piece without detection. However, in the case of a welded structure a crack that starts in a single plate or weld may progress for a large distance and cause complete failure. Another factor which should be considered when designing a welded structure is the effect of size on the energy-absorption ability of steels. A Charpy impact specimen could show a much lower brittle–ductile transition temperature than a large welded structure made of the same material. The liberty ships that were made during the Second World War are examples of monolithic structures that were made out of unsuitable steel. Many of these ships failed, some while in harbor, as a result of one crack propagating across the whole structure. Thus the notch-ductility of a steel that is to be used in large welded structures should be carefully assessed. Other rules which should be considered when designing a welded structure include:

1. Welded structures and joints should be designed to have sufficient flexibility. Structures that are too rigid do not allow shrinkage of the weld metal, have restricted ability to redistribute stresses and are subject to distortions and failure.
2. Accessibility of the joint for welding, welding position and component matchup are important elements of the design.
3. Thin sections are easier to weld than thick ones.
4. Welded sections should be about the same thickness to avoid excessive heat distortion.
5. It is better to locate welded joints symmetrically around the axis of an assembly in order to reduce distortion.
6. If possible, welded joints should be placed away from the surfaces to be machined. Hard spots in the weld can damage the cutting tools.
7. An inaccessible enclosure in a weldment, or the mating surfaces of a lap joint, should be completely sealed to avoid corrosion.
8. Where strength requirements are not critical, short intermittent welds are preferable to long continuous ones as distortion is reduced.
9. Help shrinkage forces work in the desired direction; presetting the welded parts out of position before welding so that shrinkage forces will bring them into alignment.
10. Use weld fixtures and clamps to reduce distortion.
11. Whenever possible, meeting of several welds should be avoided.
12. Balance shrinkage forces in a butt joint by welding alternately on each side.
13. Remove shrinkage forces by heat treatment or by shot peening.
14. Tolerances on the order of 1.5 mm (c. 1/16 in) are possible in welded joints. Surfaces that need closer tolerances should be finished by machining after welding and postwelding heat treatment.
15. Parts that have been designed for casting or forging should be redesigned if they are to be made by welding. The new design should take advantage of the benefits of welding and avoid its limitations.

Metal plates can be joined by welding in five main types of joints, as shown in Fig. 12.5a. Lap, tee and corner joints use fillet-type welds, as shown in Fig. 12.5b. Welding of thick plates requires edge preparation in order to ensure complete penetration. In such cases, one or both of the edges to be welded are chamfered in such a way as to minimize the amount of weld metal deposited. This is because the cost per unit weight of deposited weld metal is about 25 to 50 times as much as structural steel. In addition, the amount of shrinkage and distortion increases as the amount of deposited metal increases.

To facilitate communication between the designer and the manufacturing engineer, various welding symbols have been standardized and should be used when specifying

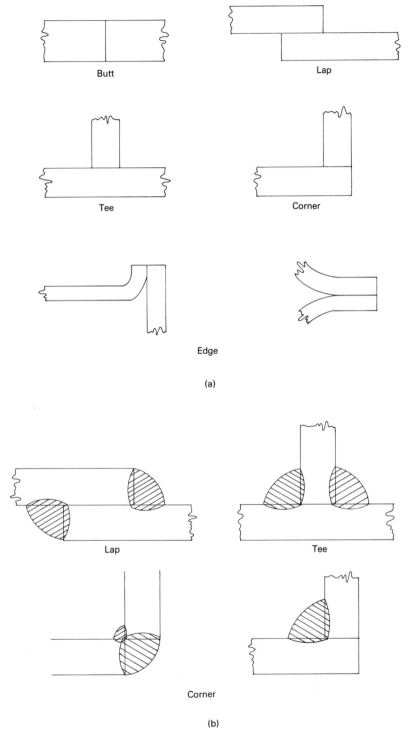

Butt

Lap

Tee

Corner

Edge

(a)

Lap

Tee

Corner

(b)

Figure 12.5 (a) Types of welded joints. (b) Use of fillet-type wells in lap, tee and corner joints.

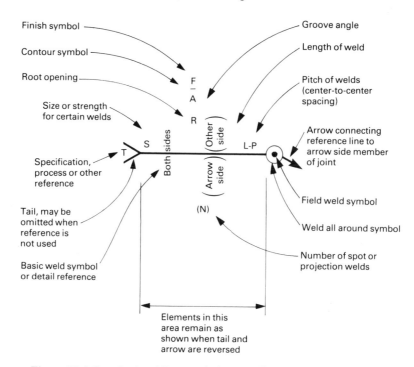

Figure 12.6 Standard welding symbols according to the AWS system.

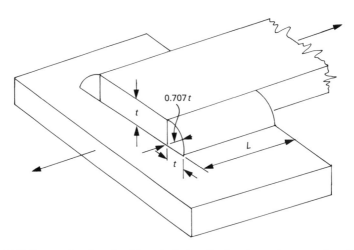

Figure 12.7 Parameters involved in calculating the load-carrying capacity of lap joints.

welded joints. Figure 12.6 shows an example of the elements of a standard welding symbol according to the American Welding Society (AWS).

Full-penetration butt welds are generally considered to have the same strength as the base metal. Hence there is no need to calculate the strength of the weld if the deposited metal is the same as the base metal. The strength of a fillet weld is inherently lower than a full-penetration butt weld. When the applied load is parallel to the weld line, the plane of rupture is at 45°, weld throat, as shown in Fig. 12.7. The AWS code gives the allowable force per unit length of the weld as 30 percent of the tensile strength, S, of the welding electrode. Thus, the load carrying capacity, P, of the two fillet welds shown in Fig. 12.7 is:

$$P = 2 \star 0.30\ S \star 0.707\ t \star L \qquad (12.1)$$

where: L = length of weld
       t = leg of weld. In this case, same as plate thickness.

Brazing

Brazing, unlike welding, uses capillary action to draw the filler material into the small clearance of the joint. As the filler material is usually weaker than the base metal, the joint should be designed differently from that of welding, as shown in Fig. 12.8. Whenever possible, a lap joint should be used in brazing and the optimum overlap is three to four times the thickness of the thinnest section in the joint. If the geometry does not allow for a lap joint, scarf or double scarf joints are preferable to butt joints. When designing a brazed assembly the following design rules should be observed:

1. The joint should be shaped in such a way that the filler material is stressed in compression. If this is not possible, it should be stressed in pure shear, and if this is not possible, then pure tension.
2. The edges of the joint should not be subjected to combined tension and bending.
3. Joint should not have features that subject the edges to high stress concentration.
4. In critical applications and when impact or fatigue loading is expected, mechanical link between the joint components should be provided to carry the load. The brazing alloy in this case should only serve to ensure rigidity in the finished joint (Fig. 12.9).

Adhesive bonding

Adhesives represent an attractive method of joining and their use is increasing in many applications. Some of the main advantages in using adhesives are:

1. Thin sheets and dissimilar thicknesses can be easily bonded.
2. Adhesive bonding is the most logical method of joining polymer matrix composites.
3. Dissimilar or incompatible materials can be bonded.
4. Adhesives are electrical insulators and can prevent galvanic action in joints between dissimilar metals.
5. Flexible adhesives spread bonding stresses over wide areas and accommodate differential thermal expansion.
6. Flexible adhesives can absorb shocks and vibrations, which increases fatigue life.
7. Preparation of bonded joins requires no fastener holes which gives better structural integrity and allows thinner gage materials to be used.
8. Adhesives provide sealing action in addition to bonding.
9. The absence of screw heads, rivet heads, or weld beads in adhesive bonded joints is advantageous in applications where interruption of fluid flow cannot be tolerated or where appearance is important.
10. Adhesive bonding can also be used in conjunction with other mechanical fastening methods to improve strength of the joint.

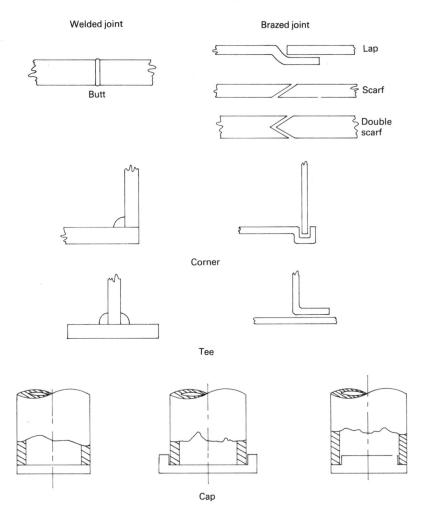

Figure 12.8 Comparison of welded and brazed joint designs.

The main limitations of adhesives are:

1. Bonded joints are weaker under cleavage and peel loading than under tension or shear.
2. Most adhesives cannot be used at service temperatures above 300°C (*c.* 600°F).
3. Solvents can attack adhesive bonded joints.
4. Some adhesives are attacked by ultraviolet light, water and ozone.
5. The designer should also be aware of the adhesive's impact resistance and creep, or cold flow, strength.

The strength of an adhesive joint depends on the joint geometry, the direction of loading in relation to the joint, the adhesive material, surface preparation and application and curing technique. As the strength of an adhesive joint is limited by the bonded area, lap and double strap joints are generally preferred to butt joints. If the geometry constraints do not allow for such joints, a scarf or double scarf joints should be made, as shown in Fig. 12.10.a. When a lap joint is used to bond thin sections, tensile shear causes deflection and this results

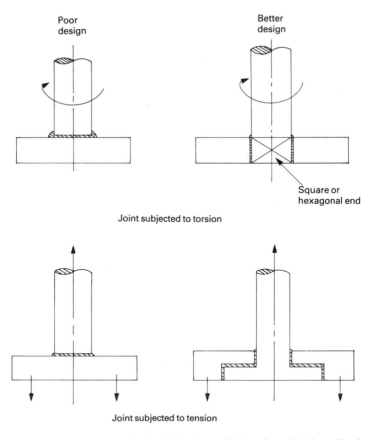

Poor
design

Better
design

Square or
hexagonal end

Joint subjected to torsion

Joint subjected to tension

Figure 12.9 Precautions taken in designing brazed joints for critical applications.

in stress concentration at the end of the lap. Tapering the ends of the joint, as shown in
Fig. 12.10b, gives more uniform loading throughout the joint. Since adhesive joints are
weaker under pealing forces, joint design should avoid this type of loading.

12.8   DESIGNS INVOLVING HEAT TREATMENT

Heat treatment represents an important step in the sequence of processes that are usually
performed in the manufacture of metallic parts. Almost all ferrous and many nonferrous
alloys can be heat treated to achieve certain desired properties. Heat treatment can be used
to make the material hard and brittle, as in the case of quench-hardening of steels, or it can be
used to make it soft and ductile, as in the case of annealing. Generally, hardening of steels
involves heating to the austenitic temperature range, usually 750 to 900°C (c. 1400 to
1650°F), and then quenching to form the hard martensitic phase. The nonuniform temperature distribution that occurs during quenching and the volume change that accompanies the
martensitic transformation can combine to cause distortions, internal stresses and even
cracks in the heat treated part. Internal stresses can cause warping or dimensional changes
when the quenched part is subsequently machined or can combine with externally applied

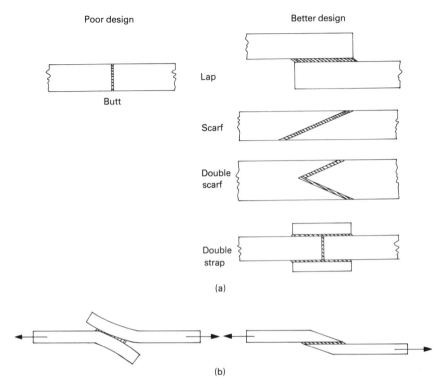

Figure 12.10 Adhesive joint design. Butt joint is weak and the bond area should be increased by using lap, scarf, double scarf or double strap joints.

stresses to cause failure. Corrosion problems can also be aggravated due to the presence of internal stresses. These difficulties can be reduced or eliminated by selecting the appropriate material, using improved techniques and eliminating undesirable design features. Undesirable design features include:

1. Nonuniform sections or thicknesses.
2. Sharp internal or external corners.
3. Nonsymmetrical shapes.

Two quenching techniques can be used to reduce the temperature gradients that can lead to internal stresses and cracking. The first technique is interrupted quench and involves quenching to a temperature above the martensitic transformation temperature, holding for a certain time to achieve more uniform temperature distribution and then cooling to form martensite (see Fig. 2.3). This process is usually called martempering or marquenching. The second technique is thermomechanical processing and is similar to martempering, but with the addition of mechanical working while the material is held above the martensitic transformation temperature (see Fig. 2.3). This process is usually called ausforming.

Selecting the appropriate steel is the most important way of avoiding heat treatment difficulties. This is because design limitations may make it difficult to adhere to the shape limitations given above. In addition, the improved quenching techniques involve extra cost and could be less economic than using a slightly more expensive but easier-to-harden steel.

Hardenability is the most important factor in the selection of steel for heat treated parts as it determines the depth and distribution of hardness induced by quenching. In addition, steels with high hardenability require a less severe cooling rate to achieve a given hardness value. Increasing the carbon content of the steel increases its hardenability. However, large amounts of carbon are undesirable as the steel becomes more prone to distortion and cracking in heat treatment. Addition of certain alloying elements to steel increases hardenability and makes it possible to harden larger sections and to use oil rather than water to minimize distortion and avoid cracking.

Manganese, chromium and molybdenum are commonly used to increase the hardenability of steels. When selecting steels for heat treated parts, it is important to specify H-steels. These are guaranteed by the supplier to meet established hardenability limits for specific ranges of chemical composition. Generally, the cost of steel usually increases as hardenability increases. In order to determine the total cost of the heat treated part, both the cost of steel and the cost of treatment must be considered together. Where only a few parts are required it may be more economical to use a steel with high hardenability and employ less accurate heating and quenching techniques. In mass production, however, using less expensive steel with lower hardenability may justify the extra cost of precision heat treating equipment.

12.9   DESIGNS INVOLVING MACHINING PROCESSES

Machining operations are the most versatile and most common manufacturing processes. Machining could be the only operation involved in the manufacture of a component, as in the case of shafts and bolts which are machined from bar stock, or it could be used as a finishing process, as in the case of cast and forged components. In all cases, it is important for the designer to ensure that the component will be machined conveniently and economically. The following discussion illustrates some component shapes and features which can cause difficulties in machining, take undue length of time to machine, call for a precision and skill that may not be available, or that may even be impossible to machine by standard machine tools and cutting tools.

1. The workpiece must have a reference surface which is suitable for holding it on the machine tool or in a fixture. This could be a flat base or a cylindrical surface. If the final shape does not have such surface, a supporting foot or tab could be added to the rough casting or forging for support purposes and removed from the part after machining.
2. Whenever possible, the design should allow all the machining operations to be completed without resetting or reclamping.
3. Whenever possible, the radii between the different machined surfaces should be equal to the nose radius of the cutting tool.
4. If the part is to be machined by traditional cutting methods, deflection under cutting forces should be taken into account. For the same cutting force, the deflection is higher the thinner the part and the lower the elastic modulus. Under these conditions, some means of support is necessary to ensure the accuracy of the machined part.
5. Twist drills should enter and exit at right angles to the drilled surface (Fig. 12.11). Drilling at an angle to the surface causes deflection of the drill and could break it.
6. Features at an angle to main machining direction should be avoided as they may require special attachments or tooling (Fig. 12.11).

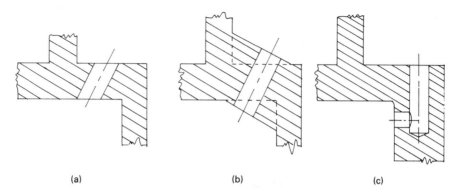

(a)   (b)   (c)

Figure 12.11 Design of drilled part. (a) Poor design as drill enters and exits at an angle to the surface. (b) Better design, but drilling the hole needs a special attachment. (c) Best design.

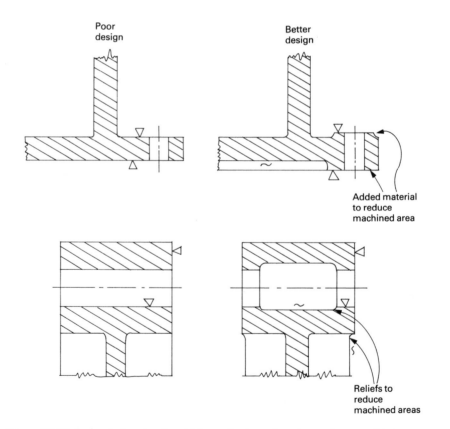

Figure 12.12 Some design details which can be introduced to reduce machining cost.

7. Flat bottom drilled holes should be avoided as they involve additional operations and the use of bottoming tools.

8. To reduce the cost of machining, machined areas should be kept to a minimum. Two examples to methods of reducing the machined area are shown in Fig. 12.12.

9. Cutting tools often require runout space, as they cannot be retracted immediately. This is particularly important in the case of grinding, where the edges of the grinding wheel wear out faster than the center. Figure 12.13 gives some examples to illustrate this point.

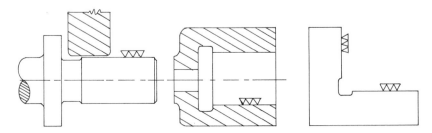

Figure 12.13 Some design details which can be introduced to give runout space for grinding wheels.

10. Thread-cutting tools normally have a chamfer on their leading edge. This chamfer means that the first two pitches do not cut a full thread. If an external diameter ends at a shoulder, the mating screwed part cannot reach the shoulder unless an undercut or a countersink are provided, as shown in Fig. 12.14. The length of the needed undercut or countersink is usually three pitches of the thread. Similar features are needed for internal screw threads, as shown in Fig. 12.15.

12.10   DESIGNING FOR CORROSIVE ENVIRONMENTS

As shown in Sections 7.7 and 7.8, corrosion represents a serious and common source of failure in many areas of engineering. Corrosion failures can be prevented or at least reduced by selecting the appropriate material, as discussed in Section 8.6; and by observing certain design rules, as will be discussed in this section.

Galvanic corrosion usually takes place as a result of design errors where dissimilar metals are placed in electrical contact. Under such conditions, corrosion occurs in the anodic material while the cathodic material is protected. The rate of corrosion and damage caused depends on the difference between the two materials in the galvanic series (Table 7.2) and on the relative areas of the exposed parts. A small anode and a large cathode will result in intensive corrosion of the anode, while a large anode and a small cathode is not as serious. For example, a steel bolt in a copper plate will be rapidly attacked in seawater, whereas a copper bolt in a steel plate may lead only to slight increase in corrosion of the steel around the bolt. The safest way of avoiding galvanic corrosion is to ensure that dissimilar metals are not in electrical contact by using insulating washers, sleeves, or gaskets. When protective paints are used, both metals or only the cathodic metal should be painted. Painting the anode only will concentrate the attack at the breaks or defects in the coating.

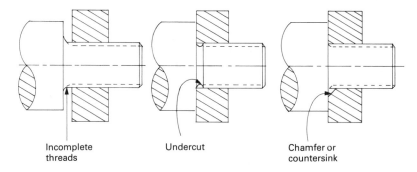

Incomplete   Undercut   Chamfer or
threads   countersink

Figure 12.14 Some design details to account for the incomplete threads at the end of external screws.

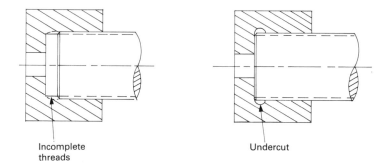

Incomplete   Undercut
threads

Figure 12.15 Some design details to account for the incomplete threads at the end of internal screws.

Severe corrosion can take place in crevices formed by the geometry of the structure. Common sites for crevice corrosion include riveted and welded joints, areas of contact between metals and nonmetals and areas of metal under deposits or dirt. Crevices can also be created between flanged pipes as a result of incorrect use of gaskets, as shown in Fig. 12.16a. Fibrous materials that can draw the corrosive medium into the crevice by capillary action should not be used as gaskets, washers, or similar applications. If crevices cannot be avoided in design, they should be sealed by welding, soldering or brazing with a more noble alloy, adhesives, or caulking compounds.

Design features that retain undrained liquids in reservoirs or collect rain water should be avoided as they cause accelerated corrosion rates. Fig. 12.16b shows examples of such features and how to avoid them. A similar problem is faced in closed tanks and sections where inadequate ventilation can cause condensation, or sweating, and accelerated corrosion rate. Closed sections are also difficult to paint and maintain. Avoiding closed sections and providing adequate ventilation can overcome this problem.

Sharp corners and convex surfaces which tend to have thinner coatings or are subject to coating cracks should be avoided. Similarly, surfaces that are exposed to direct impingement of airborne abrasive particles should be reduced. Where practicable, rounded contours and corners are preferable to angles. This feature is illustrated in Fig. 12.16c.

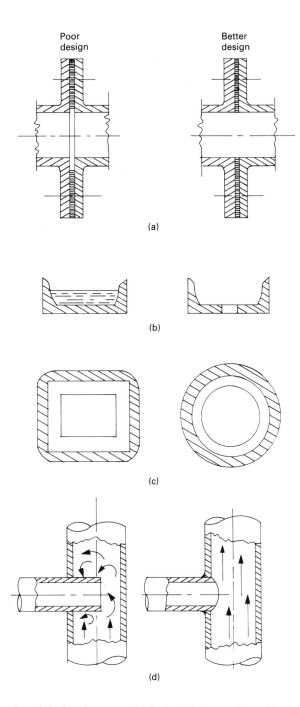

Figure 12.16 Examples of design features which should be considered in designing for corrosive environment.

Design features that cause turbulence and rotary motion in moving liquids or gases should be avoided as this can cause impingement attack and other forms of accelerated corrosion near the obstruction. Figure 12.16d shows an example of such a design feature.

Fretting corrosion can take place at the interface of two closely fitting surfaces when they are subject to slight oscillatory motion. Assemblies like shrink and press fits, bolted joints, splined couplings and keyed wheels are vulnerable to such attack. Fretting corrosion can be prevented by preventing slippage or relative motion between surfaces that are not meant to move. This can be done by roughening the surface to increase friction or eliminating the source of vibrations. Other solutions include using a soft metal surface in contact with a hard one, using low-viscosity lubricants, and phosphating.

12.11 DESIGNS INVOLVING AUTOMATED ASSEMBLY

Assembly-related tasks account for more than 50 per cent of the total time and more than 20 per cent of the labor spent in the manufacture of a typical product. The assembly process requires judgement, dexterity and flexibility and should be carried out at low cost. When carried out manually, assembly could be inefficient and inaccurate which means extra inspection, rework and cost.

Although there are many particular assembly tasks, they can be grouped into twelve basic ones, as given by Owen, T., *Engineering*, June 1983, p. 436. These tasks are: simple peg-and-hole, push and twist, multiple pegs and holes, insert peg and retainer, screws, force fit, remove location pin, flip part over, provide temporary support, remove temporary support, crimp sheet metal, weld or solder. The simple peg-and-hole are the most frequently used tasks followed by insertion of screws with the vertical downward action being the most frequently used direction in assembly operations.

In order to increase the efficiency of the assembly process and to take advantage of automation, a precedence diagram should be constructed to check the sequence of assembly and to highlight redundant assembly movements. The precedence diagram will also indicate whether or not it is possible to assemble the parts incorrectly. Symmetrical components should be used whenever possible as this requires less complicated and therefore less expensive handling and presentation equipment. If a part has to be asymmetric, then this asymmetry should be made obvious to the feeding and handling equipment. Ideally the product should be designed for unidirectional assembly so that components can be simply placed on top of each other. The number of separate parts should also be minimized by combining some of the items as this would reduce the number of assembly tasks and therefore the cost. Another important principle to be observed when designing for automated assembly is that the separate items as well as subassemblies can be handled and processed without marring and scuffing by grippers and robots. Protecting the surfaces to be gripped or providing large smooth clamping surfaces should eliminate this problem.

In automated assembly, it is important that the parts should be presented separately and in the correct attitude and orientation. Unless certain design principles have been observed problems of tangling, nesting, telescoping and shingling are likely to occur. Tangling takes place when open ended springs and rings are entangled together. Nesting takes place between components which contain self locking male and female tapers. Telescoping arises when components have pins and holes of the same diameters. Shingling problems are common with flat components as they fit together in a similar way to roof tiles. Most of these problems can be avoided by introducing slight geometrical modifications.

12.12   REVIEW QUESTIONS AND PROBLEMS

12.1   What are the problems that are likely to arise when heat treating a steel part of nonuniform sections? How are these problems overcome?

12.2   What are the advantages of casting in comparison with welding in terms of flexibility of shape design?

12.3   What are the advantages of powder metallurgy in comparison with casting when manufacturing small gears?

12.4   An AISI 1020 steel angle of dimensions 150 ∗ 100 ∗ 12 mm (6 ∗ 4 ∗ 1/2 in) is to be welded to a steel plate by fillet welds along the edges of the 150 mm leg. The angle should support a load of 270 kN (60 000 lb) acting along its length. Determine the lengths of the welds to be specified. The welding electrode used is AWS-AISI E6012 with a tensile strength of 414 MPa (60 ksi). [Answer: 128 mm on each side parallel to the axis of the angle]

12.5   Compare welding and casting as methods of fabrication of 500 mm (20 in) diameter gears. The total number required is 10 units.

12.6   Compare the use of spot welding and adhesive bonding in the assembly of the sheet metal components of motor car bodies.

BIBLIOGRAPHY AND FURTHER READING

Amstead, B. H., Ostwald, P. F. and Begeman, M. L., *Manufacturing Processes*, 7th ed., John Wiley & Sons, New York, 1979.

DeGarmo, E. P., Black, J. T. and Kohser, R. A., *Manufacturing Processes in Manufacture*, 6th ed., Collier Macmillan, New York, 1984.

Dieter, G. E., *Engineering Design: a Materials and Processing Approach*, McGraw-Hill, New York, 1983.

Doyle, L. E. *et al.*, *Manufacturing Processes and Materials for Engineers*, 3rd ed., Prentice Hall, New Jersey, 1985.

Kalpakjian, S., *Manufacturing Processes for Engineering Materials*, Addison-Wesley, Reading, Mass., 1984.

Lindberg, R. A., *Processes and Materials of Manufacture*, 3rd ed., Allyn and Bacon, Boston, 1983.

Onoda, G. Y. and Hench, L. L., *Ceramic Processing Before Firing*, John Wiley & Sons, Sussex, 1978.

Plastics Engineering Handbook, Soc. of the Plastics Industries, 4th ed., Reinhold, New York, 1982.

Smith, T., 'Ten methods to reduce corrosion', *Engineer's Digest*, vol. 43, No. 6, p. 15, 1982.

Chapter 13

Reliability of Engineering Components

13.1   INTRODUCTION

Reliability can be defined as the probability that a component or a device will perform its function adequately for the intended period of time under the specified operating conditions. In introducing a new product, it would be desirable to predict its reliability as this will allow accurate forecasts of required spares, support costs and warranty costs. Although an accurate reliability prediction can rarely be made for a new product, an analysis of the expected causes of failure and a rough estimation of reliability can provide adequate basis for forecasting life cycle costs and for comparing different design options. Reliability predictions can either be based on past experience with similar products or on statistical analysis of the loads acting on the component and the strength of the material.

In addition to service environment and operating conditions the reliability of a component is affected by its design as well as the materials and processes used in its manufacture. In a multicomponent device, reliability is also affected by the complexity of construction, reliability of the individual components in the device, efficiency of preventive maintenance and quality of repair. In addition to the above factors, the reliability of a complex system is influenced by whether or not parallel redundancy and standby systems are provided. Parallel redundancy is provided when one or more parts work in parallel such that when one fails the others share the load and continue to function. Standby systems are usually brought into operation to take over the function of the system that has failed. Utilization of the above factors to improve reliability is usually controlled by such limits as weight, space or cost. A major objective of the designer should be to maximize reliability within these limits.

The probability, P, that a device will not fail before a given time, t, i.e. its reliability, can be written as:

$$R(t) = P(T > t) \tag{13.1}$$

where T is the intended life of the device.

As reliability $R(t)$ and probability of failure $F(t)$ are mutually exclusive,

$$R(t) + F(t) = 1 \tag{13.2}$$

The above definition of reliability applies to an individual component or to a complete system made up of several components in series. To simplify mathematical analysis of the latter case it is usually assumed that failure of any component in the system is completely independent of the failure of other components and that the failure of any component causes the whole system to fail, i.e. the strength of a chain is that of its weakest link.

In Section 9.6, it was shown that the probability of failure can be represented by the area of overlap between the load and strength distributions of Fig. 9.2. An extreme case of Fig. 9.2 is when the scatter in the load distribution is very small. If the reliability of each component in the series is R, the probability that the first component will carry the load is R and the probability that both the first and second will carry the load is R^2. With n components in the series, reliability will be given by R^n or $(R_1 \star R_2 \star R_3 \ldots \star R_n)$ if the reliabilities of the different components are not similar. Another extreme case of Fig. 9.2 is when the scatter in the strength distribution is very small while the load distribution exhibits a fair amount of scatter. In this case, the reliability of the first component in the series is R as before. However, since all the components have practically the same strength, if the first component could carry the load there will be almost 100 percent certainty that all the others will not fail. Therefore the reliability of the total system is R.

The above two extreme cases can be taken as upper and lower bounds to real-life situations as shown in Fig. 13.1 which gives the variation of the overall reliability of a system with the number of components in it. Generally, the reliability of electronic devices approaches the case of almost constant loading, i.e. $\bar{R} = R^n$. This is also typical of a situation where quality control methods cannot conveniently reduce the standard deviation of the strength distribution. In this case deliberate overload can be applied to cause weak items to fail. This eliminates the tail end of the strength distribution and increases the reliability of the surviving population. This is the justification for proof-testing of pressure vessels, where the product is loaded to higher load than the expected service loads. The other extreme case of almost constant strength, i.e. $\bar{R} = R$ represents many applications in

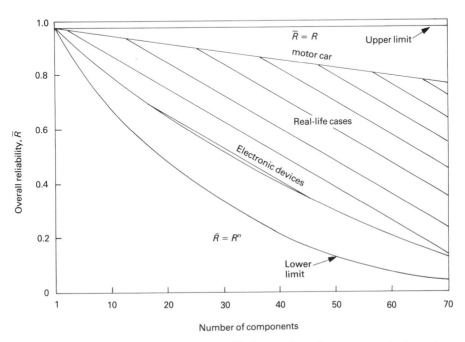

Figure 13.1 Variation of overall reliability with the number of components in the system.

mechanically loaded systems as in the case of a motor car. Reliability can be increased in these cases by increasing the safety margin by increasing the mean strength or by devising means of limiting the maximum load values. This latter option is achieved in practice by devices such as current limiters in electrical applications or by pressure relief valves in pneumatic and hydraulic systems.

13.2 ASSESSMENT OF RELIABILITY

$F(t)$ in (13.2) represents the cumulative probability function of occurrence of failure. The probability density function of occurrence of failure $f(t)$ is then given by:

$$f(t) = \frac{d}{dt}[F(t)] = -\frac{d}{dt}[R(t)] \tag{13.3}$$

As $f(t)$ is the differential of the cumulative failures with respect to time, it represents the rate at which failures occur at time t. Hazard or failure rate $h(t)$ can then be defined by relating $f(t)$ to surviving items:

$$h(t) = \frac{f(t)}{R(t)} = -\frac{dR(t)}{dt} \star \frac{1}{R(t)} \tag{13.4}$$

The value of $[h(t) \star dt]$ represents the probability that a component that has survived to time t will fail during the next short time interval (dt).

Rearranging (13.4) and integrating from 0 to time t gives:

$$R(t) = \exp[-\int_0^t h(t)\,dt] \tag{13.5}$$

This is a general mathematical description of reliability which is independent of the failure distribution. Figure 13.2 gives a schematic representation of the above parameters.

The variation of failure rate, or hazard, over the operating life of a population of homogeneous nonmaintained components is schematically represented in Fig. 13.3 and shows a bathtub-shaped curve. Initially the population will show high failure rates which occur due to design, manufacturing, materials or installation defects. Normally the failure rates rapidly decrease during this early life period which is sometimes called burn-in, debugging, shake-down, or infant mortality period. These early failures can be minimized by improving quality control or proof testing the product before sending it out of the plant. After the early life period, the failure rate stabilizes at an approximately constant value during the useful life of the component. In this period failures occur randomly and follow no predictable pattern. At the end of their useful life, the failure rate of components increases rapidly due to ageing, degrading or wearout. Analysis of failures in the wearout period is necessary to establish preventive maintenance programs. Experience has shown that wearout failures usually follow the normal distribution.

In the case of complex mechanical equipment, the failure rate exhibits the initial running-in period as in the case of nonmaintained component. Without maintenance, the failure rate will then gradually increase due to wear which takes place continuously during operation until the end of useful life, as shown in Fig. 13.4. In practice, however,

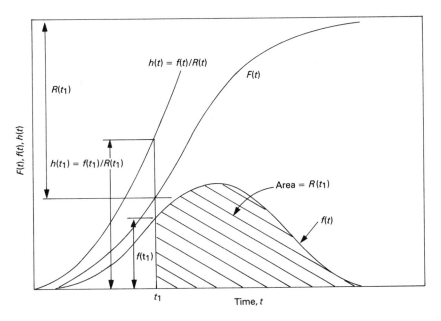

Figure 13.2 Schematic representation of the reliability parameters.

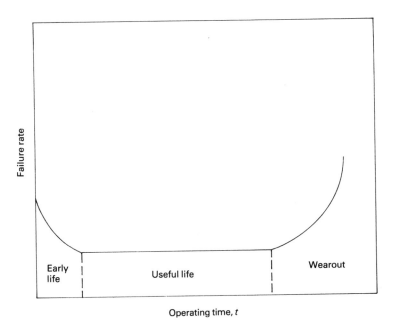

Figure 13.3 Variation of the failure rate over the operating life of nonmaintained components.

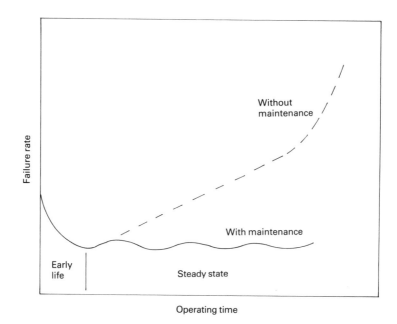

Figure 13.4 Variation of the failure rate with the operating time of well-maintained complex mechanical systems.

maintenance programs are scheduled and failed components are replaced and the equipment returned to service. The population thus remains constant and a steady state is eventually reached where the population is made up of a mixture of components in various stages of their useful lives. During this steady state, the causes of failure are either random or due to wearout or a combination of both. Random failures can be reduced by design, manufacturing or materials changes while wearout failures can be reduced by better maintenance and inspection. In practice, well maintained mechanical systems are considered to have reached the end of their useful lives when they become outmoded or uneconomical to run.

Constant failure rate (useful life)

The above discussion has shown that constant failure rates are achieved during the useful life of a component or a complex maintained equipment. Within useful life, the reliability of a device is the same for operating times of equal length. Under these conditions:

$$h(t) = \text{the constant failure rate} = \lambda \tag{13.6}$$

Equation (13.5) can be rewritten as:

$$R(t) = \exp\left[-\int_0^t \lambda dt\right] = e^{-\lambda t} \tag{13.7}$$

In practice, components which exhibit good reliability have failure rates $h(t)$ in the range of 10^{-5}–10^{-7}/h. The following example illustrates the use of (13.7). An electrical generator has a failure rate of 10^{-5} failures per hour. Calculate its reliability in an operating period of 1000 h. If 100 generators are connected to form a network, how many failures are expected in 1000 h? Assume that the generators are all within their useful life stage, i.e. constant failure rate. From (13.7):

$R(100) = \exp(10^{-5} \star 1000) = 0.99$
Number of surviving generators $= 100 \star 0.99 = 99$
Number of expected failures $= 100 - 99 = 1$

From (13.3) and (13.7):

$$f(t) = \lambda e^{-\lambda t} \tag{13.8}$$

The mean life, $\bar{t}$, is given by:

$$\bar{t} = \int_0^\infty t f(t)\, dt \qquad = \int_0^\infty t\lambda e^{-\lambda t}\, dt$$

$$= \frac{1}{\lambda}[te^{-\lambda t} + e^{-\lambda t}]_0^\infty = \frac{1}{\lambda} \tag{13.9}$$

The quantity $1/\lambda$ is known as the mean time between failures, MTBF. If the failure rate in useful life is small, the MTBF can be very long and is often much longer than the useful life. In the case of the electrical generator in the above example, the MTBF is $2 \star 10^5$ hours which is nearly 23 years. From (13.7), when a component or a complex system is operated for a period of time equal to its MTBF, its reliability, i.e. probability of survival, is $e^{-1} = 0.368$ provided it is still within the useful life period. Since life is limited by wearout, the probability of survival for a period equal to the MTBF is remote. The MTBF simply indicates how reliable a component is within its useful life.

Variable failure rate (early life or wearout stages)

Under conditions where the failure rate is a function of time, as in the cases of early life or wearout stages of Fig. 13.3, experience has shown that the Weibull distribution can be used to describe the failure pattern, as discussed in Appendix C. In this case the reliability $R(t)$ is given by:

$$R(t) = \exp\left[-\left(\frac{t - t_o}{\theta - t_o}\right)^m \right] \tag{13.10}$$

and the failure rate or hazard $h(t)$ is given by:

$$h(t) = \frac{f(t)}{R(t)} = m\frac{(t - t_0)^{m-1}}{\theta^m} \tag{13.11}$$

where $t =$ time to failure and t_0 is a locating constant defining the starting point of the distribution, i.e. time at which $F(t) = 0$.

$\theta =$ scaling constant, stretching the distribution along the time axis, called characteristic life. When $t = \theta$ the reliability is given by $e^{-1} = 0.368$, i.e. it represents the time by which 63.2 percent of the population can be expected to fail.

$m =$ shape constant which controls the failure density distribution curve.

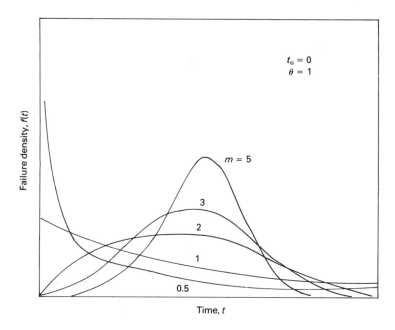

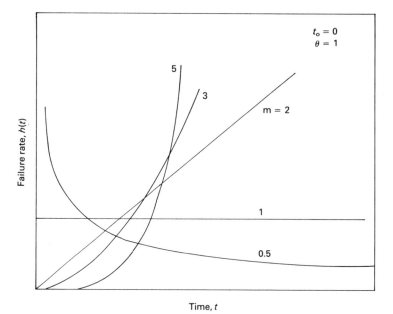

Figure 13.5 Effect of *m* on the shape of failure density and failure rate distributions.

For $m = 1$ and $t_o = 0$, the Weibull distribution reduces to the exponential distribution of (13.7). For $m < 1$ the curve takes the shape of early life failures. For $m > 1$ the curve takes the shape of wearout failures, in particular with $m = 3.44$ the Weibull distribution becomes an approximately normal distribution. Figure 13.5 illustrates the effect of m on the shape of the failure density $f(t)$ and failure rate $h(t)$ distributions. This shows that by appropriate choice of the constants, the Weibull distribution can be used to represent a wide range of distributions which cover all three stages of component life. This explains its widely spread use in practice.

13.3   SERVICE LIFE

Based on the above discussion, Fig. 13.3 can be redrawn to include the variation of reliability $R(t)$ and failure probability density $f(t)$, as shown in Fig. 13.6. The figure identifies the end of early life period, T_e, when all substandard components have all failed and the end of useful life period, T_u, when the failure rate begins to increase due to the onset of wearout and degradation. T_u represents the rated life of the component and normally only a small fraction of the population will have failed by this time. The figure identifies a time T_m which is defined as the mean wearout life of the population. About one half of the components that survive t_o, T_u, are expected to fail in the period T_u to T_m. If design and operation are correct, the wearout period should never be reached within the operating life. Extended periods of service can be achieved in complex systems by replacing components when they reach the end of their useful lives even if they have not failed. The curves of Fig. 13.6 and the values of T_e, T_u and T_m can be estimated for a

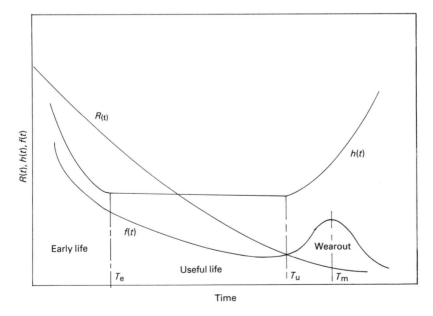

Figure 13.6 Schematic representation of the variation of reliability, failure probability density and failure rate with time.

complex system if the failure rates or reliabilities of the components in the system are available. The accuracy of the results depends on exactness of the data on which the calculations are based.

13.4   HAZARD ANALYSIS

The hazard associated with an engineering system can usually be related to its failure to perform its intended function correctly. Failure can be due to:

1. Design errors resulting from incorrect information about working stress and service conditions.
2. Poor selection of materials.
3. Manufacturing defects.
4. Neglected maintenance.
5. Subjecting the system to environmental conditions for which it was not designed.
6. Overloading the system and exceeding design limits.
7. Human errors.

A number of hazard analysis techniques have been developed to identify potential sources of failure, rate them in terms of criticality and define the conditions under which failure will most likely occur. These techniques are best applied in the design stages as tools to improving reliability but can also be employed in post-mortem analysis of failures. The analysis normally identifies the mode of operation, possible modes of failure, consequences of failure, potential sources of accidents and the events that could trigger them, severity of the hazard and measures that can be taken to prevent it or reduce the damage. Hazard analysis can be performed on the total system, subsystems, assemblies or components. The information required for hazard analysis include experience with similar systems, failure reports and interviews with operating personnel. In addition, estimations of probability of failure are required for quantitative analysis. One of the most widely used methods of analysis is failure mode, effects and criticality analysis.

Failure mode, effects and criticality analysis (FMECA)

The principle of FMECA is to consider each mode of failure of every important component of a system and to assess its effects on system operation. The analysis may be based on failure of actual components (e.g. consequences of bearing seizure or shaft fracture) or failure to function (e.g. failure of the backup unit to start or overheating of cooling system). The analysis may also be based on a combination of the two approaches. FMECA can be performed from different viewpoints, such as safety, availability or mission success. It is necessary to indicate the viewpoint adopted in the analysis since a safety related FMECA may consider an item to be more critical than an availability related study, for example. FMECA may be carried out using worksheets such as those provided by US MIL-STD-1629 (Procedures for Performing an FMECA) or using computer techniques. The information required for the worksheet includes the function of the part under consideration, expected failure modes and causes, operational mode and time, local and higher level effects of failure, failure detection method, severity rating and part criticality rating. Computer programs have been developed to replace the worksheets in performing FMECA. Like other computer-aided techniques, computerized FMECA is quicker, more accurate and easier to edit and update. The final FMECA report should rank failure effects in criticality order at different system levels and in different phases of system operation.

Human reliability

In FMECA, emphasis is placed on the malfunction of components while other categories of hazard, such as human reliability, are ignored. Human reliability describes the probability of human fallibility resulting in system failure. When people are involved in the operation and maintenance of systems, the following questions should be addressed: (a) what will be the effect if the operator fails to perform as required? and (b) are the actions required clearly specified and easy to perform even under emergency conditions? The increasing complexity of engineering systems and more emphasis on safety make the assessment of human factors in reliability and maintainability increasingly more important. Fault hazard analysis was developed by safety engineers so that human errors as well as other categories of hazards are included in the analysis. It is a qualitative method which usually considers the downstream consequences, i.e. damage to higher assemblies in the system, of the failure.

13.5   FAULT TREE ANALYSIS

Fault tree analysis (FTA) is an analysis technique which is widely used in organizing the logic in studying reliability, critical failure modes, safety, availability or advantages of design redundancy. When any of these issues is selected for analysis it is considered to be the 'top event', which forms the main trunk from which logic branches develop. The analysis proceeds by determining how the top event can be caused by individual or combined lower level events. As the fault tree grows, its logic is separated into successively smaller events until each element is sufficiently basic to be treated independently of other events. This separation is recorded using AND and OR gates in addition to other standard symbols, as shown in Fig. 13.7.

Normally, an engineering system's failure tree would have a large number of branches, gates and elements which need a computer routine to keep track of the analysis. Many FTA programs are commercially available and can be used for generating and evaluating large failure trees. A simple analysis of a gearbox failure is shown in Fig. 13.8 and is used as an example to illustrate the use of FTA.

In addition to providing a qualitative view of the impact of each element on the system, FTA can be used to quantify the top event probabilities from reliability predictions of the different events. In these calculations, an AND gate multiplies probabilities of failure and an OR gate acts additively. Two events that are connected by an OR gate will have a larger contribution to the failure of the system than two similar events that are connected by an AND gate, as shown in Fig. 13.9. Comparing events 1 and 2 with events 3 and 4 shows that the former events make a much larger contribution to the higher event and need more accurate assessment and more care in design. In performing an FTA, care should also be taken to identify common mode, or common cause, failures. These can lead to the failure

Symbol	Event

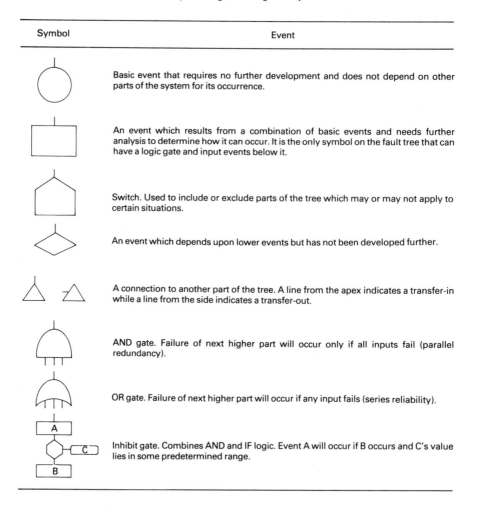

Basic event that requires no further development and does not depend on other parts of the system for its occurrence.

An event which results from a combination of basic events and needs further analysis to determine how it can occur. It is the only symbol on the fault tree that can have a logic gate and input events below it.

Switch. Used to include or exclude parts of the tree which may or may not apply to certain situations.

An event which depends upon lower events but has not been developed further.

A connection to another part of the tree. A line from the apex indicates a transfer-in while a line from the side indicates a transfer-out.

AND gate. Failure of next higher part will occur only if all inputs fail (parallel redundancy).

OR gate. Failure of next higher part will occur if any input fails (series reliability).

Inhibit gate. Combines AND and IF logic. Event A will occur if B occurs and C's value lies in some predetermined range.

Figure 13.7 Some standard symbols used in failure tree analysis.

of all paths in a redundant configuration which practically eliminates its advantage. Examples of sources of common mode failures include:

1. Failure of a power or fuel supply which is common to the main and backup units.
2. Failure of a changeover system to activate redundant units.
3. Failure of an item causing an overload and failure of the next item in series or the redundant unit.

An indirect source of failure that should be identified in performing FTA is the enabling event. This event may not necessarily be a failure in itself but could cause a higher level failure event when accompanied by a failure. Examples of enabling events include:

1. Redundant system being out of action due to maintenance.
2. Warning system disabled for maintenance.
3. Setting the controls incorrectly or not following standard procedures.

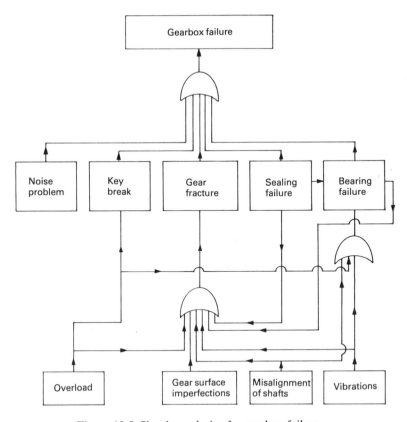

Figure 13.8 Simple analysis of a gearbox failure.

Besides their use in design and reliability assessment, FTA can also be used as a tool in trouble shooting, failure analysis and accident investigations.

13.6   THE ROLE OF DESIGN IN ACHIEVING RELIABILITY

Safe design represents one of the most important factors that can be used to ensure the reliability of a product. Reliability should be built into the design starting at the preliminary stages and maintained to the detailed design stage. Correct anticipation of working loads and service environment, adequate safety and derating factors, protection against overload and mishandling, provision of built-in system redundancy or standby units and selection of the appropriate materials that will resist degradation under working conditions are basic requirements for a reliable design. Whenever an adequate safety factor cannot be derived with reasonable accuracy from the analysis, tests should be performed to provide the necessary assurance. Protection against overload is not always possible, but should be considered whenever possible. Load analysis must take into account the effect of interactions of the various types of loads and environments, e.g. temperature and vibrations or corrosion and fatigue loading. Some combinations can have effects that are out of proportion to their separate contribution, both in terms of instantaneous effects and strength degradation effects. Strength distributions must be estimated in relation to

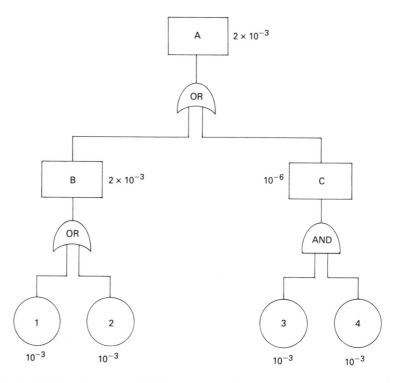

Figure 13.9 Sensitivity of system reliability to probabilities of failure of various components. Numbers beside each event represent probabilities.

material and process specifications and tolerances as well as quality control procedures. If it is necessary to use a unit with a relatively high failure rate, provision should be made to replace it easily and rapidly. This may not directly increase the reliability, but it will increase availability and may decrease possible damage to neighboring components during replacement.

Normally, simpler designs are more reliable. The greater the number of components in a system, the greater the chance of its failure and the lower is the reliability. This is particularly true for designs where the components are arranged in series, where the failure of any component causes the failure of the system. As shown in Section 13.1, the reliability of a system consisting of n components is given by:

$$R_{sys} = R_1 \star R_2 \star R_3 \star \ldots R_n \tag{13.12}$$

For example, if the reliability of individual components is 0.99, a system consisting of 10 components will have a reliability of $0.99^{10} = 0.90$, while a system consisting of 30 components will have a reliability of $0.99^{30} = 0.74$. Under constant failure rate conditions, (13.7) can be used to estimate the reliability of a two-component series system:

$$R_{sys} = R_a \star R_b = e^{-\lambda a t} \star e^{-\lambda b t} = e^{-(\lambda a + \lambda b)t} \tag{13.13}$$

The above equation shows that λ for the system is the sum of the values of λ for each component. If the electrical generator discussed in the example of Section 13.2 is driven by a diesel engine which has a failure rate of $2 \star 10^{-5}$/h, the reliability of the engine–generator system in an operating period of 1000 h can be calculated as:

$$R_{(system)} = \exp - (10^{-5} + 2 \star 10^{-5}) 1000 = 0.97$$

Higher reliability of the system can be achieved if two or more components are connected in parallel so that it would be necessary for all of them to fail in order for the system to fail. In this case the reliability of a system with n components connected in parallel is given by:

$$R_{sys} = 1 - (1 - R_1)(1 - R_2) \ldots (1 - R_n) \qquad (13.14)$$

Accordingly, if two components of reliability 0.99 each are connected in parallel, the system will have an improved reliability of 0.9999.

Under constant failure rate conditions, the reliability of a two-component parallel system is given by:

$$R_{sys} = 1 - (1 - R_a)(1 - R_b) = 1 - (1 - e^{\lambda at})(1 - e^{\lambda bt}) = e^{-\lambda at} + e^{-\lambda bt} - e^{-(\lambda a + \lambda b)t}$$
$$(13.15)$$

The concept of parallel reliability can be used to increase the reliability of critical components in a system. In this case two or more components share the load even though each one of them could carry the load alone. This means that each component is derated, which increases its life. This type of parallel system is said to have active redundancy. Another approach to redundancy is to have a standby system which is activated when the main system fails. An example is the standby electrical generator which is activated when the main power supply is interrupted. Under constant failure rate conditions, the reliability of a system with a standby unit is given by Smith, see bibliography:

$$R_{sys} = \frac{\lambda_1}{\lambda_2 - \lambda_1} (e^{-\lambda_1 t} - e^{-\lambda_2 t}) + e^{-\lambda_1 t} \qquad (13.16)$$

where λ_1 and λ_2 are the failure rates of the main system and the standby unit respectively.

It should be noted that the reliability of the sensing and switching device which activates the standby unit is critical in this case and should be included in the analysis. For example, if the diesel generator system, discussed above, is used as a standby unit for a hospital, a starting device starts the diesel engine when the main power is interrupted. If the failure rate of the main power grid is 10^{-4} and of the starting device is $5 \star 10^{-6}$ what is the reliability of the hospital power system in an operating period of 1000 h? From (13.13), λ_2 of the starter–engine–generator backup system is:

$$\lambda_2 = 5 \star 10^{-6} + 10^{-5} + 2 \star 10^{-5} = 3.5 \star 10^{-5}$$

Reliability of the backup system for 1000 h:

$$R_b = \exp - (3.5 \star 10^{-5} \star 1000) = 0.966$$

Reliability of the main power grid for 1000 h:

$$R_p = \exp - (10^{-4} \star 1000) = 0.905$$

Reliability of the hospital power system, from (13.16):

$$R_h = [1/(10 - 3.5) \star 10^{-5}][\exp - (10 \star 10^{-5} \star 1000) -$$
$$\exp - (3.5 \star 10^{-5} \star 1000)] + \exp - (10 \star 10^{-5} \star 1000) = 0.997$$

The results show that the reliability of the hospital power system is higher than that of its individual components.

The maintenance and repair of a system is an important factor in determining its reliability. If a failed component can be repaired while a redundant component has replaced it in service, then the overall reliability of the system is improved. Systems that are easy to maintain or do not need maintenance at all are less likely to fail due to poor maintenance. Systems that are likely to fail if subjected to adverse service conditions like heat, dust, moisture or vibrations should be protected. Heat shields, filters, seals and vibration damping mounts are often used by designers to increase the reliability of their designs. Ease of inspection is another reliability issue that should be built into the design. In critical components of the system provision should be made for visual inspection or other nondestructive testing techniques. If parts of the system cannot be made available for inspection, then the stress level should be lowered to a level where a defect cannot grow to a dangerous size during the expected service life of the system.

Cost of reliability

The answer to the question of how much reliability is required for a given component or system involves a large number of factors such as safety, legal considerations, expected life and availability, and economic considerations. Reliability requirements have been established for many applications, especially those which involve human life or safety. Examples are found in aerospace, automotive and defense industries. In several consumer industries, however, the level of reliability is largely determined by the producer, as in the case of electrical and electronic appliances for household use. In the latter cases economic considerations play an important role in determining the level of reliability. Introducing special design features, employing better materials and exercising more strict quality

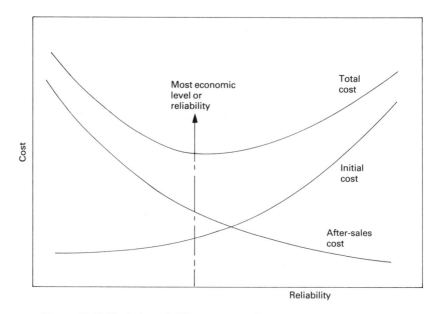

Figure 13.10 Variation of different types of system costs with reliability.

control in manufacturing is known to improve reliability but will also increase the cost. Figure 13.10 illustrates how the initial cost of the system increases as the reliability improves. On the other hand, better reliability means lower repair costs, more customer satisfaction and safety, and less warranty and replacement costs, as shown in the figure. The summation of the two types of costs produces the total cost curve which shows a minimum value at a certain level of reliability which should be aimed at for maximum economy.

It should be noted that reliability requirements, whether determined by governmental agencies or producers, should be periodically reviewed with regard to need and validity. With the increasing concern for consumer safety and protection in recent years there has been an increase in product liability cases. This has placed a burden on the designer and producer to make their products more reliable and safer not only for intended uses but also in all foreseeable, although unintended, uses. The subject of product liability will be discussed in more detail in Section 13.10.

Design review

Design review, or audit, is basically an examination of a design prior to the construction of a project and is periodically conducted during the various stages of design in order to:

1. Review the progress of the design.
2. Monitor reliability and safety to assure that their requirements will be met.
3. Ensure ease of manufacture and maintenance.
4. Provide information to all concerned.

The design review team should be multidisciplinary and is normally composed of the project engineer, design engineers, production engineers, materials engineers, research and development engineers, maintenance engineers, safety engineers, reliability engineers, packaging experts, sales personnel and management representatives. Vendor participation is also desirable. If properly conducted, design reviews can avoid serious problems which could face the product at later stages. Normally the first design review is conducted prior to the commencement of the design activity in order to review the design specifications. The second review takes place prior to the start of detail design and examines the design at the system level to eliminate any fundamental weaknesses and to identify critical areas. The issues discussed include reliability, safety, ergonomics, cost/ value, maintainability, environmental compatibility, standardization, simplicity and mod- ular design, installation, marketing and sales appeal. The main manufacturing processes and materials are identified at this stage. The third review takes place after the completion of detail design but before manufacturing is started. The purpose is to check that critical components will fulfill their required duty. Information gathered from failure mode, effects and criticality analysis and fault tree analysis represents an important input to the design reviews at the different stages of product development.

13.7   THE ROLE OF MATERIALS AND MANUFACTURING IN ACHIEVING RELIABILITY

Material behavior is one of the important factors that affect the reliability of a component in service. The behavior of materials under load is greatly affected by the loading conditions. Loading can be externally applied and can also result from component inertia or thermal effects. Knowledge of the maximum load alone is not sufficient in anticipating the behavior of materials. Loading frequency, rate and duration are also important. In addition, material behavior may also be sensitive to service conditions such as temperature, humidity, ambient chemicals, contact with other materials, radiation, etc.

Material property degradation represents an important aspect in reliability analysis. Strength degradation due to fatigue represents one of the most important sources of failure of components in service, as discussed in Chapters 7, 8 and 11. The effects of stress raisers which are caused by incorrect design or faulty manufacture, e.g. notches, sharp corners, holes and surface roughness, on the endurance limit of the material under simple loading conditions can be estimated accurately. In this case reliability analysis can be performed satisfactorily and parts can be designed for a predetermined safe life. However, the behavior of materials under complex loading conditions, similar to those usually observed in actual service, is usually more difficult to analyze and tests should be performed to provide the required data. Other factors which could make reliability predictions more difficult include corrosive environments, wear conditions, internal stresses and microstructural irregularities. If accurate prediction of fatigue life is not possible, frequent inspection procedures should be specified.

In some cases certain defects in the form of cracks are introduced in the material by the manufacturing process and are too small to be detected by inspection. Examples include porosity in castings, laps in forgings, cavities in weldments and cracks in heat-treated components. These defects could grow in service and reach a critical size which leads to failure. Under these conditions, materials with high fracture toughness should be selected for their damage-tolerance characteristics. In these materials, the propagation of cracks is slow and the critical size is large enough to be detected during routine inspection.

Estimation of reliability of components under high-temperature service conditions requires careful assessment of the long-term effects on the material behavior. These effects include microstructural changes, creep, thermal shock and oxidation. The microstructural changes, which include grain growth, coarsening of precipitates, loss of strain hardening and overtempering of steels, usually result in a reduction of strength. The loss in strength is a function of time, so that useful strength is lower after longer service. Creep results in permanent strain in the material which could eventually lead to rupture. Stress relaxation is closely associated with creep and could result in loosening and leakage in bolted joints. Service conditions which involve rapid fluctuations in temperature may cause material damage due to thermal shock or spalling. Ceramics and other brittle materials with low thermal conductivity, high thermal expansion coefficient and high elastic modulus are particularly prone to this kind of failure. Many oxidation-resistant materials are now available and should be selected for high-temperature applications. However, unexpected chemical changes of the high-temperature environment could cause catastrophic oxidation. For example, small amounts of vanadium pentoxide or lead oxide in the high-temperature atmosphere can combine with the protective oxide film to form a low-melting-point phase which accelerates the attack.

Exposing engineering materials to radiation during the course of their service can adversely affect their properties and reduce their reliability. The amount of radiation damage depends on the type of material and the energy of the radiation. Radiation damage in metallic materials usually increases their hardness and reduces their ductility which could make them prone to brittle fracture. Some polymers are also known to suffer degradation as a result of exposure to sunlight and similar sources of radiation.

13.8   THE ROLE OF THE USER IN ACHIEVING RELIABILITY

Improper installation, inadequate maintenance and abuse in service represent major sources of failure of engineering components and devices for which the user is responsible. Clear instructions indicating the correct installation, steps to be taken before the item is used for the first time, maintenance procedure, limitations of equipment and their intended use, as well as training of users are expected to reduce the occurrence of these sources of failure. In many instances, however, what is considered as normal operation by the user may be considered as abuse by the designer. This shows the importance of the role played by the user in improving the reliability of a given product. The feedback from the user would help the designer in recognizing the sources of weakness in the design and this is expected to improve performance and reliability of future models. This is particularly true in the case of large-scale users who are capable of collecting and analyzing the failure reports accumulated in service.

Failure reports are usually written by maintenance and repair engineers. Providing these personnel with specially prepared forms which only contain simple questions to answer makes it possible to obtain comparable answers which can be statistically analyzed and the results then used to improve reliability. The failure report should identify and describe the type of failure or fault and indicate whether it is primary or secondary. Secondary failures or faults are a result of another failure or fault. The report should also indicate whether failure is caused by neglect (lack of maintenance, improper storage, misuse, careless handling, etc.), or by normal wear and tear. The analysis of failure should indicate if it is random, i.e. arising from chance condition other than accident. If failure cannot be attributed to any of the above causes, it is usually caused by manufacturing, materials, or design defects. When enough failure analysis data are collected and statistically evaluated, the designer will have an invaluable tool to improve the reliability of the design.

13.9   MAINTENANCE AND CONDITION MONITORING

Maintenance and condition monitoring are two important factors that affect reliability. Generally there are two types of maintenance schemes which can be adopted:

1. Preventive or scheduled maintenance where a component which is subject to wear is replaced before it fails. Preventive maintenance is aimed at improving system reliability by minimizing its failure. Replacement before failure is based on knowledge of the statistical distribution of component life. In this case part of the useful life is traded off for increased reliability. This approach is facilitated if a condition monitoring scheme is used to detect signs of degradation or malfunction of critical components.
2. Repair maintenance which is carried out on a non-scheduled basis to restore a component to satisfactory working condition. Repairing a failed component in a series system will not improve its reliability, but decreasing the repair time will improve availability of the system, as will be discussed later. In a system with parallel redundancy, if operation continues while repair maintenance is in progress the overall reliability is improved.

Availability of a system (A) is affected by its reliability and maintainability and is usually defined as the proportion of time the system is actually working to the total scheduled working time:

$$A = \frac{\text{MTBF}}{\text{MTBF} + \text{MTTR}} \qquad (13.17)$$

where MTBF is mean time between failures which is related to the average period of time for which the system is likely to operate before failure as discussed in Section 13.2. MTTR is the mean time to repair after failure, i.e. down time of the system. The down time is made up of: the time required to determine that a failure has occurred and to diagnose the necessary action, the time of waiting for spares, the time required for repair and the time required to ascertain that the repair has been effective and that the system is operational.

Maintainability is the probability that a failed system or component is restored to operable condition in a specified downtime. Maintainability (M) is a function of MTTR, hence the probability of completing a repair in time t can be expressed as:

$$M = 1 - e^{-t/\text{MTTR}} \qquad (13.18)$$

Condition monitoring

Condition monitoring is based on the assumption that failures are normally preceded by some measurable change in the performance of the system. Thus monitoring of an appropriate variable can give an indication of an impending failure. An important design decision is to establish the methods and locations of monitoring that would be used to predict failures in the system. The simplest form of monitoring which is used in practice is visual inspection. The design should provide access covers and hatches through which the condition of the appropriate part of the system may be visually examined. Intrascopes and TV cameras are included under this form of monitoring. Visual inspection relies heavily on the experience of the inspector and can be subjective. However, it provides direct inspection and can be used to detect fatigue cracks and similar defects. More quantitative, though less direct, forms of monitoring include temperature and pressure gages, e.g. cooling water and oil pressure gages fitted to internal combustion engines. Deviation from a predetermined value is indicative of malfunction which calls for further inspection. This form of monitoring is useful in detecting wear in bearings and similar machine elements. Wear can also be estimated by analyzing the wear particles that are contained in the lubricant. The quantity, shape and size of the wear particles can be related to the condition of the wearing surfaces.

Vibration and sound measurements offer powerful tools for condition monitoring as they can be used to predict both wear failures, where small quantities of material are lost, and cracking, formation and propagation. These tools can be used for continuous monitoring of a system during its operation.

13.10   PRODUCT LIABILITY

In the past, traditional design and manufacturing practices have emphasized criteria like function, cost and marketability. Product safety was included, if at all, in the function-cost area. In recent years, however, there has been increasing concern in many countries that the public must be protected from injury by the products and by-products of technology. Governments responded to the social pressures reflecting this concern by forming commissions, such as the Consumer Product Safety Commission and the Environmental Protection Agency in the USA, and issuing liability laws, such as the Consumer Product Safety Act and the Occupational Safety and Health Act in the USA and the Consumer Protection Act in the UK. Liability is the term lawyers use to describe an obligation on one party to pay money to another party. In some countries product liability laws are a distinct category, as in the case of strict liability which is a body of law that has been developed for consumer protection in the USA; in other countries product liability arises from existing laws which have wider applicability, as in the case of contract, tort and statutory duty in the UK. The liability laws and safety regulations throughout the world are changing to reflect the increasing social awareness and to accommodate the new safety requirements as technology progresses.

Generally, the manufacturer of a defective product can now be sued directly by the injured party and is liable for bodily injury and property damage caused by its defective condition. According to the Consumer Protection Act of 1987 in the UK, it is a criminal offense to manufacture unsafe products. Under strict liability law, as defined by section 402A of the restatement of Torts in the USA, the plaintiff needs to prove that:

1. The product contained an unreasonably dangerous defect.
2. That the defect existed at the time the product left the manufacturer's hands.
3. The defect was the cause of the injury.

The fact that the injured party acted carelessly or in bad faith is not a defense under strict liability standards. Defects can be related to design, manufacturing or marketing. Common design defects are usually related to failure to include safety devices, thus causing the product to be unreasonably dangerous. Examples of safety devices are guards to protect workers from moving machine parts, shatterproof glass for windscreens of automobiles and pressure relief valves for boilers or other pressure vessels. Currently, about 40 percent of all product liability cases involve design defects. Measures to ensure safety and to guard against unforeseen design defects should be taken starting at the early design stages. Design reviews should take note of current governmental regulations, consider the different aspects of product liability and review a failure analysis, or FTA, of the design to predict the possible ways the product can become unsafe and to control consequences of failure, if not prevent it. Weinstein et al. (see bibliography) suggest applying the following procedure during a safety review of a product design:

1. Define the use and misuse of the product.
2. Define the environment in which the product is to be used.
3. Define the typical user of the product.
4. Identify all possible hazards of use and misuse and their probability of happening as best as can be anticipated.
5. Define all possible alternative design solutions to these hazards including warnings and any instructions which may help minimize the danger of safety hazards in the use of the product.
6. Evaluate how these design solutions affect overall safety, utility and cost of the product.
7. Select the optimum design solution.

In performing the last step of the above procedure, decisions have to be made on the degree of safety that gives the optimum balance between risk, utility and cost. Of course, safety features that correct obvious hazards should never be offered as an option for tradeoffs. If needed, exhaustive testing of the actual product should be performed before it is released for sale, especially in state-of-the-art products. It is also essential to document every step of the design process to explain the rationale behind the decisions and tradeoffs made.

Unsuitable materials, bad workmanship or ineffective quality control measures can constitute manufacturing defects. Examples include flawed material, omitted lubricant, loose or missing bolts, improper tolerances in critical components, defective welds and incorrect calibration or adjustment of product. It is estimated that manufacturing defects are involved in about one-third of all liability cases in the USA.

Marketing defects are associated with failure to give adequate warning and instruction for proper use of the product. For example, labels should be placed on grinding wheels indicating that they shatter or fly apart during use and that the user must use a safety guard. Products which are not foolproof in operation impose on the manufacturer a duty to warn of the dangers arising from their misapplication or misuse. Warnings and instructions must be adequate both as to content and form. It is estimated that failure-to-warn represents about 20 percent of all liability cases in the USA. A warning in the operator's manual of a machine may be adequate as to content, but inadequate as to form, if it should more properly be placed on the machine itself. Factors to consider in deciding whether or not to provide a warning are:

1. Seriousness of injury which could arise in the use of the product.
2. Probability of injury or damage.
3. Whether the intended user is professional or ordinary public. Professional users are expected to exercise their normal skill and experience in the use of products associated with their trade. However, even the professional user can get tired, become inattentive or be unduly anxious to finish a job and thus may become unaware of or even ignore danger.

13.11   REVIEW QUESTIONS AND PROBLEMS

13.1   Experience with a racing car shows that the mean distance between tire changes is 300 miles. The tires are known to have random failures. What is the probability that the car will complete a race of 100 miles without having a tire change? [Answer: 0.7165]

13.2   The failure of a group of components follows the Weibull distribution with $\theta = 100\,000$ h, $m = 4$, and $t_o = 0$. What is the probability that one of these components will have a life of 18 000 h? [Answer: 0.99895]

13.3   An electromechanical system consists of 500 elements in a series configuration. Tests on a sample of 80 elements showed that two failures occurred after 800 h. If the failure rate is assumed to be constant, what is the reliability of the system for 1000 h operation? [Answer: 0.8187]

13.4   In the system in question 13.2, if an overall system reliability of 0.98 in 1000 h is required, what would the failure rate for each element have to be? [Answer: 0.99996]

13.5   Construct a fault tree for a motor car brake system.

BIBLIOGRAPHY AND FURTHER READING

Carter, A. D. S., *Mechanical Reliability*, 2nd ed., Macmillan, London, 1986.

Carter, A. D. S. *et al.*, 'Design for reliability', in *Mechanical Reliability in the Process Industry*, The Institution of Mechanical Engineers, London, 1984.

Dieter, G., *Engineering Design: a Materials and Processing Approach*, McGraw-Hill, New York, 1983.

Muster, D. and O'Quinn, 'Products liability law in the United States', *CME*, November, 1982.

O'Conner, P. D. T., *Practical Reliability Engineering*, John Wiley & Sons, N.Y., 1985.

Smith, C. O., *Introduction to Reliability in Design*, McGraw-Hill, New York, 1976.

Stiepel, E. E., 'Product liability considerations in machinery design', *Mech. Engineering*, vol. 104, 1982, p. 64.

Vane, H. C., 'Product liability in the UK', *CME*, February, 1983.

Weinstein, *et al.*, *Product Liability and the Reasonably Safe Product*, John Wiley & Sons, New York, 1978.

Chapter 14

Computer Aided Design

14.1   INTRODUCTION

Computer aided design, CAD, can be defined as the use of computer systems to assist in the creation, modification, analysis, optimization, and storing and communication of design information. The relative importance of the roles played by the designer and computer in these activities depends on the nature of the performed task. The activity of design creation depends on experience combined with judgement and is best performed by the designer who should be free to follow his own intuitive design logic rather than a programmed computer logic. The computer can, however, provide rapid reference to previous designs which reflect the experience of other designers. The interactive nature of design requires that designs be modified to improve their performance and to eliminate possible errors. While the computer can detect errors which are systematically definable, the designer can use experience and intuition to modify the design. The iterative graphics facility offered by the CAD system allows design modifications to be performed easily and efficiently. The analysis and optimization phases of the design are easily and accurately performed by the computer while the designer will find these tasks time consuming and tedious. The role of the designer in this case is to make decisions based on the analytical calculations performed by the computer.

The above discussion shows that the designer and computer play a complementary role in performing the different activities of design. The main functions of the designer are:

1. To use creativity, experience and intuition in defining the system and in making the preliminary layout
2. To use experience and judgement in evaluating the results of computer analysis.
3. To control the modification and optimization process.

The main functions of the computer in a CAD system are:

1. To serve as an extension to the memory of the designer.
2. To perform geometric modeling through interactive graphics. This function is helpful to the designer in defining the system and making the preliminary layout.
3. To enhance the analytical and logical power of the designer.
4. To perform the repetitive and routine tasks involved in design review and evaluation.
5. To store and communicate information.
6. To produce design drawings at different levels of detail as a result of interactive computer graphics, ICG.

7. To store drawings and other information for use in manufacture and future design activities.

Computer aided design is finding increasing use in industry for the following reasons:

1. Increased productivity of the designer as the product and its component subassemblies become easier to visualize and the time required for calculations becomes shorter. This increased productivity reduces both the project completion time and cost of design.
2. Improved quality of design, as it is easy to perform more thorough analysis of a larger number of design alternatives. Design errors are also reduced through the greater accuracy provided by the system.
3. Improved communications through better documentation of the design and more consistent quality of engineering drawings.
4. Creation of a data base for future design modifications.
5. Provision of documentation for manufacturing and possibility of direct link with computer aided manufacturing, CAM, as will be discussed in Chapter 16.

The different functions that are normally performed by a CAD system will be discussed in more detail in the following sections of this chapter.

14.2 THE COMPONENTS OF CAD SYSTEMS

A typical CAD system is a combination of hardware and software. The hardware is made up of a central processing unit (CPU), data storage, input devices and output devices, as shown schematically in Fig. 14.1. The software consists of the computer programs needed for the analytical and graphic functions as well as the programs needed for running the system and controlling the peripherals.

Central processing unit

The CPU is the heart of the computer and is made up of an arithmetic and logic unit, and a control and command unit. The CPU retrieves data and programs from memory, controls

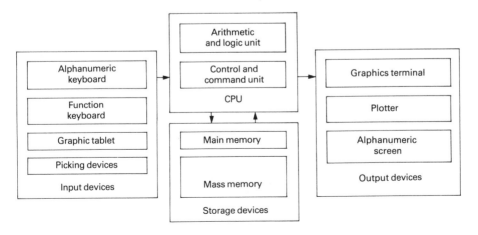

Figure 14.1 Schematic representation of the components of a typical CAD system.

programs as they work on the data and gives commands to peripheral devices. The capacity of the CPU in a CAD system can be as large as the mainframe or as small as the microcomputer, depending on the intended application, required function and expected speed of the system.

Storage devices

Data and programs are stored in the computer memory. The main memory contains the programs and data currently in use and is generally RAM (random-access memory). As this type of memory is relatively expensive, the main memory is generally a small fraction of the total memory. Mass memory is used to store the major part of the data and programs on magnetic tapes, disks, floppy disks or cassettes. Disks can be read by random access and, therefore, have a much shorter access time than tapes. On the other hand, tapes can be used to store large quantities of data at lower cost.

Input devices

Information can be given to the CAD system in the form of texts and numeric data using the alphanumeric keyboard. Preprogrammed functions can also be carried out by using function keys on the alphanumeric keyboard or by using a separate function keyboard. The graphic tablet can be used as a digitizer to input coordinates from a drawing or it can be used to select a line or character on the graphics screen using picking devices like the stylus or the puck. Other picking devices like the thumb wheel, joystick and mouse can be used independently of the graphics tablet to control a cursor on the screen. The light pen or the touch screen work directly on the face of the graphics screen.

Output devices

The graphics terminal communicates the commands between the user and the CAD system in addition to carrying out some processing of data and generating an image on the screen. Depending on the amount of data processing that the terminal performs, it is regarded as intelligent, smart, or dumb. The plotter provides a hard copy of the design either on paper (drum plotters, flatbed plotters or electrostatic plotters) or on microfilm (computer output to microfilm (COM) plotters). The alphanumeric screen has limited use in the CAD system and can be used to display the messages and commands exchanged between the user and computer, as well as results of calculations.

Software

According to Stark (see bibliography) the software used in a CAD system can be considered as made up of six levels. The first level is the operating system which controls the CPU and commands and communicates with the peripherals. The first level also includes language compilers which take a program written in a certain computer language and translate it into machine code that can be executed by the CPU. The second level of software is the graphics software which provides communication between the graphics terminal and the CPU, manipulates the images on the screen and manages the input/output devices. The third level of software includes the user interface which allows the designer to input and receive text and numeric values, digitize points and receive prompting and help messages. The fourth level is the geometric modeling software. The strength of a given CAD system largely depends on how powerful is the geometric modeler. Geometric modeling can be 2-D (two-dimensional), 2.5-D or 3-D, as will be discussed in Section 14.3. The fifth level of software is the applications software. Applications include: (a) drafting, see Section 14.4; (b) calculations of volumes, surface areas, moments of inertia, etc., of parts generated by geometric modeling; (c) kinematic analysis and animation of linkages; (d) automatic mesh generation for finite element analysis; and (e) finite element analysis. The last two applications will be discussed in Section 14.5. The above five levels of software are usually supplied by the vendor of the CAD system. However, there are instances where a user needs to perform a specific application function not covered by the vendor. In such cases, the required software is developed by the user and this forms the sixth level of software.

14.3   GEOMETRIC MODELING

Geometric modeling is an important part in computer aided design, which allows the creation of a geometric model to represent the size and shape of the component. This model can then be used for the following purposes:

1. To produce an engineering drawing of the part.
2. As a basis for a finite element model for stress analysis and design calculations.
3. As an input to numerically controlled machines when the CAD system is linked to a CAM system.

Geometric modeling usually involves:

1. Commands which generate basic geometric elements such as points, lines and circles.
2. Commands for scaling, rotation or other transformations of the geometric elements.
3. Commands which join the various elements into the desired shape of the part being created.

During the geometric modeling process, the computer converts the commands into a mathematical model, stores it in the computer data file and displays it as an image on the cathode ray tube (CRT) screen.

The geometric model may be 2-dimensional, for flat objects, 2.5-dimensional, for parts of constant section with no side-wall details, and 3-dimensional for generalized part shape. Wire-frame models of the type shown in Fig. 14.2 are the simplest and most commonly used as they are easy to create and require less computer time and memory. The image, however, is difficult to interpret, especially in complex 3-dimensional parts and contains little information about the surfaces of the part.

Surface models are more sophisticated and can overcome many of the ambiguities of wire-frame models, as shown in Fig. 14.2. Surface models define part geometry precisely and can be used to produce NC machining instructions automatically. Surface models, however, do not contain any information about the interior of the part nor whether it is solid or hollow. This means that it cannot be used as a basis for volume and weight calculations or stress and strain predictions.

The most advanced method of geometric modeling is solid modeling in three dimensions, which records the part in the computer mathematically as volumes bounded by surfaces. As a result, internal features can be defined which allows the calculations of mass

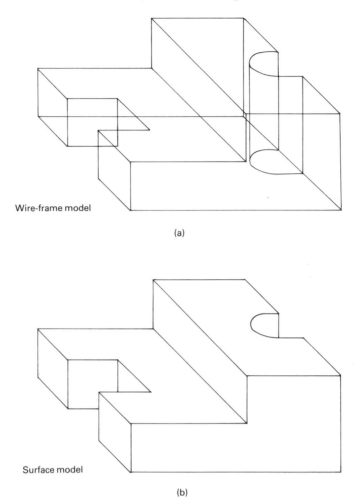

Figure 14.2 Simple geometric modeling. (a) Wire-frame model. (b) Surface model.

properties of the part which are usually needed for engineering analysis such as the finite element method. In addition, color graphics can be used to display more information and to define the different components in an assembly. Solid modeling is based on elementary, or primitive, shapes like planes, cylinders, cones and spheres. The constructive solid geometry (CSG) approach combines these primitives to create complex solid models, as shown in Fig. 14.3. Solid models, however, have the disadvantage of requiring more memory and extensive processing for their manipulation due to the more complicated data structure and associated mathematics.

Boundary modeling is another approach which starts with a planer outline of the part which then moves to create a part with the required thickness, as shown in Fig. 14.3. As the two solid modeling approaches have their relative strengths and weaknesses, maximum flexibility is achieved when the system allows the selection of the appropriate approach, depending on the part shape and complexity.

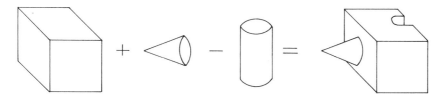

Solid modeling using primitives

(a)

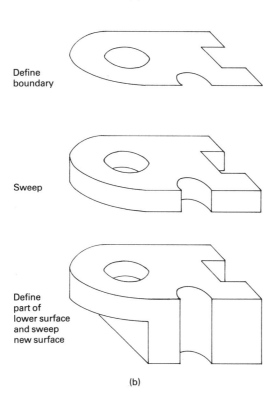

Define
boundary

Sweep

Define
part of
lower surface
and sweep
new surface

(b)

Figure 14.3 Advanced geometric modeling. (a) Solid modeling using primitives. (b) Solid modeling using boundary models.

14.4 AUTOMATED DRAFTING

Automated drafting is one of the oldest features of CAD, and in some early systems this feature represented the principal activity. Investing in such systems was justifiable on the basis that automated drafting increased productivity by more than four times over manual drafting. With automated drafting, the draftsman need not manually draw each line. Rather, the start and end need only to be defined. Circles may be drawn by specifying a center point and radius, or three points on the circumference. Other features allow features like points and lines which are drawn or removed in one view to be automatically projected into other views.

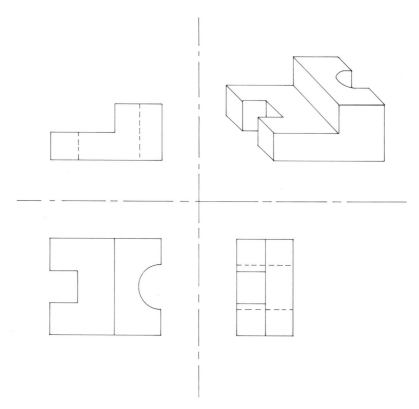

Figure 14.4 Engineering drawing with views generated automatically by a CAD system.

In addition to improved and consistent quality of drawings, automated drafting allows automatic dimensioning, generation of crosshatched areas, sectional views, scaling and enlarged views of particular part details. CAD systems usually allow isometric and perspective representation of components. These three-dimensional views can be rotated to view different sides, which can be of significant assistance in designing and drafting. This capability is related to geometric modeling which was discussed earlier. Most CAD systems are capable of automatically generating different projections of the part as it is being modeled, as shown in Fig. 14.4.

An added advantage of this feature is that drawings can be stored and recalled easily. Part classification and coding involves the grouping of similar part designs into classes and relating the similarities by means of a coding scheme. This system allows the designer to recall a particular design and to modify it to suit a different application, which may be easier than always designing new parts. This feature is also very useful in linking CAD systems with CAM systems, as will be discussed in Chapter 16.

14.5   FINITE ELEMENT ANALYSIS

Nearly all engineering design projects require some type of analysis like stress–strain calculations or heat transfer computations. The computer programs required for such analysis may be developed in-house to solve a particular design problem, or may be commercially available general-purpose programs. The finite element method (FEM) is probably the most powerful and widely used tool in computer aided design. FEM can produce accurate solutions to complex stress analysis problems with reasonable expenditure of time and resources.

In the FEM, the part is divided (idealized) into a large number of connected finite elements. Two-dimensional rectangular or triangular elements are simple and are widely used. On the other hand, 3-dimensional FEM models are best constructed from 3-D building blocks of either straight or curved sides. The points where the elements are connected are called nodes. The smaller the size of the elements used, and therefore the larger their number, the closer will be the idealized shape to the part shape. Basic stress and strain equations are used to compute the deflection at each nodal point as a result of the forces which are transmitted to it from neighboring elements. Satisfying the conditions of compatibility at the nodal points, the deflections are translated into strains and then to stresses using the appropriate constitutive equations. As the forces at each node depend on the forces at the other nodes, a complex system of simultaneous equations has to be solved in order to arrive at the FEM solution. The complexity of the simultaneous equations increases, and therefore the computing time and cost increase, as the total number of elements increases. This means that more economy is achieved by using a smaller number of elements while higher accuracy is achieved by using a larger number. A compromise solution is usually achieved by using relatively small elements in critical areas, e.g. points of stress concentration, and relatively large elements in safer areas of the component or structure under consideration, as shown in Fig. 14.5.

The first step in finite element analysis is dividing the part to be analyzed into a suitable number of finite elements and then defining the nodal point coordinates, the loads, the boundary conditions and material properties. This represents the input data for the FEM program and, if prepared manually, can be time consuming, tedious and subject to errors. This process has been automated in many commercially available general-purpose programs which has enhanced the use of FEM. Using the CAD system for automatic mesh generation allows the user to change the arrangement of the elements to suit the problem under consideration, e.g. decreasing the size of elements in critical areas to improve the accuracy of predicted stresses.

Determination of the stiffness matrix, the force–displacement characteristics, for each element is the next step in the FEM solution. The force in each element, a vector, depends on the relative displacement of its nodes, also a vector, and the element stiffness matrix. In matrix notation:

$$\{f\} = [k]\{\delta\} \tag{14.1}$$

where $\{f\}$ = column matrix, vector, of the forces acting on the element

$[k]$ = element stiffness matrix and can be constructed from the coordinate locations of the nodal points and the matrix of elastic constants of the material. The principles of statics can be used to determine $[k]$ for simple triangular elements, but more complicated elements require the use of energy principles.

$\{\delta\}$ = column matrix of deflections of the nodal points of the element.

The next step is to combine the element stiffness matrices into an overall assemblage stiffness matrix for the complete problem. The problem is then solved using the overall

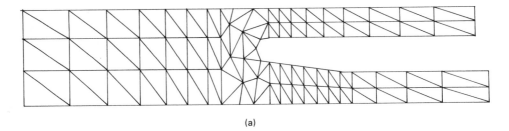

(a)

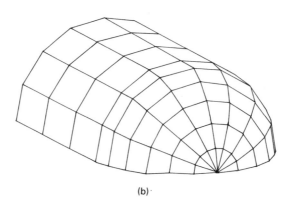

(b)

Figure 14.5 Division of a part into a number of elements in preparation for performing finite element analysis. (a) Two-dimensional part, triangular elements. (b) Three-dimensional part, rectangular elements.

stiffness matrix to obtain unknown forces and displacements from known boundary conditions. The basic matrix equation is:

$$\{F\} = [K]\{\delta\} \tag{14.2}$$

where $\{F\}$ = external forces at each node. This is known from the expected service loads and equilibrium conditions. The vector sum of the forces acting on the elements meeting at a given node must equal the sum of the externally applied forces at that node.

$[K]$ = overall stiffness matrix, assembled from $[k]$ for all elements.

$\{\delta\}$ = displacements at each node.

The displacements at the different nodes are determined from (14.2) and can then be used to determine the stresses at the different nodal points, $\{\delta\}$, using the equation:

$$\{\sigma\} = [E][B]\{\delta\} \tag{14.3}$$

where $[E]$ = matrix of elastic constants of the material

$[B]$ = matrix of coordinate locations of the nodes

The resulting nodal displacements and stresses in complex components and structures can be difficult to handle if presented as numerical values. In such cases, presenting the results graphically in the form of contours can be helpful, especially if color graphics is used. Vector representation can also be used and in this case the length of the vector is proportional to the magnitude and the direction of the arrowheads indicates whether the stresses are compressive or tensile.

14.6   DESIGN REVIEW AND EVALUATION

An advantage of using CAD systems is that checking the accuracy of the design can be accomplished on the graphics terminal. Dimensions and tolerances can be drawn semi-automatically, which reduces the possibility of error. The designer can use the zoom facility to magnify an intricate part on the graphics screen for closer inspection. CAD systems lend themselves particularly well to iterative modifications of the design and systematically definable changes in shape or material properties can be easily incorporated. For example, the computer can calculate the new torque capacity of a shaft when the diameter or material are changed.

Layering, which involves overlaying one image on top of another, allows the designer to compare visually different components on the graphics terminal. For example, the final shape of a machined part can be placed over the image of the as-forged part to ensure that sufficient machining allowance has been provided. Layering can be carried out in different stages to check each successive step in the processing of the component. Interface checking is another related procedure. This involves layering the different components of an assembly to ensure that they do not interfere with each other or occupy the same space. This is particularly useful in the design of complex piping systems and similar applications. Layering can also be used in the design of parts to be made from sheet metal to ensure proper nesting and more economic use of the material.

Animation is another important feature of CAD systems. Kinematic packages provide the possibility of simulating the motion of hinged components and linkages. Animation allows the designer to visualize the operation of the mechanism and to detect areas of possible interference between different parts of the assembly.

14.7   CREATING THE DESIGN AND MANUFACTURING DATA BASE

An important feature of a CAD system is that a design data base can be created by storing the design analysis and the engineering drawings for future use. The design data base serves as an extension to the memory and experience of the designer and can be used for design modification and review as well as for creating new designs.

Another important feature of CAD systems is that they can be used to develop the data base for part manufacture. Conventionally, engineering drawings are drawn by a draftsman and then used by the manufacturing engineer to develop the process plan. This two-step procedure is time consuming, as it involves duplication of effort by the design and manufacturing personnel. A direct link between CAD and CAM systems will allow automated transfer of information between the two activities. In this respect the manufacturing data base can be considered as an integrated CAD/CAM data base as it includes geometry and dimensions in addition to tolerances, surface finish and material specifications. Figure 14.6 shows how the data base can be related to design and manufacturing activities. Other advantages of integrating CAD/CAM systems will be discussed in Chapter 16.

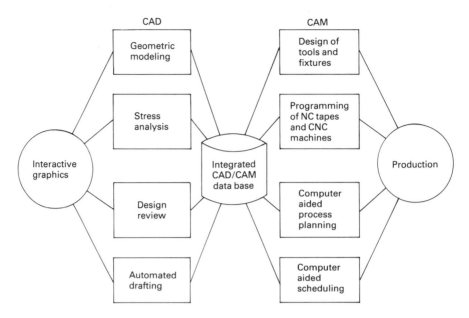

Figure 14.6 Linking of CAD and CAM systems using an integrated CAD/CAM data base.

14.8   APPLICATIONS OF CAD IN INDUSTRY

Computer aided design is now widely used in industry and in many cases it has replaced the traditional design techniques. The following three examples are selected from various industries to illustrate the use of CAD in practice.

Aerospace

Aircraft industry is one of the first users of CAD and it is now regularly used throughout the aerospace industry. As a specific example, consider the gas turbine blades for jet engines. Blade sections at different positions along the axis are drawn to conform to the fluid flow and power generation requirements and then used to build up the complex three-dimensional surface of the blade and root. This is then attached to the turbine rotor. Finite element mesh is then automatically generated in preparation for stress and thermal analysis and optimization. The geometry of the blade is also used in preparing NC tapes for machining prototypes and for making forging dies for the blades.

Automotive industry

Automotive industry is also among the first users of CAD and is now using it in all stages of production, starting from styling of the body through to the design of engine components and tires. For example, the full-scale clay model of the car which is used in design engineering is digitized and the digitized points entered into the CAD database. The data is then checked to ensure that the curves are smooth and aesthetically pleasing. The accuracy of the data can be further checked by using it to make a model of the car body in soft material on an NC machine. Having defined the external body shape, the interior body panels are then defined and the geometrical model used for: (a) structural analysis, to check the stress and strain distribution in various panels; (b) kinematic analysis, to check the action of doors and hinges; (c) aerodynamic analysis, to check air resistance and drag forces; (d) visibility analysis, for windscreen, windows and rear mirrors; and (e) space allocation for luggage compartment, engine space, seats, etc. The surface data is also used in the manufacture of the press tools for body panels.

Packaging industry

Containers and bottles for consumer goods such as toiletries and liquid detergents need to be designed in complex shapes and pleasing styles to attract customers. CAD has proved to be a convenient tool for modifying the shape of the package while maintaining a constant internal volume. Color shading under various light conditions can also be used to test the appearance of the package on the shop shelf. Structural analysis of the package will show its resistance to loading conditions during transportation and use. The geometrical model can also be used to calculate the weight of the material needed for making the container and can be used by the manufacturing engineer to generate the NC tapes for making the mold cavity. CAD can also be used to design the gating system to ensure smooth flow and uniform cooling of the plastic in the mold.

14.9   BENEFITS OF CAD

There are many benefits of computer-aided design, some of which are economic and others are related to the quality of work. Direct economic benefits include cost reduction and improved productivity. Encarnacao and Schlechtendahl (see bibliography) report that the cost of a CAD system with four terminals, $430 000, was recovered in 1.1 years when operated in two shifts. The work of 32 draftsmen working without a CAD system may be done by 8 draftsmen with the CAD system in the same amount of time. Higher productivity improvement is achieved with increasing complexity and higher level of required detail. Increased degree of repetitiveness and symmetry in the designed parts is also known to increase the gain in productivity as a result of implementing CAD.

In addition to the economic gains, employing CAD results in shorter lead time of preparing reports, component drawings and other manually performed design tasks. Shorter lead time means shorter elapsed time between receipt of a customer order and delivery of proposal or finished product.

Employing computers for design calculations improves mathematical accuracy, gives higher level of dimensional control and results in much better definition of three-dimensional curved surfaces. CAD also allows better coordination between analysis and design tasks. The same person can perform both tasks and allows the designer to interact with the design in a real-time sense. This saves time, thus allowing more design alternatives to be explored and compared in the available development time. Such an interactive system also avoids many drafting and documentation errors as it eliminates manual data compilation. In some CAD systems, a change entered on a single item can appear throughout the entire documentation package, effecting the change on all drawings which utilize that part. It is expected that a better design will result under such conditions.

As a single data base is used for all CAD work stations, design, drafting and documentation procedures are standardized. The gain is even more when the CAD and CAM systems are linked through one data base.

In addition to the above advantages, CAD systems are known to result in materials savings, improved job planning, enhanced potential for product validation, better communication between the different parties involved in product development and improved product quality. Since data storage is compact, information from previous designs can be easily retained in the system's data base, for easy access and comparison with current designs.

14.10   REVIEW QUESTIONS AND PROBLEMS

14.1   How did CAD influence engineering design?
14.2   Do you expect a CAD system to be as beneficial for a small company as for a large company?
14.3   How is automated drafting related to finite element analysis?
14.4   Give examples of the information that you would expect to find in the data base of an integrated CAD/CAM system.

BIBLIOGRAPHY AND FURTHER READING

Encarnacao, J. and Schlechtendahl, E. G., *Computer-aided Design*, Springer-Verlag, 1983.

Groover, M. P. and Zimmers, E. W., *Computer-aided Design and Manufacturing*, Prentice Hall, New Jersey, 1984.

Hearn, D. and Baker, M. P., *Computer Graphics*, Prentice Hall, London, 1986.

Jaffe, M. T. *et al.*, *Computer-aided Design, Engineering, and Drafting*, Springer-Verlag, West Germany, 1984.

Lang, J. C., *Interactive Computer Graphics Applied to Mechanical Drafting and Design*, John Wiley & Sons, New York, 1984.

Rooney, J. and Steadman, P., eds, *Principles of Computer-aided Design*, Pitman/Open University, London, 1987.

Stark, J., *What Every Engineer Should Know About Practical CAD/CAM Applications*, Marcel Dekker, New York, 1987.

Stover, R. N., *An Analysis of CAD/CAM Applications*, Prentice Hall, New Jersey, 1984.

Chapter 15

Elements of the Production Function

15.1   INTRODUCTION

Production can be defined as the processes involved in transforming raw materials into goods that have value in the marketplace. The transformation process usually involves a series of steps, each step bringing the material closer to the desired final state. Most production systems require materials, equipment, tooling, energy and labor as inputs and yield completed product, byproducts, scrap and waste as outputs. Depending on the nature of the product, industries can be identified as:

1. Process industries which produce chemicals, petrochemicals, cement, steel and similar products. These industries normally start with natural resources like iron ore and transform them into semifinished products like steel ingots, bars and sheets.
2. Manufacturing industries which use the semifinished products from the process industries to fabricate and assemble discrete items like motor cars, television sets, computers, machine tools and parts that go in these products.

The following discussion will be mainly related to the production activities in the manufacturing industries. The task of organizing and coordinating the activities of such industries is complex, because several products are usually produced simultaneously and each product is made up of many individual components. The different activities and steps involved in the manufacture of products, usually called the manufacturing cycle, may differ depending on the type of industry, nature of the product, size of the company and style of management. Generally, however, the entire production system can be represented as shown in Fig. 15.1. The figure shows that the main functions in the production system can be classified as:

1. Management function, which includes adopting and communicating the company's strategies, policies and programs and providing the necessary resources to carry them out. The information received from the different departments in the company in relation to state of the market, ideas for new products and status of the work in process form the basis upon which the resources of the company are allocated. Feasibility studies and decisions to introduce new products are also made within the management function. Some aspects of the management function are briefly discussed in Chapter 1.
2. Product development function, which includes product design, materials and process selection, and research and development. Development of new products is initiated by the management who also give the specifications and allocate the time and funds

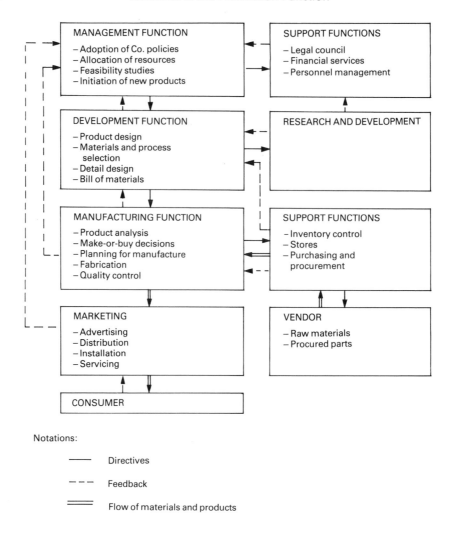

Figure 15.1 Main functions in the production system.

needed. The outputs from the product development function are layout and detail design drawings which are suitable for use in manufacture and assembly. A detailed list of parts, usually known as the bill of materials (BOM), should also accompany the drawings and identify the parts to be purchased rather than made in the company, as discussed in Chapter 9. Concepts for new products and research data represent another important output of the development function. As it is the responsibility of the product development personnel to generate functional designs, no changes in these designs should be made without their consent. The activities of the development function – design and materials selection – are discussed in detail in Parts II and IV respectively.

3. Manufacturing function, which converts the procured materials and parts into finished products according to the designs supplied by the development function. Planning for

manufacture shows how each product is to be made and how each part of each product is to be made, assembled and tested for quality. The development function is kept informed about the status of the processing capabilities to ensure that the generated designs can be manufactured within the company and to take advantage of any new additions. The manufacturing processes used to convert the different materials into products are discussed in Part I. The different activities of the manufacturing function occupy a major part of the discussions in this part.

4. Marketing function, which includes advertising, distributing, selling, delivering, in- stalling and servicing of the company products. It is also the responsibility of marketing to perceive the changing needs of the market place for larger or smaller quantities of the current products and for new or improved products. In addition, marketing contributes to the pricing structures, assessing the competition and market condition. The outcome of such activities represents an important input to the management of the company. Some aspects of the marketing function are briefly discussed in Chapter 1.

5. Support function, which includes legal council, financial services and accounting, personnel management and manufacturing support. The latter activity covers inven- tory control, stores and purchasing and procurement.

15.2   TYPES OF MANUFACTURING SYSTEM

Manufacturing systems have traditionally been classified into job shop and mass produc- tion. Job shops are flexible and allow small to medium lot sizes of a wide variety of products to be manufactured following different processing routes. This type of activity requires flexible general-purpose machines, skilled labor, much indirect labor, a great deal of material handling, large inventory and long in-process waiting times. A typical job shop may produce 2000 to 10 000 different components per year on 100 to 500 machine tools in lot sizes ranging from 10 to 100 units. Scheduling and planning in the job shop environment is difficult and this could lead to uncertainties in delivery times, cost estimation and product quality.

Mass production represents the other extreme in being intended for large numbers of products which undergo the same sequence of operations using specialized, more dedi- cated equipment. The most automated examples of mass production employ transfer lines as in motor car assembly. Transfer lines are efficient, need less skilled labor and are much easier to schedule.

In spite of the low productivity and high cost of job shops, social and technological trends show that there is an increasing need for small-lot production systems. This is due to the increased variety of product models, as in the case of consumer products; and faster development in technology, as in the case of electronic-based industries. On the other hand there is increasing pressure to improve productivity and quality, to reduce lead time and to halt inflation. This has led in recent years to the introduction of new types of manufacturing systems, such as group technology and flexible manufacturing systems, which combine the flexibility of the job shop and the efficiency of the transfer line. Group technology will be discussed in Section 16.5 and flexible manufacturing systems will be discussed in Section 16.7.

The following discussion of the different elements of the production function applies equally well to the different types of manufacturing systems discussed above.

15.3   PRODUCTION PLANNING AND CONTROL

Planning for manufacture is the first step in the manufacture of a new batch of an ongoing product or a new product. Planning is usually carried out on two levels of detail:

1. Planning for manufacture which involves the creation of an overall strategy for manufacturing the product.
2. Process planning which involves the processing of individual parts and the methods of assembling them into end products.

In the case of large companies producing complex products, e.g. the motor car industry, both levels of planning will be needed. On the other hand, the two levels of planning may not be easily distinguishable in a small company producing simple products.

Planning for manufacture

Planning for manufacture starts when the detail design and bill of materials are received from the development function. The design consists of layout drawings and detail drawings for every part to be manufactured. Specifications of the parts to be purchased are also supplied. The planning function is based on directives from the management function giving the number of units to be made, allocating the budget for manufacture and specifying the expected delivery schedule. The major objectives of the planning function are: (a) to optimize the use of available facilities, including machines, tools, test facilities and manpower; (b) to minimize the cost materials, fabrication, inspection, assembly and testing of the product; and (c) to minimize the time required for start-up and production of the required batch.

Planning for manufacture starts with a breakdown of the product into its major subassemblies which are then divided into subassemblies, sub-subassemblies and then to separate parts. As an example, consider the case of a medium-size manufacturer of household refrigerators. Figure 15.2 shows how the refrigerator can be divided into major subassemblies, subassemblies and sub-subassemblies. For each sub-subassembly or part, a make-or-buy decision is made. Some parts are obviously going to be made in-house, others are equally obviously best procured from other companies which specialize in making them. In the case of the refrigerator, the body, door and cooler and condenser subassemblies are among the items that will most likely be fabricated in-house, although the raw materials such as steel sheets, plastic powder and copper tubes will be bought from other suppliers. The motor–pump unit, ball bearings, electrical cables, switches, light bulbs and connectors will most likely be bought from specialized companies.

Some parts, however, may present an option. In such cases, a decision has to be made whether or not company facilities should be used. In some cases such a decision is influenced by factors like quality and reliability of supply, control of trade secrets, patents, flexibility and availability of alternative supply sources. A company policy to specialize and to concentrate its efforts and skill in one basic line rather than to diversify may also affect make-or-buy decisions. In most cases, however, the major factor that influences the make-or-buy decision is cost. If a part can be bought cheaper than it can be made, the decision is usually to buy it. In the case of the household refrigerator, the items which may represent the option for make-or-buy decision include expansion valves, temperature selector and rubber seals. Although the actual procurement of the bought items will be

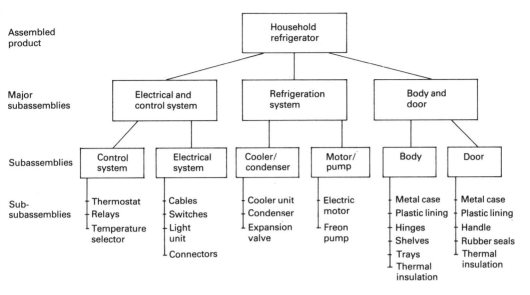

Figure 15.2 Breakdown of a household refrigerator into major subassemblies, subassemblies and sub-subassemblies.

conducted by the purchasing department, quantities, specifications and delivery dats are determined by the planning function.

For parts that are to be made in-house, processing times and assembly times are then roughly estimated. Some of these times may run concurrently and others sequentially. Parallel with this activity, several support functions must also be planned. These include material requirements, tooling, quality assurance facilities and personnel. These estimates are then used to prepare a production master schedule, which is a listing of the products to be produced or purchased, when they are to be delivered and in what quantities. Based on the master schedule, raw materials that will be used to make the various parts and all the items that are to be purchased must be ordered. Delivery times must be planned to ensure that they are available when needed. This task is usually called materials requirement planning, MRP.

In addition to the above scheduling activities, cost estimates and budgets are set up to control manufacturing costs. Costs in this context include all the expenditures necessary to manufacture the product. The different elements of cost include direct and indirect labor and materials in addition to overheads, as will be discussed in Chapter 18.

Process planning

The main objective of process planning is to convert the detail design into detailed manufacturing instructions. A sequence of operations is listed showing every act that is necessary to convert a raw material into a finished part. Such documents are usually called process sheets, route sheets or methods planning sheets. Specifications and dimensions of the material from which the part is to be made should also be included in the process sheet, as shown in Fig. 15.3. Every operation on every part has to have machines, tooling, gages and supplies planned out in advance, ordered, procured and delivered according to

	PROCESS SHEET			Page 1 of 3 pages		
Written by M.M.F.		*Order no.* 1844			*Dwg no.* 12	
Date 1/3/88		*Date* 1/1/88		*Pcs req'd* 30	*Patt. no.* 5	
Enters assembly at stage	x-23 Loader 6		*Part name* 250 mm Pulley			
Material condition Gray CI, ASTM A48-74 35, 245 BHN			*Rough weight* 15 kg		*Finish weight* 12 kg	
Oper. no.	*Description*			*Set-up time (h)*	*Cycle time (h)*	*Mach. no.*
10	Turn O.D. of body and flanges, face hub Speed 200 rpm, feed 0.25 mm per rev tool no. TT-25			0.5	0.5	L-2
20	Turn bore and face other side of hub Speed 200 rpm, feed 0.25 mm per rev tool no. TT-25			0.6	0.3	L-2
30	Drill and tap 2 holes, 10 and 12 mm diameter standard metric thread M10 and M12			0.3	0.2	D-1

Figure 15.3 Example of a simple process sheet.

schedule. If the need for special production aids is economically justified, requests for special jigs and fixtures are given to the tool design department.

In addition to their use in manufacturing, process sheets also serve as a basis on which to predict times and to estimate costs. Virtually everything that takes place in manufacturing takes time and creates costs, and the more detailed the process sheets, the better the estimates will be. Computer programs are now available to assist the planner in this tedious and time-consuming activity. Computer aided process planning, CAPP, will be discussed in Chapter 16.

In the course of this detailed planning, some changes in part design may be requested to facilitate manufacture. If the designer is aware of the available manufacturing facilities, and if communications are maintained between design and manufacturing personnel, there should be little need for such design changes.

Following process planning, assembly operations are planned and the necessary tools and equipment are listed. Assembly proceeds by assembling a few parts in a subassembly, and then assembling the different subassemblies until a finished product is obtained. Even at this stage, design changes may still be necessary because of the difficulties encountered in assembly and problems due to tolerance build-ups. Testing and inspection may be necessary during assembly and appropriate facilities should be provided.

The outputs of the planning function include instructions to the manufacturing shop and lists of types and quantities of materials to the purchasing department. The shop instructions include for each part a work packet containing the work order, process sheet, NC tapes if appropriate and all the necessary drawings. It also includes a phasing list to help in assembly operations. Testing instructions for the final product are also included in the manufacturing shop instructions.

15.4   THE MANUFACTURING SHOP

The main activities that take place in the manufacturing shop include actual conversion of raw materials into parts, assembly, testing and inspection. The master schedule and the detailed instructions received from the production planning function represent the main inputs to the manufacturing control department in the shop. Manufacturing control will then make detailed assignment schedules of all the separate production tasks to the available resources. It determines the priorities of work at each work station, follows the progress of work from task to task and ensures that the different materials are available when needed. The different tasks should be scheduled so that the different parts are delivered to the assembly station on time with the highest possible efficiency. This can be a difficult task in a shop that manufactures widely different products. For example, in a job shop some products may be manufactured on a long-term basis in a continuous stream of units simultaneously with other products which are produced in batches at random intervals. With the increasing use of computers in the manufacturing industries, software programs have been developed to help production engineers in performing this task more effectively. This subject will be discussed in Chapter 16.

Production control

When instructions to manufacture a new product are first received, the inventory files are searched in case some parts are already in stock. Some items may have been made or bought in economical lot sizes and stored for later use. Knowing the total number to be produced, the next step is to set economical lot sizes for the production, and to determine the time when each lot must be entered on the shop floor to assure assembly of the product at the due date. The economic lot size is determined by a balance between the cost of setup and the costs of storage and of the capital that is tied up in the finished product.

Reference to the process sheet for a given part will show what machines will be involved in manufacture, the setup time and the cycle time. From the number of parts, the production controller determines the length of time where the machine is tied up with each operation and calculates the loading schedules for the different machines. The use of computers in manufacturing has made this task much easier.

Fabrication steps

The first step in fabricating a given part is to set up the machine and to equip it with the necessary dies, tools, jigs or fixtures. Usually the first part is checked for accuracy before the rest of the batch is processed. Samples may be periodically removed for inspection. Inspection may have to be completed before the setup is dismantled, so that if some parts are rejected, they may be reworked or replaced without another setup cost. When the required number of parts is completed, the setup is dismantled. Tools and other production aids are then returned to their storage areas. The machine will then be ready for a new assignment.

Assembly, testing and shipping

Assembly usually proceeds in stages, with small subassemblies joined to form larger ones until the final product is achieved. The sequence of assembly normally proceeds according to phasing instructions given by production planning. Subassemblies are usually tested before moving on to higher levels of assembly. Rejected subassemblies may have to be disassembled and the defective parts replaced or reworked. Recurrent problems in assembly may indicate the need for adjustment of procedures or even of detail design. For example, in the case of the household refrigerator, Fig. 15.2, the refrigeration system as well as the electrical system need to be tested before final assembly.

The fully assembled product is normally tested and adjusted to achieve full operation capability. Tests are normally carried out by quality control personnel who operate independently from the shop personnel.

Accepted products are then moved from test to finishing and packaging, then stored in readiness for shipping. The type of package depends on the nature of the product and the distance it will have to travel before reaching the customer. In all cases, it should provide protection against weather and handling.

15.5   MODELING OF PRODUCTION SYSTEMS

The above discussion has shown that manufacturing consists of a series of steps, each step bringing the material closer to the desired final state. In routing the material through the shop, it must be transported between successive operations. Delays often occur while the part waits in a queue for its turn to be processed on each machine or transported from one operation to another. This series of events can be represented by an operation sequence model, as shown in Fig. 15.4, for a hypothetical part. It has been estimated that of the total time an average part spends in a batch-type metal cutting shop, only about 5 percent is spent on machine tools, while the rest of the time spent in the shop is nonproductive. Of the 5 percent time spent on the machine, only about 30 percent, i.e. 1.5 percent of the total time in the shop, represents the actual cutting time. This shows that there is great need for reducing nonproductive time in order to improve productivity. It is expected that better productivity is achieved in automated and computer aided manufacturing systems.

The total manufacturing time, T_t, for a batch of n parts can be represented by the following simple mathematical model:

$$T_t = m(T_s + nT_o + T_w) \tag{15.1}$$

where T_s = average setup time per machine
m = number of machines involved in making the part
T_w = average waiting and temporary storage time per machine.
T_o = average operation time per machine
$\quad = T_1 + T_2 + T_3$
T_1 = actual time of the process
T_2 = workpiece handling time
T_3 = tool handling time, e.g. time to change worn or damaged tools

The times in (15.1) are taken as averages for simplicity. In actual shop situations, however, these times will vary with the type of process, the required operations in each process and the size and material of the part. If the part is in production, the exact times can be measured and added to give better estimates of the time elements in the above equation. From (15.1) the total batch time per machine, T_b, is given as:

$$T_b = T_s + nT_o \tag{15.2}$$

The average production time per machine, T_p, is then:

$$T_p = \frac{T_s + nT_o}{n} \tag{15.3}$$

The average production rate per machine, R_p, is:

$$R_p = \frac{1}{T_p} \tag{15.4}$$

The above equations show that the average production time is reduced, and the average production rate is increased, as the batch size is increased.

To illustrate the use of the above model, consider the manufacture of a batch of 50 parts of the hypothetical case given in Fig. 15.4. Using the terms of (15.1), $n = 50$. As the actual times for each operation are known in this case, the time elements can be estimated by addition. Rewriting (15.1) as $T_t = mT_s + mnT_o + mT_w$, then:

$$mT_s = T_{s1} + T_{s2} + T_{s3} + T_{s4}$$
$$= 1.0 + 0.5 + 1.5 + 2.0 = 5.0 \, \text{h}.$$
$$mT_o = T_{o1} + T_{o2} + T_{o3} + T_{o4}$$
$$= 0.40 + 0.20 + 0.01 + 0.012 = 0.622 \, \text{h}.$$

As shown in Fig. 15.4:

Tr-1 is transportation to first operation
Tr-2 is transportation to second operation
TS-1 is temporary storage (wait for operation 1)
TS-2 is temporary storage (wait for operation 2)

$$mT_w = (\text{Tr-1}) + (\text{TS-1}) + (\text{TS-2}) + (\text{Tr-2}) + (\text{TS-3}) + (\text{TS-4}) + (\text{Tr-3}) + (\text{Tr-4})$$
$$+ (\text{TS-5}) + (\text{Tr-5})$$
$$= 0.16 + 0.35 + 0.12 + 0.02 + 0.30 + 0.50 + 0.30 + 0.03 + 1.5 + 0.25 = 3.53 \, \text{h}.$$
$$T_t = 5.0 + 50 \star 0.622 + 3.53 = 39.63 \, \text{h}.$$

The average production time per machine, T_p, can be calculated by rewriting (15.2) as:

$$T_p = \frac{mT_s + mnT_o}{mn} = \frac{5.0 + 0.622 \star 50}{4 \star 50} = \frac{36.1}{200} = 0.18 \, \text{h}$$

15.6   IMPROVING PRODUCTIVITY IN MANUFACTURING

Based on the discussion in this chapter, several methods can be suggested to improve the productivity of manufacturing systems:

1. Use specialized machines rather than general-purpose machines, if the number of parts required is sufficiently large. This method is used in automatic transfer lines, as in the case of motor car industry.

Distance m (ft)	Time (0.1 h)	Symbol	Description of operation
–	48.00	PS	Permanent storage
50 (164)	0.16	Tr-1	Transportation to first operation (milling machine)
–	0.35	TS 1	Temporary storage (wait for operation 1)
–	$T_o1 = 0.40$ $T_s1 = 1.00$	0-1	First operation
–	0.12	TS 2	Temporary storage (wait for move man)
5 (16.4)	0.02	Tr-2	Transportation to second operation (drill press)
–	0.3	TS 3	Temporary storage (wait for drilling)
–	$T_o2 = 0.2$ $T_s2 = 0.5$	0-2	Second operation (Drill 3 holes)
–	0.5	TS 4	Temporary storage (wait for move man)
100 (330)	0.3	Tr-3	Transportation to third operation (washing section)
–	$T_o3 = 0.01$ $T_s3 = 1.50$	0-3	Third operation (wash)
10 (33)	0.03	Tr-4	Transportation to fourth operation
–	$T_o4 = 0.012$ $T_s4 = 2.00$	0-4	Fourth operation (Apply paint)
–	1.50	TS 5	Temporary storage (wait for paint to dry)
75 (246)	0.25	Tr-5	Transportation to store room
–	24.00	PS	Permanent storage

Notation T_o = operation time, T_s = setup time

Figure 15.4 Operation sequence model.

2. Reduce setup time by scheduling similar parts on the same machine. This will allow the use of common tools and fixtures. This strategy has been effectively used in group technology as will be discussed in the following chapter.

3. Plan processing so that similar operations are performed sequentially on the same machine. This has the advantage of reducing the times for setup, work handling and waiting. This concept is utilized in numerically controlled (NC) machines. Multiple spindle machines will also allow several processes to be performed simultaneously, thus improving the productivity of manufacturing. NC machines will be discussed in the following chapter.

4. Integrate manufacturing processes by linking them by automatic work handling devices. This reduces transportation and handling times and provides smoother flow of work pieces in the different work stations; transfer lines and flexible manufacturing systems are examples.

5. Use improved process plans by adopting the computer software which optimizes process planning. Computer aided process planning (CAPP) will be discussed in the following chapter. Process planning can also be made more efficient by employing computerized manufacturing data base to automate the generation of manufacturing data, e.g. cutting speeds and feeds for cutting a given material on a given machine tool.

6. Employ computers in managing manufacturing operations on the company level, in product design and in shop floor control. The concept of automated factories will be discussed in the following chapter.

15.7 REVIEW QUESTIONS AND PROBLEMS

15.1 Discuss the differences between a job shop and mass production systems.

15.2 How would you classify the maintenance and repair shop of a medium-size manufacturing company? How would you improve the efficiency of operation of such shop?

15.3 Explain the reasons why a large proportion of manufacturing is done according to the batch productions system.

15.4 The manufacture of a given component requires eight operations in the machine shop. The average setup time per operation is 2.5 h. Average operation time per machine is 7 min. The average nonoperation time needed for handling of work pieces and tools, inspection, delays, temporary storage, etc., is 50 min per operation. Estimate the total time required to get a batch of 100 components through the shop. [Answer: 120 h]

15.5 Prepare a process sheet for machining of 10 bronze sleeves for a journal bearing from a cast hollow cylinder. An engine lathe will be used. Casting dimensions are: 65 mm outer diameter, 45 mm inner diameter, and 74 mm length. Finished dimensions are: outer diameter 60 mm, inner diameter 50 mm, and length 70 mm. All surfaces are to be given finishing cuts.

15.6 A component needs six operations on an engine lathe in a batch production shop. The different times associated with each operation are shown in Table 15.1.

Ignoring the setup time and taking the workpiece handling time as 2 min, find the production time (T_p) and production rate (R_p) for this part. [Answer: $T_p = 24.5$ min, $R_p = 2.45$/h]

15.7 The component in question 15.4 is required in batches of 500, and a turret lathe is being considered for manufacturing the component. The machining time will be unchanged, but the total tool handling and changing time will be reduced to 1.5 min per operation. If the setup time of the turret lathe is estimated as 6 h, calculate:

(a) Total time to process the batch on the engine lathe.

(b) Total time to process the batch on the turret lathe.
(c) Average production time in both cases.

[Answers: (a) 493.3 h, (b) 478.87 h, (c) 59.2 min for engine lathe, 57.5 min for turret lathe]

Table 15.1 Batch production shop times (see Problem 15.6)

Operation	Operation time (min.)	Handling and changing of tools (min.)
1	0.5	2.5
2	2.0	3.0
3	1.5	2.5
4	0.5	1.5
5	1.0	2.0
6	2.5	3.0

BIBLIOGRAPHY AND FURTHER READING

Besant, C. B. and Lui, C. W. K., *Computer-aided Design and Manufacture*, 3rd ed., Ellis Horwood, London, 1986.
Buffa, E. S., *Modern Production Management*, 4th ed., John Wiley & Sons, New York, 1973.
Dallas, D. B., *Tool and Manufacturing Engineering Handbook*, 3rd ed. SME, McGraw-Hill, New York, 1976.
Groover, M. P., *Automation, Production Systems, and Computer-aided Manufacturing*, Prentice Hall, N.J., 1980.
Harrington, J., *Understanding the Manufacturing Process*, Marcel Dekker, New York, 1984.
Koren, Y., *Computer Control of Manufacturing Systems*, McGraw-Hill, London, 1983.
Riggs, J. L., *Production Systems: Planning, Analysis, and Control*, 3rd ed., John Wiley & Sons, New York, 1981.

Chapter 16

Computer Aided Manufacture

16.1   FUNDAMENTALS AND APPLICATIONS OF CAM

Computer aided manufacture, CAM, can be defined as the use of computer systems to plan, manage and control the operations of manufacture through either direct or indirect computer interface with the production facilities. The direct applications involve connecting the computer directly to the manufacturing process with the purpose of monitoring and/or controlling the process. Monitoring involves observing the process and associated equipment and collecting data from the process. In this case the computer is not used to control the process directly. Control remains in the hands of the operator, who is guided by the information compiled by the computer. Computer process control goes further than monitoring by not only observing but also controlling the process, based on the observations. Indirect applications of computers in manufacture include support operations like planning, scheduling, forecasting, inventory control, programming of machine tools and supplying information.

The gap between CAD and CAM is decreasing and integrated CAD/CAM systems are now commercially available. In such integrated systems, the link between design and manufacture is accomplished by the use of a computer. This link allows the information and data from the CAD process to be used directly in the CAM procedures, thus avoiding the independent generation of data for computer programs in the manufacturing area.

16.2   NUMERICAL CONTROL

A numerical control (NC) system is one in which instructions for performing a job are given to an automatic machine as numbers. NC has mostly been applied to machine tools including lathes, drilling and boring machines, milling, broaching and grinding machines. NC has also found applications in punching presses, arc welding, flame cutting, riveting operations, filament winding of composites and assembly. NC machines may be classified according to the number of axes of numerically controlled movements with respect to Cartesian x-y-x coordinates. There may be other movements not numerically controlled. A two-axis machine would have the table moved in the x and y directions in one plane, while a three-axis machine would have an additional movement in the z direction. Four-, five, or six-axis machines provide additional linear or rotary movements. Machining centers are NC systems of high versatility and have been developed to perform a wide variety of machining operations. Such machines are capable of milling, drilling, boring, reaming and tapping multiple faces in one setup. Machining centers carry a large number of tools, as many as 60 tools in some models, and each cutting tool is preset for diameter and depth in a quick-change holder. This allows cutting of programmed dimensions without any adjustment on the machine. In unmanned machining centers, parts are loaded and unloaded automatically.

The instructions to an NC machine consist of a series of numbers assembled into a program. The program lists the steps and conditions for each operation to produce the part. Programs are fed into the machine control unit (MCU) as punched tapes, magnetic tapes, signals directly from computer logic or computer peripherals, such as coated-metal disks or floppy disks, or voice command. In integrated CAD/CAM systems, product specifications are computer generated by the designer and transmitted directly to a computer that prepares the NC program. Systems which have an external computer connected directly to several satellite NC controls are called direct numerical control (DNC). In addition to preparation and storing NC programs, the external computer can also be used to perform management reporting and production control functions, as in the case of CAM systems.

The MCU converts the numbers in the program into signals to activate the machine components. For example, the position commands are commonly converted into a series of pulses, with each pulse representing an incremental movement of about 2.5 μm (0.0001 in) in the x, y, or z direction. This value represents the resolution of a typical high-quality NC system. The rate of pulses can be used to control the feed rate and to coordinate the feed rates of the various axes if contouring is to be done. On the other hand, spindle speed commands are translated into electric currents to the gear shifters in the drive. The accuracy of most NC systems, which signifies how closely a machine member can be moved to any specified point, lies in the range of ±5 to 25 μm (±0.0002 to 0.001 in). Of equal importance is the repeatability or precision of an NC machine, which is how closely the table or any other member can be returned to an initial position repeatedly. This value is usually in the range of ±8 μm (±0.0003 in).

Modern controls in NC systems are also capable of interpolation between discrete points established along the tool path. When the control functions are performed by a computer, the system is called computer numerical control (CNC). The program in the CNC memory can easily be updated and revised and standard subroutines can be used to carry out whole operations like drilling, tapping, boring, etc.

An NC machine may be controlled through an open-loop or a closed-loop circuit. The open-loop system is the simplest and cheapest but does not assure accuracy. A signal for a certain action travels through the drive mechanism which imposes the action upon the appropriate member of the machine. There is no feedback to report whether the action was successfully completed and the accuracy is not assured. The most commonly used type of closed-loop method of control is the indirect feedback system where the linear travel is inferred from the rotary position of the lead screw or pinion. The feedback signal is continually compared to the order until the two are equal. This indicates that the movement has been completed and the power is turned off.

An added feature of NC systems is adaptive control (AC) which is a means of continuously monitoring the critical parameters of an operation to maintain optimum conditions. AC can be used to detect faults, such as worn or broken cutting tools or hard spots in material. Sensors constantly measure the critical parameters in the operation and report to a computer that is programmed to compare the current status to standard values. If the difference persists or exceeds a given value, the computer orders corrections to be made. In metal cutting, the common parameters that are measured include cutting forces or resulting deflection, speed and torque or power at the spindle. If deflections and torque indicate cutting forces more than the tool can stand, the computer may order an override of the originally programmed feed rate to reduce it to a tolerable level, or may call for a change of tools for resharpening. The computer may also be programmed to adjust the feed or speed to obtain optimum tool life. Many modern NC systems have diagnostic capabilities to check the mechanical and electrical parts of the system, to find the causes of trouble that may arise in operation and display instructions to correct the faults.

In spite of the high initial cost of NC systems, which cost two to five times as much as conventional machines of like size, and in spite of the higher skill needed for their programming and maintenance, they are finding increasing use in many industries. There is a number of reasons for the increased use of NC system which can be summarized as follows:

1. Increased productivity as a result of shorter cycle time, shorter idle time, no fatigue and no human errors. As reported by Doyle *et al.*, 'Manufacturing processes and materials for engineers', Prentice Hall 1985, a study showed that NC machines are actually engaged in cutting about 80 percent of the time, as opposed to 25 percent for conventional machine tools.
2. Employing NC systems is known to decrease scrap and rework by about 25 percent, reduce materials handling by 20 to 50 percent, and reduce inspection cost by 30 to 40 percent.
3. Assembly costs are reduced by 10 to 20 percent as a result of reduced operator errors and increased accuracy.
4. Tooling costs, tool storage charges, setup costs and the cost of making changes are much less in NC systems. This means that smaller lot sizes are economical and less floor space is needed for materials in process and storage.

The above advantages which NC systems give can be exploited best for medium-size production. When one or very few relatively simple pieces are required, it may be more economical for skilled workers to perform the job on conventional machines. Also, NC cannot compete with special-purpose machines and tools when very large numbers are required.

16.3   COMPUTER AIDED PROCESS PLANNING

Process planning is the activity which determines the sequence of individual manufacturing and assembly operations needed to produce a given part or product. The resulting listing of the production operations and associated machine tools is usually called a route sheet. Traditionally, the different functions of process planning and the related activities of determining machining conditions and setting of time standards for the different operations have been carried out manually which is tedious and time consuming. In addition, traditional process planning procedures are dependent on experience and judgement which often lead to large variability among different planners. With computers these tasks can be carried out more accurately and efficiently. Based on the characteristics of a given part, the computer aided process planning (CAPP) system automatically generates the manufacturing operation sequence. Such systems integrate logic, judgement and experience into a viable problem-solving method. CAPP systems have the advantage of producing consistent plans because the same computer software is used by all planners. Experience with the use of CAPP has shown increased productivity and reduced manufacturing lead time.

Two alternative approaches to CAPP have been developed:

1. Retrieval or variant process planning systems depend on the parts classification and coding that will be discussed in the section on group technology. The parts are grouped into families according to their manufacturing characteristics and a standard process plan is established for each family. The standard process plan can be slightly modified to accommodate the different manufacturing requirements of a new part. The machine routing may be the same for the new part, but the specific operations required at each machine may be different. The complete process plan must document the operations as well as the sequence of machines through which the part must be routed. The system may also include programs to compute machining conditions, production times, work standards and standard costs for pricing purposes.

2. Generative computer aided process planning systems involve creating an individual process plan for each new product. This is done automatically using a set of algorithms to progress through the various decisions to reach a final process plan. The system synthesizes the optimum sequence of processes based on part initial shape and geometry, final shape and geometry, dimensions, required accuracy, material and other relevant parameters. Inputs to the system normally include a detailed description of the workpart. This may involve the use of some form of part code number to summarize the data, but it does not involve the retrieval of existing standard process plans. Some interaction between the planner and the computer is often utilized during decision making for checking purposes. As in the case of retrieval-type systems, generative process planning systems often calculate production times and costs in addition to the optimum sequence of operations.

An important activity which is closely related to process planning is the specification of the cutting conditions to be used in the various machining operations. The cutting conditions, which consist of the speed, feed and depth of cut, can be selected to satisfy one or more of the following criteria:

1. Tool life criterion, where the cutting tool is expected to last a specified length of time.
2. Surface finish criterion, where a specified smoothness is achieved and maintained.
3. Accuracy criterion, where deflections and vibrations should be kept below a specified maximum.
4. Power consumption criterion, which restricts the power to a value below a given level.
5. Economic criteria, where a maximum production rate or minimum cost per piece is achieved.

The criteria used to select the cutting conditions will depend on the type of process and the status of the part. For example, power consumption and economic criteria are important for roughing operations, whereas accuracy and surface finish are important for finishing operations. In addition, the selection of the optimal cutting conditions is affected by a multitude of different parameters which include:

1. Type of machining operation, whether roughing or finishing and whether it is turning, milling, drilling, grinding, etc.
2. Machine tool parameters, such as power, size and rigidity, available spindle speeds and feed rates, accuracy, etc.

3. Cutting tool parameters, including tool material (high-speed steel, cemented carbide, ceramic, etc.), type of tool (single point, drill, milling cutter, etc.), geometry (nose radius, rake angle, number of teeth, etc.).
4. Workpiece parameters, including hardness and strength, machinability, size and shape, rigidity, required accuracy, initial surface condition, etc.
5. Other cutting parameters, e.g. cutting fluids, available jigs and fixtures.

The complexity of the parameters that affect the cutting conditions has led to the development of many computerized machinability data systems. These systems can be generally classified into:

1. Data base systems, which are based on experimentally collected information for different cutting tool materials, machine tools and workpiece materials.
2. Mathematical model systems, which are based on empirical relationships like the Taylor tool-life formula:

$$vT^n = C$$

where v is the cutting speed, T is the tool life in minutes, n is an empirical tool-life exponent, and C is a constant of the material.

The Taylor formula is a simple model that gives fairly good predictive data over a limited range of conditions. In practice, a computerized machinability system often employs complex models that are coupled with optimization procedures so that the cutting conditions may be selected for the best economy, accuracy or other criteria.

Computer-generated time standards represent another important activity which is related to process planning. Work measurement can be defined as the development of a time standard to indicate the value of a work task. Work measurement is used to estimate:

1. Time needed for a given operation.
2. Cost of the job production scheduling and capacity planning.
3. Measurement of worker performance.
4. Wage incentives.

The time required for a given job depends not only on the activity but also on the type of machine and the characteristics of the workpiece in terms of weight, size, etc. To use the computerized system, the job to be timed must first be divided into its elements. The computer than calculates the element times, sums the times and applies the necessary allowances to determine the standard time for the total cycle.

16.4 COMPUTER AIDED QUALITY CONTROL

Quality can be defined as the degree to which a product or its components conform to certain standards. Quality in manufacturing is usually related to materials, dimensions and tolerances, appearance, performance, reliability and any other measurable characteristic of the product. Quality control (QC) is concerned with the activities related to testing and inspection of a product to assess its quality. These activities also include corrective action needed to eliminate poor quality. Quality control is traditionally performed manually on a sample of the product and the results are presented on control charts for record keeping. This is usually time-consuming and often leads to delays as parts have to be taken to a separate inspection area. Another disadvantage of manual inspection is that it can involve subjective judgement on the part of the inspector which can lead to variability in results. When quality control is performed after the fact, its usefulness is limited, as defective parts may need to be scrapped or reworked at a high cost. With the increasing use of computers in industry, computer aided quality control (CAQC) is being increasingly used in manufacturing. The following points represent some of the advantages of CAQC:

1. Increases productivity in the inspection process.
2. Allows 100 percent inspection, if technically possible.
3. Allows on-line inspection during production and can be integrated in the production process, rather than being performed after the fact.
4. CAQC system can be used to make adjustments to the process variables based on analysis of the data collected by the inspection sensors. The data analysis can include statistical trend analysis to account for variables like tool wear.
5. Robots can be used in completely automated QC cells.

The sensors used in CAQC systems can be classified into contact and noncontact types. The contact methods usually involve the use of coordinate measuring machines (CMM). Most of these machines are controlled by NC or computers and are capable of accuracies in the range of ± 5 μm (± 0.0002 in). Programs for CMM can be downloaded from a central computer in a similar manner to DNC. If the CMM is located away from the production area, time will still be wasted in transporting samples. Noncontact inspection sensors offer several attractive features in comparison with contact-type sensors. Noncontact methods are usually much faster and avoid the danger of damage of the inspected surface. Noncontact inspection methods can be divided into optical and nonoptical. The optical methods usually involve some sort of machine vision system, although scanning laser beam devices and photogrammetry techniques are also used. The nonoptical techniques are usually based on the use of electrical fields, ultrasonics and radiation to sense the desired characteristics of the part.

The full advantage of CAQC is achieved when it is integrated with CAD/CAM. In this case it is essential that the systems use the same data base. The programs that are used to develop the NC tapes can also be downloaded from the central computer to the CMM. Similar sort of downloading is possible for some of the noncontact inspection devices.

16.5   GROUP TECHNOLOGY

Introduction

Although mass production plays an important role in modern industry, about 75 percent of all metal working is still done in batches of 50 parts or less. Such work is normally done in job shops where it is difficult efficiently to plan the processes and predict delivery times as a result of the large differences in part shapes and batch sizes. Each batch is usually treated with no reference to previous work, which results in lack of consistency in process planning and pricing. Traditionally, job shops have used functional layout for their machines and all lathes are grouped in one section, milling machines in another, etc. During the machining of a given part, the workpiece is moved back and forth between sections, with perhaps the same section visited more than once. This results in extensive material handling, large in-process inventory, long manufacturing lead time and high cost. One approach to overcome these difficulties is to apply group technology and the associated computer aided manufacturing techniques.

Group technology concepts

Group technology (GT) is a manufacturing philosophy in which similar parts are identified and grouped together to take advantage of their similarities in design and manufacturing. A part family may be defined as a collection of related parts which are similar either because of geometric shape and size or because similar processing steps are required in their manufacture. The parts within one family may look different, but their similarities are close enough to merit their identification as members of the part family. This GT concept is an essential part of computer aided design retrieval systems and most computer aided process planning schemes.

An important factor in the successful application of GT in manufacturing is that the machine tools should be grouped on the basis of manufacturing parts families. The machines should be organized into cells so that a given component is completed before it leaves the cell. Manufacturing cells are usually made up of about 10 machines and are self-contained and self-regulating. A conveyor system may be used to connect the different machines. Figure 16.1 shows a schematic example of a manufacturing cell.

Parts classification and coding systems

Many parts classification and coding systems have been developed and there are several commercially available packages on the market. The selection of one system in preference to others depends on the nature of the company and the type of work. The classification of parts families can be done according to:

1. Design features: such features include internal and external shape, major and minor dimensions, length/diameter ratio, type of material, part function, surface finish and tolerances. This method is useful for design storing and retrieval systems and for design standardization. However, parts classified according to this method may not have similar manufacturing processes.
2. Manufacturing features: such features include major and minor operations, operation sequence, production time, batch size, annual production, surface finish and tolerances, major dimensions, length/diameter ratio, machine tools, cutting tools and fixtures. There is a certain amount of overlap between the design features and manufacturing features although the latter method is preferred from the manufacturing point of view.
3. Production flow analysis (PFA): parts with common operations and routes are grouped and identified as a manufacturing family. This method is based on the analysis of existing production data, such as operation sheets and route cards, instead of part drawings data.

Regardless of the criteria used for classification of parts families, a coding system has to be used to identify the parts. Coding systems consist of a sequence of symbols that identify the part geometry, size, processing information, material, etc. There are three basic code structures in use in group technology applications:

1. Hierarchical, monocode or tree, structure. The information represented by each subsequent digit is dependent on the preceding digit and the code is assigned by a step-by-step examination of part characteristics. This method provides a relatively compact structure which conveys much information about the part in a limited number of digits.

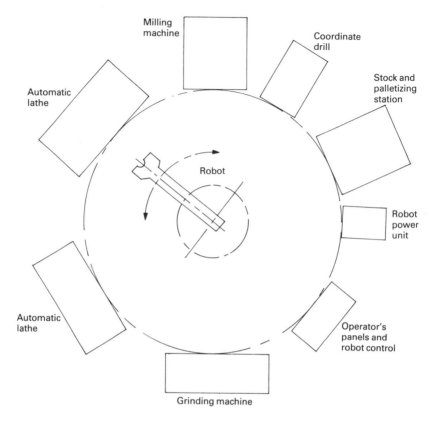

Figure 16.1 Schematic illustration of the elements of a manufacturing cell.

2. Chain-type, or polycode, structure. A given digit value always represents the same feature. Although such codes tend to be relatively long, they are helpful in identifying parts with similar processing requirements.
3. Hybrid structure. Most of the commercial parts coding systems used in industry are a combination of the two systems, with hierarchical codes being used for the primary numbers and the chain code for secondary numbers. Examples of hybrid codes include Opitz, MICLASS and CODE systems.

The Opitz coding system

The Opitz coding system consists of nine digits, which describe both design and manufacturing data. The code can be extended by adding four more digits. The first five digits, 12345, are called the 'form code' and describe the primary design attributes, as shown in Fig. 16.2. The next four digits, 6789, constitute the 'supplementary code' and describe some of the attributes that are needed for the manufacture of the part. The extra four digits are called 'secondary or supplementary code' and can be added by the company to serve their particular needs and are intended to describe the production operation type and sequence. As an illustration of the use of the Opitz coding system, consider the part shown in Fig. 16.3. The overall length/diameter ratio, $L/D = 4$, which makes the first digit $= 2$. The part is stepped on one end with a thread, so that the second digit of the code is 2. The

Figure 16.2 The Opitz coding system.

Digit	Category	0	1	2	3	4	5	6	7	8	9
1st Digit	Component class	L/D ≤ 0.5	0.5 < L/D < 3	L/D ≥ 3	*(Rotational parts)*	*(Rotational parts)*	*(Rotational parts)*	*(Nonrotational parts)*	*(Nonrotational parts)*	*(Nonrotational parts)*	*(Nonrotational parts)*
2nd Digit	External shape, external shape elements	Smooth, no shape elements	No shape elements *(Stepped to one end or smooth)*	Thread	Functional groove	No shape elements *(Stepped to both ends)*	Thread	Functional groove	Functional cone	Operating thread	All others
3rd Digit	Internal shape, internal shape elements	No hole, no break-through	No shape elements *(Smooth or stepped to one end)*	Thread	Functional groove	No shape elements *(Stepped to both ends)*	Thread	Functional groove	Functional cone	Operating thread	All others
4th Digit	Surface machining	No surface machining	Surface plane curved in one direction	External plane surface	External groove and/or slot	External spine (polygon)	External plane surface or slot	Internal plane surface and/or slot	Internal spline (polygon)	Internal and external polygon	All others
5th Digit	Auxiliary holes and gear teeth	No auxiliary hole *(No gear teeth)*	Axial, not on P C D	Axial on P C D	Radial not on P C D	Axial, radial other direction	Axial, radial on P C D or other directions	Spur gear teeth *(With gear teeth)*	Bevel gear teeth	Other gear teeth	All others

Secondary code:

6	Dimensions
7	Material
8	Original shape of raw material
9	Accuracy

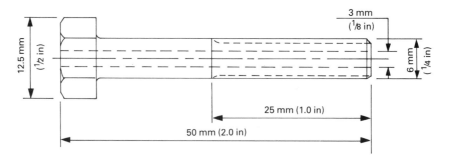

Figure 16.3 Dimensions of the workpiece used to illustrate the use of the Opitz coding system. Opitz code: 22140.

third digit is 1 because of the through hole. The fourth digit is 4 because of the hexagonal head. The fifth digit is 0 because no auxiliary holes or gear teeth are required. The Opitz code for the shown part is 22140. To complete the supplementary code the sixth through the ninth digits with data on dimensions, material, starting workpiece shape and accuracy would have to be added.

The MICLASS system

MICLASS stands for Metal Institute Classification System and was developed in the Netherlands. The system uses 12 digits to describe and classify workpiece characteristics. The company may use up to 18 additional digits to code specific data like lot size, piece time, cost data and operation sequence. The first 12 digits can be assigned manually or interactively with a computer. The number of questions asked by the computer varies with the complexity of the workpiece. Once all the questions have been answered, the computer will assign a classification number to the part as follows:

Digit 1	Main shape
Digits 2 and 3	Shape elements
Digit 4	Position of shape elements
Digits 5 and 6	Main dimensions
Digit 7	Dimension ratio
Digit 8	Auxiliary dimensions
Digits 9 and 10	Tolerance code
Digits 11 and 12	Material code

Having classified the part, the computer can then be asked to obtain numbers of similar drawings or related parts that are in the company files.

The CODE system

The CODE system was developed by Manufacturing Data System, Inc, Michigan, USA, to describe the part's design and manufacturing characteristics. It has 8 digits, and for each digit there are 16 possible values, 0 through 9 and A through F. The digits are identified as follows:

Digit 1	The basic shape, e.g. cylindrical

Digit 2	Further definition of basic shape, e.g. uniform diameter, stepped, conical, etc.
Digit 3	Internal details, e.g. central hole, threaded, etc.
Digits 4, 5, 6	Secondary manufacturing processes other than digit 3, e.g. grooves, slots, flats, etc.
Digits 7, 8	Main dimensions, e.g. largest outside diameter or section and overall length.

Tables containing the different shapes for each digit and its 16 possible values make it easy to select the correct code numbers for a given part.

The second step of classification is to record the code number on a source data collection card for input to the computer. Other information may be added to this card, such as material code, heat treatment specification, tolerance class or surface coating.

Benefits of group technology

When group technology is successfully applied, benefits in several areas may be realized. The area of product design benefits from GT as the coding system may avoid redesigning a part which has been designed before. Even if an exact part cannot be found, alteration of a part from the same family may still save designer time. Using GT will also help standardize design features such as corner radii, chamfers, tolerances, etc. Grouping of machine tools yields several benefits to the areas of tooling, as a result of using group jigs which can accommodate different members of the parts family, and setups, as a result of savings in setup and idle times. Materials handling and in-process inventory are other areas which benefit from the arrangement of machine tools in cells. Grouping of parts into families reduces the complexity and size of the parts scheduling problem.

In spite of these benefits, GT has not yet achieved the widespread application which might be expected. This is because rearranging the machines into cells is expensive and disruptive. This also requires reorganization in almost all departments, including accounting, engineering, management and shop personnel. Another reason is the difficulty and expense encountered in parts classification and coding.

16.6   INDUSTRIAL ROBOTS

In a manufacturing context, a robot has been defined as a programmable device capable of performing complex actions in a wide variety of operations. It is a manipulator that can normally be reprogrammed to do various repetitive actions without human intervention. As shown in Fig. 16.4, the robot is generally composed of an arm that swivels on a base through a total arc of 90 to 360°, depending on the model. Usually the arm also lifts in a vertical plane. A wrist on the end of the arm provides rotary motion for positioning the end-effector, which may grasp with finger-like grippers or have a tool fastened to it. The tool can be a welding head, a spray gun or a machining tool, depending on the specific application of the robot. Basically the robot needs six axes of motion, or degrees of freedom, to reach a point with a specific orientation in the space. Typically the arm has three degrees of freedom, in linear or rotary motions, and the wrist section contains three rotary motions. Each axis of motion is separately driven by an actuator which may be a d.c. servomotor, a stepping motor, a pneumatic actuator or a hydraulic motor or actuator. The

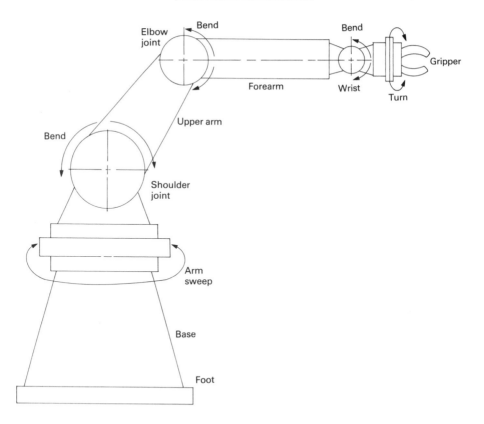

Figure 16.4 Schematic illustration of the main elements of an industrial robot.

robot is directed by an information processor following instructions programmed in its memory.

The first generation robots were used to replace human beings in dangerous operations, such as loading and unloading hot parts from processing furnaces or handling toxic materials. Robots can lift heavy loads and work tirelessly, keeping up a steady pace and making less scrap.

Depending on their capability, robots can be classified as:

1. Pick and place, first generation, robots. The drive elements of these robots are usually pneumatic or hydraulic cylinders. The feedback devices are limit switches and stoppers for each axis of motion of the arm. Although these robots are relatively inexpensive, they have limited control flexibility.
2. Point-to-point transfer robots have the capability of moving to and from a number of points and not necessarily in the same sequence. Reprogramming is done by the operator leading the robot through its routine initially. Some robots can be programmed off-line by input of coded instructions like NC machines. Typical applications of such robots include spot welding, material handling, simple assembly tasks and drilling operations.
3. Continuous-path robots are the same as point-to-point ones but have a much larger memory and a processor capable of coordinating the movements of the joints and

interpolating a given path between end points. When programmed, the robot is led not only from point to point but also along the paths between points. Typical applications of such robots include arc welding, spray painting, deburring and assembly. The robot is an important element of the manufacturing cell in flexible manufacturing. It loads and unloads the machines and transfers the parts between them.

4. Intelligent, third generation, robots use sensors, such as TV cameras for vision or force transducers for detecting the disturbing forces acting on the robot end-effector. The addition of sensors allows the robot to deviate from its programmed path and respond to changes in its working environment, i.e. to make decisions during its operation. Among the capabilities of intelligent robots is perception and pattern recognition, where one shape can be distinguished from another, with the aid of a vision system. Typical applications of such robots include complicated assembly tasks and quality inspection.

The decision whether or not a robot should be used for a certain operation is usually based on economic considerations. In addition to the economic considerations, there are the safety and environmental issues. The use of robots to load and unload presses, for example, eliminates the possibility of operator injury. The robot can also take operators out of hazardous or noisy environments. A basic equation for determining the payback period for a robot installation is:

$$P = \frac{R}{L - M}$$

where P = payback period in years
R = total initial investment for robot and accessories
L = annual labor savings
M = annual maintenance and programming expense per year.

As an example take $R = \$40\,000$, $L = \$20\,000$, $M = \$3000$. Substitution in the above equation gives $P = 2.35$ years. If the production rate increases as a result of employing the robot, then the payback period is even shorter. Another aspect is material saving due to the more consistent and accurate operation compared to human operation.

16.7   FLEXIBLE MANUFACTURING SYSTEMS

Flexible manufacturing systems (FMS) incorporate many of the individually developed systems of modern manufacture (such as NC and CNC, robots, group technology principles, automatic inspection and automatic materials handling) into a single large-scale system. An FMS consists of a group of processing stations, usually CNC machines, connected by an automated workpiece handling system and controlled with the aid of a central computer, as shown in Fig. 16.5. Most of the machines in FMS utilize randomly selectable heads to perform functions such as drilling, tapping, turning, etc. Some FMS systems may incorporate some special machine tools to perform specific functions. Materials handling is an important feature of FMS with automated guided vehicles and robots playing an increasingly important role as they reduce the requirement for special loading fixtures, which are expensive and reduce flexibility. The FMS is capable of processing a variety of different types of parts simultaneously at the various workstations.

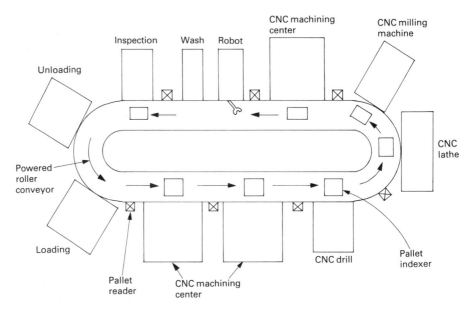

Figure 16.5 Elements of a flexible manufacturing system.

If only parts of the same type are loaded onto the system, the loading of the workstations will not be even.

The functions performed by the operators include loading unmachined workpieces onto the system and unloading finished ones, changing of tools and tool setting, equipment maintenance and repair, input data and change part programs. Once a workpiece is loaded onto the handling system, it is automatically routed to the appropriate workstations. The routing will be different for different workpieces and the operations and tooling may also be different. The coordination and control of the different activities involved in FMS is accomplished under the control of the computer. These activities include:

1. Control of NC, CNC and DNC machines.
2. Reporting on system performance by monitoring and collecting production information like number of workpieces, tool changes and machine utilization.
3. Storing of CNC part programs with editing facilities.
4. Control of materials handling system.
5. Supervisory functions like assigning alternative routing when a part of the system breaks down, controlling traffic of workpieces in the system and monitoring and control of the tools and their status.

The FMS provides the efficiency of mass production for batch production and is designed to fill the gap between high-production transfer lines (more than about 50 000 units/year) and low-production job-shop approach (less than about 200 parts/year). Batch production is applied to parts manufactured in lots ranging from several units to more than 50, for which the total demand is fewer than about 50 000 units. FMS is particularly suitable for manufacturing complex parts requiring considerable machining in the range of 10 to 1000 kg (*c.* 20 to 2000 lb). Transfer lines are most efficient when used in the production of large numbers of identical parts at high output rates, as in the case of the motor car industry. However, they are inflexible and cannot tolerate variations in part design. On the other hand, the job-shop approach provides the flexibility needed for small batches and many parts can be machined concurrently with the correct planning. The job-shop approach is costly since the output per machine is low owing to setting-up time, materials handling and dependence on direct labor. In-process inventory also tends to be high.

Flexible manufacturing systems offer considerable benefits to manufacturing industry because changes can rapidly be made through computer software rather than changes through hardware. Other benefits include:

1. Reduced direct and indirect labor since machines, inspection and materials handling are all done automatically.
2. Reduced lead time and inventory since FMS reduces waiting time and provides for a rapid throughput of work on the part.
3. Reduced need for special tooling since the machines used are universal in nature and the software substitutes for specialized tooling.

16.8   THE AUTOMATED FACTORY

The concept of the automated factory is not new to industry and is well established in the process industries, petrochemicals, petroleum refineries, basic metals, etc. In such plants, a relatively small crew manages the entire production operations. Their role is mainly to supervise, maintain and repair the equipment but not to participate directly in the processing of the product. The manufacturing industries, which have to produce a diverse mix of discrete products, have not achieved the same level of factory automation because of the difficulties encountered in managing, handling, assembling and inspecting these products. It is not difficult, however, to visualize the combination of the many production systems discussed in this chapter into an integrated computer automated factory.

The principal functions that must be integrated in the automated factory include processing, assembly, inspection, materials handling, parts transport, packaging and storage. As discussed earlier in this chapter, most of these functions have been automated individually. What remains is to integrate them into one coordinated system under computer control. When this is accomplished, raw materials entering the factory will be automatically transferred and stored. The identification and location of each item in storage will be retained in computer files. When needed, the materials will be moved to the production location by an automated materials handling system. The parts will then be processed according to a closely coordinated schedule through the proper sequence of machines. The optimal routing for the different workpieces will be decided by the computer. In-process inspection will ensure the quality of the different parts and adaptive control feedback will ensure that all the tools are in good working order. Robots will be used to aid in many of these operations. When the processing of individual parts is satisfactorily completed, they will then be routed to automated assembly stations. After automatic testing of the assembled products, they will be transported to an automated warehouse for storage until shipment to the customer.

The role played by the operators in the automated factory will be upgraded from manual and tedious tasks to functions requiring higher skill and intellect. These functions will include computer aided management and supervision, computer aided planning, programming and operating computers, computer aided engineering analysis and design, and maintenance and repair. Computer aided diagnosis of machine breakdowns will help the operators in determining the cause of failure. Preventive maintenance schedules will also be prepared by the computer.

The main benefits of increasing automation in the manufacturing industries include:

1. Increased productivity.
2. More efficient use of machines and production facilities.
3. More efficient use of raw material and reduced scrap.
4. Faster throughput of the material and less in-process inventory.
5. Shorter lead times and faster response to market demands.
6. More efficient planning.
7. Possibility to manufacture small quantities at competitive costs.
8. Less direct and indirect labor costs.
9. Less repetitive and tedious tasks and more meaningful work.

16.9   REVIEW QUESTIONS AND PROBLEMS

16.1   Compare a turret lathe, an automatic lathe and an NC lathe in terms of rate of production and economics.

16.2   Compare NC machines with CNC and DNC machines.

16.3   How do NC, CNC, and DNC machines fit in a CAM system?

16.4   What is the main difference between a functional layout and a cellular layout?

16.5   What are the main advantages of group technology?

16.6   How are NC machines related to group technology?

16.7   Develop the Opitz code for the following parts:
       (a) A washer of 20 mm outer diameter, 10 mm inner diameter, and 2 mm thickness.
       (b) A bearing sleeve of 60 mm outer diameter, 46 mm inner diameter, and 80 mm long.
       (c) A bolt similar to that in Fig. 16.3 but without the central hole and with the total length of the bolt being threaded.

16.8   Compare the MICLASS and Opitz coding systems.

16.9   Do you expect the flexible manufacturing system to be as beneficial to a small manufacturer as it is to a large one?

16.10 What are the main differences between a transfer line and a flexible manufacturing system?

BIBLIOGRAPHY AND FURTHER READING

Besent, C. B. and Lui, C. W. K., *Computer-aided Design and Manufacture*, 3rd ed., Ellis Horwood, London, 1986.

Doyle, L. E. *et al.*, *Manufacturing Processes and Materials for Engineers*, 3rd ed., Prentice Hall, New Jersey, 1985.

Gallangher, C. C. *et al.*, 'Group technology in the plastics moulding industry,' *The Production Engineer*, 1973, p. 127.

Groover, M. P., *Automation, Production Systems, and Computer-aided Manufacturing*, Prentice Hall, New Jersey, 1980.

Groover, M. P. and Zimmers, E. W., *CAD/CAM: Computer-aided Design and Manufacturing*, Prentice Hall, London, 1984.

Koren, Y., *Computer Control of Manufacturing Systems*, McGraw-Hill, Aukland, 1983.

Lindberg, R. A., *Processes and Materials of Manufacture*, 3rd ed., Allyn and Bacon, Boston, 1983.

Opitz, H. and Wiendahl, H. P., 'Group technology and manufacturing systems for small and medium quantity production,' *Int. J. Prod. Res.*, 1971, vol. 9, No. 1, p. 181.

Stover, R. N., *An Analysis of CAD/CAM Applications*, Prentice Hall, New Jersey, 1984.

PART III

ECONOMIC CONSIDERATIONS

Engineers make their living walking the sometimes narrow line
between the physical sciences and the dismal science of economics, between the do-able and the affordable
Doug McCormic

In addition to designing and manufacturing products at the required level of quality, organizations must also be able to sell their products at competitive prices and to make profits. In order to achieve these objectives, it is important that the different materials and processes involved in the design and manufacture should be evaluated in terms of their economic as well as their technical merits. An understanding of the elements that make up the cost of a finished product is vital in making these evaluations. However, it is not enough for engineers to be cost conscious, they should also be provided with accurate cost information for use in decision making. The introduction of computers in industry and the development of cost accounting software systems have proved to be of value in this respect.

Part III of this book discusses the different factors that are normally involved in evaluating the economic worth of an engineering product.

Chapter 17

Concepts of Economic Analysis

17.1   TYPES OF COSTS IN MANUFACTURING

There are several methods of classifying the different elements of cost that are encountered in manufacturing. A common method is to divide costs into fixed and variable costs. Fixed costs are constant and do not depend on the level of production output, while variable costs change in close proportion with production rate. When fixed and variable costs are combined, the total cost of production is obtained. Deducting total cost from the selling price gives the profit (if any); the selling price as such is determined by the market.

Fixed costs include:

1. Investment costs, such as interest on cost of the factory buildings, depreciation of production equipment, insurance and property taxes.
2. Administrative expenses, such as share of executive and legal staff and share of corporate research and development staff.
3. Sales and services expenses, such as sales personnel, delivery and warehouse costs, technical service staff and nontechnical service staff.

Variable costs include:

1. Direct labor costs, including fringe benefits.
2. Materials, including raw materials, operating supplies, scrap, lubricants, etc.
3. Cost of power needed to operate production facilities.
4. Maintenance costs.
5. Quality control costs.
6. Royalty payments.
7. Packaging and storage costs.

Some of the above costs, e.g. direct labor and maintenance, are sometimes considered as semivariable costs as they do not increase in direct proportion to production rate. At zero production, a portion of the semivariable costs, about 20 to 40 percent of the costs at full production, continues to be incurred.

Rather than classifying costs into variable and fixed cost, accountants and finance personnel usually find it more convenient to think in terms of direct labor cost, material cost and overhead costs. The direct labor cost is the sum of the wages paid to personnel who operate the production facilities and perform the process and assembly operations. The material cost is the cost of all materials used to produce the finished product. Overhead costs are all the other costs associated with running the company. Overhead can be divided into factory overhead and corporate overhead.

Factory overhead includes the costs of operating the factory, other than direct labor and materials, and can amount to several times the cost of direct labor. The share of a given product in the factory overhead can be allocated in proportion to the direct labor cost, direct labor hours, space, material cost, etc. The elements of cost that are usually considered as factory overheads include:

1. Plant supervision, foremen and security.
2. Maintenance crew, materials handling crew, shipping and handling, and custodial services.
3. Taxes and insurance.
4. Cost of interest on equipment and factory buildings.
5. Cost of depreciation of equipment and factory.
6. Cost of power for machines, lighting, heating, etc.

It should be noted that basing the share of a product in the factory overhead on direct labor cost alone could lead in some cases to inconsistencies. For example, a machine operator who runs a small inexpensive engine lathe should not be costed at the same overhead rate as an operator who runs a much more expensive NC machining center. The time on the NC machine should be valued at a higher rate. The appropriate rate can be obtained by dividing production costs into direct labor and direct machine costs. Each of these costs will then be charged an appropriate portion of the factory overhead. The direct labor cost would consist of the wages paid to operate the machine. The applicable factory overheads include fringe benefits and supervision. The machine cost is the capital cost of the machine apportioned over its life at the appropriate rate of return used by the company. The machine overhead rate is based on those factory expenses which are directly applicable to the machine, e.g. power, floor space, maintenance and repair expenses. It may be difficult in some cases to separate the applicable factory overhead items given above between direct labor and machine. In such cases, some judgement is needed.

Corporate overhead cost is the cost of running the company other than its manufacturing activities and can be determined in a similar way to the factory overhead. Separating factory overheads from corporate overhead is particularly important in the case where the manufacturing organization operates more than one plant. The elements of cost which are usually considered as corporate overhead include:

1. Share in corporate management.
2. Share in accounting, finance and legal personnel.
3. Share in sales and other support personnel.
4. Share in design and research and development personnel.
5. Share in cost of office space, lighting, heating, etc.

17.2 BREAK-EVEN ANALYSIS

Break-even analysis is a method of assessing the effect of changes in production output on costs, revenues and profits. Break-even analysis makes use of the concept of dividing manufacturing costs into fixed and variable costs to calculate the range of production volumes necessary for profitable operation. Figure 17.1 illustrates a simple break-even chart where the semivariable costs are reduced to fixed and variable components. The change in the variable cost per unit change of production volume is taken to be linear in this

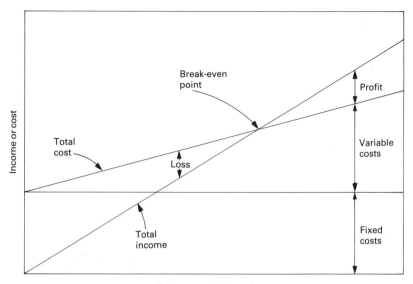

Figure 17.1 A simple break-even chart.

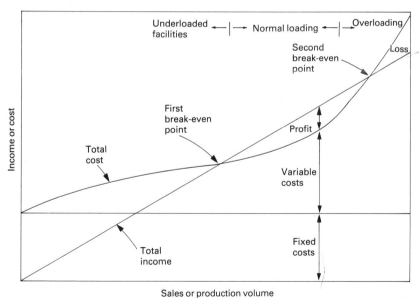

Figure 17.2 A break-even chart with nonlinear total cost.

simple case. However, in actual industrial cases the total production costs are usually nonlinear functions of production volume, as shown in Fig. 17.2. The least-cost load usually determines the level of operation for maximum economy and profit. Generally, loading production facilities to their full rated capacity corresponds to maximum efficiency of operation. Lack of efficiency when the facilities are underloaded, and the increased

Table 17.1 Economics of volume production of a turned part

	Engine lathe	Capstan	Automatic
a. Tooling cost ($)	30	30	30
b. Cost of cams ($)	–	–	120
c. Material cost/part ($)	0.5	0.5	0.5
d. Direct labor cost ($/h)	8.0	6.0	3.0
e. Cycle time/part (h)	0.3	0.1	0.05
f. Setting up labor cost ($/h)	10.0	10.0	10.0
g. Setting up time (h)	0.2	1.0	8.0
h. Machine overheads (% of item d)	100%	300%	1000%

Table 17.2 Results of calculations using data in Table 17.1

	Engine lathe	Capstan	Automatic
FC = fixed costs ($)	33.6	58	350
VC = variable costs/part ($)	5.3	2.9	2.15

depreciation and maintenance costs when the facilities are overloaded, contribute to the nonlinearity of the total costs. In such cases it is possible to have more than one break-even point.

The break-even chart is an important tool for the analysis of many production problems. It can be used to show how the profits, or losses, will vary for different output levels. The break-even point is the output level at which total costs equal revenues and the profit is zero. Other uses of break-even analysis are in plant capacity planning, equipment replacement studies and make or buy decisions. The break-even chart can also be used to show the effect of changes of output level on the costs of different methods of production. The break-even point in this case is the output level at which the costs for two production methods are equal. In this form, the break-even chart can be of help in equipment and process selection for a given product, as will be illustrated in the following example:

A turned part can be produced on an engine lathe, capstan lathe or automatic lathe. Find the most economical volume of production in each case using the information in Table 17.1.

Method of calculation

$$\text{overheads} = d \star h = O$$
$$\text{fixed costs} = a + b + g(f + O)$$
$$\text{variable costs/part} = (d \star e) + c + (O \star e)$$

Performing the calculations according to the above procedure gives the results in Table 17.2.

Break-even quantities can be determined graphically, as shown in Fig. 17.3, or calculated analytically from:

$$(FC_1) + (VC_1)n = (FC_2) + (VC_2)n \qquad (17.1)$$

where n is the break-even point between machine 1 and 2.

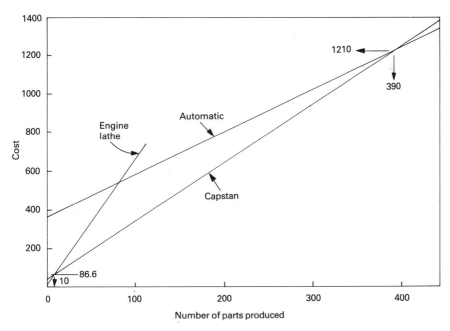

Figure 17.3 Break-even chart for a turned part, see example in Section 17.2.

The break-even results show that the engine lathe is more economical for quantities less than 10 parts, and the automatic lathe is more economical for quantities more than about 390 parts. The capstan lathe is more economical if the required number of parts is in the range 10 to 390.

Another illustration of the use of break-even analysis in selection of tooling materials is given in the following example:

A pattern for sand casting can be made from either wood or aluminum. Find the break-even number of castings using the information in Table 17.3.

Table 17.3 Comparison of sand casting pattern materials

	Wood	Aluminum
Fixed cost (initial cost of pattern) ($)	40	150
Variable cost/part (cleaning and preparation)	$0.02	0.02
Life of pattern, no. of parts	500	10 000

The costs of the two patterns are shown in Fig. 17.4 as a function of the number of parts to be cast. The figure shows that the break-even number is 1500 parts.

17.3 TIME VALUE OF MONEY

Money is considered to have a time value because when it is borrowed for a period of time, it is expected that the amount paid back will be greater than that which was borrowed. The difference is referred to as interest. The amount of interest is determined by the length of time the money was borrowed and the rate of interest. If an amount of money, P, is

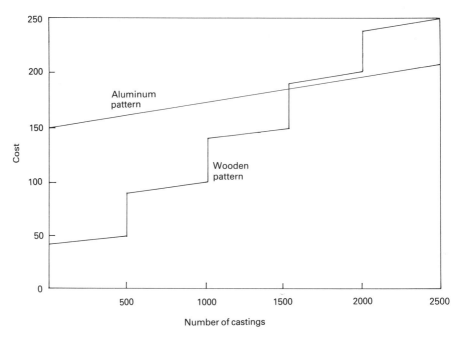

Figure 17.4 Use of break-even analysis in tool-material selection, see example in Section 17.2.

borrowed for a period of n years at an annual rate of interest i, the amount of money that has to be paid at the end of the period, F, with simple interest is given by:

$$F = P(1 + ni) \qquad (17.2)$$

P is usually called present worth and F is future worth.

Usually, financial transactions use compound interest and the parameters P, F, i, and n are related by the equation:

$$F = P(1 + i)^n \qquad (17.3)$$

The quantity $(1 + i)^n$ is usually called single-payment compound-amount factor, SPCAF, and tables are available for its value at various i and n. SPCAF can be used to calculate the future worth, F, when the present worth, P, is known. The inverse of SPCAF is usually called single-payment present worth factor, SPPWF, and is used to calculate the present worth, P, of some future value, F.

Instead of paying off a borrowed sum with a single future payment, another common practice is to pay annual installments, A, at the end of each of n years. The amount A is the annual payment needed to return the initial borrowed sum, P, plus interest on that investment at a rate i over n years and can be calculated from the relationship:

$$A = P\left[\frac{i(1 + i)^n}{(1 + i)^n - 1}\right] = P\,[\text{CRF}] \qquad (17.4)$$

where CRF is capital recovery factor. As an example to the use of (17.4), calculate the annual payment needed to return the capital of $100 000 and interest at 10 percent in 15 years.

$$A = 100\,000 \left[\frac{0.1\,(1+0.1)^{15}}{(1+0.1)^{15} - 1} \right] = 100\,000\,[0.1315] = \$13\,150$$

Another important parameter is the sinking fund factor, SFF, which is related to the sum of money put aside at the end of each year, A, so that after n years the accumulated fund, with interest compounded, will be worth F. The value of A is calculated according to the relationship:

$$A = F \left[\frac{i}{(1+i)^n - 1} \right] = F\,[SFF] \qquad (17.5)$$

This relationship is often used to set aside money in a sinking fund to provide funds for replacing worn-out equipment. As an example to illustrate the above concept, calculate the amount of money that should be set aside each year to replace a $100 000 plant in 15 years at 10 percent interest rate.

$$A = 100\,000 \left[\frac{0.1}{(1+0.1)^{15} - 1} \right] = 100\,000\,[0.1315] = \$3150$$

The results of the above examples show that the capital recovery factor, CRF, can be related to the sinking fund factor, SFF, by the relation CRF = SFF + i.

Although the above discussion was based on 1-year periods, other periods can be used provided the appropriate rate of interest is used.

17.4   COMPARING ALTERNATIVES ON COST BASIS

Economic considerations usually represent an important factor in comparing the various technically feasible alternative solutions to an engineering problem. In many cases a company or an organization may not accept a project unless it satisfies a minimum attractive rate of return, MARR. The value of MARR may be determined on the basis of past financial performance of the company, or may be related to the cost of capital. A company may use different MARR for different planning horizons or different magnitudes of initial investment. There are several methods of comparing investment opportunities and making economic decisions. These can be summarized as:

1. Present worth method (PW) converts all cash flows to a single sum equivalent at time zero using i = MARR.
2. Annual worth method (AW) converts all cash flows to an equivalent uniform annual series of cash flows over the planning horizon using i = MARR.
3. Future worth method (FW) converts all cash flows to a single sum equivalent at the end of planning horizon using i = MARR.
4. Rate of return method (RR) which goes slightly beyond either PW, AW, or FW by actually calculating the rate of return. If the calculated rate of return exceeds MARR the investment is acceptable.

5. Payback period method (PBP) determines how long at a zero interest rate it will take to recover the initial investment.
6. Capitalized worth method (CW) determines the single sum at time zero that is equivalent at i = MARR to a cash flow pattern that continues indefinitely.

The first four methods yield the same results and selection of which method to use will depend on the preference of the company. Only commonly used methods will be discussed here. For a more detailed discussion and for the derivation of the formulas that will be used in the following paragraphs, the reader should refer to one of the engineering economy books listed in the bibliography.

Present worth method

The present worth method, PW, is probably the most commonly used approach for comparing alternatives which have a common planning horizon or time periods. The PW method uses the equivalent present value of all current and future cash flows to evaluate the different alternatives. The following example will illustrate the use of this method.

The two machines detailed in Table 17.4 each have a useful life of 5 years. If MARR is 12 percent, which machine is more economical?

$$\text{PW for A} = -16\,000 - 4\,000 \frac{1}{\text{CRF}} + 3000 \frac{1}{(1+i)^n}$$

From (17.4), CRF = 0.277
From (17.3), $(1+0.12)^5 = 1.76$
PW for A = $-\$28\,735$

$$\text{PW for B} = -25\,000 - 2000 \frac{1}{\text{CRF}} + 600 \frac{1}{\text{CRF}} + 4000 \frac{1}{(1+i)^n}$$

$$= -\$27\,781$$

Machine B is more economical because it has the lower cost on a present worth basis.

The present worth method can be used to compare alternatives that have different planning horizons or lives, by taking n as a common multiple of the individual lives.

Table 17.4 Comparison by present worth

	A	B
Initial cost ($)	16 000	25 000
Maintenance cost per year ($)	4000	2000
Salvage value ($)	3000	4000
Annual extra income for higher productivity ($)	–	600

Annual worth method

Annual worth, AW, method is convenient to use when alternatives have different planning horizons or lives because all comparisons are made on an annual basis. The AW is related to PW by the constant multiplier CRF such that AW = PW * CRF. The two machines discussed in the above example will be used to illustrate the use of the annual worth method.

AW for A = $-16\,000$ CRF $-4000 + 3000$ SFF

From (17.5), SFF = 0.1574, and CRF is the same as in previous example.

AW for A = $-\$7960$/year

AW for B = $-25\,000$ CRF $-2000 + 600 + 4000$ SFF = $-\$7695$/year.

Machine B has a lower annual cost and is, therefore, more economical. This result is in agreement with the PW method in the above example.

Payback period method

The payback period, PBP, method is different from the PW and AW methods in that it does not consider the time value of money nor the MARR. The PBP method involves the calculation of the length of time required to recover the initial capital investment based on zero interest rate. For example, consider a plant of initial cost of $100 000 and output worth $25 500 per year. After 4 years, the total value of output would be $102 000 which is greater than the initial cost. The PBP in this case is 4 years as it is the minimum number of years of operation that are required to recover the initial cost of the plant. There are several obvious deficiencies of PBP method, but it continues to be one of the most popular methods of judging the desirability of investing in a project. The reasons are that it does not require interest rate calculations, nor MARR decisions. The PBP gives a rough measure of the liquidity of an investment and hedges against uncertainty of future cash flows. It is suggested that the PBP method be used in addition to one of the methods that employ the time value of money.

Capitalized worth method

Capitalized worth, CW, method is a special case of present worth method. The capitalized cost of a project is the present value of providing for that project in perpetuity. The concept was developed for use with long-term investment projects, such as bridges, dams, highways, or endowment funds where the estimated life is 50 years or more. The CW method has been more broadly used in economic decision making because it provides a method that is independent of the time period of various alternatives.

If a present value P is deposited into a fund at interest rate i per period so that a payment of amount A may be withdrawn at each period indefinitely, then the following relation holds:

$$A = Ci \qquad\qquad (17.6)$$

This means that a deposit of $100 000 at 8 percent annually will be required to fund a payment of $8000 each year indefinitely. A broader use of the method is illustrated in the following example. Consider a machine that costs $10 000 and has an average life of 5 years. If $i = 12$ percent, then the capitalized cost, C, is related to the machine cost, P, by combining (17.4) and (17.6). Thus:

$$P\,[\text{CRF}] = Ci \qquad\qquad (17.7)$$

CRF at 12 percent for 5 years is calculated from (17.4) as 0.2774

$$C = 10\,000 \,\frac{0.2774}{0.12} = \$23\,120$$

It is noticed that the excess over the cost of the machine is $13 120. From (17.3), it can be shown that if this amount is invested at 12 percent for the 5 years life of the machine,

$$F = 13\ 120\ (1 + 0.12)^5 = 13\ 120\ (1.762) = \$23\ 120$$

Thus when the machine is to be replaced, $23 120 has been generated. $10 000 will be taken to buy a new machine, ignoring inflation, and the difference is invested again.

The CW method can be used to compare alternatives, as illustrated in the following example. Compare the above machine with another one that costs $14 000, lasts for 6 years, and has a salvage value of $3000.

For 6 years at 12 percent, CRF = 0.2432

$$C = 14\,000\ \frac{0.2432}{0.12} - 3000 \left(\frac{0.2432}{0.12} \right) \left(\frac{1}{1.974} \right) = \$25\ 293$$

The term $(1/1.974)$ is used to bring the salvage value to present worth, as given by (17.3). The first machine is better as it requires a lower CW.

17.5   DEPRECIATION AND TAX CONSIDERATIONS

In manufacturing and many other industries, depreciation of facilities may occur by wear, deterioration, or obsolescence. Therefore a company must set aside enough funds each year to accumulate the capital required to replace the depreciated facility. Although depreciation allowances are not actually cash flows, their magnitudes and timing do affect taxes. Taxes are based on net income that results by deducing certain items such as expenses and depreciation from gross income. This can be written as:

$$TI = GI - E - D \tag{17.8}$$

where TI = annual taxable income
      GI = gross or total annual income
       E = allowable annual expenses
       D = annual depreciation charge

As taxes are cash flows, they should be included in economic analysis in the same way as wages, equipment, materials and energy. The different types of taxes that are imposed on a company are:

1. Property taxes which are based on the value of property owned by the company. These taxes do not vary with profits and are usually not large.
2. Sales taxes which are imposed on sales of products and are usually paid by the retail purchaser. These taxes are not relevant to engineering economy studies.
3. Excise taxes which are imposed on the manufacture of certain products like tobacco and alcohol are usually passed on to the consumer.
4. Income taxes which are imposed on company profits or personal income; gains resulting from the sale of capital property also are subject to a special form of income tax. The rate of income tax depends on the annual taxable income and is subject to change over time. During periods of recession and inflation there is a tendency for tax laws to be changed in order to improve the state of the economy. As a rough guide, the income tax rate in the USA can be taken as 46 percent of taxable incomes in excess of $100 000.

The time period over which depreciation is calculated, the method of spreading the total depreciation charge over the period, and the salvage value of the asset are the main factors that determine the value of the annual depreciation, D. The period of calculation is normally related to the useful life of the asset and is usually given by the tax authorities. There are several models for calculating the depreciation charge, depending on the company policy and the local tax laws.

The straight-line depreciation model assumes that the value of an asset decreases at a constant rate and the depreciation charge for each year, D, is represented by:

$$D = \frac{V_o - V_s}{N} \qquad (17.9)$$

where V_o and V_s are original and salvage values of the asset respectively and N is estimated life of the asset.

The book value of the asset, V_b, is defined as the original value less the accumulated depreciation at a given time. Thus the book value after n years is:

$$V_b = V_o - nD = V_o - \frac{n}{N}(V_o - V_s) \qquad (17.10)$$

The declining-balance method is another model for calculating depreciation. In this method a fixed percentage, R, is multiplied times the book value of the asset at the beginning of the year to determine the depreciation charge for that year. Thus as the book value of the asset decreases through time so does the value of the depreciation charge. The depreciation charge in year n is:

$$D_n = R(R-1)^{(n-1)} V_o \qquad (17.11)$$

and the book value is:

$$B_n = (1-R)^n V_o \qquad (17.12)$$

The maximum rate that may be used for income tax purposes is double the straight-line rate. Thus for an asset of life N years, the maximum rate is $(2/N)$. This method is commonly referred to as the double-declining-balance.

The Accelerated Cost Recovery System (ACRS) is applied in the USA to any depreciable property that is placed in service after 1980. According to the ACRS, the recovery period is simply defined as the time over which the cost basis can be recovered, and is 3, 5, 10 or 15 years depending upon the type of property. The recovery period is usually shorter than the inherent physical life of the asset. Neither the salvage value nor the useful life is used explicitly in the ACRS. Examples of the 3-year property include motor cars and light general-purpose trucks, while 5-year property includes the bulk of equipment, machinery, tooling, computers and office furniture. 10-year property includes residential manufactured homes, while 15-year property includes public utilities, buildings, fixtures and fences.

The ACRS deduction is calculated by applying the appropriate percentage, $d_t * 100$ percent, for a specified recovery property class and year to the cost basis, P. These percentages are given in Table 17.5.

The Table shows that the percentages add up to 100 percent for each recovery property class. This means that the entire cost basis of a property can be recovered over its recovery

Table 17.5 ACRS percentages ($dt * 100\%$) for different property classes

Year, t	3-year	5-year	10-year	15-year
1	25	15	8	5
2	38	22	14	10
3	37	21	12	9
4		21	10	8
5		21	10	7
6			10	7
7			9	6
8			9	6
9			9	6
10			9	6
11				6
12				6
13				6
14				6
15				6

Table 17.6 ACRS deduction and unrecovered investment on computer system

End of year, t	ACRS deduction, D_t	Unrecovered investment, B_t
0	–	$100 000
1	$15 000	85 000
2	22 000	63 000
3	21 000	42 000
4	21 000	21 000
5	21 000	0
6	0	0
7	0	0

property class life, even if its useful life is much longer. The allowable deduction in year t, D_t, is given by:

$$D_t = d_t * P \qquad (17.13)$$

The unrecovered investment, B_t, at the end of year t is given by:

$$B_t = P - \sum_{j=1}^{t} D_j = P(1 - \sum_{j=1}^{t} d_j) \qquad (17.14)$$

This represents the amount of the cost basis yet to be recovered.

As an example to illustrate the use of the ACRS method, consider a company that purchased a computer system for $100 000 with an estimated salvage value of $10 000 at the end of a projected useful life of 7 years. This is a 5-year recovery property. The allowable ACRS deduction and unrecovered investment for each year are given in Table 17.6.

The example shows that the estimated salvage value does not enter into the ACRS calculations. Further, the entire cost basis was recovered during the class life of 5 years, rather than over the 7-year useful life.

17.6   BENEFIT-COST ANALYSIS

Benefit-cost analysis is a commonly used method of evaluating projects, especially in the case of public sector projects. As the name implies, the method consists of comparing the annual worth of the benefits of a proposed project to the annual worth of the costs. For example, a proposed project with estimated benefits of $B = \$10\,000$ per year and estimated costs of $C = \$5000$ per year has a benefit-cost ratio of $B/C = 2.0$. This ratio can be taken as an indicator of the price paid for improvements. Benefits in the case of public sector projects, like a new highway or bridge, can be taken as any savings in costs or time as well as improved safety for the users. Costs can be divided into two categories: capital costs and maintenance costs. Capital costs are usually considered as the construction, acquisition or other similar costs of the facility. The capital cost is usually expressed on an annualized basis using the CRF of (17.4) and a suitable rate of return, i, and life, n. Maintenance costs are the owner's costs for operating and maintaining the facility. When the benefits are expressed in the same units as cost, a B/C ratio of unity represents the minimum justification for adopting a new project in place of the present facility.

Considerable care should be exercised in deciding which items to include in benefits and which items to include in costs. As a result, considerable judgement is required in performing B/C analysis. In the conventional B/C analysis, the net benefits, B_n, are considered to be equal to the user annual costs of the present facility minus the user annual costs of the proposed facility, i.e. net savings to the user. The costs consist of the owner's net annualized capital cost, C_c, plus owner's net operating and maintenance costs, C_m. This can be written as:

$$\text{Conventional } B/C = \frac{B_n}{C_c + C_m} \tag{17.15}$$

The modified B/C analysis uses the input data but the operating and maintenance costs, C_m, are treated as negative benefits rather than as costs. The resulting equation is:

$$\text{Modified } B/C = \frac{B_n - C_m}{C_c} \tag{17.16}$$

To illustrate the difference between the two methods, consider the following example. Two proposals, A and B, are suggested to replace an existing road. Taking $i = 10$ percent and $n = 25$ years, compare the two proposals given the information in Table 17.7.

Table 17.7 Benefit-cost analysis

	Existing road	Proposal A	Proposal B
Construction cost ($)	–	100 000	100 000
User's cost ($/year)	200 000	160 000	180 000
Owner's maintenance cost ($/year)	250 000	260 000	240 000
Total annual cost ($/year)	450 000	420 000	420 000
Savings over present road ($/year)	–	30 000	30 000

The first step is to calculate the annual equivalent of the construction cost using (17.4) CRF = 0.1102 and A = \$11 020.
Substitution in (17.16) gives:
Proposal A compared to existing road:

$$\frac{200\,000 - 160\,000}{(11\,020 - 0) + (260\,000 - 250\,000)} = \frac{40\,000}{21\,020} = 1.9$$

Proposal B compared to existing road:

$$\frac{200\,000 - 180\,000}{(11\,020 - 0) + (240\,000 - 250\,000)} = \frac{20\,000}{1020} = 19.6$$

By comparison, the modified B/C gives the following results:
Substituting in (17.16),
Proposal A compared with existing road:

$$\frac{(200\,000 - 160\,000) - (260\,000 - 250\,000)}{(11\,020 - 0)} = \frac{30\,000}{11\,020} = 2.72$$

Proposal B compared with existing road:

$$\frac{(200\,000 - 180\,000) - (240\,000 - 250\,000)}{(11\,020 - 0)} = \frac{30\,000}{11\,020} = 2.72$$

The results of the modified method show that with equal capital investment of \$100 000 and equal annual savings of \$30000/year, proposals A and B have equal B/C ratios which is more reasonable than the results of the conventional method. Similar comparisons indicate that the modified B/C method usually gives more consistent results. However, if no or very small capital investment is required C_c = 0, (17.16) gives infinite B/C ratio and the conventional method is more suitable.

When several acceptable proposals exist, each with $B/C > 1$, the best solution is not always the one with the highest B/C ratio. This is illustrated in Table 17.8.

Table 17.8 Multiple benefit-cost analysis

Solution	a	b	c	d
Annual benefits (\$/year)	35 000	40 000	50 000	58 000
Annual costs (\$/year)	15 000	19 000	30 000	38 000
B/C ratio	2.3	2.11	1.67	1.53

The first step in comparing the different solutions is to arrange them in increasing order of costs, which has already been done here. The next step is to calculate the incremental B/C ratio, beginning with the lowest cost pair:

$$\frac{B_b - B_a}{C_b - C_a} = \frac{40\,000 - 35\,000}{19\,000 - 15\,000} = \frac{5\,000}{4\,000} = 1.25$$

Each additional dollar invested in (b) over (a) gives a benefit of $1.25. Therefore solution (b) is preferred to (a). Solutions (b) and (c) are then similarly compared, which yield:

$$\frac{B_c - B_b}{C_c - C_b} = \frac{50\,000 - 40\,000}{30\,000 - 19\,000} = \frac{10\,000}{11\,000} = 0.91$$

This means that solution (b) is preferable to solution (c). Solutions (b) and (d) are then compared, thus:

$$\frac{B_d - B_b}{C_d - C_b} = \frac{58\,000 - 40\,000}{38\,000 - 19\,000} = \frac{18\,000}{19\,000} = 0.95$$

This again shows that solution (b) is the better solution and should be selected, even though it does not have the highest B/C ratio. But it offers the greatest benefit for the total expenditure.

17.7 COST-EFFECTIVENESS ANALYSIS

In the benefit-cost analysis discussed in the above section it was assumed that the effects of a project are measurable, either directly or indirectly, in monetary terms. There are circumstances, however, when project outputs are not measurable monetarily and must be expressed in physical units appropriate to the project. For example, it would be difficult to give monetary values to the visual appearance of the different methods of finishing the exterior surfacing for a building wall. Similarly, how would different models of motor cars be evaluated in terms of comfort, noise level, styling and prestige? In such cases, cost-effectiveness analysis has proven to be a useful technique for objective comparison of different alternatives.

There are certain steps which constitute a standardized approach to cost-effectiveness analysis which can be summarized as:

1. Define the goals, purpose and missions of the system. This analysis should lead to a list of desirable features or performance goals.
2. Establish evaluation criteria for both cost and effectiveness aspects of the system under study. Among the categories of cost are those arising throughout the system life cycle. These include costs associated with research and development, engineering, testing, production, operation, maintenance and salvage. The effectiveness parameters are usually more difficult to establish. Also many systems have multiple purposes, which complicates the problem further. Some general effectiveness categories are utility, merit, worth, benefit and gain. These are difficult to quantify and, therefore, such criteria as mobility, availability, maintainability and reliability are normally used. Although precise quantitative measures are not available for all of these evaluation criteria, they are useful as a basis for describing system effectiveness.
3. Select the fixed-effectiveness or fixed-cost approach. The fixed-cost criterion is used to identify the alternative that gives the maximum amount of effectiveness at a given cost. This approach is useful in the case of public projects, where the available budget is fixed and the project that gives maximum utility is selected. The fixed effectiveness criterion is used to identify the least expensive alternative which achieves the specified goals or effectiveness levels. Alternatives failing to achieve these levels may either be eliminated or given penalty costs. When multiple alternatives, which provide the same service, are compared on the basis of cost, the fixed effectiveness approach is being

used. This approach is more appropriate when selecting designs, materials or processes for a given application.

4. Develop alternative solutions and designs and evaluate each individual feature or goal in terms of its contribution to the overall objective.

5. Conduct a sensitivity analysis to see if minor changes in assumptions or conditions cause significant changes in the order of preference of the different alternatives.

To illustrate the use of cost-effectiveness analysis in selection from different alternatives, consider the case of buying a hard hat for use on a construction site. Four brands A, B, C and D are available on the market at the costs of $50, 45, 40 and 35 respectively. The main function of the hard hat is to provide safety but other factors should also be considered in comparing the different brands. These factors include comfort, appearance and weight of the hat. The relative importance values of the different parameters are given by:

Performance goals:	Safety	Comfort	Weight	Appearance
Relative importance:	0.55	0.20	0.15	0.10

It is noted that the sum of the relative importance of the different parameters is unity.

The next step is to rate the different brands in terms of the different goals. A scale of 10 to 1 is used, with 10 being excellent rating and 1 being very poor rating (Table 17.9).

The next step is to evaluate the effectiveness, E, of each brand by multiplying the rating times the relative importance and adding the products.

$$E_A = 9*0.55 + 7*0.2 + 8*0.15 + 5*0.1 = 8.05$$
$$E_B = 7*0.55 + 5*0.2 + 8*0.15 + 4*0.1 = 6.45$$
$$E_C = 7*0.55 + 6*0.2 + 6*0.15 + 7*0.1 = 6.65$$
$$E_D = 6*0.55 + 4*0.2 + 5*0.15 + 7*0.1 = 5.55$$

Having estimated the effectiveness for each brand of hard hats, the next step is to calculate the effectiveness/cost (EC) ratio (Table 17.10).

The results show that brand C has the highest E/C ratio and is, therefore, a better buy in

Table 17.9 Cost-effectiveness analysis

Brand	A	B	C	D
Safety	9	7	7	6
Comfort	7	5	6	4
Weight (the lighter the better)	8	8	6	5
Appearance	5	4	7	7

Table 17.10 Effectiveness/cost ratios

Brand	A	B	C	D
Effectiveness, E	8.05	6.45	6.65	5.55
Cost, C, ($)	50	45	40	35
E/C	0.161	0.143	0.166	0.159

terms of effectiveness rating points per dollar expended. However, brand A has the highest effectiveness rating. An incremental E/C analysis shows that:

$$\frac{E_A - E_C}{C_A - C_C} = \frac{8.05 - 6.65}{50 - 40} = 0.14$$

This result means that the extra effectiveness points of brand A are purchased at a rate of 0.14 effectiveness points/dollar, which is more expensive than the 0.166 points/dollar paid for brand C. Brand C is, therefore, a better choice.

In the above example, it was easy to assign numerical values to the relative importance of the performance goals. In other cases where the number of performance goals is large or when the relative importance is not clear, the digital logic procedure can be used to simplify decision making. In this procedure evaluations are arranged such that only two alternatives are considered at a time. Every possible combination is compared and no shades of choice are required, only a yes or no decision for each evaluation. To determine

Table 17.11 Determination of the relative importance of performance goals using the digital logic method

Goals	Number of possible decisions [N = n(n − 1)/2]										Positive decisions	Relative emphasis coefficient (α)
	1	2	3	4	5	6	7	8	9	10		
1	1	1	0	1							3	$\alpha_1 = 0.3$
2	0				1	0	1				2	$\alpha_2 = 0.2$
3		0			0			1	0		1	$\alpha_3 = 0.1$
4			1			1		0		0	2	$\alpha_4 = 0.2$
5				0			0		1	1	2	$\alpha_5 = 0.2$
	Total number of positive decisions										= 10	$\Sigma\alpha = 1.0$

the relative importance of each performance goal or requirement a table is constructed, the goals are listed in the left-hand column and comparisons are made in the columns to the right as shown in Table 17.11.

In comparing two goals or requirements, the more important goal is given numeral one (1) and the less important is given zero (0). The total number of possible decisions $N = n(n - 1)/2$, where n is the number of goals or requirements under consideration. A relative emphasis coefficient, α, for each goal is obtained by dividing the number of positive decisions for each goal (m) into the total number of possible decisions (N). In this case $\Sigma\alpha = 1$. This procedure can be modified by giving the more important goal a number larger than one, depending on its importance, and the less important goal a number larger than zero if it is relevant to the application under consideration.

17.8 MINIMUM COST ANALYSIS

When the cost of a product is a function of a variable that may take on a range of values, it may be useful to determine the value of the variable for which the cost of the product is a minimum. Alternative products whose costs depend upon the same variable can be compared on the basis of their minimum costs. This situation usually arises when the alternatives contain two or more cost factors that are modified differently by the common variable, x. Certain cost factors may vary directly with an increase in the value of x, while others may vary inversely. The general solution for the situation outlined above may be represented as:

$$T_c = ax + \frac{b}{x} + c \qquad (17.17)$$

where a, b and c are constants. The condition for minimum cost can be determined by differentiating (17.17) with respect to x, equating to zero, and solving for the value of x:

$$\frac{dT_c}{dx} = a - \frac{b}{x^2} = 0 \qquad (17.18)$$

$$x = \sqrt{\left(\frac{b}{a}\right)} \qquad (17.19)$$

It may be noted that, for the condition of straight-line variation of the variable costs assumed, when minimum cost is obtained, the increasing costs are equal to the decreasing cost.

$$\text{Increasing costs} = \text{decreasing costs} = \sqrt{(a * b)} \qquad (17.20)$$

Figure 17.5 graphically illustrates the minimum cost analysis with straight-line variation.

A more general solution is where some costs increase with some production or size function, and other costs decrease, but not in direct or inverse proportion as shown in Fig. 17.6. In this case, the minimum cost point does not have to occur at the point where the two product costs are equal. In many cases, the material of a product can be considered as the main independent variable that effects the cost, and a relationship similar to (17.18) can be written to relate the product cost, T_c, to its service life, L, weight, W, cost of development effort and pilot runs, R, processing cost, P, assembly and similar costs, A, and material cost, M:

$$\frac{dT_c}{dM} = \left(\frac{\partial T_c}{\partial L}\right)\left(\frac{\partial L}{\partial M}\right) + \left(\frac{\partial T_c}{\partial W}\right)\left(\frac{\partial W}{\partial M}\right) + \left(\frac{\partial T_c}{\partial R}\right)\left(\frac{\partial R}{\partial M}\right) + \left(\frac{\partial T_c}{\partial P}\right)\left(\frac{\partial P}{\partial M}\right)$$

$$+ \left(\frac{\partial T_c}{\partial A}\right)\left(\frac{\partial A}{\partial M}\right) + \frac{T_c}{M} \qquad (17.21)$$

When (17.21) equals zero, the minimum product cost is obtained. In the above equation, the total product cost is assumed to have some functional relationship of the form:

$$T_c = f(L, W, R, P, A, M) \qquad (17.22)$$

It is also assumed that the service life, L, weight, W, research and development costs, R, and processing costs, P, are related to the material costs. This is usually true for many engineering applications. However, (17.21) can only be solved if equations relating each variable to the material costs are formulated. Usually these relations are known only for large increments. Even so, a term-by-term examination of (17.21) will be instructive. Depending upon the nature of the product, the analysis is made on a per unit basis or on annual volume basis, whichever is more convenient.

The first term on the right-hand side of (17.21) is the product of two rates of change, $\partial T_c/\partial L$ and $\partial L/\partial M$. Knowledge of what the consumer will pay for long life or what penalty

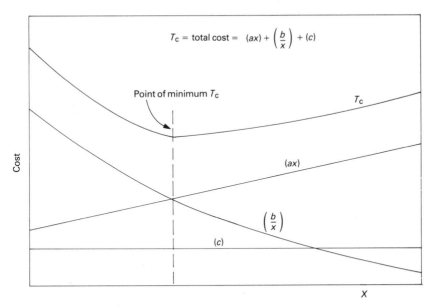

Figure 17.5 Minimum cost analysis with straight-line variation.

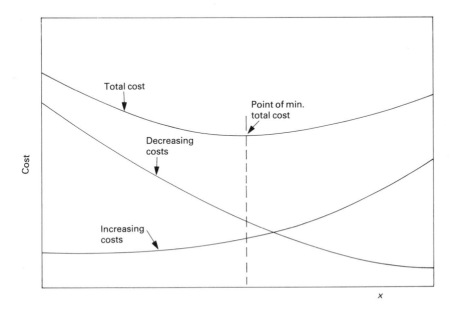

Figure 17.6 Minimum cost analysis with nonlinear cost variation.

attaches to short life is necessary to determine $\partial T_c/\partial L$. Data on $\partial L/\partial M$ can be obtained from information on the material's behavior under expected service conditions. Signs of both parts of this term and the other terms of (17.21) are determined by increases or decreases in the variables.

The second term in (17.21) may require two or more complete sets of design calculations, in order to establish the relation of total cost to weight $\partial T_c/\partial W$ and the dependency of weight on material cost $\partial W/\partial M$. Cost of development necessary before the material is adopted for manufacturing may be spread over several years, but it must be known in order to find the third terms $\partial T_c/\partial R$ and $\partial R/\partial M$.

The processing and assembly terms are important, especially in mass-produced products where they represent a large proportion of the cost. The remaining term $\partial T_c/\partial M$ is the change of total cost with respect to material cost, all other variables being excluded. As is true of the other derivatives, $\partial T_c/\partial M$ represents the change of cost in both numerator and denominator, not the ratio of the total material cost.

In most practical situations, analytical solution of (17.21) requires data that are not available to the engineer, and graphical solutions based on this equation are usually easier to achieve. Prior experience, model tests, good engineering judgement and accurate cost data are necessary for a sound analysis. The above fundamentals also apply to decisions in designs using the same materials in different proportions.

17.9   VALUE ANALYSIS

The objective of value analysis is to determine the most economical way of providing the use values which the consumer requires in a product. Value analysis does not have as its objective the cheapening of products or services by diminishing or eliminating the essential or desirable qualities of proper function and safety. This definition of value analysis involves two types of value. The first is 'use value', which is related to the characteristics that accomplish a use, work or service. The second type of value is 'esteem value', which is related to the characteristics that make the consumer want to possess the product or service. These characteristics include appearance and aesthetic acceptability, reliability, durability and ease of servicing.

A three-step approach is often used in value analysis work. The first step involves identification of primary or secondary functions. The second step is to evaluate the worth of the functions by comparison, and the third step is to develop value alternatives. Grading the value alternatives requires simple economic studies involving selection between alternative designs, materials and manufacturing processes. At this stage esteem values become relevant. An alternative, which might be completely satisfactory from a functional viewpoint, might be lacking in sales appeal. Changes and additions may be necessary in order to provide the required esteem value at a minimum cost.

Depending on the type of industry and on the product, the value analysis assessment will be performance oriented or value oriented. Military and aerospace applications usually involve performance oriented value analysis, at least in the initial designs. Materials and production methods are used almost entirely to assure the required ultimate performance. The value oriented approach has greater applications in the consumer industries. In fact, the mass-production industries depend on being able to produce appliances that satisfactorily fulfil a required function at a price that the general public can afford to pay.

In the field of quality control much can be done by applying the principles of value analysis to inspection procedures. In fields where a part failure would endanger human life, 100 percent inspection is necessary; whereas in less vital fields, it may be found that better value is obtained by testing the finished assembly and neglecting the inspection of components or subassemblies. In the latter case, the cost saved could be more than the cost of the occasional finished appliance that is rejected at the final inspection. Generally, it is useful to review at frequent intervals the cost of an inspection operation and compare it with the cost that would have been incurred if the operation had not been carried out and the rejects had been found at a later stage.

17.10 REVIEW QUESTIONS AND PROBLEMS

17.1 What sum of money must be invested now if its value in 5 years from now is to be $9000 at an interest rate of 10 percent. [Answer: $5590]

17.2 An automatic lathe has a first cost of $50 000, a service life of 5 years, and an anticipated salvage value at the end of this period of $8000. Annual maintenance cost is $2500. The machine will be operated for 2000 h/year at a labor rate of $10/h. Calculate the equivalent present worth of the system if the expected rate of return is 12 percent. [Answer: $126 512]

17.3 For the lathe in question 17.2, calculate the equivalent uniform annual cost. [Answer: $35 119]

17.4 Calculate the appropriate hourly rate for the following worker-machine system:

Direct labor rate = $10.00
Labor factory overhead rate = 60%
Machine factory overhead rate = 40%
Capital investment in the machine = $70 000
Service life = 7 years
Salvage value = $8000
Number of working hours per year = 2000
Rate of return = 10%
[Answer: $22.6]

17.5 If the machine in question 17.4 is used 6000 h/year, what would be the hourly rate? [Answer: $18.2]

BIBLIOGRAPHY AND FURTHER READING

Au, T. and Au, T. P., *Engineering Economics for Capital Investment*, Prentice Hall, New Jersey, 1983.
DeGarmo, E. P. and Canada, J. R., *Engineering Economy*, 5th ed., Macmillan, London, 1973.
Fabrycky, W. J. and Thuesen, G. J., *Economic Decision Analysis*, Prentice Hall, New Jersey, 1980.
Humphreys, K. K. and Katell, S., *Basic Cost Engineering*, Marcel Dekker, New York, 1981.
Thuesen, H. G., Fabrycky, W. J. and Thuesen, G. J., *Engineering Economy*, 6th ed., Prentice Hall, New Jersey, 1984.
White, J. A., Agee, M. H. and Case, K. E., *Principles of Engineering Economic Analysis*, 2nd ed., John Wiley & Sons, New York, 1984.

Chapter 18

Economics of Manufacturing Processes

18.1 INTRODUCTION

The cost of selling a product is in the background of almost every decision that is made in industry, starting with the initial stages of feasibility study of a project, designing the product, selecting the materials and processes, manufacturing the different parts and assembling them, testing and ending with packaging and selling the product. In almost all cases the value of engineering products lies in their utility measured in economic terms. There are numerous examples of products that exhibit excellent performance but possess little economic merit and consequently have not been adopted commercially. An understanding of the elements that make up the cost of a product is, therefore, essential in ensuring that the product will be competitive economically as well as technically. Estimating all the quantities and values for the parts and assemblies that comprise the final product is essential in:

1. Establishing the selling price.
2. Determining the most economical method, process or material for manufacturing.
3. Deciding whether or not a part or assembly is to be purchased from outside or manufactured in the plant.
4. Initiating programs of cost reduction.
5. Testing the economies due to the use of new materials or processes.
6. Determining the standards of production performance that may be used to control costs.
7. Providing input concerning the profitability of a new product.

In some cases a product is meant to compete directly with existing products. The selling price is already established and the problem is to work backward from it to determine the likely cost of each of the product elements. This procedure is sometimes called design to cost.

18.2 METHODS OF COST ESTIMATION IN THE PROCESS INDUSTRIES

Cost estimates in the process industries usually involve estimates of building a new plant or installing a process within an existing plant to produce a new product. This type of cost estimate is normally based on past costs of a similar project or design, with due allowance for cost escalation and difference in size. Such estimates require a backlog of experience or published cost data if reliable cost estimates are to be made. These methods are also useful in estimating the cost of buying capital equipment like electric motors, transformers, pumps, compressors, gas-holding tanks, heat exchangers, crushers, ball mills, dust collectors and conveyor belts. Published charts give the variation of cost with capacity for a given type of equipment. An example and the use of such charts are discussed by Humphreys and Katell (see bibliography). The cost and capacity can also be related by:

$$C_1 = C_2 \left(\frac{Q_1}{Q_2}\right)^x \qquad (18.1)$$

where C_1 and C_2 are capital costs associated with the capacity Q_1 and Q_2 respectively. The exponent x varies from about 0.4 to 0.8, and is approximately 0.6 for much process equipment. The cost-capacity relation can also be used to estimate the effect of plant size on capital cost.

Rapid preliminary estimates of the total project cost in the process industries can also be built up by applying various factors to the quotations for major items of equipment. The equipment costs to which the factors are applied are called the base cost. In this case:

$$C_t = fC_b \qquad (18.2)$$

where C_t = total installed cost of the plant
$\quad C_b$ = base cost
$\quad\quad f$ = Lang factor which depends on the nature of the plant
$\quad\quad\quad$ = 3.1 for solids-processing plant
$\quad\quad\quad$ = 3.6 for a solid-liquid-processing plant
$\quad\quad\quad$ = 4.7 for a fluid-processing plant.

The above method is described in detail by Lang, H. J.: 'Simplified approach to preliminary cost estimate' (*Chem. Eng.*, **55**, pp. 112–13, June 1948). Other cost factors for process plants are available in the literature, e.g. Bassel, W. D.: *Preliminary Chemical Engineering Plant Design*, Elsevier Scientific Publishing 1976.

Because inflation can cause published cost data to be out of date, cost indices have been introduced to compensate for this shortcoming. Cost indices are used to convert past costs to present costs. Examples of engineering cost indices include, Marshall and Swift (M&S), Engineering News Record (ENR), Chemical Engineering and Bureau of Labor Statistics (BLS). The use and composition of important cost indices are discussed by Humphreys and Katell (see bibliography).

As an illustration of the above method, estimate the cost of a conveyor belt to carry 2000 ton/h of coal. Past company records show that in 1977 a conveyor of capacity 1500 ton/h cost \$95 000. The 1977 M&S index = 505.4. Take the present value of M&S index as 600. Assume a cost-capacity factor $x = 0.62$.

$$\text{Present cost of conveyor} = \$95\,000 \left(\frac{2000}{1500}\right)^{0.62} \left(\frac{600}{505.4}\right)$$

$$= \$95\,000\,(1.195)\,(1.187) = \$134\,754.$$

18.3   METHODS OF COST ESTIMATION IN MANUFACTURING INDUSTRIES

Methods of cost estimation in the manufacturing industries are usually based on a particular sequence of manufacturing steps which are applied to the stock material to produce a finished part. Costs can be estimated on different levels of detail depending on the type of product, available information and required accuracy of the estimate.

Experience (guesstimating)

This type of cost estimation is usually performed by an experienced employee based on his acquaintance with similar products and a few superficial parameters, such as cost per unit weight or cost per unit length. This method is fast but also inaccurate, because two products may look alike when examined superficially but may involve widely different manufacturing problems. This method may be usable for simple routine parts but it is unsuitable for making an accurate estimate of a complex new product. The reliability of the estimates can be improved by employing a committee of representatives of the design, production and sales departments.

Engineering analysis

In this method the separate elements of work are identified in great detail in order to develop step by step material requirements and processing times for the different steps of manufacture. Such estimates may be based upon predetermined standards of output, usage and costs. The materials, labor and overhead costs for the different parts are prepared, they are added to give the material, labor and overhead costs for all parts and assembly operations involved in a subassembly. The total cost of the product is then obtained by adding the costs for material, labor and overheads for all subassemblies and final assembly operations, as well as for engineering, selling and administration costs. Using the engineering analysis method on complex products requires a great deal of effort and computation, which limits its use in practice. However, there are trends toward putting material and processing costs into a computer data base and using a computer to search out the optimum processing sequence and calculate the costs.

Statistical method

The statistical method of cost estimation applies regression analysis techniques to past cost data in order to identify the main variables that affect the cost of a product and to find the relationship that produces the best fit. With multiple linear regression analysis, the basic relationship is usually written as:

$$C = a + b_1 X_1 + b_2 X_2 + \ldots b_n X_n \qquad (18.3)$$

where C is the predicted cost
$\quad$ a is a constant (intercept) value
$\quad$ $X_1, X_2, \ldots X_n$ are the variables that affect the cost
$\quad$ $b_1, b_2, \ldots b_n$ are coefficients that determine the contribution of the variables to the cost.

Successful application of this method to the cost estimation of a given product depends on the availability in the company of reliable past cost data on similar products. This method can be computerized by using a coding system similar to that used in group technology to allow the operator to retrieve the relevant information from the data bank. If this system is used as a part of a CAD/CAM system, cost estimating information can be automatically generated.

Lowe and Walshe (see bibliography) applied this method to the manufacture of injection molding tools and found that the important variables that affect the cost are box volume, number of shape primitives, projected area, depth from split line, split line perimeter, shape/split line complexity, number of side cores and splits and shape category.

18.4   MANUFACTURING TIME

The manufacturing time represents an important parameter in most of the major elements of manufacturing costs. For example, the direct labor cost for a given process is usually calculated by multiplying the time required for the process by a labor rate. Overhead costs are also commonly calculated by multiplying operation time by an overhead rate.

Elements of manufacturing time

The total time required to perform an operation may be divided into four parts as follows:

1. Setup time. This is the time required to prepare for operation and may include the time to get tools from the crib and arrange them on the machine. Setup time is performed one for each lot of parts and should, therefore, be listed separately from the other elements of the operation time. If 45 min are required for a setup and only ten parts are made, an average of 4.5 min must be charged against each part. On the other hand, if 90 parts are made from the same setup, only 0.5 min is charged per part. Setup time is usually estimated from previous performance on similar operations.
2. Man or handling time. This is the time the operator spends loading and unloading the part, manipulating the machine and tools and making measurements during each cycle of the operation. Personal and fatigue allowances as well as time to change tools, etc., are also included in this part. Fatigue allowances depend on the type of work and generally can be considered as proportional to the energy exerted in performing the job. Personal time allows a break from both the physical and psychological stresses that a job may contain and is, in a sense, a minimum fatigue allowance. The minimum allowance is normally 5 percent of the total available time.
3. Machine time. This is the time during each cycle of the operation that the machine is working or the tools are cutting. Many organizations have developed standard data for various machine classes based on accumulated time studies. In some cases the machine time can be calculated from the process parameters as will be discussed later in this section.
4. Down or lost time. This is the unavoidable time lost by the operator because of breakdowns, waiting for tools and materials, etc. The actual down time is difficult to predict for a specific operation and is usually based on the average currently lost in the plant.

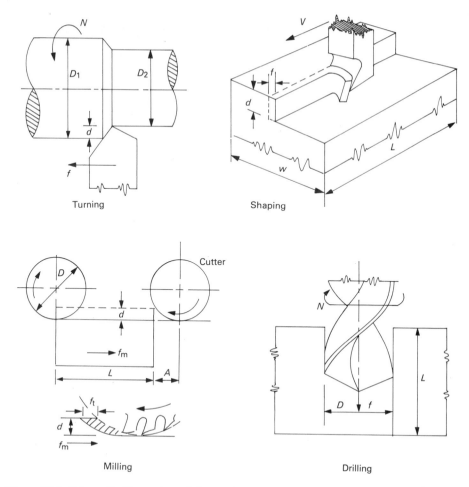

Figure 18.1 Schematic illustration of the cutting parameters in some metal cutting processes.

Floor-to-floor time (FFT) is the time which elapses between picking up a part to load on the machine and depositing it after unloading from the machine. FFT includes loading time, manipulation, machining or processing and unloading the part. Allowances for tool set-up and changes, fatigue and delays are added to FFT to give the basic production time for the operation.

Metal cutting time

Calculation of cutting time depends on the correct selection of values for the cutting speed and feed rate, each of which depends on other factors such as depth of cut, rigidity of set-up, power of the machine, etc. Tool and machine-tool manufacturers publish tables of recommended speeds and feeds to be used with their equipment and in practice such tables should be consulted when estimating a cutting time.

Figure 18.1 shows the different cutting parameters for commonly used metal cutting processes and Table 18.1 gives the different formulas for calculating the cutting time. The

Table 18.1 Formulas for calculating the cutting time and metal removal rate for different cutting operations

Process	Relation between V and N	Cutting time (Ct)	Metal removal rate (MRR)
Turning	$V = \pi DN$	$Ct = \dfrac{(L+A)}{fN} = \dfrac{\pi D(L+A)}{fV}$	$\dfrac{\pi L(D_1^2 - D_2^2)}{4Ct}$
Milling	$V = \pi DN$	$Ct = \dfrac{(L+A)}{f_m} = \dfrac{(L+A)}{f_t Nn}$	$\dfrac{Lbd}{Ct}$
Shaping (mech.)	$N' = \dfrac{7.4V}{L+A}$		
		$Ct = \dfrac{W}{N'f}$	$\dfrac{LWd}{Ct}$
Shaping (hydraulic)	$N' = \dfrac{8V}{L+A}$		
Drilling	$V = \pi DN$	$Ct = \dfrac{(L+A)}{fN}$	$\dfrac{\pi D^2 L}{4Ct}$

V = Cutting speed in units of length/time, e.g. fpm, ipm, m/min.
N = Spindle speed in units of revolutions/time, e.g. rev/min.
N' = Strokes per time.
f = Feed in units of length/rev., length/stroke, length/time, length/tooth.
f_m = Table feed in milling = $f_t Nn$; f_t = Feed/tooth in milling.
d = Depth of cut in units of length.
A = Approach or overrun allowance in units of length, taken as $D/2$ in milling and drilling.
L = Length of cut in units of length.
n = Number of teeth in milling cutter.
b = Width of cut or width of surface whichever is narrower.
Refer to Fig. 18.1 for other symbols.

metal removal rate (MRR) is an important parameter that affects the economics of metal cutting. MRR is calculated by dividing the total volume of metal removed by the cutting time, C_t. Care should be taken to allow for the difference in the units used for the different parameters that are involved in calculating C_t and MRR.

In practice, time estimates for cutting processes should be based on the speeds and feeds that are available on the machine that will be used. This means that it is necessary to calculate an approximate spindle speed, N', using the appropriate value of V and then select the nearest, N, which is available on the machine. In the absence of values of cutting speeds for milling, V can be taken as 0.75 the turning speed, and in the absence of values for cutting speeds in drilling, V can be taken as 0.9 the turning speed.

The following simple example illustrates the use of the above equations in calculating the standard manufacturing time. Determine the standard milling time for a steel block 250 mm long and 80 mm wide, taking one cut 6 mm deep. Use a slab milling cutter 100 mm diameter, 100 mm wide, and 10 teeth. Feed per tooth is 0.2 mm. Cutting speed is 18 m/min. Allow 25 s for loading and 25 s for unloading, 6 s to wind table to starting position, 3 s for start/stop spindle, 6 s for approaching work to cutter, 10 s for gaging, and 15 percent of cutting and handling time for fatigue allowance.

$N' = 18 \star 1000/\pi \star 100 = 57.3 \text{ rpm}$. Nearest available N on the machine is $N = 60 \text{ rpm}$.

From Table 18.1,

$$C_t = \frac{250 + 50}{0.2 \star 60 \star 10} = 2.5 \text{ min} = 150 \text{ s}$$

From Table 18.1,

$$\text{MRR} = \frac{250 \star 80 \star 6}{2.5 \star 1000} = 48 \text{ cc/min}$$

Standard time Load   = 25 s
               Wind table to starting position   = 6 s
               Spindle start/stop   = 3 s
               Approach work to cutter   = 6 s
               Cut by face milling   = 150 s
               Unload   = 25 s
               Gage   = 10 s
                                                                      _____
               Basic time   = 225 s
               Fatigue allowance at 15 percent of 225 = 33.75 s
               Standard time   = 258.75 s

18.5 MANUFACTURING COSTS

Elements of manufacturing costs can be grouped into three major categories: materials, labor and manufacturing expenses or overheads.

Material costs

Direct materials include all materials which enter into and form part of the product in measurable quantities. The quantities of direct materials required as well as sizes are normally recorded on bills of materials or on process sheets. The quantities should be gross, which includes necessary allowance for waste and scrap. For example, the quantity of bar stock required for a part should be indicated in units of length to cover the length of the part plus the length of the cut-off. If the material is purchased on the basis of unit length, the price is determined directly; otherwise the weight of bar stock is determined in order to estimate the material cost. Costing data can be determined from quotations and previous purchases. Such prices should be FOB plant or should include transportation to plant. Since prices of certain commodities, like steel and nonferrous alloys, vary with the quantity ordered, care should be taken to base the prices on the correct quantities.

Indirect materials include materials used in quantities too small to be readily identified with units of product. Examples include cutting oils, solders and adhesives. Such items are considered as shop supplies.

A credit should be included to account for recycled scrap or by-products that are sold elsewhere.

Labor costs

Labor may be defined as the employment costs, wages and other associated payments of the employees whose effort is involved in the fabrication of the product. Labor may be further classified as direct and indirect labor. Direct labor is required to change the form or advance the stage of manufacture or assembly of the product and includes machine operators, tool setups and process inspection. Direct labor costs are based on direct labor time needed for processing the product. Estimating the time involves the determination of the sequence of operations and knowledge of the standard time required to perform each operation. This time is determined in two parts: setup time for the job and cycle time to process each part. Indirect labor costs include overtime, nightwork premiums, paid vacations and holidays, insurance, old-age benefits, unemployment taxes and other fringe benefits. Indirect labor may be included in cost either as a markup of direct labor or as part of manufacturing expense.

The development of standard times requires time and motion study and involves a number of departments in the company. However, the expense of such activity is justifiable in most manufacturing situations in which cost control is vital for profitability. The standard time of production is compared with the actual hours worked to measure efficiency. With a system of standard times and costs, increasing or decreasing labor costs are reflected in variance accounts. This information is important for cost control. Standard times are also useful in cost reduction activities which are aimed at changing the design, the material, or the process in a way that will reduce the cost without compromising the quality.

Manufacturing expense (overhead costs)

Manufacturing expense, or overhead costs, covers all the other costs associated with running a manufacturing company. Examples of manufacturing expense include salaries and fringe benefits of company executives, accounting personnel, purchasing department and R&D. Other overhead expenses include depreciation, taxes, insurance, heating, light, power and maintenance. In some companies, indirect materials and labor are included as manufacturing expense or overhead costs. In other companies, the indirect labor is recovered as a markup of the direct labor rate. When the manufacturing company operates more than one plant, overheads should be divided into plant overhead and corporate overhead. In such cases, plant overhead includes the costs of operating the plant other than direct materials and labor while corporate overhead is the cost of running the company other than manufacturing activities.

The method of allocating manufacturing expenses to individual products depends on the nature of the manufacturing activity. A single overhead rate can be used in the case of plants manufacturing a single end product involving the use of diverse processes, e.g. a refrigerator, or in the case of plants manufacturing a variety of items using a common process, e.g. forgings. In such cases, the total manufacturing expense is allocated either on the basis of the units produced or on the basis of direct labor hours. In the case of plants manufacturing a variety of products using a variety of processes, several rates will be required to apportion the overheads equitably. In such cases, cost centers may be set up for each production unit doing one type of work, e.g. machining, assembly or painting. The total manufacturing expense is divided equitably among the cost centers. Each cost center then distributes its share of overheads among the different products either on the basis of direct labor hours or on the basis of the units processed. The time on machines of different degrees of sophistication and productivity should be valued differently even though they are in the same production unit or a cost center. For example, time on a small engine lathe should not be costed at the same overhead rate as a modern NC machine. The time on the automated machine should, obviously, be valued at a higher rate. The rate of overhead for each machine should be related to its floor space, power, maintenance and repair expenses, etc.

18.6   ECONOMIC JUSTIFICATION OF JIGS AND FIXTURES

Jigs and fixtures are production tools that are specially designed for quick and accurate location of the workpiece during manufacture. A fixture is a special work-holding device that holds the workpiece during machining, welding, assembly, etc. It is usually designed to facilitate setup or holding of a particular part or shape. A jig, on the other hand, not only holds the workpiece but also guides the tools, as in drill jigs, or accurately locates the parts of the work relative to each other, as in welding jigs. Such production tools are expensive and their costs add to the total production cost. It is, therefore, important to make sure that they can be justified economically by the saving in production time that will result from their use. The following factors must be considered when considering the economics of special tooling:

1. The cost of the special tooling.
2. Interest rate on the cost of the tooling.
3. Savings in labor cost as a result of using the tooling.
4. Savings in machine cost as a result of increased productivity.
5. The number of units that will be produced using the tooling.

From the economic point of view, the use of special tooling can only be justified if the saving in production costs per piece is greater than, or at least equal to, the tooling cost per piece. The savings per piece, S_p, can be calculated from the following relationship:

$$S_p = (R \star t + R_o \star t) - (R' \star t' + R_o \star t') \tag{18.4}$$

where R = labor rate per hour without tooling
      R' = labor rate per hour using tooling
       t = production time per piece without tooling (h)
      t' = production time per piece with tooling (h)
     R_o = machine cost per hour, including overheads

The total tooling cost, C_T, is the sum of the initial tooling cost, C_t, plus the interest on the tooling cost, I_t. Taking the number of years over which the tooling will be used as n, the rate of interest, or MARR, as i, and assuming straight-line depreciation:

$$C_T = C_t + \frac{C_t \star n \star i}{2} \tag{18.5}$$

This relationship is only approximate, but it is sufficiently accurate for our purpose because the tooling life is usually relatively short. When the time over which the tooling will be used is less than one year, the interest on the tooling may be ignored.

The tooling cost per piece, C_p, is calculated as:

$$C_p = \frac{C_T}{N} \qquad (18.6)$$

where N = number of pieces that will be produced with the tooling.

For the tooling to be economically justified, $S_p > C_p$.

The following example illustrates the use of the above procedure. Using a drill jig is expected to reduce the drilling time from 30 min to 12 min. If a jig is not used, a skilled worker will be needed with an hourly rate of $12.00. Using the jig makes it possible for a less skilled worker to do the job at an hourly rate of $10.00. The hourly rate for using the drilling machine is $8.00. The cost of designing and manufacturing the jig is estimated as $1200.00. The MARR is 12 percent and the expected life of the jig is 2 years. It is estimated that the jig will be used for the production of 400 parts during its useful life. Is the use of the jig economically justifiable? How many parts need to be produced for the jig to break even?

The saving per piece, S_p, as a result of using the jig is calculated from (18.4) as:

$$S_p = \left(12.00 \star \frac{30}{60} + 8.00 \star \frac{30}{60} \right) - \left(10.00 \star \frac{12}{60} + 8.00 \star \frac{12}{60} \right)$$

$$= 10.00 - 3.60 = \$6.40$$

The total jig cost, C_T, is calculated from (18.5) as:

$$C_T = 1200.00 + \frac{1200.00 \star 2 \star 0.12}{2} = \$1344.00$$

The jig cost per part is calculated from (18.6) as:

$$C_p = \frac{1344}{400} = \$3.36$$

As the cost of jig per part is less than the savings per part, the jig is economically justifiable. The break-even number of parts, N', is calculated as:

$$6.40 = \frac{1344}{N'}. \text{ Thus } N' = 210 \text{ parts.}$$

This means that at least 211 parts need to be manufactured in order to justify the use of the jig.

18.7 ECONOMICS OF METAL CUTTING

Manufacturing parts by metal cutting is usually wasteful of material, slow and expensive. In spite of these drawbacks, metal cutting is still a major process in the engineering industry and extensive work has been done to improve its efficiency. The main cutting conditions that are known to affect the economics of manufacturing a certain material are: (a) cutting speed; (b) feed; (c) depth of cut; and (d) tool material. Other factors like tool shape and cutting fluids have a minor effect on economics and will not be discussed here.

It is known that as the cutting speed increases, the cutting time and the cutting cost decrease. This improvement is, however, accompanied by a decrease in tool life and an increase in tool cost. This means that the total cost will show a minimum at a certain intermediate value of the cutting speed, as shown in Fig. 17.6. F. W. Taylor found that, for many workpiece materials and different cutting conditions, the tool life, T, changes with the cutting speed, V, according to the empirical relationship:

$$VT^n = C \qquad (18.7)$$

where C and n are constants related to the tool material, workpiece material, tool form and shape, size and shape of cut and cutting fluid. Experience has shown that n is closely related to the tool material. For high-speed steel tools $n = 0.1$–0.15, for cemented carbide tools $n = 0.2$–0.3, and for ceramic tools $n = 0.4$–0.6.

The elements of manufacturing cost that are affected by the cutting speed are the tool costs and the cutting costs. Setup and workpiece handling times are not affected and will not be included. The following analysis will be based on cost per unit volume removed. The cost to supply a cutting-tool edge for an operation, C_t, is considered to be made up of three components:

1. Cost of paying for the tool, c_o.
2. Cost of indexing or changing the tool, c_i.
3. Cost of grinding or sharpening the tool, c_s.

Thus $C_t = c_o = c_i + c_s$

$$= \frac{C_T}{N_c * N_g} + t_i * R + \frac{t_g * R'}{N_c} \qquad (18.8)$$

where C_T = original cost of cutting tool, ($\$$).
$\quad N_c$ = number of cutting edges obtained per grind or number of cutting edges on a throwaway tool.
$\quad N_g$ = number of times the tool can be ground, including when the tool was made.
$\quad t_i$ = time needed to index or change the tool, (min).
$\quad R$ = labor and overhead rate for metal cutting, ($\$$/min).
$\quad t_g$ = time needed to grind or sharpen the tool, (min).
$\quad R'$ = labor and overhead rate for tool grinding, ($\$$/min).

For a total tool life of T min, the labor cost over the life of the tool is $C_c = R * T$. During this time, the volume of metal removed, Q, is:

$$Q = VTfd \qquad (18.9)$$

where f = feed in units of length
$\quad d$ = depth of cut in units of length.

From (18.7) and (18.9) and combining all the constants as K,

$$Q = \frac{K}{T^{n-1}} \qquad (18.10)$$

The total cost per unit volume of metal removed, C_m, is:

$$C_m = \frac{C_c + C_t}{Q} = \frac{T^{n-1}(R \star T + C_t)}{K} \qquad (18.11)$$

Differentiating C_m with respect to T and setting the result = zero to find the minimum, the most economical tool life, T_e, is:

$$T_e = \left(\frac{1}{n} - 1\right)\frac{C_t}{R} \qquad (18.12)$$

The cutting speed that corresponds to the economical tool life can be found from (18.7). This is the speed at which the operation should be run for the lowest total cost.

Another criterion which may be used in evaluating metal cutting operations is maximum rate of production. Assuming ready availability of sharpened tools, the only tool factor that affects the rate of production is the tool changing or indexing time, t_i. The average rate of production, P_r, is:

$$P_r = \frac{Q}{T + t_i} = \frac{K \star T^{n-1}}{T + t_i} \qquad (18.13)$$

Differentiating with respect to T and equating the result to zero, the tool life for maximum rate of production, T_p, is:

$$T_p = \left(\frac{1}{n} - 1\right)t_i \qquad (18.14)$$

Since t_i is always less than C_t/R, the tool life for maximum production is always less than the economical tool life for a given operation. This means that the cutting speed for maximum production is always higher than that for the lowest cost. The range between the two speeds is called high-efficiency, Hi-E, range. Experience shows that the cutting speed for maximum profit always lies within the Hi-E range.

As an example to illustrate the use of the above relationships, determine the Hi-E range for the following case: Tool changing time is 25 s, labor rate including overheads for cutting and grinding is $12.00/h, time to grind tool 20 min. Tool costs $10.00 and can be ground 6 times, and 8 cutting edges are produced each time. The Taylor constants for the cutting tool-workpiece combination can be taken as $n = 0.25$, $C = 150$ for metric units and 492 for Imperial units.

From 18.8,

$$C_t = \frac{10.00}{8 \star 6} + \frac{25}{60} \star \frac{12}{60} + 20 \star \frac{12}{60} \star \frac{1}{8} = \$0.79$$

From (18.12),

$$T_e = \left(\frac{1}{0.25} - 1\right)\frac{0.79 \star 60}{12} = 11.85 \text{ min}$$

$$V_e = \frac{150}{(11.85)^{0.25}} = 80.86 \text{ m/min, or} = \frac{492}{(11.85)^{0.25}} = 265.2 \text{ ft/min}$$

From (18.14),

$$T_p = \left(\frac{1}{0.25} - 1\right)\frac{25}{60} = 1.25$$

$$V_p = \frac{150}{(1.25)^{0.25}} = 141.9 \text{ m/min, or } = \frac{492}{(1.25)^{0.25}} = 465.5 \text{ ft/min}$$

The Hi-E range is 80.86–141.9 m/min (265.2–465.5 ft/min).

18.8 STANDARD COSTS

Allocating the different costs to a product can be based on either historical or standard costs. Historical costs provide a record of the actual incurred costs as calculated from job orders after the job is completed and are, therefore, of little control value. Standard-cost systems are based on the proposition that there is a certain amount of material in a part and a given amount of labor goes into the part's manufacture. Such systems involve the preparation in advance of standard rates for materials, labor and expense. Standard time is defined as the total time in which a job should be completed at standard performance. It includes an allowance for relaxation, known as relaxation allowance, which in turn includes a fatigue allowance. Standard performance is defined as the rate of output which qualified workers will naturally achieve without overexertion as an average over the working day or shift. Many companies collect their own data for a particular process using time studies taken with a stop-watch. A large amount of data for special or general work has also been published in predetermined motion time systems.

Comparing the actual materials, labor and times against the corresponding standard values makes it possible to isolate the reason for any deviation from the overall standard cost. The variance is expressed as the difference between the actual performance against the standard base. The variance can be related to materials, labor or cost. For example, performance variance of labor may be calculated by comparing actual time worked with standard time. The ratio of actual performance to standard is sometimes called labor–effectiveness ratio or labor–efficiency ratio.

As an illustration to the use of variance in analyzing performance, consider the following example. Standard costs for a cast product are:

Labor	= $1.00
Materials	= $0.50
Overheads	= $2.00
Total cost	= $3.50

Standard production rate is 2000 castings per month. A scheme for increasing labor efficiency caused the production to increase to 2200 per month. The actual costs during this month are given in Table 18.2. Standard costs for producing 2200 castings are also shown in Table 18.2.

The results show that the increased labor effectiveness caused a favorable variance of $200. However, the decreased efficiency of using materials caused an unfavorable variance of $700, which is more than the gain due to increased labor effectiveness. The net result is unfavorable variance of $500.

Table 18.2 Performance analysis by variance method

	Actual costs	Calculated according to standard costs	Variance
Labor	$2000	$2200	−$200
Materials	$1800	$1100	+$700
Overheads	$4400	$4400	−
Total cost	$8200	$7700	+$500

18.9 LEARNING CURVE

It is commonly observed that the performance of workers improves as they gain experience in doing a certain job. In a manufacturing situation, a worker can produce more parts in a given time as he becomes more skilled in doing the work. The extent and rate of improvement depend on the nature of the process, the length of production run, and the working environment. Usually, the higher the human input to the process the greater the improvement. Processes performed on automated machines show a lower improvement rate than those performed on manually operated machines. Figure 18.2 gives a schematic representation of a learning curve which shows how the performance varies with time in a typical manufacturing situation. After a slow start in stage I, the performance shows progressive improvement until it reaches a maximum at the end of stage II. Stage III represents the optimum performance range, after which performance could decline if the job becomes monotonous. If the performance is measured in terms of the production time, stage II can be represented by an exponential curve as shown in Fig. 18.3. For an

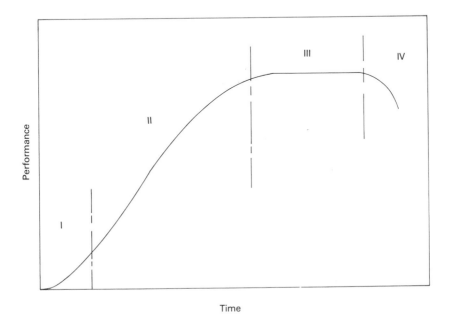

Figure 18.2 Schematic representation of the change of performance with time.

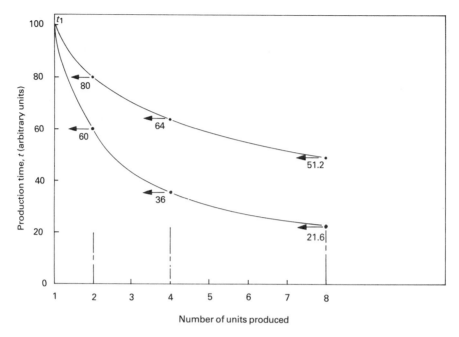

Figure 18.3 Comparison of the 80 percent and 60 percent learning curves.

80 percent learning curve, each time the total number of produced units doubles the production time is 80 percent of what it was before the doubling occurs. For a 60 percent learning curve, the production time is 60 percent of the time before doubling.

The learning curves can be expressed as

$$t = t_1 N^n \tag{18.15}$$

where t = production time per unit.
  t_1 = production time of the first unit
  N = total number of units produced
  n = negative slope of the learning curve.

It can be shown that learning curve percentage $P = 2^n$, thus $n = -0.322$ for an 80 percent curve and $n = -0.737$ for a 60 percent curve.

As an example to illustrate the use of the learning curve, calculate the time required to manufacture the 10th and 30th parts if the first part took 4 h to manufacture. Assume a 75 percent learning curve.

For $P = 75$ percent, $n = -0.415$
Time to manufacture the 10th part $= 4 \times 10^{-0.415} = 1.538$ h
Time to manufacture the 30th part $= 4 \times 30^{-0.415} = 0.975$ h.

18.10   SELLING PRICE OF A PRODUCT

In a free enterprise system, the price of goods and services is ultimately determined by supply and demand. Typical supply and demand curves are illustrated in Fig. 18.4. The

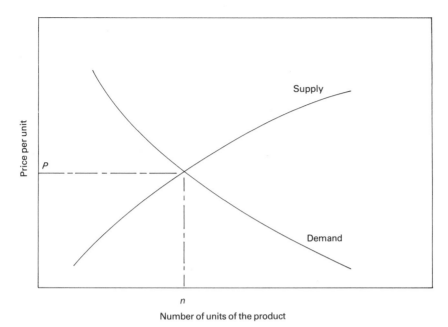

Figure 18.4 Schematic representation of typical supply and demand curves.

demand curve shows the relationship between the quantity of a product that customers are willing to buy and the price of the product. The supply curve shows the relationship between the quantity of a product that vendors will offer for sale and the price of the product. The intersection of the two curves determines both the quantity, n, and the price, p, of the product in the free market. The concept of supply and demand is important in engineering economy studies since proposed ventures frequently involve action that will increase the supply of a product or influence its demand. The effect of such action upon the price at which the product can be sold is an important factor to be considered in evaluating the desirability of the venture. A key issue in establishing a price is the reaction of the competition and the customer. Prices are set by the corporate management with the help of personnel from marketing, accounting, product engineering and manufacturing.

Figure 18.5 shows the elements of cost that add together to establish the selling price of a product. The cost of product engineering covers the direct labor and materials costs required to prepare the design and drawings, to perform development work and experiments and to make tools, jigs, fixtures, or dies for the different parts of the product. Each department has a direct-labor rate per hour for each class of work charged to the product. Similar techniques of cost estimates to those described for manufacturing may be used here. Overheads may also be charged on a similar basis. The costs of product engineering are usually charged to the number of products to be sold within a given period.

The actual manufacturing cost is estimated as outlined in previous sections. Figure 18.5 shows that standard costs are used to estimate the direct material, direct labor and overheads. In order to convert these costs to actual costs, the estimating department is supplied with monthly operating ratios or variances which give the relationships between actual costs and standard costs of each element included in the estimate. The estimating

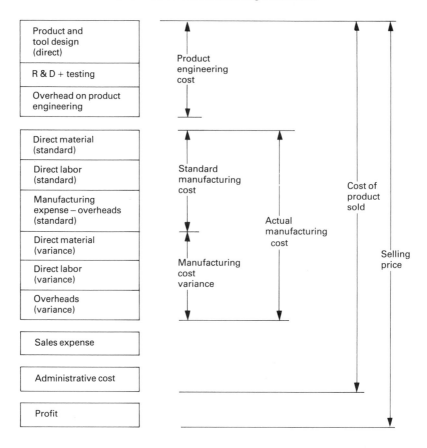

Figure 18.5 Elements of cost that add together to establish the selling price of a product.

department must exercise sound judgement in the selection of the unit prices. The standard-cost prices, as adjusted by operating ratios, are satisfactory for job estimates that are to be produced in the near future. Estimates of major projects that require longer periods of design and planning must be based upon a careful forecast of the price situation expected to prevail at the time of manufacture.

The sales expense covers the cost of distribution as well as the cost of advertising and marketing of the product. In some cases the sales expense is charged as a percent of the combined costs of product engineering and manufacturing. Administrative expenses can be treated in a similar manner to manufacturing overhead.

The cost of product sold, as shown in Fig. 18.5, is the total cost of product engineering, manufacture, selling and administration. An amount of profit is then added to this total cost to arrive at the selling price. The net profit to the company that results from selling a certain number of units of the product per year can be calculated as:

$$\text{Net profit} = (\text{Number of units} \star \text{profit per unit}) - \text{income taxes}$$

Income taxes are usually calculated as a certain percent of the total profits, as discussed earlier.

18.11   LIFE CYCLE COSTING

The total cost of a product throughout its life includes the costs associated with feasibility studies, research and development, design, production, maintenance, replacement and disposal. Other life-cycle costs include support, training, site preparation and operating costs generated by the acquisition. Life-cycle costs can be estimated for new projects at either total plant level or at equipment item level. The main aims of life cycle costing (LCC) are:

1. To provide a comprehensive understanding of the total commitment of asset ownership.
2. To identify areas of the life cycle where improvement can be achieved through redesigning or relocation of resources.
3. To improve profitability and industrial efficiency.

The applications of LCC in industry include comparing mutually exclusive projects, checking future performance of an asset, and tradeoffs between design and operation parameters. Because LCC deals to a large extent with future costs, its successful application depends on the availability of accurate information on:

1. All future cost potential cost elements.
2. Expected length of useful life.
3. Cost of lost production which is affected by the availability, reliability, and maintainability characteristics.
4. Rate of discounting future cash flows to present value. This rate depends on interest rate and inflation over the expected life.

Historical data and unit prices form the basis for predicting the most likely future costs. Cost models are available for LCC predictions. The net present worth (NPW) is usually used in such models. NPW discounts future cash flows to present values. A suitable discount rate is used in determining whether the NPW of the future cash flows will exceed the NPW outlayed.

18.12   REVIEW QUESTIONS AND PROBLEMS

18.1 Estimate the total production time for machining 100 units of the sleeve given in question 15.5. Available spindle speeds are: 550, 660, 800, 960, 1150, 1380, 1650 and 2000 rev/min. Available feeds are: 0.1, 0.15, 0.2, 0.25, 0.3 mm/rev. Recommended cutting speeds 150 m/min for roughing and 200 m/min for finishing. Take a tool life of 120 min. Average time for tool change 7 min/tool, average time for handling and changing workpiece 5 min. No setting up time is required. [Answer: 11.87 h]

18.2 Estimate the total production time for a batch of 500 parts to be machined on a horizontal milling machine using a fixture that takes four parts at a time. The parts are held end-to-end in the fixture. The width of the available milling cutter is larger than the width of the surface to be machined. Use the following information:

Length of cut surface on each part	= 75 mm
Approach distance	= 30 mm
Overrun	= 25 mm
Feed per tooth	= 0.25 mm

Number of teeth in the cutter	= 10
Rotational speed of the cutter	= 80 rev/min
Loading and unloading time per cycle	= 2 min
Other allowances per cycle	= 1 min

[Answer: total time = 9.95 h]

18.3 Design a simple jig for drilling three holes in a rectangular workpiece of 250 mm long, 150 mm wide and 50 mm thick.

18.4 Calculate the number of parts that will justify introducing the jig discussed in question 18.3 given the following information:

Labor rate without jig	= $12/h
Labor rate with jig	= $10/h
Production time per part without jig	= 15 min
Production time with jig	= 10 min
Machine cost per hour	= $20
Cost of designing and making the jig	= $500
The expected length of production run	= 7 months
Rate of interest	= 12%

[Answer: number of parts = 173]

BIBLIOGRAPHY AND FURTHER READING

Coates, J. B., 'Tool costs and tool estimating', *The Prod. Eng.*, vol. 55, No. 4, pp. 202–8, 1976.

Dallas, D. B. ed., *Tool Manufacturing Engineers Handbook*, Soc. Manufacturing Engr., McGraw-Hill, N.Y., 1976.

DeGarmo, E. P., Black, J. T. and Kohser, R. A., *Materials and Processes in Manufacturing*, Collier Macmillan, London, 1984.

Dieter, G., *Engineering Design: a Materials and Processing Approach*, McGraw-Hill, N.Y., 1983.

Doyle, L. E. *et al.*, *Manufacturing Processes and Materials for Engineers*, 3rd ed., Prentice Hall, New Jersey, 1985.

Haslehurst, M., *Manufacturing Technology*, 3rd ed., Hodder and Stoughton, London, 1981.

Humphreys, K. K. and Katell, S., *Basic Cost Engineering*, Marcel Dekker, New York, 1981.

Lowe, P. H. and Walshe, K. B., 'Computer-aided tool cost estimating', *Int. J. Prod. Res.*, vol. 23, No. 2, pp. 371–80, 1985.

Ludema, K. C., Caddell, R. M. and Atkins, A. G., *Manufacturing Engineering: Economics and Processes*, Prentice Hall, London, 1987.

Thabit, S. S., 'Life cycle costings: a decade of progress', *CME*, No. 5, pp. 46–9, 1983.

White, J. A. *et al.*, *Principles of Engineering Economic Analysis*, 2nd ed., John Wiley & Sons, N.Y., 1984.

Chapter 19

Economics of Materials

19.1 INTRODUCTION

Generally, engineering materials can be classified into two main categories, depending on their cost. The first category contains the commonly used materials, like plain carbon steels and polyethylenes, which are manufactured by large-scale processes to produce them cheaply and more competitively. The second category contains the special or high-performance materials, like the superalloys and silicones, which are manufactured to meet special needs. The materials in the second category are more expensive than the materials in the first, and are only used in order to meet special requirements that cannot be met by the less expensive, commonly used materials. This division must not be too rigid because a material developed for a particular application may prove to have properties that eventually lead to its widespread use. For example, aluminum and titanium moved in a few decades from being special and expensive materials to being moderately priced items of everyday industrial use.

Regardless of whether the material used in making a product is common and inexpensive or special and expensive, its cost is expected to have a considerable influence on the final cost of the product. This is because materials cost usually represents a high proportion of the total product cost, and also because material processability affects manufacturing costs. In a wide range of engineering industries, the direct cost of materials represents 30 to 70 percent of the value of production. As the cost of materials is so important, efforts should be made to optimize their use in order to achieve an overall reduction of the product cost.

Selecting a cheaper material may not always be the answer to a less expensive product. For example, as materials are usually priced on the basis of cost per unit weight, it may be more economical to pay the extra cost of higher strength since less material will be needed to carry the load. Another example is the case where the manufacture of a product involves a large amount of machining. In such cases, it may be more economical to select a more expensive material with better machinability than to select a cheaper material which is difficult to machine. To illustrate this point consider the production of large numbers of bolts on a high-speed turret lathe. Compare the economics of manufacturing the bolts from AISI 1112 steel, cartridge brass, and 2014 aluminum. The dimensions of the bolt are such that it will need 10 cc of the stock material for its manufacture, the differences in material utilization due to differences in stock sizes being ignored. Table 19.1 gives the characteristics of the different materials and the estimated cost of manufacture.

The calculations show that the superior machinability of aluminum more than offsets

Table 19.1 Characteristics and estimated cost of manufacture of some materials

	AISI 1112 steel	Cartridge brass	2014 aluminum
Cost of stock material ($/kg)	0.6	4.2	3.6
Machinability index	100	200	400
Density of material (g/cc)	7.8	8.5	2.8
Weight of material required (g)	78	85	28
Cost of material required ($)	0.047	0.36	0.10
Number machined bolts per hour	52	85	90
Labor and overhead rate ($/h)	15.00	15.00	15.00
Labor cost per bolt ($)	0.29	0.18	0.17
Cost of material and labor ($)	0.337	0.54	0.27

the savings in material cost of steel, which makes the aluminum bolt less expensive than steel bolt.

The dimensions and shape of the stock material can affect the amount of resulting scrap and, consequently, the final cost of the product. This means that the possibility of recycling the scrap is another material factor that can influence the cost of product, especially when using an expensive material.

19.2 ELEMENTS OF THE COST OF MATERIALS

As the cost of materials represents a high proportion of the total cost of a product, it is necessary to analyze its various elements in order to find possible means of minimizing it. This can be done by considering the sequence of operations in which raw materials are progressively converted into final products. As an example, Fig. 19.1 shows the build-up of cost of steel and aluminum with the progress of processing operations from ore to finished product. The main elements of the cost of materials can be grouped as follows.

Cost of ore preparation

The main elements of the cost of ore preparation are the cost of ore at the mine and the cost of beneficiation. The cost of ore depends on its location and the method of mining it, while the cost of beneficiation depends on the concentration of the required material in the ore, the degree of complexity in mineralogical association, and the type of gangue materials. For example, the cost of iron ore preparation is relatively low because commercial ores are usually mined in open pit mines and contain 50 to 65 percent Fe. On the other hand, preparation of copper and gold ores is much more expensive as the concentration of the metal is 1–1.5 percent and 0.0001–0.001 percent respectively. The cost of transporting the ore from the mine to the extraction site can be considerable and can be reduced by performing the beneficiation process at the mine. Even so, ores are known to be transported across countries and continents and the cost can be considerable.

Cost of extraction from the ore

The main elements of the cost of extraction include the cost of power and the cost of auxiliary materials. In metals, the more stable the compound in which the element is found, the greater will be the amount of energy and cost needed for reduction. For

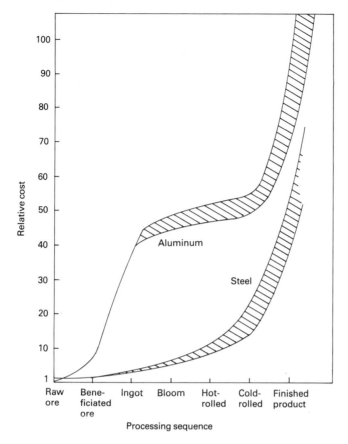

Figure 19.1 Build-up of material cost with the progress of manufacturing processes from ore to finished product.

example, the large amount of electric power and the high cost of auxiliary materials needed for extraction of aluminum are the main reasons for its high cost, as can be shown from Fig. 19.1.

Cost of purity and alloying

Impurity level is known to have an important influence on the cost of materials and, in general, the higher the permissible impurity level the lower the cost. For example, the relative cost of aluminum ingots nearly doubles as the purity increases from 99.5 to 99.99 percent. The cost of an alloy is not only affected by the purity of the elements used, but also by their nature and degree of complexity of alloy structure. The cost of an alloy is not simply the cost of constituents, because in the majority of cases more sophisticated techniques of production have to be employed in order to make full use of the alloying elements. For example, the less demanding specifications of SAE 326 (LM4) aluminum alloy permit the use of lower purity aluminum as the base metal, which makes its cost about 50 percent less than the cost of SAE 324 (LM10) alloy which specifies at least 99.7 percent purity for its base aluminum. Another example is the case of AA 7075 aluminum alloy whose complexity makes it at least 30 percent more expensive than the simpler 5083 binary alloy. The more complex alloys usually require expensive melting procedures and heat treatment, especially if they contain small quantities of readily oxidizable elements.

Cost of conversion to semifinished product

The cost of converting ingots to semifinished products ready for delivery to the manufacturers of end products includes the costs of casting, forging, rolling, etc. The main cost elements in this case are: labor, energy, overheads and the cost of material losses. This latter element can be considerable in metal industries, where material losses range from 25 to more than 50 percent. The size, complexity, surface finish and dimensional accuracy of the product as well as the required number influence the method of production and the cost. Table 19.2 gives an example of the costs in a cast iron foundry. Although this is a hypothetical case, the figures can be considered as representative of those found in industry.

Table 19.2 Cost analysis in a cast iron foundry

Item	Cost $/casting	Percent of total cost
Metal [weight of casting 15 kg (33 lb), 55% yield, 12% scrap, sprue]	4.40	35.9
Mold and core	1.80	14.7
Clean and sort	0.53	4.3
Heat treat	0.56	4.5
Hand tool finish and grind	1.30	10.6
Inspect	0.12	1.0
Ship	0.05	0.4
Total direct cost	$8.76	71.4
Overhead and administration	$3.50	28.6
Total cost	$12.26	100.0

Cost of conversion to finished product

The final stage in production is to convert the semifinished material into a finished product ready for delivery to the end user. Manufacturing processes that are usually involved in this stage are pressing, machining, surface finishing, assembly and packaging. Many of these processes are expensive and wasteful of materials. Material losses of 200 to 300 percent are not uncommon at this stage. With mass production, the costs of direct labor and overheads are usually small in proportion to the material cost. In such cases, it becomes even more important to reduce the material costs and to optimize material utilization.

19.3   FACTORS AFFECTING MATERIAL PRICES

As with most other commodities, the price of engineering materials is affected by a variety of factors. Inflation, economic recessions, supply and demand, amount of material purchased, inventory costs and quality of material are among the major factors that can influence the price and will be discussed in this section.

General inflation and price fluctuations

Generally, the prices of most established engineering materials show a steady increase when considered over a relatively long period of time. The main reasons for such price increases are rising costs of raw materials, energy and labor. Governmental anti-pollution policies and similar legislations have also contributed to the increase in prices in recent years. Over the past two decades, most metallic and polymeric materials showed an average price inflation rate of about 5 to 15 percent per year. In addition, material prices are also known to suffer short-term price fluctuations. Political factors, local wars, industrial strikes and world recessions are mostly responsible for such price fluctuations, which occur as a result of changes in supply and demand.

Supply and demand

In a free market economy, the price of a commodity is fixed by the equilibrium between supply and demand. This price is given by the point of intersection of the supply and demand curves of Fig. 19.2. Prices vary as a result of horizontal shifts in one or the other of the supply and demand curves. When the supply increases, S1 to S2, prices should decrease as competing producers pare their profit margins to maintain their market share.

Order size

The cost of material is usually affected by the size of order. The larger the size of an order for a given material the smaller will be the unit cost. This is because administrative expenses and delivery charges remain almost unchanged and tend to represent a higher proportion of the total cost of a smaller order. For example, the unit prices of most popular plain carbon steels can nearly double as the amount of material purchased

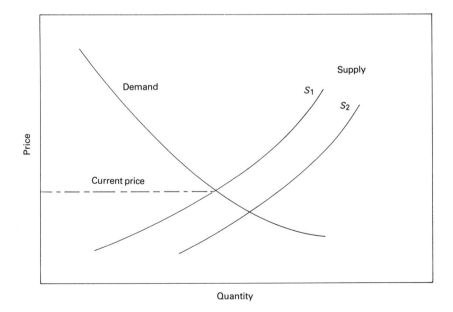

Figure 19.2 Effect of supply and demand on the prices of commodities.

decreases from 5-ton lots to about 1/2 ton. The unit price can be even higher for smaller order sizes.

Standardization of grades and sizes

If selection of material were based entirely on picking the most suitable grade and size for each part, almost as many materials, grades and sizes as parts would be chosen. If all parts were produced in equal and large quantities, it might be practical to make each part from a different material. However, production is seldom in equal and large quantities, and selecting a different material for each part can lead to serious problems in material cost, storage, inventory and equipment required by each material. The other extreme is to make all parts from the same grade and size of one material. This is obviously not practical, except in rare instances. A compromise is to standardize on the smallest number of material grades and sizes that will satisfy plant-wide needs at the lowest cost. Fabrication requirements must be checked to ensure that cost savings in eliminating grades and sizes are not more than offset by added manufacturing costs. Minor design changes can often be made to permit standardization.

Inventory costs

It has been shown that significant savings may be realized by ordering in large quantities. Inventory costs, however, can quickly offset the initial saving in large-quantity purchases. For example, the cost of storing steel for one year can range from 10 to 25 percent of the initial cost. Inventory costs include interest on investment, storage, taxes, insurance, obsolescence, deterioration, inventory taking, record keeping, rehandling and reinspection.

Cost extras for special quality

In the case of most steels, a base price which represents the lowest cost is usually given to the most often used quality, e.g. merchant quality for hot-rolled bars and commercial quality for sheets. Cost extras are then added to account for the customer's special requirements. Some of the special requirements that usually increase the base price, cost extras, are:

1. Grade: Standard AISI quality is supplied as the base material. Semikilled or killed grades are extra.
2. Restrictions: Standard quality usually allows for wider chemistry limits. Extra cost is usually charged for specifying a narrower range of carbon, a minimum manganese content, etc. Resulfurized or free-machining steel is supplied at extra cost. Specifying the grain size or hardenability is also an extra.
3. Size and form: Special sections or closer tolerances than specified in standards are cost extras.
4. Treatment: Specifying treatments like annealing, normalizing, quenching and tempering, stress relieving or pickling is usually considered as a cost extra.
5. Length: Specifying a certain length of the stock is usually considered as a cost extra.
6. Cutting: Hot cut and cold shear are standard. Machine cutting is an extra.
7. Packaging: Wrapping, burlap or boxing are extras.

The price after adding the extras can be more than twice the base price in some cases.

Geographic location

Material prices are usually given FOB of the supplier. The customer pays the cost of transportation. This cost item can be considerable for longer distances and smaller quantities.

19.4   COMPARISON OF MATERIALS ON COST BASIS

As discussed in the preceding section, the cost of engineering materials varies over a wide range depending on their relative abundance, ease of extraction, volume of production, supply and demand, quality and quantity purchased. Timber, concrete and mild steel are the most widely used engineering materials because of their low cost and availability. The use of plastics and aluminum is expanding rapidly and, compared on a volume basis, the use of polymers is similar to that of steel.

Most engineering materials are sold on the basis of cost per unit weight, although some semifinished and finished products are sold on other bases. For example, timber is sold on the basis of cost per unit volume, while window glass is sold on the basis of cost per unit area. The prices of plumbing, heating, sewer pipe and tubing are usually given on a unit length basis while the prices of paint are given on a unit liquid volume basis. Figure 19.3 compares some metallic and plastic materials on the basis of relative cost per unit weight. The cost of hot-rolled plain carbon steel is taken as unity and all other costs are given relative to it. All forms of plain carbon steels, cast irons and low-alloy steels are less expensive than other materials, which explains their widespread use. The figure shows the wide difference between the prices of the different materials. For example, nickel and inconel 600 are about 20 times as expensive as plain carbon steel, tin and titanium 30 times as expensive, while zirconium and tungsten are 50 times as expensive.

In many applications, engineering materials are not highly stressed, largely because the amount of material used is determined by the size and shape, method of production, or rigidity of the part. Examples include machine frames and motor bodies, conduits and sewer pipes, furniture, household appliances and fittings. In such cases it may be more appropriate to compare materials on the basis of their cost per unit volume, as shown in Fig. 19.4. As in Fig. 19.3, the different materials are related to hot-rolled plain carbon steel, whose cost per unit volume is taken as unity. Some plastics now appear to be cheaper than steel because of their low density. Aluminum also becomes close to steel.

19.5   VALUE ANALYSIS OF MATERIAL PROPERTIES

One of the important applications of value analysis is to assess the value of any product by reference to the cheapest available or conceivable product which will perform the same function. This technique can be adapted to material selection. In the case of steel, for example, plain carbon steels should be considered as a reference point for estimating the value. Additional steel prices above those of plain carbon steel should be critically analyzed. The value of each item of cost extras needs to be examined in relation to the function that the part has to perform in service.

In comparing materials in order to select the one which will perform the required function at the least cost, the engineer has two basic alternatives: (a) to select the least

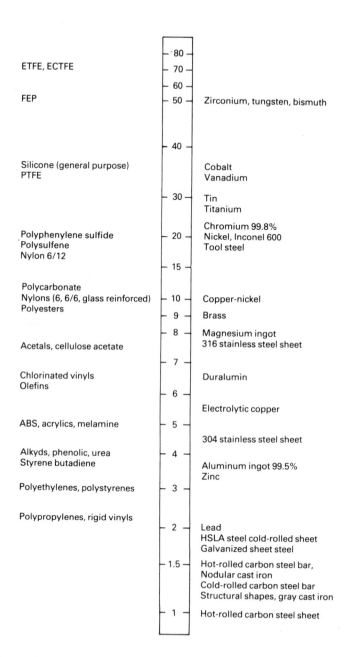

Figure 19.3 Comparison of some engineering materials on the basis of cost per unit mass. Comparison is made relative to the cost of hot-rolled low carbon steel sheet.

Source: Farag, M. M., *Materials and Process Selection in Engineering*, Applied Science Publishers, London, 1979.

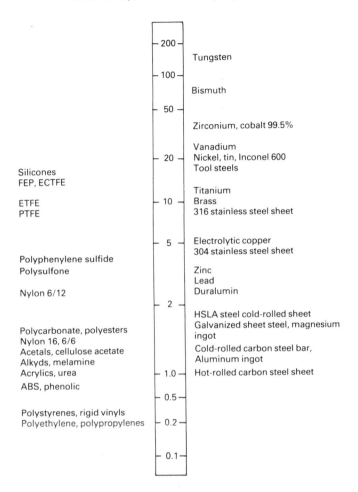

Figure 19.4 Comparison of some engineering materials on the basis of cost per unit volume. Comparison is made relative to the cost of hot-rolled low carbon steel sheet.

Source: Farag, M. M., *Materials and Process Selection in Engineering*, Applied Science Publishers, London, 1979.

expensive material; and (b) to select a more expensive material that will simplify processing or eliminate steps in manufacturing. An example of cheaper materials that perform the job of more expensive alternatives are EX steels. These steels are designed to replace standard AISI-SAE steels of similar hardenability, but they contain less expensive alloying elements, which make them less expensive than standard steels. Thus in situations where hardenability is the main requirement needed to assure equivalent strength, an EX steel is usually a suitable replacement. Another example is the manganese-containing grades of stainless steels like 201, 202, 203 and 216, which are less expensive than their type 300 counterparts because they contain less nickel. Cladding, galvanizing and tinning offer cheaper alternatives to using stainless steels if the corrosion conditions are not severe. Surface hardening or cold working can be used to increase the wear and fatigue resistance of plain carbon steel parts and the use of expensive alloy steels can thus be avoided.

An example of a higher-cost material resulting in a lower-cost component, because of savings in processing, is prehardened steel sheets. The use of this material makes it possible for users to eliminate heat treatment and the resulting part distortion, as well as the associated handling and salvage operations. Another example is precoated coils which enable fabricators to omit the finishing step. The coating can be alkyd, polyester, acrylic, vinyl, epoxy or phenolic paint, or it can be a plastic, such as vinyl, fluorocarbon or polyethylene. The coatings can be applied to most metals, e.g. steel, tin, zinc and aluminum, and each type of coating provides a different set of properties. Aluminized and chromized coatings on carbon steel resist moderate heat and do not need expensive finishing by enameling.

19.6   ECONOMICS OF MATERIAL UTILIZATION

Manufacturing a part or a product at a competitive cost can only be accomplished when materials and processes are used as effectively as possible. Ideally, the manufacturer should use the cheapest material and not pay for properties that are not needed for successful performance of the part. As discussed earlier, this could lead to cost and inventory problems as a result of stocking a very large number of materials, grades and sizes. Standardization of material grades and sizes as well as judicious design were given as an answer to this problem. Where production involves more than one size, it may become necessary to buy base quantities of large size stock and use it for a variety of smaller size parts. Break-even analysis can be useful in determining the number of parts at which the cost of extra machining involved in reducing the large stock is equal to the extra cost of buying and stocking the correct size, as shown in Fig. 19.5. Line (a) represents the case where an extra cost is paid to buy and stock the correct size which needs less fabrication costs, while line (b) represents the case where cheaper standard stock is used to fabricate

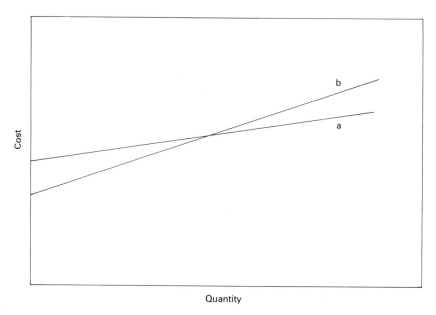

Figure 19.5 Application of break-even analysis to the selection of economic stock size.

the part at a higher cost. The position of the break-even point is expected to vary with the type of material and the available manufacturing facilities.

Another area of cost saving material utilization is the reduction of scrap. A material utilization factor, m, can be defined as:

$$m = \frac{\text{weight of the finished part}}{\text{weight of material used to make the part}} \qquad (19.1)$$

The nearer the value of m to unity, the less waste will be incurred and hence the lower will be the direct material cost. The cost of scrap material, C_s, that results from manufacturing N components is given by:

$$C_s = (1 - m)\, W\, C_w\, N \qquad (19.2)$$

where W = weight of the component
C_w = cost of stock material per unit weight

Figure 19.6 shows an example of how the value of m can be increased by a change in the blanking die design. The proposed change involves a more costly die and a higher capacity

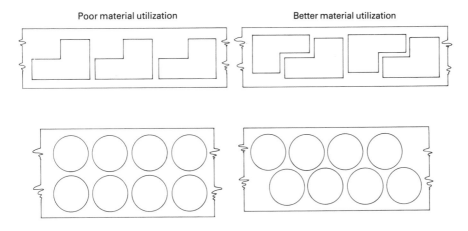

Figure 19.6 Effect of blanking die design on material utilization.

press as two parts instead of one are punched in each stroke. This problem can also be solved using the break-even analysis. Taking line (a) of Fig. 19.5 to represent the modified blanking die, it can be seen that, at a large enough number of parts, the higher initial cost is more than offset by the savings due to the more economic utilization of the material.

Another form of inefficiency in material utilization is encountered in metal casting. The yield in casting is expressed as the percentage of the weight of good castings obtained from the charged metal weight. With high yield there is less return metal to remelt, which in turn reduces the net cost of conversion, melting loss and molten metal treatments. The ultimate metal cost of castings with a high yield is therefore less than that of low-yield castings. Typical yields in cast iron foundries range from 40 to 70 percent.

19.7   COMPETITION IN THE MATERIALS FIELD

Although the average world consumption of engineering materials is increasing with time, the consumption of some materials is increasing at a much faster rate than others. Generally, the consumption of the older materials like steel, copper, concrete and timber is growing at a slower rate than aluminum and plastics. The slow growth of the older materials reflects the increasing efficiency in their utilization, as well as competitive inroads made by aluminum and plastics. The more efficient utilization of materials is illustrated by the fact that the use of copper in electrical generators has been reduced from about 100 kg (220 lb) per megawatt to 25 kg (55 lb) per megawatt during the past decade. The more efficient design and better alloys used in present-day aircraft have reduced the use of metals from 3.5 kg (7.7 lb) per passenger mile in the Boeing stratocruiser to about 1.4 kg (3 lb) in the Boeing 707.

The substitution of a new material for an established one usually involves overcoming the inertia which helps to preserve the existing structure of the industry, i.e. the investment in capital plant, labor skills, etc. A good example of this is the difficulty some car companies might find in introducing all plastics injection- or blow-molded car seats in place of the current labor-intensive tubular metal frame construction. The main forces which can overcome the inertia against change are: (a) legislation; (b) cost saving; and (c) performance.

Legislation can be a major driving force for change, as in the case of the motor car industry. For example, legislation on crash padding in car interiors led to a sudden increase in the amount of plastics employed in European cars. Similarly, the introduction of legislation in the US requiring that new cars should average 29 miles to the gallon resulted in the initiation of development programs to reduce the weight of the car. This caused the average amount of plastics in the motor car to increase from about 55 kg (120 lb) in 1980 to about 75 kg (165 lb) in 1985, and is expected to reach more than 100 kg (220 lb) by 1990. Legislation, however, can also oppose change, as in the case of side-impact resistance legislation which restricted the design and introduction of all-plastics car doors and other structural members.

An example of how cost savings can be the driving force for change is the case of beer cans and oil containers. Aluminum has taken about one-sixth of this business from steel, but the steel industry has now reduced the thickness of steel in a can to two-thirds in the hope of holding the market. Aluminum has also taken a large proportion of the market share from copper in the electrical industry. Overhead cables for power transmission down to 10 kVA, and even lower, are much cheaper when made of aluminum, thus leaving copper to be used mainly for generators and motors because of its higher conductivity per unit cross-section and, therefore, its greater compactness in design. Aluminum, in turn, is facing challenges in the construction of aircraft, and may now yield in critical areas to high-strength steels and to the newer titanium alloys and fiber reinforced plastics. Although some plastics raw materials are more expensive than common metals when compared on a weight basis, they have captured large proportions of the market share which was traditionally served by other materials. This is because their low density and ease of manufacture has made it possible to produce inexpensive products with attractive colors and aesthetic appeal.

The superior performance of new materials is also a major cause of change. The sports-equipment industry offers many examples, like tennis rackets (Chapter 24) and vaulting poles, where fiber reinforced composites have replaced traditional materials.

However, substitution of a new material for an established one is not always easy. For example, the introduction of the strong aluminum alloys with their high strength/weight ratios made it possible to use them in structural applications. To make use of this improved performance, considerable changes in design procedures had to be made to allow for their lower elastic modulus. In some areas, however, the low modulus can be regarded as a virtue. This is illustrated by the successful combination of aluminum and steel in making the top deck and life boats of ocean liners and cargo vessels. Apart from the obvious advantage of low density, low modulus helps to keep stresses low and allows the use of medium strength alloys with good weldability and corrosion resistance.

19.8   REVIEW QUESTIONS AND PROBLEMS

19.1   Explain why stainless steels are more expensive than plain carbon steels.

19.2   Why has the price of titanium decreased in the last 20 years?

19.3   What are the economic reasons for the increased use of carbon fiber reinforced materials?

19.4   Although plastics are more expensive than many metals, on the basis of cost per unit weight, plastic products are generally less expensive. Discuss this statement.

19.5   Although powdered alloys are more expensive than similar solid alloys, powder metallurgy products are competitive on a cost basis.

BIBLIOGRAPHY AND FURTHER READING

Crane, F. A. and Charles, J. A., *Selection and use of Engineering Materials*, Butterworths, London, 1984.
Farag, M. M., *Materials and Process Selection in Engineering.*, Applied Science Publishers, London, 1979.
Sharp, H. J. ed., *Engineering Materials: Selection and Value Analysis*, Heywood Books, London, 1966.

INTEGRATION OF DESIGN AND ECONOMIC ANALYSIS WITH MATERIALS AND PROCESS SELECTION

Discussions in earlier parts of this book have illustrated the interdependence of the various activities involved in developing a concept into a finished product. It was shown that the materials and processes used in making a component have a large influence on its design, cost and performance in service. This means that these aspects should not be considered in isolation of each other and that an integrated approach will make it easier to arrive at the optimum solution. This is not an easy task, especially in the context of today's technical and social climate where a large number of factors, not all of which are necessarily compatible, have to be taken into consideration. The procedures which will be presented in this part are meant to make it easier to analyze the large amount of data involved in selection in order to arrive at the optimum solution. Case studies drawn from widely different areas are used to illustrate the use of the selection procedures.

Chapter 20

The Selection Process

20.1 INTRODUCTION

One of the most important requisites for the development of a satisfactory product at a competitive cost is making a sound economic choice of engineering designs, materials and manufacturing processes. Earlier discussions in various parts of this book have shown that the product function, design, selected material properties and manufacturing processes are closely related. The relationship between design and product function was discussed in Chapter 9, while the effects of material properties and manufacturing processes on design were discussed in detail in Chapters 11 and 12 respectively.

It has been reported that there are more than 40 000 currently useful metallic alloys and probably close to that number of nonmetallic engineering materials like plastics, ceramics and glasses, composite materials and semiconductors. This large number of materials and the many manufacturing processes available to the engineer, coupled with the complex relationships between the different selection parameters, often make the selection process a difficult task. If the selection process is carried out haphazardly, there will be the risk of overlooking a possible attractive alternative solution. This risk can be reduced by adopting a systematic selection procedure. Several quantitative selection procedures have been developed to analyze the large amount of data involved in the selection process so that a systematic evaluation can be made. Most of the quantitative procedures can be adopted to computer aided selection from a data bank of material properties and process characteristics. Rigorous and thorough approach to materials selection is, however, often not followed in industry and much selection is based on past experience. What worked before is obviously a solution, but it may not be the optimum solution. It is often said 'when in doubt make it stout out of the stuff you know about'. While it is unwise totally to ignore past experience, the frequent introduction of new materials and manufacturing processes, in addition to the increasing pressure to produce more economic and competitive products, makes it necessary for the engineer to be always on the lookout for possible improvement.

Although the materials and process selection is most often thought of in terms of new product development, there are many other reasons for reviewing the type of materials and processes used in making an existing product. These reasons include:

1. Taking advantage of new materials or processes.
2. Improving service performance, including longer life and higher reliability.
3. Meeting new legal requirements.
4. Accounting for changed operating conditions.
5. Reducing cost and making the product more competitive.

If a decision is taken to substitute a new material for an established one, care must be taken to ensure that all the characteristics of the new material are well established. A large number of product failures have resulted from new materials being used before their long-term properties were fully known. Another source of failure results from substituting a new material without reviewing the design. As an example, consider the case where thinner high-strength low-alloy (HSLA) steel sheets are substituted for the thicker steel sheets currently used in motor car bodies. Having overcome the processing problems, the substitution appears attractive in view of the weight saving. It should be remembered, however, that while the strength of HSLA steel is higher, corrosion resistance and elastic modulus are essentially the same as those for low carbon steel. Thinner HSLA sheets would be damaged by corrosion in a shorter time and undesirable vibrations would be more of a problem. Design modifications might be necessary to overcome these problems.

20.2   THE NATURE OF THE SELECTION PROCESS

Discussions in various parts of this book have emphasized the iterative nature of the design process and the interdependence of design, material properties and method of fabrication. This means that selecting the optimum combination of material and process is not a simple task which can be performed at one certain stage in the history of a project, but should gradually evolve during the different stages of product development.

The selection process steps

While each material and process selection decision has its own individual character and its own sequence of events, there is a general pattern common to the selection process. In the first stages of development of a new product, such questions as the following are posed: What is it? What does it do? How does it do it? Answering these questions will help in specifying the functions of the product. This is then followed by the questions: what are the important, or primary, design and material requirements? What are the secondary requirements and are they necessary? Answering these questions makes it necessary to specify the performance requirements of the different parts involved in the design and to broadly outline the main materials and processing requirements. On this basis certain classes of materials and manufacturing processes may be eliminated and others chosen as likely candidates.

Having specified the performance requirements of the different parts, the materials property requirements can be established. These requirements may be quantitative or qualitative, essential or desirable. For example, the function of a connecting rod in an internal combustion engine is to connect the piston to the crankshaft. The performance requirements are that it should transmit the power efficiently without failing during the expected life of the engine. The essential material properties are tensile and fatigue strengths, while the desirable properties that should be maximized are processability, weight, reliability and resistance to service conditions. All these properties should be achieved at a reasonable cost.

The major part of materials and process selection is performed at the stage of final design development. Having identified the relevant material properties for the different parts, the importance of each property is determined and a short-list of candidate materials is prepared. Candidate materials are then graded according to their expected performance and cost. Processing details are also examined at this stage. Optimization techniques may then be used to select the optimum design, material and processing route. This may result in design modifications to achieve production economy or to fit available production facilities and equipment. These modifications could require significant changes in materials, or the design may be refined to conform with the exact properties of the selected material.

In the stage of product manufacture, some changes in materials may still be necessary. Processing problems may arise causing the replacement of an otherwise satisfactory material. For example, heat treating, joining or finishing difficulties may require material substitution which, in turn, may result in different service performance characteristics requiring some redesign.

General procedure for selection

A general procedure for selection can be summarized in the following steps:

1. Analysis of the performance requirements.
2. Development of alternative solutions to the problem.
3. Evaluation of the different solutions.
4. Decision on the optimum solution.

The above steps can be followed in selection among different designs, products, materials or manufacturing methods. Application of this general procedure to the problem of materials selection will be discussed in Sections 20.3 and 20.4.

20.3   ANALYSIS OF THE MATERIAL PERFORMANCE REQUIREMENTS

The outcome of the design stage in product development is a detailed specification of the material performance requirements, which are based on the desired function and expected performance of the component, as well as the environment in which it will operate.

Analysis of performance requirements

The material performance requirements can be divided into five broad categories, namely functional requirements, processability requirements, cost, reliability and resistance to service conditions.

1. Functional requirements. Functional requirements are directly related to the required characteristics of the part or the product. For example, if the part carries a uniaxial tensile load, the yield strength of a candidate material can be directly related to the load-carrying capacity of the product. However, some characteristics of the part or product may not have simple correspondence with measurable material properties, as in the case of thermal shock resistance, wear resistance, reliability, etc. Under these conditions, the evaluation process can be quite complex and may depend upon predictions based on simulated service tests or on the most closely related mechanical, physical or chemical properties. For example, thermal shock resistance can be related to thermal expansion coefficient, thermal conductivity, modulus of elasticity, ductility and tensile strength. On the other hand, resistance to stress corrosion cracking can be related to tensile strength, KISCC and electrochemical potential.

2. Processability requirements. The processability of a material is a measure of its ability to be worked and shaped into a finished part. With reference to a specific manufacturing method, processability can be defined as castability, weldability, machinability, etc. Ductility and hardenability can be relevant to processability if the material is to be deformed or hardened by heat treatment respectively. The closeness of the stock form to the required product form can be taken as a measure of processability in some cases.

It is important to remember that processing operations will almost always affect the material properties, so that processability considerations are closely related to functional requirements.

3. Cost. Cost is usually the controlling factor in evaluating materials because in many applications there is a cost limit for a material intended to meet the application requirements. When the cost limit is exceeded, the design may have to be changed to allow the use of a less expensive material.

The cost of processing often exceeds the cost of the stock material. In some cases, a relatively more expensive material may eventually yield a less expensive product than a low-priced material that is more expensive to process. The economics of manufacturing processes and materials were discussed in detail in Chapters 18 and 19 respectively.

4. Reliability requirements. As discussed in Chapter 13, reliability of a material can be defined as the probability that it will perform the intended function for the expected life without failure. Material reliability is difficult to measure, because it is not only dependent upon the material's inherent properties, but is also greatly affected by its production and processing history. Generally, new and nonstandard materials will tend to have lower reliability than established, standard materials.

Despite difficulties of evaluating reliability, it is often an important selection factor that must be taken into account. Failure analysis techniques are usually used to predict the different ways in which a product can fail, and can be considered as a systematic approach to reliability evaluation. The causes of failure of a part in service can usually be traced back to defects in materials and processing, to faulty design or to unexpected service conditions. The subjects of failure analysis and reliability were discussed in detail in Chapters 7 and 13 respectively.

5. Resistance to service conditions. The environment in which the product or part will operate plays an important role in determining the material performance requirements. Corrosive environments, as well as high or low temperatures, can adversely affect the performance of most materials in service, as discussed in Chapter 7. Whenever more than one material is involved in an application, compatibility becomes a selection consideration. In thermal environments, for example, the coefficients of thermal expansion of all the materials involved may have to be similar in order to avoid thermal stresses. In wet environments, materials that will be in electrical contact should be chosen carefully to avoid galvanic corrosion.

In applications where relative movement exists between different parts, wear resistance of the materials involved should be considered. The design should provide access for lubrication, otherwise self-lubricating materials have to be used. The functional requirements of engineering materials under different service conditions were discussed in Chapter 8.

Classification of material requirements

After the above analysis of the engineering material performance requirements, the next step is to classify these requirements into two main categories:

1. Rigid, or go-no-go, requirements.
2. Soft, or relative, requirements.

Rigid requirements are those which must be met by the material if it is to be considered at all. Such requirements are usually used for the initial screening of materials to eliminate the unsuitable groups. For example, metallic materials are eliminated when selecting materials for an electrical insulator. If the insulator is to be flexible, the field is narrowed further as all ceramic materials are eliminated. Examples of the material rigid requirements include behavior under operating temperature, resistance to corrosive environment, ductility, electrical and thermal conductivity or insulation and transparency to light or other waves. Examples of process rigid requirements include batch size, production rate, product size and shape, tolerances and surface finish. Whether or not the equipment or experience for a given manufacturing process exists in a plant can also be considered as a hard requirement in many cases. Compatibility between the manufacturing process and the material is also an important screening parameter. For example, cast irons are not compatible with sheet metal forming processes and steels are not easy to process by die casting. In some cases, eliminating a group of materials results in automatic elimination of some manufacturing processes. For example, if plastics are eliminated because service temperature is too high, injection and transfer molding should be eliminated as they are unsuitable for other materials.

Soft, or relative, requirements are those which are subject to compromise and trade-offs. Examples of soft requirements include mechanical properties, specific gravity and cost. Soft requirements can be compared in terms of their relative importance, which depends on the application under study.

20.4   DEVELOPMENT AND EVALUATION OF ALTERNATIVE SOLUTIONS

Having specified and classified the material requirements, the rest of the selection process involves the search for the material or materials that best meet those requirements. The starting point for materials selection is the entire range of engineering materials.

In this phase of the selection process different approaches to the solution of the materials problems, as distinct from simply choosing candidate materials for evaluation, should be considered. At this stage, creativity is essential in order to open up channels in different directions and not to let traditional thinking interfere with the exploration of ideas. The importance of this phase is that it creates alternatives without much regard to their feasibility. After all the alternatives have been suggested, the ideas that are obviously unsuitable are eliminated and attention is concentrated on those that look practical. At the end of this phase, the number of candidate materials is narrowed down to a manageable number for subsequent detailed evaluation. The classification of the material requirements into rigid and soft requirements can also simplify the initial screening process.

Having narrowed down the field of possible materials to those that do not violate any of the rigid requirements, the search starts for the material, or materials, that best meet the soft requirements. The objective of the evaluation stage is to weigh the candidate materials against the specified material requirements in order to select the optimum one for the application. Unlike the exact sciences, where there is normally only one single correct solution to a problem, materials selection requires the consideration of conflicting advantages and limitations, necessitating compromises and decision making; and as a consequence different solutions are possible. This is illustrated by the fact that similar components performing similar functions, but produced by different manufacturers, are often made from different materials and even by different manufacturing processes.

Several quantitative methods for selection optimization will be discussed in the following sections. It should be emphasized at this stage that none of the proposed quantitative methods is meant to replace the judgement and experience of the engineer. The methods are only meant to help the engineer in ensuring that none of the viable solutions are neglected and in making sounder choices and trade-offs for a given application.

20.5 COST PER UNIT PROPERTY METHOD

In the simplest cases of optimizing the selection of materials, one property stands out as the most critical service requirement. In such simple cases, the cost per unit property can be used as a criterion for selecting the optimum material. Considering the case of a bar of a given length, L, to support a tensile force, F. The cross-sectional area, A, of the bar is given by:

$$A = \frac{F}{S} \qquad (20.1)$$

where S = working stress of the material, which is related to its yield strength by an appropriate factor of safety.

The cost of the bar (C') is given by:

$$C' = C \star p \star A \star L = \frac{C \star p \star F \star L}{S} \qquad (20.2)$$

where C = cost of the material per unit mass
p = density of the material

in comparing different candidate materials, only the quantity $[(C*p)/S]$, which is the cost of unit strength, needs to be compared as F and L are constant for all materials. The material with the lowest cost per unit strength is the optimum material.

As shown earlier, manufacturing costs are a significant factor in evaluating materials. The life, installation cost, maintenance cost and salvage costs are also important factors which should be taken into consideration when comparing different alternatives. This means that the material cost C in (20.2) should be taken as the total life-cycle cost.

The working stress of the material in (20.1) and (20.2) was related to the static yield strength of the material, since the applied load was static. If the applied load is alternating, it would be more appropriate to use the fatigue strength of the material. Similarly, the creep strength should be used under loading conditions that cause creep.

When one material is being considered as a substitute for an existing material, the two materials, 'a' and 'b', can be compared on the basis of relative cost per unit strength, RC':

$$RC' = \frac{C'_a}{C'_b} = \frac{C_a \star p_a \star S_b}{C_b \star p_b \star S_a} \tag{20.3}$$

RC' less than unity indicates that material 'a' is preferable to material 'b'.

Equations similar to (20.2) and (20.3) can be used to compare materials on the basis of cost per unit stiffness when the important design criterion is deflection in the bar. In such cases, S_w is replaced by the Young's modulus of the material. The above equations can also be modified to allow comparison of different materials under loading systems other than uniaxial tension. Table 20.1 gives some formulas for the cost per unit property under different loading conditions based on either yield strength or stiffness.

The following example illustrates the use of the cost per unit property method. Consider a structural member in the form of a simply supported beam of rectangular cross-section. The length of the beam is 1 m (39.37 in), width is 100 mm (3.94 in), and there is no restriction on the depth of the beam. The beam is subjected to a concentrated load of 20 kN (4409 lb) which acts in the middle of the beam. The main design requirement is that the beam should not suffer plastic deformation as a result of load application. Use the information given in Table 20.2 to select the least expensive material for the beam.

Based on Table 20.2 and the appropriate formula from Table 20.1, the cost of unit strength for the different materials is calculated and the results are given in the last column of Table 20.2. The results show that steels AISI 1020 and 4140 are equally suitable, while aluminum 6061 and epoxy-glass are more expensive. This answer is reasonable since steels are usually used in structural applications unless special features, like light weight or

Table 20.1 Formulas for estimating cost per unit property

Cross-section and loading conditions	Cost of unit strength	Cost of unit stiffness
Solid cylinder in tension or compression	$\dfrac{C \star p}{S}$	$\dfrac{C \star p}{E}$
Solid cylinder in bending	$\dfrac{C \star p}{S^{2/3}}$	$\dfrac{C \star p}{E^{1/2}}$
Solid cylinder in torsion	$\dfrac{C \star p}{S^{2/3}}$	$\dfrac{C \star p}{G^{1/2}}$
Solid cylindrical bar as slender column	$-$	$\dfrac{C \star p}{E^{1/2}}$
Solid rectangle in bending	$\dfrac{C \star p}{S^{1/2}}$	$\dfrac{C \star p}{E^{1/3}}$
Thin-walled cylindrical pressure vessel	$\dfrac{C \star p}{S}$	$-$

S = working yield strength of the material. C = cost per unit mass. p = density. E = Young's modulus. G = shear modulus.

Table 20.2 Characteristics of candidate materials for the beam

Material	Working stress[a]		Specific gravity	Relative cost[b]	Cost of unit strength
	(MPa)	(ksi)			
Steel AISI 1020, normalized	117	17	7.86	1	0.73
Steel AISI 4140, normalized	222	32	7.86	1.38	0.73
Aluminum 6061, T6 temper	93	13.5	2.7	6	1.69
Epoxy + 70% glass fibers	70	10.2	2.11	9	2.26

[a] The working stress is computed from yield strength using a factor of safety of 3.
[b] The relative cost per unit weight is based on AISI 1020 steel as unity. Material and processing costs are included in the relative cost.

corrosion resistance, are required. If the weight of the beam that can carry the given load is calculated for the different materials, it can be shown that, in a descending order, the 1020 steel beam is the heaviest, followed by 4140 steel, then aluminum, with epoxy-glass being the lightest. In some cases, as in aerospace applications, it is usually worth paying the extra cost of the lighter structural member.

Although the cost per unit property method can be useful in optimizing simple cases of selection, it has the drawback of considering one requirement only as the most critical and ignoring other requirements. In many engineering applications, the situation is more complicated than this and material requirements usually specify more than one property as being important. In such cases, the weighted properties method may be a useful tool for optimizing materials selection.

20.6 WEIGHTED PROPERTIES METHOD

The weighted properties method can be used in optimizing materials selection when several properties should be taken into consideration. In this method each material requirement, or property, is assigned a certain weight, depending on its importance. A weighted property value is obtained by multiplying the numerical value of the property by the weighting factor (α). The individual weighted property values of each material are then summed to give a comparative materials performance index (γ). The material with the highest performance index (γ) is considered as the optimum for the application.

In its simple form, the weighted properties method has the drawback of having to combine unlike units, which could yield irrational results. This is particularly true when different mechanical, physical and chemical properties with widely different numerical values are combined. The property with higher numerical value will have more influence than is warranted by its weighting factor. This drawback is overcome by introducing scaling factors. Each property is so scaled that its highest numerical value does not exceed 100. When evaluating a list of candidate materials, one property is considered at a time. The best value in the list is rated as 100 and the others are scaled proportionally. Introducing a scaling factor facilitates the conversion of normal material property values to scaled dimensionless values. For a given property, the scaled value, B, for a given candidate material is equal to:

$$B = \text{scaled property} = \frac{\text{numerical value of property} * 100}{\text{maximum value in the list}} \qquad (20.4)$$

For properties like cost, corrosion or wear loss, weight gain in oxidation, etc., a lower value is more desirable. In such cases, the lowest value is rated as 100 and B is calculated as:

$$B = \text{scaled property} = \frac{\text{minimum value in the list} * 100}{\text{numerical value of property}} \qquad (20.5)$$

For material properties that can be represented by numerical values, application of the above procedure is simple. However, with properties like corrosion and wear resistance, machinability and weldability, etc., numerical values are rarely given and materials are usually rated as very good, good, fair, poor, etc. In such cases, the rating can be converted to numerical values using an arbitrary scale. For example, corrosion resistance rating: excellent, very good, good, fair and poor can be given numerical values of 5, 4, 3, 2 and 1 respectively.

$$\text{Material performance index} = \gamma = \sum_{i=1}^{n} B_i \alpha_i \qquad (20.6)$$

where i is summed over all the n relevant properties.

In the cases where numerous material properties are specified and the relative importance of each property is not clear, determinations of the weighting factors (α) can be largely intuitive, which reduces the reliability of selection. This problem can be solved by adopting a systematic approach to the determination of α. The digital logic approach described for the cost-effectiveness analysis, Section 17.7, can be used as a systematic tool to determine α. To increase the accuracy of decisions based on the digital logic approach, the yes–no evaluations can be modified by allocating gradation marks ranging from 0 (no difference in importance) to 3 (large difference in importance). In this case, the total gradation marks for each selection criterion are reached by adding up the individual gradation marks. The weighting factors are then found by dividing these total gradation marks by their grand total.

Cost (stock material, processing, finishing, etc.) can be considered as one of the properties and given the appropriate weighting factor. However, if there is a large number of properties to consider, the importance of cost may be emphasized by considering it separately as a modifier to the material performance index (γ). In the cases where the material is used for space filling, cost can be introduced on a per unit volume basis. A figure of merit, M, for the material can then be defined as:

$$M = \frac{\gamma}{C * p} \qquad (20.7)$$

where C = total cost of the material per unit weight (stock, processing, finishing, etc.)
p = density of the material.

When an important function of the material is to bear stresses, it may be more appropriate to use the cost of unit strength instead of the cost per unit volume. This is because higher strength will allow less material to be used to bear the load and the cost of unit strength may be a better representative of the amount of material actually used in making the part. In this case, (20.7) is rewritten as:

$$M = \frac{\gamma}{C'} \qquad (20.8)$$

Where C' is determined from Table 20.1 depending on the type of loading.

This argument may also hold in other cases where the material performs an important function like electrical conductivity or thermal insulation. In these cases the amount of the material, and consequently the cost, are directly affected by the value of the property.

The weighted properties method can be used when a new material is considered as a substitute of an existing one. This is done by computing the relative figure of merit, RM, which is defined as:

$$RM = \frac{M_n}{M_e} \tag{20.9}$$

where M_n and M_e are the figures of merit of the new and existing materials respectively. If RM is greater than unity, the new material is more suitable than the existing material.

When the large number of materials with a large number of specified properties are being evaluated for selection, the weighted properties method can involve a large number of tedious and time-consuming calculations. In such cases, the use of a computer would facilitate the selection process. The steps involved in the weighted properties method can be written in the form of a simple computer program to select materials from a data bank. An interactive program can also include the digital logic method to help in determining the weighting factors.

The selection of a material for a large cryogenic storage tank to be used in transporting liquid nitrogen gas will be used to illustrate the use of the weighted properties method. An important rigid requirement for materials used in cryogenic applications is that the material must not suffer ductile–brittle transition at the operating temperature, which is about $-196°C$ ($-320.8°F$) in this case. This rules out all carbon and low-alloy steels and other body centered cubic (BCC) materials (Section 8.5). Face-centered cubic materials (FCC) do not usually become unduly brittle at low temperatures. Many plastics are also excluded on this basis. Processability is another rigid requirement. As welding is normally used in manufacturing metal tanks, good weldability becomes a rigid requirement. Availability of materials in the required plate thickness and size is also another screening factor. As a first step, the performance requirements of the storage tank should be translated into material requirements. In addition to having adequate toughness at the operating temperature, the material should be sufficiently strong and stiff. With stronger material, thinner walls can be used which means lighter tank and lower cool-down losses. Thinner walls are also easier to weld. Lower specific gravity is also important as the tank is used in transportation. Lower specific heat reduces cool-down losses, lower thermal expansion coefficient reduces thermal stresses, and lower thermal conductivity reduces

Table 20.3 Application of digital logic method to cryogenic tank problem

Property	\multicolumn{21}{c}{Decision number}

Property	1	2	3	4	5	6	7	8	9	10	11	12	13	14	15	16	17	18	19	20	21
Toughness	1	1	1	1	1	1															
Yield strength	0						1	0	0	1	1										
Young's modulus		0					0					0	0	0	1						
Density			0					1				1				1	1	1			
Expansion				0					1				1			0			1	1	
Conductivity					0					0				1			0		0		0
Specific heat						0					0				0			0		0	1

heat losses. The cost of material and processing will be used as a modifier to the material performance index, as given in (20.8).

The digital logic method is used to determine the weighting factors. With seven properties to evaluate, the total number of decisions = $N(N-1)/2 = 7\,(6)/2 = 21$. The different decisions are given in Table 20.3.

The weighting factor can be calculated by dividing the number of positive decisions given to each property by the total number of decisions. This gives the following weighting factors given in Table 20.4.

Toughness is given the highest weight followed by density. The least important properties are Young's modulus, thermal conductivity and specific heat. Other properties are in between.

Table 20.4 Weighting factors for cryogenic tank

Property	Positive decisions	Weighting factors
Toughness	6	0.28
Yield strength	3	0.14
Young's modulus	1	0.05
Density	5	0.24
Thermal expansion coefficient	4	0.19
Thermal conductivity	1	0.05
Specific heat	1	0.05
Total	21	1.00

The properties of a sample of the candidate materials are listed in Table 20.5. The yield strength and Young's modulus correspond to RT which is conservative as they generally increase with decreasing temperature.

The next step in the weighted properties method is to scale the properties given in Table 20.5. For the present application, materials with higher mechanical properties are more desirable and highest values in toughness, yield strength and Young's modulus are considered as 100 and other values in Table 20.5 are rated in proportion. On the other hand, lower values of specific gravity, thermal expansion coefficient, thermal conductivity and specific heat are more desirable in this case. Accordingly, the lowest values in the table were considered as 100 and other values rated in proportion according to (20.5). The scaled values are given in Table 20.6. The table also gives the performance index which is calculated according to (20.6).

The performance index shows the technical capability of the material without regard to the cost. In this case, stainless steels are the optimum materials. It now remains to consider the cost aspects by calculating the figure of merit, M. In the present case it is more appropriate to use (20.8), as the primary function of the tank material is to bear stresses. The formula for a thin-wall pressure vessel is given in Table 20.1 as:

$$\text{Cost of unit strength} = \frac{C * p}{S}$$

where S is the yield strength.

The values of the relative cost, cost of unit strength, performance index, figure of merit M and the ranking of the different materials is shown in Table 20.7. The results show that full hard stainless steel grade 301 is the optimum material followed by A1 2014-T6.

Table 20.5 Properties of candidate materials for cryogenic tank

Material	1 Toughness index[a]	2 Yield strength (MPa)	3 Young's modulus (GPa)	4 Specific gravity	5 Thermal expansion[b]	6 Thermal conductivity[c]	7 Specific heat[d]
Al 2014-T6	75.5	420	74.2	2.8	21.4	0.37	0.16
Al 5052-O	95	91	70	2.68	22.1	0.33	0.16
SS 301-FH	770	1365	189	7.9	16.9	0.04	0.08
SS 310-3/4H	187	1120	210	7.9	14.4	0.03	0.08
Ti-6 Al-4 V	179	875	112	4.43	9.4	0.016	0.09
Inconel 718	239	1190	217	8.51	11.5	0.31	0.07
70 Cu-30 Zn	273	200	112	8.53	19.9	0.29	0.06

[a] Toughness index, TI, is based on UTS, yield strength Y, and ductility e, at $-196°C$ ($-321.8°F$). $TI = (UTS + Y)e/2$.
[b] Thermal expansion coefficient is given in $10^{-6}/°C$. The values are average between RT and $-196°C$.
[c] Thermal conductivity is given in $cal/cm^2/cm/°C/s$.
[d] Specific heat is given in $cal/g/°C$. The values are average between RT and $-196°C$.

Table 20.6 Scaled values of properties and performance index

Material	Scaled properties							Performance index (γ)
	1	2	3	4	5	6	7	
Al 2014-T6	10	30	34	96	44	4.3	38	42.2
Al 5052-O	12	6	32	100	43	4.8	38	40.1
SS 301-FH	100	100	87	34	56	40	75	70.9
SS 310-3/4H	24	82	97	34	65	53	75	50.0
Ti-6 Al-4 V	23	64	52	60	100	100	67	59.8
Inconel 718	31	87	100	30	82	5.2	86	53.3
70 Cu-30 Zn	35	15	52	30	47	5.5	100	35.9

Table 20.7 Cost, figure of merit and ranking of candidate materials

Material	Relative cost[a]	Cost of unit strength ⋆ 100	Performance index	Figure of merit	Rank
Al 2014-T6	1	0.67	42.2	62.99	2
Al 5052-O	1.05	3.09	40.1	12.98	6
SS 301-FH	1.4	0.81	70.9	87.53	1
SS 310-3/4H	1.5	1.06	50.0	47.17	3
Ti-6 Al-4 V	6.3	3.20	59.8	18.69	4
Inconel 718	5.0	3.58	53.3	14.89	5
70 Cu-30 Zn	2.1	8.96	35.9	4.01	7

[a] The costs include stock material and processing cost. The relative cost is obtained by considering the cost of Al 2014 as unity and relating the cost of other materials to it.

In the above procedure, the strength and density were considered twice, once in calculating the performance index (γ) and another time in calculating the cost of unit strength. This procedure may have overemphasized their effect on the final selection. This could be justifiable in this case as higher strength and lower density is advantageous from the technical and economic points of view.

20.7 INCREMENTAL RETURN METHOD

This method is suitable for materials substitution studies where new improved materials or processes are introduced to provide better properties or service at an increased cost. In such cases, the benefit-cost analysis described in Chapter 17 can be used to evaluate the new material or process. The first step is to calculate the performance index of the new material (γ_n), as described in Section 20.6, and to compare it with the performance index of the presently used material (γ_b) which is considered as the basis. The incremental relative performance index is taken as:

$$\Delta\gamma_r = \frac{\gamma_n - \gamma_b}{\gamma_b} \qquad (20.10)$$

The next step is to calculate relative incremental cost, which is taken as:

$$\Delta C_r = \frac{C_n - C_b}{C_b} \qquad (20.11)$$

where C_n and C_b are the total costs of the new and basis materials respectively.

If $\Delta\gamma_r/\Delta C_r$ is greater than unity, the new material gives better value and the substitution is beneficial. When more than one new material is available, they are arranged in the order of increasing cost, $n_1, n_2, n_3, \ldots$, etc. The comparison is made between the basis material and n_1. If n_1 is better, the comparison then proceeds between n_1 and n_2. If n_2 gives better value, n_1 is rejected and the comparison continues between n_2 and n_3. If n_2 gives better value, n_3 is rejected and the comparison continues between n_2 and the next material. The procedure is repeated until all the alternatives have been compared and the surviving material is the optimum.

To illustrate the use of this method consider the case of cryogenic tank discussed in Section 20.6. The results of the analysis show that SS 301-FH is the optimum material and is, therefore, used in making the tank. Suppose that at a later date a new fiber reinforced material is introduced and it is proposed to manufacture the tank from the new material by the filament winding technique. The properties of the new fiber reinforced material are given in Table 20.8 together with the properties of SS 301-FH.

Following the procedure in Section 20.6, the properties are first scaled. Using the same weighting factors as in Table 20.4, the performance index is calculated and the results are given in Table 20.9.

Following the same procedure of Section 20.6, the cost of unit strength will be used, as shown in Table 20.10.

Table 20.8 Properties of candidate materials for cryogenic tank[a]

Material	1 Toughness index	2 Yield strength	3 Young's modulus	4 Specific gravity	5 Thermal expansion	6 Thermal conductivity	7 Specific heat
Composite	175	1500	200	2.0	12	0.005	0.1
SS 310-FH	770	1365	189	7.9	16.9	0.04	0.08

[a]For definition and units of properties see Table 20.5.

Table 20.9 Scaled values of properties and performance index

Material	Scaled properties							Performance index (γ)
	1	2	3	4	5	6	7	
Composite	23	100	100	100	100	100	80	77.4
SS 301-FH	100	91	95	25	71	12.5	100	70.6

Table 20.10 Cost, and cost of unit yield strength, of candidate materials

Material	Relative cost	Cost of unit strength $\star 100$
Composite	5	0.67
SS 301-FH	1	0.57

From (20.10), the incremental relative performance index is:

$$\Delta\gamma_r = \frac{77.4 - 70.6}{70.6} = 0.096$$

From (20.11), the relative incremental cost is:

$$\Delta C_r = \frac{0.67 - 0.57}{0.57} = 0.175$$

$$\frac{\Delta\gamma_r}{\Delta C_r} = \frac{0.096}{0.175} = 0.549$$

As the ratio $\Delta\gamma_r/\Delta C_r$ is less than unity, the basis material still gives better value than the new material and no substitution is required. If, however, the increasing use of the new material causes its cost to decrease to 4.5 times the cost of stainless steel, the cost of unit strength becomes 0.60 instead of 0.67. In this case, ΔC_r becomes 0.053 and the ratio $\Delta\gamma_r/\Delta C_r$ becomes greater than unity which means that the new material gives better value.

20.8 LIMITS ON PROPERTIES METHOD

In the limits on properties method, the performance requirements are divided into three categories: lower limit properties, upper limit properties, and target value properties. For example, if it is desired to have a strong light material, a lower limit on the strength and an upper limit on the density are specified. When compatibility between materials is important, a target value for the thermal expansion coefficient or for the position in the galvanic series may be specified to control thermal stresses or galvanic corrosion respectively. Whether a given property is specified as an upper limit or lower limit may depend upon the application. For example, when selecting materials for an electrical cable, the electrical conductivity will be specified as a lower limit property for the conductor and as an upper limit property for the insulation.

The limits on properties method is usually suitable for optimizing material and process selection when the number of possible alternatives is relatively large. This is because the limits which are specified for the different properties can be used for eliminating unsuitable materials from a data bank. The remaining materials are those whose properties are above the lower limits, below the upper limits and within the limits of target values of the respective specified requirements. After the screening stage, the limits on properties method can then be used to optimize the selection from among the remaining materials.

As in the case of the weighted properties method, each of the requirements or properties is assigned a weighting factor, α, which can be determined using the digital logic method, as discussed earlier. A merit parameter, m, is then calculated for each material according to the relationship:

$$ m = \left[\sum_{i=1}^{n_l} \alpha_i \frac{Y_i}{X_i} \right]_l + \left[\sum_{j=1}^{n_u} \alpha_j \frac{X_j}{Y_j} \right]_u + \left[\sum_{k=1}^{n_t} \alpha_k \left| \frac{X_k}{Y_k} - 1 \right| \right]_t \qquad (20.12) $$

where l, u, and t stand for lower limit, upper limit, and target value properties respectively.
n_l, n_u and n_t are the numbers of lower limit, upper limit and target value properties.
α_i, α_j and α_k are the weighting factors for the lower limit, upper limit and target value properties.
X_i, X_j and X_k are the candidate material lower limit, upper limit and target value properties.
Y_i, Y_j and Y_k are the specified lower limits, upper limits and target values.

According to (20.12), the lower the value of the merit parameter, m, the better the material.

As in the weighted properties method, the cost can be considered in two ways:

1. Cost is treated as an upper limit property and given the appropriate weight. When the number of properties under consideration is large, this procedure may obscure its importance.
2. Cost is included as a modifier to the merit parameter as follows:

$$ m' = \frac{CX}{CY} m \qquad (20.13) $$

where CY and CX are the specified cost upper limit and the candidate material cost.
m is the merit parameter calculated without taking the cost into account.

In this case the material with the lowest cost-modified merit parameter, m', is the optimum.

The following example will be used to illustrate the use of the limits on properties method in selection. Consider the case of selecting an insulating material for a flexible electrical conductor for a computer system. Space saving and adaptability to special configurations are important. Service temperature will not exceed 75°C (167°F). Cost is an important consideration because large quantities of these cables will be used in installing the system. Rigid requirements in this case are flexibility, or ductility, of the insulating material and operating temperature. The requirement for ductility eliminates all ceramic insulating materials and the operating temperature eliminates some plastics like low-density polyethylene. The next step is to analyze the electrical and physical design requirements. These are:

1. Dielectric strength which is related to the breakdown voltage is a lower limit property in this case. Due to space limitations, the dielectric strength should be more than 10 000 V/mm.

2. Insulating resistance depends on both the resistivity of the material and geometry of the insulator and is a lower limit property. The minimum acceptable value in this case is 10^{14} ohm/cm.
3. Dissipation factor affects the power loss in the material due to the alternating current and is an upper limit property. The maximum acceptable value in this case is 0.0015 at 60 Hz.
4. Dielectric constant is a measure of the electrostatic energy stored in the material and affects the power loss. This property is an upper limit requirement in this case, although it is taken as a lower limit property in applications like capacitors. The maximum allowable value in this case is 3.5 at 60 Hz.
5. Thermal expansion coefficient is a target value to ensure compatibility between the conductor and insulator at different temperatures. As the conductor is made of aluminum, the target value for the expansion coefficient is $2.3 * 10^{-5}/°C$.
6. Specific gravity is an upper limit property to ensure light weight. This property will not be considered here because weight is not critical.

Table 20.11 gives the properties of some candidate materials which do not violate the rigid requirements of ductility and operating temperature and also satisfy the upper limits or lower limits of design.

The first step is to determine the weighting factors for the different properties. Because the number of properties is relatively small, the cost will be included as one of the properties. The digital logic method is used as shown in Table 20.12. The number of properties under consideration is 6 and the total number of decisions $= 6(6-1)/2 = 15$.

Table 20.11 Properties of some candidate insulating materials

Material	Dielectric strength (V/mm)	Volume resistance (ohm/cm)	Dissipation factor (60 Hz)	Dielectric constant (60 Hz)	Thermal expansion $10^{-5}/°C$	Relative cost[a]
PTFE	14 820	10^18	0.0002	2.1	9.5	4.5
CTFE	21 450	10^18	0.0012	2.7	14.4	9.0
ETFE	78 000	10^16	0.0006	2.6	9.0	8.5
Polyphenylene oxide	20 475	10^17	0.0006	2.6	6.5	2.6
Polysulfone	16 575	10^14	0.001	3.1	5.6	3.5
Polypropylene	21 450	10^16	0.0005	2.2	8.6	1.0

[a] Cost includes material and processing cost. Relative cost is based on the cost of material and processing for polypropylene.

Table 20.12 Weighting factors for an electrical insulator

Property	Decision number															Total	Weighting factor
	1	2	3	4	5	6	7	8	9	10	11	12	13	14	15		
Dielectric strength	0	1	1	0	1											3	0.20
Volume resistance	1					1	1	1	1							5	0.33
Dissipation factor		0				0				1	1	0				2	0.13
Dielectric constant			0				0			0			1	0		1	0.07
Thermal expansion				1				0			0		0		0	1	0.07
Cost					0				0			1		1	1	3	0.20
																15	1.0

Table 20.13 Evaluation of insulating materials

Material	Merit parameter (m)	Rank
PTFE	0.78	3
CTFE	1.07	6
ETFE	0.81	5
Polyphenylene oxide	0.66	1
Polysulfone	0.78	3
Polypropylene	0.66	1

The next step in the selection process is to calculate the merit parameter m using (20.12) for the different materials. In the present case the dielectric strength and volume resistance are lower limit properties, dissipation factor, dielectric constant and cost are upper limit, and thermal expansion coefficient is a target value. In calculating the relative merit of the different materials, the log value of the volume resistivity was used. As no upper limit is given to the cost, the cost of the most expensive material in Table 20.11 is taken as the upper limit. The relative merit parameter m and the rank of the different materials is given in Table 20.13.

The results show that polypropylene and polyphenylene oxide have equal merit parameter. The final selection between the two materials may depend on availability, possibility of coloring, etc.

20.9   COMPUTER AIDED MATERIALS AND PROCESS SELECTION

As discussed earlier, the materials and process selection starts by narrowing down the range of available alternatives to a manageable number. This preliminary selection of materials and processes can be a tedious task if performed manually from handbooks and supplier catalogs. This difficulty has prompted the introduction of several computer-based systems for materials and/or process selection, as shown in the bibliography. As illustrative examples, the systems proposed by Dargie *et al.* (MAPS-1) and by Swindells (PERITUS) (see bibliography) will be briefly described here.

MAPS-1 system

Dargie *et al.* proposed a part classification code similar to that used in group technology for their computer aided preliminary selection system, MAPS-1. The first five digits of the code are related to the elimination of unsuitable manufacturing processes. The first digit is related to the batch size and varies between 0, for 10 parts and less; and 9, for over 300 000 parts. The second digit characterizes the bulk and varies between 0 and 9 depending on the major dimension and whether the part is long, flat or compact. The third digit characterizes the shape and also varies between 0 and 9. Shapes are classified on the basis of being prismatic, axisymmetric, cup-shaped and nonaxisymmetric and nonprismatic. The fourth digit is related to tolerance with values between 0 and 9 depending on the tolerance and nominal dimensions of the part. The fifth digit is related to surface roughness with 0 given to the roughest surface and 9 given to the smoothest.

The next three digits of the MAPS-1 code are related to the elimination of unsuitable materials. The sixth digit is related to service temperature and varies between 0, for temperatures less than $-100°C$, and 9 for temperatures above $1100°C$. The seventh digit is related to the acceptable corrosion rate with 0 given to rate over 1 mm/year and 3 for rates less than 0.1 mm/year. The eighth digit characterizes the type of environment to which the part is exposed.

The system proposed by Dargie *et al.* uses two types of data bases for preliminary selection: the suitability matrices and the compatibility matrix. The suitability matrices deal with the suitability of processes and materials for the part under consideration. Each of the code digits has a matrix. The columns of the matrix correspond to the value of the digit and the rows correspond to the processes and materials in the data base. The elements of the matrix are either 0, indicating unsuitability, or 2 indicating suitability. The compatibility matrix expresses the compatibility of the different combinations of processes and materials. The columns of the matrix correspond to the materials while the rows correspond to the processes. The elements of the matrix are either 0 for incompatible combinations, 1 for difficult or unusual combinations, or 2 for combinations used in usual practice. Based on the part code, the program generates a list of candidate combinations of materials and processes to produce it. This list helps the designer to identify possible alternatives early in the design process and to design for ease of manufacture.

The above procedure can be performed on a purely qualitative basis with no need to specify actual values of the properties. When the requirements are easy to meet, the resulting list of possible materials and processes may be too large for detailed analysis and optimization. In such cases, it may be necessary to perform an intermediate selection process where some quantitative values of the properties are used to eliminate some of the alternatives and to reduce the list to a manageable number. For example, when the weight of a highly-stressed structure is important, it may be necessary to screen the materials on the basis of a specified minimum specific strength (strength/density) or specific stiffness (Young's modulus/density). This intermediate selection process may be combined with the optimization stage of selection, as was discussed in the limits on properties method of Section 20.8.

PERITUS system

Swindells and Swindells (see bibliography) proposed a knowledge-based system for use in selection of materials at the innovation stage of design when the general feasibility of a new design concept needs to be established. The system is based on the observation that material properties, manufacturing route and shape, dimensions and failure mode of a component are interactive variables. The system has three main stages:

1. The director stage, where the user of the system can start from knowing in general outline only what has to be designed and what it has to do. Inputs in relation to component category (pipe, pressure vessel, gear, bolt, bearing, spring, etc.), shape category (complex solid, hollow concentric, hollow nonconcentric, cup, cone, spiral, flat, etc.) and operational factors (low temperatures, high temperatures, environment resistance, etc.) are used to broadly select possible groups of materials. Subsequent steps in the director stage examine the materials characteristics (toughness, stiffness, strength, density, cost, etc.) and process characteristics (quality, size, precision, complexity, cost, etc.) help the designer identify materials classes and fabrication processes which should suit his requirements in general terms.

2. The pre-sort stage provides a detailed match between the functional requirements and the material characteristics for a preferred manufacturing route in order to generate a short list of candidate materials. The user specifies the level of importance for each general characteristic, such as weldability, corrosion resistance, etc., on a scale from 0 (not important) to 3 (very important). The same procedure is used to specify the required manufacturing route, such as extrusion, sand casting, etc., depending on which processes are appropriate for the particular class of materials. A branching algorithm then identifies which materials in the class meet or exceed these requirements.

3. Evaluation and optimization stage evaluates the materials in the short list from the pre-sort stage using two alternative approaches. The simpler approach gives a priority order to the materials in the short list by comparing each one against a set of standard values chosen by the user. The second approach requires a more complex optimization process involving failure modes, component dimensions and the properties of the materials in the short list. Each engineering situation requires a separate module, such as beams in bending, tubes in torsion, etc.

An important feature of the PERITUS system is that it presents a short list of possible solutions for the user to make a final decision or to carry out more detailed investigation if required.

20.10 MATERIALS DATA BASES

As discussed in previous sections, computers can be of great help in solving many of the problems associated with selection. A major part of any computer aided system for selection is the materials properties data bank and in recent years many data bases have been established in various countries and by different organizations. Examples include the high-temperature data bank of the Commission of the European Communities and the Metsel system which is developed by the American Society for Metals, ASM (see bibliography). As the latter system is developed for microcomputers and is easily available at an affordable price it will be discussed later in more detail.

An obvious advantage of a computerized system of storage and retrieval of materials properties is that it replaces the multiplicity of handbooks, card files, manufacturer catalogs, etc. and provides the user with a single source for most information. The simplest level of data storage is a file containing m properties of n materials. At a higher level of sophistication, the computerized data base providing a selective listing of materials that have certain properties is useful. This facility corresponds to the role of an automatic filing clerk. However, using the data base as a mere computerized tabulation of data would be only a marginal improvement on the traditional sources of information. The computerized data base should be capable of providing other services to the designer and the manufacturing engineer. For example, the system should provide graphics, file inversion or profile matching and unit conversion. It is also desirable that the system should be able to manipulate the raw data into various formats appropriate to the required levels of detail and purpose. Integrating the data base with design algorithms and CAD/CAM programs will have many benefits including homogenization and sharing of data in the different departments, decreased redundancy of effort and decreased cost of information storage and retrieval.

Data banks can be classified into two main categories:

1. General, or centralized, banks which contain all the available information. Both standard handbook data and those experienced by the users themselves through production, tests and failure analysis are included. Such data banks are large and complex and are not economic for small industries. Generally the cost of supplying new data and updating existing information is small compared with the total cost of the system.
2. Local data banks which are designed to satisfy the needs, especially of a certain group or small company. Such banks are easier to initiate and use than the centralized ones. However, one of the difficulties with local data banks is to define which data is to be included. Data that seems to be of no interest at present may be important in the future. Another problem is that the expense of supplying new data and updating the existing information may be beyond the means of the group. Connecting the two types of bank makes it possible to update the information at the central bank and to transfer it periodically to the local one.

In some cases the required information is not available in textbooks, supplier's catalogs or company standards. Ideally, the required information should be generated by testing prototypes or laboratory samples. In many cases, however, time and cost limitations do not allow testing and the required information has to be obtained by interpolation or extrapolation of existing results. Although many data banks have sufficient mathematical and statistical facilities for generating new data from existing information, the results are not as reliable as data generated experimentally. This problem can be overcome by using the appropriate factor of safety or derating factor, as discussed in Chapter 9.

ASM METSEL software system

The Metal Selector software was introduced by ASM in 1985 as a companion to *Metals Handbook*, desk edition. In addition to an expandable data base, the system provided graphics, unit conversion and the possibility of communication with ASM's bibliographical data base. An enhanced version, MetSel/2 was introduced in 1987. MetSel/2 is a menu-driven materials data base management program that allows the user to screen, or sift through, the data base to eliminate unlikely candidates and to narrow down the field to a small number of possible materials.

In addition to the basic data base that comes with the system, specialized data bases compiled by ASM and other societies and organizations participating in the MetSel/2 project can be added to widen the scope of selection. The user can also add his own materials information to the system. In addition to tabulated information, graphs such as end-quench hardenability bands, fatigue curves, or tempering curves are also included. The interactive nature of the program helps the user to define and redefine the selection criteria to isolate gradually the materials that meet the requirements. The selection criteria can combine properties, processability, chemical composition, product forms, material groups, manufacturer and country of origin.

The selection process starts with the whole range of materials in the data base. A variety of sift criteria are then used to define the requirements and to direct the search. These criteria include:

1. Sifting by designations. UNS numbers, AISI numbers, common names, material

group, country of origin and user identification are among the designations that can be used to sift the data base.

2. Sifting by specifications allows the operator to select the materials that are acceptable to organizations like ASTM and SAE.
3. Sifting by composition allows the operator to select the materials that have certain min. and max. values of alloying elements.
4. Sifting by forms using options like rod, wire, sheet, tube, cast, forged, welded, etc.
5. Sifting by class, e.g. fatigue resistant, corrosion resistant, heat resistant, electrical materials, etc.
6. Sifting by rankings allows the operator to eliminate materials that do not conform to certain levels of machinability, weldability, formability, availability, processing cost, etc.
7. Sifting by properties, where selection can be performed according to specified numeric values of a set of properties.

More than one of the above sifting criteria can be used to identify suitable materials. Sifting can be performed in the AND or OR modes. The AND mode narrows the search, since the material has to conform to all the specified criteria. The OR mode broadens the search, since materials that satisfy any of the requirements are selected.

The number of materials that survive the sifting process depends on the severity of the criteria used. At the start of sifting, the number of materials shown on the screen is the total in the data base. As more restrictions are placed on the materials the number of surviving materials gets smaller and could reach 0, i.e. no materials qualify. In such cases, some of the restrictions have to be relaxed and the sifting restarted. This interactive mode of operation is useful in the initial stages of design as it helps the designer in determining the limitations of the part under consideration. Ideally, a limited number of materials survives the sifting process. A hard copy of characteristics of these materials is then printed. The hard copy includes designations, specifications, composition, forms, properties, temper, processability, cost, etc. This information will then be used to optimize the selection, as described in earlier sections of this book.

20.11   SOURCES OF INFORMATION ON MATERIALS PROPERTIES

One important requisite to successful materials and process selection is a source of reliable and consistent data on materials properties. There are many sources of information which include governmental agencies, trade associations, engineering societies, textbooks, research institutes and materials producers. Table 20.14 lists some of the widely available sources of information on materials properties. An article published in *Materials Engineering*, **11** (1975), pp. 20–32, lists many addresses for materials information sources in the USA.

Assembling, analyzing and classifying the very large amount of available data on materials properties into a useful form is not easy. Usually, tabulated data give a single value for each property and it must be assumed that the value is typical. When the scatter is large, maximum and minimum values may be given. It is rare, however, to find a property data presented in a statistical manner by a mean value and the standard deviation. Using such widely different information for design and selection needs judgement and understanding of the nature and properties of materials and their processing. For applications

Table 20.14 Selected sources of information on materials properties

Source	Description
Materials selector	A yearly issue of *Materials Engineering*
	Contains tabulated data and advertisements on common engineering materials
Volumes 1, 2 and 3 of Metals Handbook	Properties and selection of metals
	American Society for Metals, ASM, USA
	Contains large amount of data on metals and alloys in tabulated and graphical form
ASM Metals reference book	Provides extensive information on metals and metalworking in tabular and graphical forms
Databook	A yearly issue of *Metal Progress*. Contains tabulated data on materials and processes
SAE Handbook	Annually by the Society of Automotive Engineers, USA. Contains information on materials and components
ASTM Standards	American Society for Testing and Materials
	An authoritative source of information on materials
ASTM information retrieval system	Available free of charge from ASTM information center
ASM Computerized database	Comprehensive system for access to metals literature from 1966 to present day
American metal market newspaper	Provides costs of metallic materials
American machinist manufacturing cost guide	Provides a good source for processing cost estimation
Engineering properties of selected ceramic materials	American Ceramic Society, Columbus, Ohio
Fulmer materials optimizer	Fulmer Research Institute, England. Information system for the selection of engineering materials (metals and nonmetals). Cost data and processing are also included
Alloy index of Metals Abstracts	Metals Society UK. Lists all references to specific alloys mentioned in *Metals Abstracts*
Rapra Technology Ltd, Shrewsbury, UK	Abstracts, review reports, new trade names, on rubber and plastics, Pergamon Press, London
Ross, R. B.	E&FN Spon Ltd, London
	Metallic materials specification handbook
Modern plastics encyclopedia	McGraw-Hill Book Company. A yearly issue of *Modern Plastics*
Encyclopedia of polymer science and technology	John Wiley and Sons Inc. vol 1–16, 1964–72

where reliability is critical, the behavior of the materials under the expected service conditions is usually determined experimentally using prototypes and laboratory tests.

20.12 REVIEW QUESTIONS AND PROBLEMS

20.1 What are the main functional requirements and corresponding material properties for the following products:
   (a) Milk containers.
   (b) Gas turbine blades.
   (c) Sleeve for sliding journal bearing.
   (d) Piston for an internal combustion engine.
   (e) Air plane wing structure.

20.2 It is required to design and select materials for a suitcase for traveling on air lines:
   (a) What are the main structural elements of the suitcase?
   (b) What are the main functional requirements of each structural element?
   (c) Translate the functional requirements into material properties.
   (d) Give weighting factors to the different properties.
   (e) Suggest possible materials for each structural element of the suitcase.

20.3 It is required to design and select materials for an overhead pedestrian crossing to connect two parts of a company.
  (a) What are the main design features and structural elements?
  (b) What are the main functional requirements of each structural element?
  (c) What are the corresponding material requirements for each element?
  (d) Use the digital logic method to determine the relative importance of each property for the different structural elements.
  (e) Recommend the possible materials that may be used as candidates for the final selection of the optimum material for each structural element.

20.4 The following three materials are being considered for manufacturing a welded structure in an industrial atmosphere. The structure is expected to be subjected to alternating stresses in addition to the static loading.

Material	A	B	C
Relative weldability (0.15)	5	1	3
Relative tensile strength (0.15)	3	5	2
Relative fatigue strength (0.25)	5	3	3
Relative corrosion resistance (0.20)	3	5	3
Relative cost (0.25)	2	5	3

Taking the weighting factors for the different properties as shown in brackets after each property, what would be the best material? [Answer: material B]

20.5 If the weighting factors used in question 20.4 are changed to 0.25, 0.10, 0.2, 0.25, and 0.2 for weldability, tensile strength, fatigue strength, corrosion resistance and cost respectively, find the new optimum material. [Answer: material A]

20.6 A walking aid is a device for helping people who have lost partial control of their legs. Users of walking aids usually have at least 25 percent of their leg power, otherwise a wheelchair will be necessary. Figure 20.1 shows an example of a walking aid in common use by elderly people and those recovering from illness.
  In order to perform its function effectively, the walking aid should fulfill the following requirements:
1. The walking aid should be adjustable in order to serve users of different heights. It is recommended that the height of the walking aid be adjusted between 700 and 1000 mm (c. 27 and 40 in).

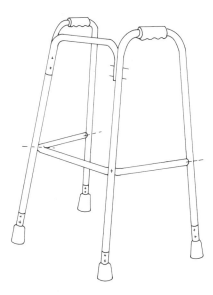

Figure 20.1 An example of a walking aid.

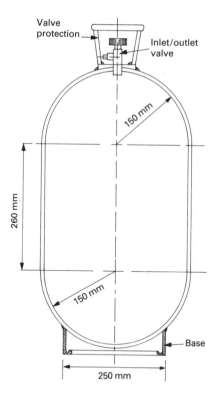

Figure 20.2 Shape and main dimensions of the butane container discussed in the case study.

2. In view of the nature of the health of the users, it is recommended that the total weight of the walking aid should not exceed 2 kg (4.4 lb).
3. As the company proposing to make the walking aid depends on mail orders for a large proportion of its sales, the design should allow easy packaging and assembly by the users.
4. As failure by buckling under the user's weight may have serious consequences, special effort should be made to avoid this mode of failure.
5. Cost is an important consideration.

It is required to design the walking aid and select the appropriate materials and processes for its manufacture.

20.7 Liquified butane gas is frequently used as a source of energy on camp sites and remote settlements. The liquified gas is usually supplied in portable containers of various sizes. In the present case study it is required to perform the necessary design calculations and to select the optimum material for a container of the inner dimensions shown in Fig. 20.2. These dimensions allow the transportation of about 16 kg (c. 35 lb) of butane. The internal pressure in the container is expected to be 2 MPa (290 lb/in^2). As the thickness of the container is expected to be less than one-twentieth of the radius of the container, the container can be treated as a thin-walled cylinder.

The available manufacturing facilities include a variety of sheet metal forming processes, fusion welding methods and metal turning lathes. The weld efficiency can be taken as 70 percent.

A major requirement in the present case is safety. The design should ensure that the container should not burst as a result of general yielding of the material. In addition, brittle fracture as a result of an impact loading should not be allowed. Leakage of the gas can be avoided by applying appropriate quality control techniques. Other requirements for the container include light weight, dent resistance, corrosion resistance and cost. As a rough

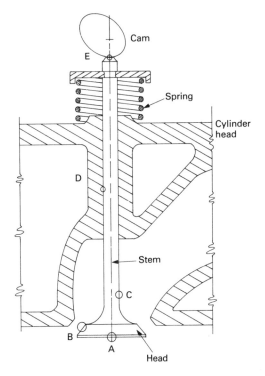

Figure 20.3 Exhaust valve assembly showing the main parts and critical areas.

estimate, the dent resistance can be taken as proportional to $(YS * t^2)$, where YS is yield strength and t is thickness of the material.

20.8 The exhaust valve in the internal combustion engine functions as a high-temperature seal which maintains the high pressure inside the cylinder during the compression and power strokes. Figure 20.3 is a schematic representation of a typical exhaust valve assembly. Good contact between the valve head and cylinder block is essential for efficient functioning of the engine. The exhaust valve, especially the head, is subjected to high temperatures, thermal stresses, wear and corrosive gases. The exhaust valve has several critical locations as shown in Fig. 20.3. The service conditions at the critical areas depend on the type of the engine, size, load and speed. The temperature values given in the following discussion are based on the published results for a V-8 car engine running at 3500 rev/min and full throttle by Kocis, J. F. and Matlock, W. M.; 'Alloy selection for exhaust valves', *Metal Progress*, August, 1975, pp. 58–62.

The highest operating temperatures are encountered in areas A and C and can reach about 810°C (1490°F). Good corrosion resistance is also required in these areas. Increasing the operating temperature causes a reduction in the mechanical strength and an increase in the corrosion rates of the valve. In addition to high temperature and corrosive environment, the valve underhead (area C) experiences cyclic loading as a result of pressure variations inside the cylinder during the different strokes of the piston. For the sake of the present analysis, the maximum pressure which occurs in the cylinder at the beginning of the ignition stroke is assumed to be about 4 MPa.

The valve face (area B) normally operates at a lower temperature, about 700°C (1290°F), because of heat conduction into the valve seat. It must, however, withstand indentation from loose fuel ash deposits and also resist the high temperatures which may occur if contact with the seat is disrupted by ash deposits. This area must also resist the thermal fatigue which results from repeated heating by exhaust gases and cooling by the valve seat.

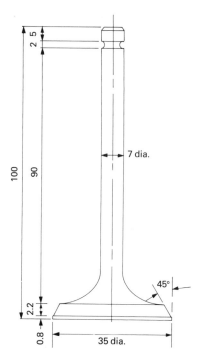

Figure 20.4 Dimensions of the exhaust valve for the case study (mm).

The valve stem (areas D and E) operates at lower temperatures, less than 600°C (1112°F), but they could suffer wear and fatigue failures.

The above discussion shows that the valve has to satisfy several difficult, and in some cases conflicting, requirements. To satisfy these requirements at a practical cost, different areas of the exhaust valve are often made of different materials. The static and fatigue stresses acting on the different parts of the valve normally dictate the required strength of the valve material at the operating temperature.

In the present case study, it is required to estimate the stresses, evaluate the relative importance of the different requirements, and to select the appropriate materials for the different areas of the exhaust valve shown in Fig. 20.4. Similar working conditions to those given above can be assumed. In addition the force exerted by the spring on the stem of the valve can be assumed to vary from 200 N, when the valve is closed, to 0, when the valve is open. This loading cycle occurs once for each two revolutions of the engine.

20.9 Packaging is an important and fast growing industry which utilizes modern design, materials and manufacturing technology. The shape of the package can range from a simple box, as in the case of industrial packages, to a complex design, as in the case of cosmetics. The materials used in making the package cover a wide variety of engineering materials which include paper, wood, glass, metal, plastic and composites of various materials. Manufacturing techniques used in making the package range from manual cutting and assembly to fully automatic processing and filling. The cost of the package can represent a considerable portion of the total cost of the product, especially in the case of consumer products. For example, the cost of packaging is about 30 percent of the selling price of cosmetics, 25 percent for drugs and pharmaceuticals, 20 percent for foods and 10 percent for toys.

A well-designed package in the consumer industry should satisfy the following require-ments:

1. Does not adulterate the contents, especially in the case of food and pharmaceutical packages.

2. Maintains quality of the contents after it has been opened and until the consumer finishes the contents.
3. Protects contents against environment and handling during shipping from manufacturer to consumer.
4. Provides a convenient and efficient means of storage and handling at the wholesaler's warehouse, retailer's store room and consumer's home.
5. Conforms to the specifications of the transportation company or post office.
6. Allows clear labeling and identification of the type, composition and amount of the contents.
7. Provides an attractive visual appearance and a high value as a sales tool.
8. Does not endanger public safety at any stage of its life.
9. Is easy to dispose of and recycle after the contents have been consumed.
10. Has reasonable cost.

The relative importance of the above requirements depends on the contents of the package, expected shelf life, distance between manufacturer and consumer and method of delivery, as well as local and international laws.

In the present case study, a large manufacturer of instant coffee is in the process of reviewing the packaging policy. It is required to analyze the requirements, design the package, select the materials and propose the method of manufacturing of coffee packages for the following cases:

1. Package for a single cup of coffee, 2 g (0.07 oz)
2. Small container for household use, 50 g (1.75 oz)
3. Medium-size container for household use, 100 g (3.5 oz)
4. Large-size container for household use, 200 g (7 oz)
5. Commercial-size container for cafeteria and restaurant use, 1000 g (2.2 lb).

BIBLIOGRAPHY AND FURTHER READING

Alexander, W. O. and Appoo, P. M., 'Material selection: the total concept', *Design Engineering*, November 1977, pp. 59–66.

Boyer, H. E. and Gall, T. L., *Metals Handbook Desk Ed.*, ASM, Ohio, 1985.

Budinski, K., *The Selection Process, Engineering Materials: Properties and Selection*, Reston, N.Y. 1979.

Buffa, E. S., *Modern Production Management*, 4th ed., John Wiley & Sons, New York, 1973.

Crane, F. A. and Charles, J. A., *Selection and Use of Engineering Materials*, Butterworths, London, 1984.

Dargie, P. P. *et al.*, MAPS-1: computer-aided design system for preliminary material and manufacturing process selection, *Trans. ASME, J. Mech. Design*, Vol. 104, 1982, pp. 126–36.

Dieter, G. E., *Engineering Design: a Materials and Processing Approach*, McGraw-Hill, New York, 1983.

Farag, M. M., *Materials and Process Selection in Engineering*, Applied Science Publishers, London, 1979.

First Demonstration Report on the High-Temperature Materials Data Bank of JRC, Commission of the European Communities Report EUR 8817 ENfs.

Gall, T. L., Project Manager: 'Metsel2' Engineered by PSI/Systems, ASM, Ohio 1987.

Gillam, E., 'Materials selection: principles and practice', *The Metallurgist and Materials Technologist*, Sept. 1979, pp. 521–5.

Graham, J., 'Computerized materials properties storage, retrieval, and use', in *Computers in Materials Technology*, ed. T. Ericsson, Pergamon Press, 1981, pp. 3–10.

Gutteridge, P. A. and Turner, J., 'Computer-aided materials selection and design', *Materials and Design*, vol. 3, Aug. 1982, pp. 504–10.

Hanley, D. P., *Selection of Engineering Materials*, Van Nostrand Reinhold, N.Y., 1980.

Heller, M. E., *Metalselector*, ASM, Ohio, 1985.

Kusy, P. F., *Plastics Selection Guide*, ASM-Engineering Bookshelf, Source Book on Materials Selection, vol. II, USA, 1976, pp. 353–83.

Lenel, U. R., *Materials Selection in Practice: The Institute of Metals Handbook*, London, 1986, p. 165.

Olsson, L., Bengtson, U. and Fischmeister, H., 'Computer-aided materials selection', in *Computers in Materials Technology*, ed. T. Ericsson, Pergamon Press, 1981, pp. 17–25.

Ostberg, G., 'A paradox in the development of computerized materials data systems', *Materials and Design*, vol. 5, 1984, pp. 15–19.

Sandstrom, R., 'An approach to systematic materials selection', *Materials and Design*, vol. 6, 1985, pp. 328–38.

Svinning, T., 'A data bank for small and medium-sized industries', in *Computers in Materials Technology*, ed. T. Ericsson, Pergamon Press, 1981, pp. 11–16.

Swindells, N. and Swindells, R. J., 'System for engineering materials selection', *Metals and Materials*, May 1985, p. 301.

Waterman, N. A., 'Materials optimization', *Proc. Conf. on Selection of Materials in Machine Design*, Institution of Mech. Engr., 1973.

Chapter 21

Design and Selection of Materials for a Turnbuckle

21.1   INTRODUCTION

Objective

The objective of this case study is to make the design and to select the appropriate materials and processes for a turnbuckle.

Functional requirements

A turnbuckle is a loop with opposite internal threads in each end for the threaded end of two ringbolts, forming a coupling that can be turned to tighten or loosen the tension in the members attached to the ringbolts. Figure 21.1 shows an assembly of a typical turnbuckle. The turnbuckle is used in different applications involving widely different requirements of forces, reliability and service conditions. Examples include guy wires for telegraph poles, ship rigs, sports equipment and camping gear. The main functional requirements of a turnbuckle are to apply and maintain tensile forces to the members attached to the ringbolts. It should be possible for an operator to release and reapply the tensile forces when needed.

Factors affecting performance in service

The forces acting on the turnbuckle are usually tensile, although fatigue and impact loading can be encountered. Corrosion becomes a problem in aggressive environments, especially if the loop is made from a different material other than the ringbolt material. The possible modes of service failure and their effect on the performance of the turnbuckle are:

1. Yielding of the loop or one of the ringbolts. This will release the tensile forces in the system and could make the operation unsafe.
2. Shearing, or stripping, of threads on the loop or on one of the ringbolts. This will release the tensile forces in the system and would make it impossible to reapply the required forces.
3. Fatigue fracture of the loop or one of the ringbolts. Fatigue fracture could start at any of the points of stress concentration in the turnbuckle assembly.

4. Creep strain in the loop or one of the ringbolts. This will relax the tensile forces in the system and could make operation unsafe.
5. Fracture of the loop or one of the ringbolts. This could take place as a result of excessive loading of the system or as a result of impact loading if materials lose their toughness in service.
6. Corrosion as a result of environmental attack and galvanic action between ringbolt and loop if they are made of widely different materials. Excessive corrosion will make it difficult to apply and release the tension in the system and could reduce the cross-sectional area to dangerous limits. Stress corrosion cracking can also occur in this system.

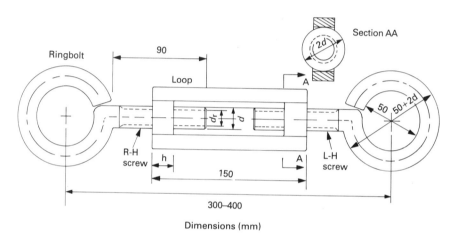

Figure 21.1 Assembly and material-independent dimensions of the turnbuckle in the case study.

One or more of the above failure modes could prove to be critical depending on the materials used in making the turnbuckle components, type of loading and service environment. For example, fatigue is expected to be critical if the load is fluctuating, while creep should be considered for high-temperature service.

21.2   DESIGN OF THE TURNBUCKLE

General design specifications

The tensile force to be applied by the turnbuckle consists of a static component, $L_m = 20$ kN and an alternating component, $L_a = 5$ kN
Inner diameter of the rings at the end of ringbolts $= 50$ mm
Shortest distance between centers of rings on ringbolts $= 300$ mm
Longest distance between centers of rings on ringbolts $= 400$ mm
Other dimensions of the turnbuckle, which are material-independent, are shown in Fig. 21.1
Service environment is industrial atmosphere.

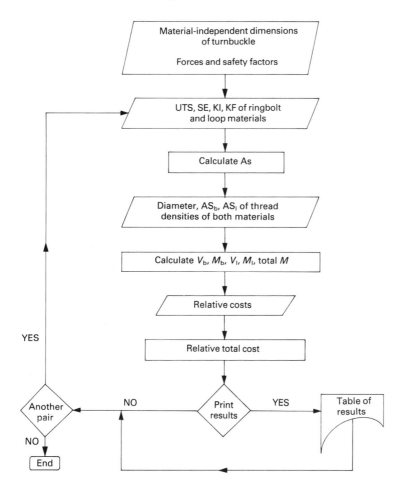

Figure 21.2 Flow chart for turnbuckle program.

Design calculations

Figure 21.1 shows that the threaded length of the ringbolt is critical in view of the reduction in diameter involved in manufacturing the thread and the stress concentration at the roots of the teeth. It will, therefore, be assumed that if the threaded part can carry the service loads the rest of the ringbolt will be safe.

In calculating the axial stress carried by a threaded bolt, an effective cross-sectional area, called tensile stress area (A_s), is used. The tensile stress area for standard metric threads is given by:

$$A_s = 0.25\pi(d - 0.9382p)^2 \qquad (21.1)$$

where d = major diameter of the bolt, see Fig. 21.2
p = pitch of the thread.

Table 21.1 gives the values of A_s for some standard metric threads.

Design for static loading

The tensile stress on the threaded part of the ringbolt, S_t, is given by:

$$S_t = \frac{LK_t}{A_s} = \frac{YS_b}{n_b} \qquad (21.2)$$

where L = Applied load
YS_b = Yield strength of the ringbolt material
n_b = Factor of safety used for the ringbolt calculations
K_t = Stress concentration factor.

If a ductile material is used for making the ringbolt, there will be no need to introduce a stress concentration factor in (21.2) and K_t can be ignored. However, with brittle materials, like cast iron, the stress concentration at the roots of the threads should be taken into consideration.

In the case of the loop, the combined area of the two sides $(2 \star A_1)$ should withstand the applied load L without failure. Thus:

$$\frac{L}{2 \star A_1} = \frac{YS_1}{n_1} \qquad (21.3)$$

where YS_1 = Yield strength of the loop material.
n_1 = Factor of safety used for the loop calculations.

Stripping of threads on the ringbolt will occur when its threads fail in shear at the minor diameter (d_r), Fig. 21.1. The shear stress in the ringbolt threads (τ) is given by:

$$\tau = \frac{L}{AS_b \star h} \qquad (21.4)$$

Where AS_b = Shear stress area per unit length of the ringbolt
h = Length of engagement between the ringbolt and loop.

The values of AS_b for some standard metric threads are given in Table 21.1.
Similarly, stripping of the internal threads in the loop will occur when its threads fail in shear at the major diameter of the ringbolt, d. The shear stress in the loop threads, τ, is given by:

$$\tau = \frac{L}{AS_1 \star h} \qquad (21.5)$$

where AS_1 = Shear stress area per unit length of the loop thread.
The values of AS_1 for some standard metric threads are given in Table 21.1.
As Table 21.1 shows, the shear area of internal threads on the loop is greater than that of the external threads on the ringbolt. This means that, if the ringbolt and loop materials have the same shear strengths, stripping of the ringbolt threads will normally occur before stripping of the internal threads on the loop. For this reason, the material selected for the loop can be weaker than that of the ringbolt.

Table 21.1 Stress areas for some standard metric threads

Major diameter and pitch (mm)	Tensile stress area (As) (mm²)	Thread shear area per mm of engaged threads	
		AS_b (mm²)	AS_1 (mm²)
M4 $\star$ 0.7	8.78	5.47	7.77
M5 $\star$ 0.8	14.20	7.08	9.99
M6 $\star$ 1.0	20.10	8.65	12.20
M8 $\star$ 1.25	36.60	12.20	16.80
M10 $\star$ 1.5	58.00	15.60	21.50
M12 $\star$ 1.75	84.30	19.00	26.10
M14 $\star$ 2	115	22.40	31.00
M16 $\star$ 2	157	26.10	35.60
M18 $\star$ 2.5	192	29.70	40.50
M20 $\star$ 2.5	245	33.30	45.40
M22 $\star$ 2.5	303	37.00	50.00
M24 $\star$ 3	353	40.50	55.00
M27 $\star$ 3	459	46.20	62.00
M30 $\star$ 3.5	561	51.60	69.60
M36 $\star$ 4	817	61.30	84.10
M42 $\star$ 4.5	1120	74.30	99.20

Design for fatigue loading

When the turnbuckle is subjected to fatigue loading, the procedure outlined in Section 11.6 can be used. Equation 11.23 can be used to calculate the ringbolt tensile stress area A_s. Thus:

$$\frac{n_m K_t L_m}{UTS A_s} + \frac{n_a K_f L_a}{S_e A_s} = 1 \tag{21.6}$$

where n_m and n_a are factors of safety for static and fatigue strengths respectively.
L_m and L_a are static and alternating loads respectively.
K_t and K_f are static and fatigue stress-concentration factors respectively. K_t can be ignored for ductile materials.
UTS and S_e are tensile strength and modified endurance limit respectively.

According to (11.21), S_e can be calculated by multiplying the endurance limit of the material by a set of modifying factors. For the present case study, all the modifying factors will be grouped as one modifying factor k_i.

A similar procedure can be used to calculate the length of engagement between the ringbolt and loop h.

For the ringbolt:

$$\frac{n_m K_t L_m}{\tau_{bu} AS_b h} + \frac{n_a K_f L_a}{\tau_{be} AS_b h} = 1 \tag{21.7}$$

where τ_{bu} and τ_{be} are the static and fatigue shear strengths of the ringbolt material respectively.
As in the case of tensile stress calculations, K_t can be ignored for ductile materials.

For the loop:

$$\frac{n_m K_t L_m}{\tau_{lu} A S_l h} + \frac{n_a K_f L_a}{\tau_{le} A S_l h} = 1 \tag{21.8}$$

where τ_{lu} and τ_{le} are the static and fatigue shear strengths of the loop material respectively.

The values of h calculated from (21.7) and (21.8) are compared and the larger one is selected.

For the present case study, the maximum-shear-stress theory will be used to predict the shear strengths of the ringbolt and loop materials. Thus $\tau_u = 0.5$ UTS and $\tau_e = 0.5\, S_e$.

21.3   CANDIDATE MATERIALS AND MANUFACTURING PROCESSES

As shown in Section 21.2, the strength of the loop material need not be as high as the ringbolt material. At the same time, the two materials should not be far apart in the galvanic series in order to avoid failure due to galvanic corrosion, as discussed in Section 7.7. For the present case, the ringbolt and loop materials will be selected from the same group, see Table 7.2.

The ringbolt can be manufactured by several methods including:

1. From bar stock by first threading and then forming the ring by bending. Threading can be done by cutting or rolling.
2. From bar stock by upset forging to form a head, flattening the head and forming the ring by forging, then threading as above.
3. Sand casting, then thread cutting.
4. Shell molding, then thread cutting.
5. Die casting, then thread cutting.

For the present case study, it is assumed that the available facilities favor the first method of manufacturing the ringbolt where a bar stock is threaded by rolling and then bent round a die at room temperature to form the ring. The main processing requirement in this case is ductility. A minimum elongation of 15 percent is assumed to be necessary. This figure can be arrived at from experience with similar products or by performing development experiments.

The loop can be manufactured by several methods, including:

1. Sand casting using wooden or metal pattern, then thread cutting.
2. Shell molding using metal pattern, then thread cutting.
3. Die casting, then thread cutting.
4. Die forging of a bar stock, then thread cutting.
5. Machining from a bar stock, then thread cutting.
6. Welding of the threaded ends to round or square bars.

For the present case study, it will be assumed that the available facilities and the required accuracy and surface finish favor the shell molding method.

Table 21.2 lists the properties of some candidate wrought alloys for the ringbolt and cast alloys for the loop materials. The values of K_f in the table are calculated according to

Table 21.2 Candidate materials for the ringbolt and loop materials

Material	UTS (MPa)	YS (MPa)	Se' (MPa)	ki[a]	q	K_f	p[b]	C[c]	Relative cost[d] Mat.	Mfr.
RINGBOLT MATERIALS (minimum elongation 15%)											
Steels											
AISI 1015	430	329	195	0.7	0.1	1.15	7.8	1	1	1
AISI 1040	599	380	270	0.6	0.2	1.30	7.8	1	1.1	4
AISI 1340	849	567	420	0.5	0.7	2.05	7.8	3	2.2	4
AISI 4820	767	492	385	0.55	0.6	1.90	7.8	3	2.5	8
Aluminum alloys											
AA 3003 O	112	42	56	0.7	0.1	1.15	2.73	4	5	1
AA 5052 O	196	91	90	0.65	0.4	1.60	2.68	4	7	1
AA 6061 O	126	56	55	0.6	0.3	1.45	2.7	3	7	1
Copper-base alloys											
Al bronze	420	175	147	0.7	0.4	1.60	8.1	4	12	6
Si bronze	441	210	175	0.6	0.5	1.75	8.5	4	12	6
70/30 brass	357	133	145	0.75	0.3	1.45	8.5	4	10	4
LOOP MATERIALS (cast alloys)											
Gray CI ASTM A48-74											
grade 20	140	140	70	0.5	0.2	1.30	7.5	4	1.2	1
grade 40	280	280	130	0.45	0.2	1.30	7.5	4	1.25	1
grade 60	420	420	168	0.4	0.2	1.30	7.5	4	1.3	1
Nodular CI ASTM A536											
60-40-18	420	280	210	0.6	0.2	1.30	7.5	4	1.9	4
80-55-06	560	385	280	0.55	0.2	1.30	7.5	4	2	4
120-90-02	840	630	420	0.5	0.2	1.30	7.5	4	2.1	4
Aluminum alloys											
AA 208.0	147	98	44	0.6	0.4	1.60	2.8	3	5	1
AA 356.0 T6	231	168	79	0.5	0.5	1.75	2.68	4	5	1
AA B443.0	133	56	40	0.6	0.7	2.05	2.7	4	4	1
Copper-base alloys											
Al bronze	590	210	200	0.55	0.4	1.60	8.1	4	12	6
Si bronze	420	125	120	0.6	0.4	1.60	8.3	4	12	6
Mn bronze	640	340	300	0.5	0.5	1.75	8.3	4	11	6

[a]ki is endurance limit modifying factor.
[b]p is specific gravity.
[c]C is corrosion resistance: 1 = poor, 2 = fair, 3 = good, and 4 = very good.
[d]Relative materials and processing costs are based on the cost of steel AISI 1015 which is taken as unity.

(8.8) using the given values of q and assuming that $K_t = 2.5$, which is reasonable for coarse threads.

21.4   SAMPLE CALCULATIONS

The procedure for calculating the dimensions, weight and cost of the turnbuckle will be illustrated in this section. The calculations will be based on steel AISI 1015 as the ringbolt material and nodular cast iron ASTM A536 60-40-18 as the loop material. According to Table 7.2, these two materials are compatible from the galvanic corrosion point of view.

As the turnbuckle is subjected to combined static and fatigue loading, (21.6) will be used to calculate a preliminary value for A_s. According to Shigley and Mitchell (see bibliography) the factors of safety can be taken as $n_m = 1.5$ for static loading, and $n_a = 3.0$ for fatigue loading. As the ringbolt material is ductile, the static stress concentration factor K_t can be taken as unity. Using the values in Table 21.2 for steel AISI 1015, (21.6) can be written as:

$$\frac{1.5 \star 1 \star 20\,000}{430 \star A_s} + \frac{3.0 \star 1.15 \star 5000}{195 \star 0.7 \star A_s} = 1 \qquad (21.9)$$

which gives,

$$A_s = 196.14 \text{ mm}^2$$

From Table 21.1, it can be seen that the M18 standard metric thread has an A_s of 192 mm² which is close to the calculated A_s. This means that the major diameter of the bolt is 18 mm.

The mass of the ringbolt, w_b

    = density (volume of the straight part + volume of the ring).
    = 603 g.

The next step is to calculate the length of engagement between the ringbolt and the loop using (21.7) and (21.8). According to (21.7), and using the same values n_m, n_a, K_t and K_f as before:

$$\frac{1.5 \star 1 \star 20\,000 \star 2}{430 \star 29.7 \star h} + \frac{3.0 \star 1.15 \star 5000 \star 2}{195 \star 0.7 \star 29.7 \star h} = 1 \qquad (21.10)$$

This gives $h = 13.2$ mm

In the case of the loop, K_t will be taken as unity since the nodular cast iron used is ductile. Factors of safety similar to those used for the ringbolt calculations will be used for the loop. According to (21.8):

$$\frac{1.5 \star 1 \star 20\,000 \star 2}{420 \star 40.5 \star h} + \frac{3.0 \star 1.3 \star 5000 \star 2}{210 \star 0.6 \star 40.5 \star h} = 1 \qquad (21.11)$$

This gives $h = 11.14$ mm, which is smaller than the value given by (21.10).

The larger value of h is taken as the design value.

The two webs that connect the threaded ends of the loop have to resist the combined effects of the static and alternating loads. Equation 21.6 can be used to calculate the total cross-sectional area of the webs, A_w. Using the same factors of safety as above,

$$\frac{1.5 \star 1 \star 20}{420 \star A_w} + \frac{3.0 \star 1.3 \star 5}{210 \star A_w} = 1 \qquad (21.12)$$

This gives $A_w = 164.3$ mm²

The mass of the loop w_1

  = density (volume of threaded ends + volume of webs)
  = 336 g.

Total mass of the turnbuckle = w_{tb}

$\qquad$ = 2 ⋆ weight of ringbolt + weight of loop
$\qquad$ = 1542 g = 1.542 kg.

Relative cost of materials and processing = C_m

  = 2 ⋆ relative cost of each ringbolt + relative cost of loop
  = 2 ⋆ 0.603(1 + 1) + 0.336(1.9 + 4)
  = 4.39 units of relative cost.

21.5   SELECTION OF OPTIMUM MATERIALS

In order to arrive at the optimum pair of materials for the turnbuckle, the calculations given in Section 21.4 should be repeated for all possible combinations of ringbolt materials and loop materials given in Table 21.2. For example, each one of the three aluminum ringbolt alloys represents a possible candidate for each one of the three aluminum loop alloys, which involves nine combinations. On the other hand, aluminum alloys do not represent possible candidates for copper-base or ferrous alloys as they are too far apart in the galvanic series, which could cause galvanic corrosion. From Table 21.2, it can be shown that there are 42 possible materials combinations (24 combinations for ferrous alloys, nine combinations for aluminum alloys, and nine combinations for copper alloys). It would be tedious and time consuming to perform the required calculations manually. A computer program has been written in BASIC for this purpose. The flow chart of the program is given in Fig. 21.2 and the program listing is given in Fig. 21.3. For each possible pair of materials, the program calculates the total weight of turnbuckle and its relative cost. A sample of the computer output for the AISI 1015 bolt–ASTM A536 60-40-18 loop, discussed in Section 21.4, is shown in Fig. 21.4.

In order to select the optimum pair of materials for the turnbuckle under consideration, the weighted properties method which was discussed in Section 20.6 is used. When the corrosion resistances of the two materials in a turnbuckle are different, the lower value is taken to represent the corrosion resistance of the turnbuckle.

For the present case study, the weighting factors are taken as 0.5, 0.3 and 0.2 for the cost, corrosion resistance and weight respectively. The performance index (γ) of a turnbuckle made of a pair of materials is calculated as follows:

γ = 0.5 (scaled relative cost) + 0.3 (scaled corrosion resistance) + 0.2 (scaled total weight)

Scaling was performed such that a turnbuckle with lower weight and cost was given a lower scaled value. To be consistent, material combinations with higher corrosion resistance were given a lower scaled value. With this method of scaling, turnbuckles with lower performance index are preferable to those with higher performance index.

The calculated relative total weight, relative cost, and performance index for 10 turnbuckles with the lowest performance indexes are given in Table 21.3. The results show that, with the present selection criteria, ferrous alloys are preferable. The main reasons for

```
                            PROGRAM FOR DESIGN OF TURNBUCKLE''
 10 CLS
 20 PRINT ''
 30 PRINT ''
 40 PRINT ''
 50 PI = 22/7
 60 INPUT ''STATIC FORCE kN, LM'';LM
 70 INPUT ''ALTERNATING FORCE kN, LA'';LA
 80 PRINT ''MAT-INDEP DIMENSIONS OF TURNBUCKLE:''
 90 INPUT ''       INNER DIA. mm'';ID
100 INPUT ''       LENGTH OF RINGBOLT LEG mm'';RL
110 INPUT ''       LENGTH OF LOOP mm'';LL:PRINT
120 PRINT ''                   CALCULATING TENSILE STRESS AREA, AS'':PRINT
130 INPUT ''STATIC FACTOR OF SAFETY NM'';NM
140 INPUT ''FATIGUE STRENGTH FACTOR OF SAFETY NA'';NA
150 INPUT ''STATIC STRESS-CONC. FACTOR KT'';KT
160 INPUT ''FATIGUE STRESS-CONC. FACTOR FOR RINGBOLT MATERIAL, KFB'';KFB
170 INPUT ''FATIGUE STRESS-CONC. FACTOR FOR LOOP MATERIAL, KFL'';KFL
180 INPUT ''RINGBOLT MAT. END. LIMIT MOD. FACTOR, KIB'';KIB
190 INPUT ''LOOP MAT. END. LIMIT MOD. FACTOR, KIL'';KIL
200 INPUT ''UTS OF SELECTED RINGBOLT MATERIAL, MPa'';UTS
210 INPUT ''SE MODIFIED ENDURANCE LIMIT FOR RINGBOLT, MPa'';SE
220 AS=((NM*KT*LM)/UTS)+((NA*KFB*LA)/(SE*KIB)):PRINT :AS=AS*1000
230 PRINT '' ** TENSILE STRESS AREA, AS = '';AS;''mm2'':PRINT
240 PRINT ''     SELECT STANDARD METRIC THREAT'':PRINT
250 INPUT ''MAJOR DIA. FOR SELECTED STANDARD'';D
260 V=((PI/4)*D^2*RL) + ((PI*PI/4)*D^2*(ID+D)):PRINT
270 INPUT ''SPECIFIC GRAVITY FOR RINGBOLT MATERIAL'';R:PRINT
280 VV=V/1000
290 PRINT ''   **   VOL. OF RINGBOLT = '';V;''cc''
300 WB = R*VV
310 PRINT ''  ==   MASS OF THE RINGBOLT = '';WB;''gm''
320 TBU=UTS/2
330 TBE=SE/2:PRINT
340 INPUT ''UTS FOR LOOP MATERIAL'';UTSL
350 INPUT ''END. LIMIT FOR LOOP, SE'';SEL
360 TLU=UTSL/2
370 TLE=SEL/2
380 INPUT ''FROM STD. ENTER ASB'';ASB
390 INPUT ''FROM STD. ENTER ASL'';ASL
400 HH=((NM*KT*LM)/(TLU*ASL)) + ((NA*KFL*LA)/(TLE*ASL*KIL)):HH=HH*1000
410 H=((NM*KT*LM)/(TBU*ASB)) + ((NA*KFB*LA)/(TBE*ASB*KIB)):H=H*1000
420 IF H>HH THEN 440
430 HO=HH:GOTO 450
440 HO=H:PRINT
450 PRINT ''    **     H= '';HO;''mm''
460 AW=((NM*KT*LM)/UTSL)+((NA*KFL*LA)/(SEL)):AW=AW*1000:PRINT
470 PRINT ''    **     AREA OF THE WEB'';AW;''mm^3'':PRINT
480 VL=((3*PI/2)*D^2*HO) + (LL*AW)
490 VL=VL/1000
500 PRINT ''     **     VOL. OF LOOP = '';VL; ''mm^3'':PRINT
510 INPUT ''SP. GRAVITY OF LOOP MATERIAL'';RR
520 WL=RR*VL:PRINT
530 PRINT ''     **     MASS OF LOOP = '';WL;''gm'':PRINT
540 WTB= 2*WB + WL:PRINT :PRINT:WTB=WTB/1000
550 PRINT ''     **     MASS OF TURNBUCKLE'';WTB;''Kg'':PRINT
560 INPUT ''REL. COST OF MAT. OF RINGBOLT'';MCB
570 INPUT ''REL. COST OF MFR. OF RINGBOLT'';PCB
580 INPUT ''REL. COST OF MAT. OF LOOP'';MCL
590 INPUT ''REL. COST OF MFR. OF LOOP'';PCL
600 C=(2*WB*(MCB+PCB)) + (WL*(MCL+PCL)):PRINT :PRINT:C=C/1000
610 PRINT ''           RELATIVE COST OF THE TURNBUCKLE = '';C:PRINT :PRINT
620 INPUT ''PRINT RESULTS (Y/N)'';B$
630 IF B$=''Y'' THEN 650
640 GOTO 940
650 LPRINT ''                      DESIGN OF A TURNBUCKLE''
660 LPRINT ''                      ************************
670 LPRINT:LPRINT :LPRINT
680 LPRINT ''STATIC FORCE '';LM;''kN''
690 LPRINT ''ALTERNATING FORCE '';LA;''kN''
700 LPRINT ''STATIC FACTOR OF SAFETY '';NM
```

```
710 LPRINT ''FATIGUE STRENGTH FACTOR OF SAFETY '';NA
720 LPRINT ''STATIC STRESS-CONC FACTOR '';KT:LPRINT
730 LPRINT ''INNER DIAM.'';ID;''mm''
740 LPRINT ''LENGTH OF RINGBOLT LEG '';RL;''mm''
750 LPRINT ''LENGTH OF LOOP '';LL;''mm'':LPRINT
760 LPRINT,,''RINGBOLT'',''LOOP''
770 LPRINT,,''————————'',''————'':PRINT
780 LPRINT''UTS MPa'',,UTS,UTSL
790 LPRINT ''SE MPa'',,SE,SEL
800 LPRINT ''Ki'',,KIB,KIL
810 LPRINT ''Kf'',,KFB,KFL
820 LPRINT ''SP.GTY.'',,R,RR
830 LPRINT ''THREAD AS mm^2'',ASB,ASL
840 LPRINT ''AREA mm^2'',,AS,AW
850 LPRINT ''VOLUME cm^3'',,V,VL
860 LPRINT ''MASS gm'',,WB,WL
870 LPRINT ''REL. MAT. COST'',MCB,MCL
880 LPRINT ''REL. MANU. COST'',PCB,PCL
890 LPRINT ''——————————————————————————————————————''
900 LPRINT
910 LPRINT ''H USED = '';HO;''mm''
920 LPRINT ''TOTAL MASS OF TURNBUCKLE= '';WTB; ''Kg''
930 LPRINT ''REL. COST OF TURNBUCKLE = '';C
940 INPUT ''IS THERE ANOTHER PAIR TO BE SELECTED? (Y/N)'';A$
950 IF A$=''Y'' THEN 160
960 PRINT
970 PRINT
980 PRINT ''            END''
990 PRINT ''            ***''
```

Figure 21.3 Listing of turnbuckle program.

DESIGN OF A TURNBUCKLE

STATIC FORCE 20 kN
ALTERNATING FORCE 5 kN
STATIC FACTOR OF SAFETY 1.5
FATIGUE STRENGTH FACTOR OF SAFETY 3
STATIC STRESS-CONC.FACTOR 1

INNER DIA. 50 mm
LENGTH OF RINGBOLT LEG 90 mm
LENGTH OF LOOP 150 mm

	RINGBOLT	LOOP
UTS MPa	430	420
SE MPa	195	210
Ki	.7	.6
Kf	1.15	1.3
SP.GTY	7.8	7.5
THREAD AS mm^2	29.7	40.5
AREA mm^2	196.1411	164.2857
VOLUME cm^3	77316.98	44.81737
MASS gm	603.0725	336.1303
REL. MAT. COST	1	1.9
REL. MFR. COST	1	4

H USED = 13.20815 mm
TOTAL MASS OF TURNBUCKLE= 1.542275 Kg
RELATIVE COST OF TURNBUCKLE= 4.395458

Figure 21.4 Sample of computer output for turnbuckle program.

Table 21.3 Comparison of turnbuckle materials

Material pair		Relative total weight	Relative cost	Merit value
Ringbolt	Loop			
AISI 1340	Nod. CI 120-90-02	1.082	2.264	1.748
AISI 1340	Gray CI Grade 60	1.221	2.504	1.896
AISI 1015	Gray CI Grade 60	1.220	1.000	1.944
AISI 1340	Gray CI Grade 40	1.313	2.580	1.952
AISI 1340	Nod. CI 80-55-06	1.117	2.717	1.982
AISI 1015	Nod. CI 120-90-02	1.091	1.157	1.997
AISI 1015	Gray CI Grade 40	1.314	1.076	2.001
AISI 1340	Nod. CI 60-40-18	1.152	2.790	2.025
AISI 1015	Nod. CI 80-55-06	1.126	1.232	2.041
AISI 1015	Nod. CI 60-40-18	1.161	1.305	2.085

this are their lower cost and higher strengths. The higher strengths are reflected in the total weight of the turnbuckle as shown in Table 21.3. If the weighting factor given to corrosion resistance was increased at the expense of that given to cost, turnbuckles made of nonferrous alloys would be expected to be preferable.

BIBLIOGRAPHY AND FURTHER READING

Black, P. H. and Adams, O. E., *Machine Design*, 3rd ed., McGraw-Hill, 1968.
Farag, M. M., *Materials and Process Selection in Engineering*, Applied Science Pub., London, 1979.
Mohamedein, A. A., Mehenny, D. S. and Abdel-Dayem, H. W., *Turnbuckle Project*, American University in Cairo, 1986.
Parmley, R. O., *Standard Handbook of Fastening and Joining*, McGraw-Hill, New York, 1977.
Shigley, J. E. and Mitchell, L. D., *Mechanical Engineering Design*, 4th ed., McGraw-Hill, New York, 1983.

Chapter 22

Design and Selection of Structural Parts of a Cargo Trailer

22.1 INTRODUCTION

The rising cost of energy, uncertainty of oil supply, and government legislation in some countries have made it important for motor car manufacturers to achieve the highest possible fuel economy for cars and trucks. This can be achieved by two main routes:

1. Improving engine efficiency. As modern engines are already efficient, it would be difficult to greatly improve fuel consumption without changing the present technology and materials used in building the engine. This, however, remains as a long-range target for many motor car manufacturers.
2. Reducing the weight of the vehicle. Ashby and Jones (see bibliography) have shown that there is a linear correlation between fuel consumption and car weight. Halving the weight almost halves the fuel consumption. The continuing introduction of stronger and lighter engineering materials makes this route an attractive one to explore. Unfortunately, however, these new materials are also more expensive.

The present case study will discuss the design and selection of materials and processes that will give the optimum combination of weight saving and cost. The case of a trailer used for transporting cargo will be used to illustrate the procedure.

22.2 DESIGN OF THE CARGO TRAILER

For the sake of simplicity of discussion, it will be assumed that the main load-carrying structure of the cargo trailer is in the form of a platform 7 m long and 3 m wide, as shown in Fig. 22.1. The main structural members used to build the platform are assumed to be beams of I-section or box-section and flat panels. The maximum mass of the goods that can be carried by the trailer is 5000 kg and is assumed to be uniformly distributed on the panels which are supported by the beams. From the shape of the trailer, it would be reasonable to assume that the limiting criterion for designing the frame beams is plastic yielding and the limiting criterion for designing the panels is elastic deflection.

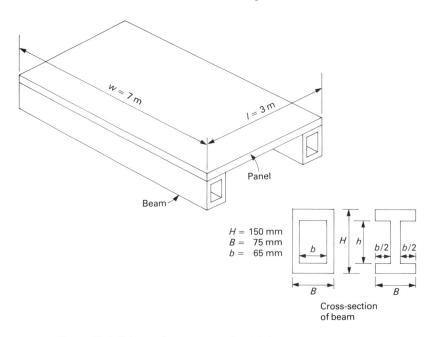

$H = 150$ mm
$B = 75$ mm
$b = 65$ mm

Cross-section
of beam

Figure 22.1 Schematic representation of the cargo trailer platform.

Design of the frame beams

The main frame elements can be approximated to simply supported beams subjected to uniformly distributed loads along their lengths. In this case, the maximum stress, s, will take place in the middle of the beam and can be calculated from the relationship:

$$s = \frac{M}{Z} \tag{22.1}$$

where M = maximum bending moment = $\dfrac{Lw^2}{8}$

    L = the uniformly distributed load per unit length
    w = length of the beam

    Z = section modulus = $\dfrac{BH^3 - bh^3}{6H}$

H, B, h and b are as shown in Fig. 22.1.

If failure by plastic yielding is to be avoided, s should be less than the working strength of the material. Taking a factor of safety of 2, $s = $ YS/2, where YS is the yield strength of the material.

    If s, H, B, h and b are considered as variables, there would be a very large number of possible material and dimension combinations which satisfy a given set of design requirements. In order to limit these combinations to a practical number, it will be assumed that

H, B and b will be kept constant while h will vary with the material strength. For the present application, the fixed dimensions of the beam cross-section will be taken as those of a steel standard I-beam of section index 10 I 35. For this I-beam, $H = 150$ mm, $B = 75$ mm, and $b = 65$ mm. From the yield strength of a given material, the value of h can be calculated which will allow the calculation of the cross-sectional area, mass and cost of the beam.

Design of the body panels

The body panels can be assumed to be subjected to uniformly distributed load. The elastic deflection y of a panel of length l, width w and thickness t under uniformly distributed load can be calculated according to the relationship:

$$y = \frac{Kl^3 F}{Ewt^3} \qquad (22.2)$$

where E = Young's modulus of the material
$\quad\; F$ = the total force acting on the panel
$\qquad = L' \star l \star w$
$\quad L'$ = distributed load per unit area of the panel
$\quad K$ = a constant which depends on how the panel is held
$\qquad = 5/32$ when the panel is simply supported as shown in Fig. 22.1. The value of K will not influence the selection process.

From Fig. 22.1, $l = 3$ m, $w = 7$ m. Taking the maximum allowable deflection $y = 15$ mm the value of t for a given material can be calculated from (22.2). Knowing the thickness will allow the calculation of the mass and cost of the panel.

22.3 CANDIDATE MATERIALS AND MANUFACTURING PROCESSES

In this case study it will be assumed that the structural members of the trailer platform are currently being fabricated from AISI 1015 steel. This material will be referred to as the base material in the following discussion. Although the shape and requirements of the beams are different from those of the panels, similar materials can be considered for the fabrication of both types of components. The manufacturing processes used in making the beams are, however, different from those used for the body panels. Possible candidate materials which may be used to replace the base material in order to reduce the weight of the trailer include steels, aluminum alloys, magnesium alloys and fiber reinforced resins.

Candidate materials for the beams

In the case of metallic materials, the cross-section can be either I-beam or box-section. In the case of steel, it may be possible to use standard hot-rolled structural sections. In the case of the other metallic alloys, the section can be manufactured from sheets by welding. For uniformity of analysis, the steel sections will also be assumed to be made by welding. In the case of composite resins the box-section is easier to manufacture than the I-section. Fiber fabric can be wrapped round a core of the required dimensions to form the

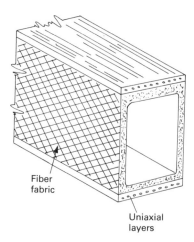

Figure 22.2 Arrangement of reinforcement in the cargo trailer beams.

Table 22.1 Properties of some candidate materials for cargo trailer

Material	YS (MPa)	E (GPa)	Specific gravity	Total cost ($/kg)[a]
STEELS				
AISI 1015	329	207	7.8	1.1
AISI 1040	380	207	7.8	1.2
AISI 1340	567	207	7.8	2.25
AISI 4130	436	207	7.8	2.2
AISI 4820	492	207	7.8	2.75
HSLA ASTM A572 grade 60	415	207	7.8	2.0
ALUMINUM ALLOYS				
AA 2014 T6	42	70	2.8	4.5
AA 3003 O	42	70	2.73	5
AA 5052 O	91	70	2.68	7
AA 6061 O	56	70	2.7	7
MAGNESIUM ALLOYS				
M1A hard	18.2	44.8	1.76	8.5
AZ31B	22.1	44.8	1.77	10
Zk60A F	26.2	44.8	1.8	10
FIBER REINFORCED PLASTICS				
Epoxy-70% glass fabric	680	22	2.1	20
Epoxy-35% glass fabric + 35% glass fibers	1400	40	2.1	25
Epoxy-63% carbon fabric	560	56	1.61	90
Epoxy-33% carbon fabric + 30% carbon fibers	2000	100	1.61	100
Epoxy-62% aramid fabric	430	29	1.38	40
Epoxy-32% aramid fabric + 30% aramid fibers	860	55	1.38	45

[a]The total approximate cost represents the cost of material on the job, which includes processing and finishing costs. Differences in price for the different shapes are ignored in these approximate estimates.

box-section. The fabric is then impregnated with the resin, which is subsequently cured by heating. The fabric may be supplemented by uniaxial layers in the top and bottom faces of the box-section, as shown in Fig. 22.2. Table 22.1 gives the properties of the base material (AISI 1015) and some candidate materials.

Candidate materials for the body panels

As the body panels are in the form of flat sheets, standard sheet metals may be used. In this case the only manufacturing processes needed are cutting the sheets to size and then joining them to the frame. In the case of fiber reinforced resins, the panels can be made by laying prepregs in various orientations or by using layers of fiber fabrics to obtain equal properties in different directions. The matrix resin can then be cured by hot pressing.

22.4   EVALUATION OF CANDIDATE MATERIALS

Having decided on the design procedure and located the possible candidate materials, the next step is to calculate the dimensions of the main structural elements of the trailer when the different materials are used in their manufacture. From the dimensions, the weight and cost are calculated for the different materials.

Evaluation of materials for the beams

From (22.1) and using the fixed dimensions given in Section 22.2, it can be shown that:

$$s = \frac{YS}{2} = \frac{M}{Z}$$

$$h^3 = \frac{1}{b}\left[BH^3 - 6H\left(\frac{M \star 2}{YS}\right)\right]$$

$$M = \frac{1}{8}\left(\frac{5000 \star 9.8}{2 \star 7000}\right)(7000)^2 = 21\,437\,500\ \text{N mm}$$

$$h^3 = \frac{1}{65}\left[2.53 \star 10^8 - 900\left(\frac{21.437 \star 10^6 \star 2}{YS}\right)\right] \qquad (22.3)$$

This shows that as the yield strength of the material decreases, the value of h decreases, which means that the flange thickness increases. A limiting value of YS is when $h = 0$, i.e. the beam is of a solid rectangular cross-section. From (22.3), when $h = 0$ YS = 152.5 MPa. Materials with yield strengths less than this limiting value will fail under the applied load even when made in the form of a solid rectangular section of $150 \star 75$ mm. This means that all the aluminum and magnesium alloys in Table 22.1 should be eliminated from the selection process.

Equation 22.3 also shows that as the strength of the material increases, the value of h increases. This means that for relatively strong materials, the flange thickness could become too thin to resist other modes of failure, such as buckling. A reasonable assumption is that flange thickness should not be less than the side-wall thickness in a box-section or half the web thickness in the case of an I-section. This means that the maximum allowable value of $h = 140$ mm. From (22.3) it can be shown that this limiting condition corresponds to YS = 514.6 MPa. Materials of higher strength will not be utilized to their maximum potential.

Table 22.2 gives the calculated values of h according to (22.3) and Table 22.1. The value of h is taken as 140 mm for all materials whose strengths are higher than the limiting value of 514.6 MPa.

Having calculated the values of h for the different materials, it is then possible to calculate the cross-sectional area, mass and cost of beams made from the different materials, as shown in Table 22.2.

Evaluation of materials for the panels

From (22.2) and using the given panel dimensions and limiting deflection, it can be shown that:

$$y = 15 = \frac{5 \star (3000)^3 \star 5000 \star 9.8}{32 \star 7000 \star E \star t^3}$$

$$t^3 = \frac{1.97 \star 10^9}{E} \qquad (22.4)$$

where t is in mm and E is MPa.

Equation 22.4 is used to calculate the panel thickness for the different materials in Table 22.1 and the results are given in Table 22.3. As the stiffness requirements are equal in the plane of the panel, only the fabric reinforced composites are considered. Equation 22.4 shows that the thickness of the panel is only affected by the Young's modulus. This means that materials of the same group, i.e. same Young's modulus, give the same panel thickness regardless of their yield strength. Knowing the thickness, the mass and cost of the panels made from the different materials are then calculated, as given in Table 22.3. In

Table 22.2 Comparison of different candidate materials for use in making beams for cargo trailer frame

Material	YS (MPa)	h (mm)	Area (mm²)	Mass (kg)	Cost ($)
STEELS					
AISI 1015 (base material)	329	127.8	2943	160.69	176.76
AISI 1040	380	132.6	2631	143.65	176.38
AISI 1340	567	140	2150	117.39	264.13
AISI 4130	436	136.3	2391	130.55	287.21
AISI 4820	492	139.0	2215	120.94	332.59
HSLA ASTM A572 grade 60	415	135.0	2475	135.14	270.28
FIBER REINFORCED PLASTICS					
Epoxy-70% glass fabric	680	140	2150	31.61	632.20
Epoxy-35% glass fabric + 35% glass fibers	1400	140	2150	31.61	790.25
Epoxy-63% carbon fabric	560	140	2150	24.23	2180.70
Epoxy-33% carbon fabric + 30% carbon fibers	2000	140	2150	24.23	2423.00
Epoxy-62% aramid fabric	430	135.9	2475	23.91	956.40
Epoxy-32% aramid fabric + 30% aramid fibers	860	140	2150	20.77	934.65

Table 22.3 Comparison of candidate materials for use in making panels of cargo trailer

Material	E (GPa)	t (mm)	Mass (kg)	Total cost ($)
STEELS				
AISI 1015 (base material)	207	21.2	3473	3820.3
AISI 1040	207	21.2	3473	4167.6
AISI 1340	207	21.2	3473	7814.25
AISI 4130	207	21.2	3473	7640.6
AISI 4820	207	21.2	3473	9550.75
HSLA ASTM A572 grade 60	207	21.2	3473	6946
ALUMINUM ALLOYS				
AA 2014 T6	70	30.4	1788	8046
AA 3003 O	70	30.4	1743	8715
AA 5052 O	70	30.4	1711	11977
AA 6061 O	70	30.4	1724	12068
MAGNESIUM ALLOYS				
M1A hard	44.8	35.3	1305	11092
AZ31B	44.8	35.3	1312	13120
Zk60A F	44.8	35.3	1334	13340
FIBER REINFORCED PLASTICS				
Epoxy-70% glass fabric	22	44.7	1971	39420
Epoxy-63% carbon fabric	56	32.8	1109	99810
Epoxy-62% aramid fabric	29	40.8	1182	47280

the case of steels in Table 22.1, as the density is the same, the mass of the panel is the same in all cases and there is no need to consider steels other than the least expensive one.

22.5   SELECTION OF OPTIMUM MATERIALS

In the present case study, savings in the weight of the trailer platform can be looked at in two different ways:

1. Savings in weight are translated into savings in fuel consumption and savings in suspension and other components which can be reduced in size as a result of the reduction in mass of the platform.
2. Savings in the weight of platform can also mean a corresponding extra mass of cargo to be carried over the entire life of the platform. In this case, there is no change in the fuel consumption or the design of the suspension or other components of the trailer.

In this study it will be assumed that any reductions in the weight of the platform will allow a corresponding increase in the mass of the cargo. Assuming that the owner of the trailer can charge $7.00 for transporting 1000 kg of goods over a distance of 100 km, and assuming that the life of the platform is 2 years during which it will travel a distance of 100 000 km, it can be shown that the extra income which results from transporting 1 kg of cargo over the entire life of the platform, i, is:

$$i = \frac{7.00 \star 100\ 000}{1000 \star 100} = \$7.00/kg$$

Assuming that the added cost of labor and overheads required to load and unload the extra cargo is 40 percent of the extra income i, the net benefit of weight saving is $0.6\,i$. This means that a maximum of $0.6\,i$ can be spent towards paying the extra cost of using the new lighter materials in fabricating the platform.

The benefit-cost analysis and incremental return methods, as discussed in Sections 17.6 and 20.7 respectively, will be used to select the optimum materials for fabricating the beams and panels.

Based on the cost values of Tables 22.2 and 22.3, the candidate materials are arranged in the order of increasing component cost as shown in Tables 22.4 and 22.5. The benefit, B, that results from substituting a new material for the base material is given by:

$$B = (M_b - M_n) \star 0.6\,i \qquad (22.5)$$

where M_b = mass of the component when made of the base material

M_n = mass of the component when made of the new material

Table 22.4 Selection of the optimum candidate material for use in making beams for cargo trailer frame

Material	Mass (kg)	Cost ($)	B	C	B/C
AISI 1015 (base material)	160.69	176.76			
AISI 1040	143.65	176.38	71.57	−0.38	>1
AISI 1340	117.39	264.13	181.86	87.37	2.08
HSLA ASTM A572 grade 60	135.14	270.28	107.31	93.58	1.15
AISI 4130	130.55	287.21	126.63	110.45	1.15
AISI 4820	120.94	332.59	166.95	156.83	1.07
Epoxy-70% glass fabric	31.61	632.20	542.14	455.44	1.19
Epoxy-35% glass fabric + 35% glass fibers	31.61	790.25	542.14	613.49	reject
Epoxy-32% aramid fabric + 30% aramid fibers	20.77	934.65	587.66	757.89	reject
Epoxy-62% aramid fabric	23.91	956.40	574.48	779.64	reject
Epoxy-63% carbon fabric	24.23	2180.70	573.13	2003.94	reject
Epoxy-33% carbon fabric + 30% carbon fibers	24.23	2423.00	573.13	2246.24	reject

Table 22.5 Comparison of candidate materials for use in making panels of cargo trailer

Material	Mass (kg)	Total cost ($)	B	C	B/C
AISI 1015 (base material)	3473	3820.3			
AISI 1040	3473	4167.6	0	347.3	reject
HSLA ASTM A572 grade 60	3473	6946	0	3125.7	reject
AISI 4130	3473	7640.6	0	3820.3	reject
AISI 1340	3473	7814.3	0	3993.95	reject
AA 2014 T6	1788	8046	7077	4215.7	1.68
AA 3003 O	1743	8715	7266	4894.7	1.49
AISI 4820	3473	9550.75	0	5730.45	reject
M1A hard	1305	11092	9106	7271.7	1.25
AA 5052 O	1711	11977	7400	8156.7	reject
AA 6061 O	1724	12068	7346	8247.7	reject
AZ31B	1312	13120	9076	9299.7	reject
Zk60A F	1334	13340	8984	9519.7	reject
Epoxy-70% glass fabric	1971	39420	6308	35599.7	reject
Epoxy-62% aramid fabric	1182	47280	9622	43459.7	reject
Epoxy-63% carbon fabric	1109	99810	9929	95989.7	reject

The extra cost, C, that will have to be paid as a result of using the new material is given by:

$$C = C_n - C_b \qquad (22.6)$$

where C_n = cost of component when made of the new material
C_b = cost of component when made of the base material

The values of B and C are calculated for the different candidate materials for the case of beams and panels and the results are shown in Tables 22.4 and 22.5 respectively.

For a material to be acceptable as a substitute to the base material, its B/C should be >1. Materials with $B/C<1$ are rejected, as shown in Tables 22.4 and 22.5. Materials with $B/C>1$ are then evaluated according to the procedure outlined in Sections 17.7 and 20.7.

Final selection of panel material

In the case of the panel materials, Table 22.5, the first and second materials with $B/C>1$ are compared as follows:

$$\frac{B(3003) - B(2014)}{C(3003) - C(2014)} = \frac{7266 - 7077}{4894.7 - 4215.7} = 0.278$$

This means that each additional dollar invested in making the panel from AA 3003 over making the panel from AA 2014 yields a return of \$0.278, which is not economical. Therefore AA 2014 is preferred to AA 3003. AA 2014 is then compared with M1A hard as follows:

$$\frac{B(M1A) - B(2014)}{C(M1A) - C(2014)} = \frac{9106 - 7077}{7271.7 - 4215.7} = 0.664$$

As this ratio is less than unity, AA2014 is preferred to M1A hard.

From the above analysis it becomes clear that aluminum AA2014 T6 is the optimum material for making the panels for the cargo trailer.

The selection of the panel material is likely to be affected if sandwich materials were considered as candidates. The example discussed in Section 6.6 shows that such materials can be considerably lighter than solid plates.

Final selection of the beam material

Using a similar procedure to that followed in the case of the panel, steel AISI 1040 is compared with AISI 1340 as follows:

$$\frac{B(1340) - B(1040)}{C(1340) - C(1040)} = \frac{181.86 - 71.57}{87.37 + 0.38} = 1.26$$

This means that steel AISI 1340 is preferred to AISI 1040. The above calculations are not necessary in the case of HSLA ASTM A 572, AISI 4130 or AISI 4820 since they give lower

B for a higher *C* than AISI 1340 and are therefore rejected. The comparison is then continued between AISI 1340 and Epoxy-70 percent glass fabric:

$$\frac{B(\text{Epoxy-70\% glass}) - B(1340)}{C(\text{Epoxy-70\% glass}) - B(1340)} = \frac{542.14 - 181.86}{455.44 - 87.37} = 0.979$$

This shows that steel AISI 1340 is preferred to Epoxy-70 percent glass fabric composite. However, this latter material is a strong competitor and would be economically feasible if it became less expensive as production volume increases or as more efficient production technology is introduced.

BIBLIOGRAPHY AND FURTHER READING

Ashby, M. F. and Jones, D. R. H., *Engineering Materials 2: an Introduction to Microstructure, Processing and Design*, Pergamon Press, London, 1987.

Chang, D. C. and Justusson, J. W., 'Structural requirements in material substitution for car-weight reduction', *SAE Trans.*, vol. 86, 1977, pp. 66–8.

Kennedy, F. E. and Hooven, F. J., 'An analysis of automobile weight reduction by aluminum substitution', *SAE paper* No. 770805, 1977.

Selwood, P., Wolton, S. and Stoddart, T., 'Advanced materials in racing cars', *Metals and Materials*, vol. 3, No. II, Nov. 1987, pp. 655–8.

Chapter 23

Design and Materials Selection for Lubricated Journal Bearings

23.1 INTRODUCTION

A journal bearing is a machine element designed to transmit loads or reaction forces from a rotating shaft to the bearing support. Besides carrying loads, the bearing material is subjected to the sliding movement of the shaft. The friction forces that result from the sliding motion are normally reduced by lubrication. In the case of lubricated journal bearings, a continuous oil film is formed between the shaft and the bearing as shown in Fig. 23.1. When the shaft is at rest, metal-to-metal contact occurs at point x. As the shaft rotates slowly, the point of contact moves to position y. A thin adsorbed film of lubricant may partially separate the surfaces, but a continuous film will not exist because of the slow speed. With increasing speed of rotation, a continuous lubricant film is established and the

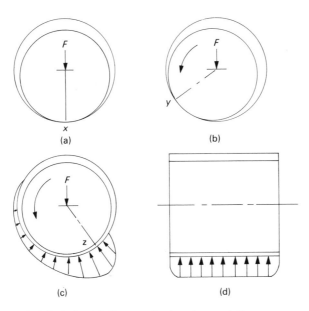

Figure 23.1 Position of the shaft in relation to the bearing and the pressure distribution in a lubricated journal bearing. (a) Shaft at rest. (b) Shaft rotating at a slow speed. (c) Shaft rotating at high speed. (d) Distribution of pressure in the axial direction.

center of the shaft is moved so that the minimum film thickness is at z. The wedge shape of the film helps to build up sufficient pressure to support the external force. The pressure distribution in both the radial and axial directions is shown in Fig. 23, c and d. The drop in pressure at the edges of the bearing in the axial direction is due to the leakage of the lubricant from the sides.

Experience shows that higher lubricant pressure builds up in the bearing as the clearance between the shaft and bearing decreases. In addition, the bearing gives better guidance to the shaft as the clearance decreases. However, allowance must be made for manufacturing tolerances in the journal and sleeve, for deflection of the shaft and for space to permit foreign particles to pass through the bearing. The clearance, c, is usually related to the diameter of the shaft, d and c/d usually ranges between 0.001 and 0.0025.

The bearing length should be as long as possible in order to reduce the compressive stresses on the bearing material. Also a longer bearing will reduce the side leakage of the lubricant from the bearing, thus allowing higher loads to be supported without suffering metal-to-metal contact between the journal and bearing. However, space requirements, manufacturing tolerances and shaft deflection are better met with a shorter bearing length. The length of the bearing is usually related to the diameter of the shaft and L/d usually ranges between 0.8 and 2.0.

In the present case study it is required to design and select the materials for the two journal bearings that support the rotor of a centrifugal pump in its casing. The mass of the rotor is equally distributed between the two bearings which carry a load of 7500 N each. The diameter of the rotor shaft is 80 mm and its speed of rotation is 1000 rpm.

23.2   DESIGN OF THE JOURNAL BEARING

The magnitude and nature of the load are the main factors that affect the bearing design and the performance of the bearing material. Heavy loads require high compressive yield strength to avoid plastic deformation, while cyclic loads require high fatigue strength to avoid fracture. An approximate value of the mean compressive stress, S_c acting on the bearing material is given by:

$$S_c = \frac{F}{L * D} \qquad (23.1)$$

where F = the force acting on the bearing = 7500 N
       L = axial length of the bearing
       D = diameter of the bearing
          = $d + c$
       d = shaft diameter = 80 mm
       c = clearance

For the present case study it is reasonable to assume that L/d is 1.25, i.e. L = 100 mm, and c/d is 0.001. These values are within the practical limits discussed in Section 23.1. Based on these assumptions, it can be shown that the compressive stress acting on the bearing material is about 0.93 MPa. This value is within 0.6 and 1.2, which are the limits recommended by Shigley and Mitchell (see bibliography) for centrifugal pumps.

The coefficient of friction, f, is an important parameter in bearing design as it affects both the power loss and temperature rise due to friction. The coefficient of friction for a

well lubricated journal bearing can be expressed by the McKee empirical relation
(reported by Black and Adams, see bibliography):

$$f = K \frac{ZN}{S_c} \frac{d}{c} + K' \qquad (23.2)$$

where Z = viscosity of the lubricant, centipoises
N = rotational speed of the shaft (rpm)
K = Constant whose value depends on the system of units used.
     = 473×10^{-10} when S_c is in lb/in^2, and d and c in inches
     = 3.3×10^{-10} when S_c is in MPa, and d and c in mm.
K' = constant to account for end leakage of the lubricant
     = 0.002 for L/d in the range 0.75 to 2.8

The quantity ZN/S_c is called the bearing characteristic number and is dimensionless if the
quantities are expressed in a consistent system of units. Higher values of this parameter
ensure the continuity of the lubricant film and avoid metal-to-metal contact between the
bearing material and the journal. The value of Z depends on the grade of the oil and its
temperature at the bearing–journal interface. This temperature is a function of the heat
balance between the heat generated due to friction, H_1, and the heat dissipated by the
bearing to the surroundings, H_2. Under thermal equilibrium, heat will be dissipated at the
same rate that it is generated in the lubricant.

The usual procedure in bearing design is to assume a bearing temperature and then
estimate H_1 and H_2. If thermal equilibrium is indicated, then the assumed bearing
temperature is correct. If H_1 and H_2 are not approximately equal, the designer must
assume a different operating temperature, a different oil or different values of L and d. For
the present case study, it is reasonable to assume an ambient temperature, t_a, of 20°C
(68°F) and a bearing surface temperature, t_b, of 70°C (158°F). Black and Adams report
that experimental work shows that:

$$t_b - t_a = 0.5(t_o - t_a) \qquad (23.3)$$

where t_o = temperature of the oil film.

Using the above assumptions and (23.3), t_o = 120°C (248°F). Selecting a lubricating oil of
grade SAE No. 30, gives Z = 5.5 cP at 120°C.

Under these conditions, the bearing characteristic number is:

$$\frac{ZN}{S_c} = \frac{5.5 \star 1000}{0.93} = 5914$$

This value is within the range of 4300–14 300 which is recommended by Jain (see
bibliography) for centrifugal pumps.

From (23.2), $f = 0.00395$

The heat generated due to friction, H_1, is given by:

$$H_1 = fFV \qquad (23.4)$$

where F = force acting on the bearing = 7500 N

V = rubbing velocity (m/s)

$\quad = \pi \star d \star N/60$

$\quad = 4.187$ m/s.

From (23.4), $H_1 = 0.00395 \star 7500 \star 4.187 = 124$ W

According to Black and Adams, the heat dissipated by the bearing, H_2, is given by the empirical relationship:

$$H_2 = CA(t_b - t_a) = \frac{1}{2}CA(t_o - t_a) \qquad (23.5)$$

where 'A = projected area of the bearing = $L \star d$

C = heat-dissipation coefficient

$\quad = 3.75$ ft lb/min/in^2/°F (244 W/m^2/°C) for average industrial unventilated bearings

$\quad = 5.8$ ft lb/min/in^2/°F (377 W/m^2/°C) for well ventilated bearings

Using the assumed values of temperature and (23.5) it can be shown that:

under average industrial unventilated conditions $H_2 = 97.6$ W

under well ventilated conditions $H_2 = 150.8$ W

Comparing the calculated values of H_1 and H_2 shows that thermal balance can be achieved under moderate ventilation conditions.

Having determined the bearing loads, lubricant and operating temperatures, it is now possible to select the optimum material for the bearing as will be discussed in the following sections.

23.3 ANALYSIS OF BEARING MATERIAL REQUIREMENTS

The above discussion shows that the compressive strength of the bearing material at the operating temperature (120°C or 248°F in the present case) must be sufficient to support the load acting on the bearing. If the material is not strong enough, it could suffer considerable plastic deformation by extrusion. Fatigue strength also becomes important under conditions of fluctuating load. Both compressive and fatigue strengths are known to increase as the thickness of the bearing material decreases. This is achieved in practice by bonding a thin layer of the bearing material (0.05–0.15 mm or 0.002–0.006 in) to a strong backing material to form a bimetal structure. Common examples include lead and tin alloys on steel or bronze backs. An intermediate layer of copper or aluminum alloys may also be introduced between the bearing material and the steel back to produce a trimetal structure. In such cases, the bearing material can be made as thin as 0.013 mm (0.0005 in).

Conformability of the bearing material allows it to change its shape to compensate for slight deflections, misalignments and inaccuracies in the journal and bearing housing. Bearing materials with lower Young's modulus will undergo larger deflections under lower loads and are, therefore, more desirable.

Embeddability is the ability of the bearing material to embed grit, sand, hard metal particles or similar foreign materials and thus prevent them from scoring and wearing the journal. Such foreign materials can be introduced with the lubricant or ventilating air. Materials with lower hardness are expected to have better embeddability.

Wear resistance of bearing materials is an important parameter in cases where the position of the journal is to be kept within narrow tolerances, or where the bearing material is a thin layer on a hard backing. Generally, the wear rate depends on the tendency of the system towards adhesive weld formation, and on the resistance of the bearing material to abrasion by asperities on the journal surface. The rate of wear can be an important factor in determining the bearing life.

Thermal conductivity becomes an important factor in selecting bearing materials under conditions of high speeds or high loads, where heat is generated at high rates, see (23.4). The ability of the bearing material to conduct heat away from friction surfaces reduces the operating temperature, and thus reduces the possibility of lubricant-film failure, melting of the bearing material and seizure of the bearing.

Corrosion resistance of the bearing material becomes an important parameter when the lubricating oil is likely to contain acidic products or to be contaminated by corrosive materials.

As many of the bearing materials contain expensive elements, the cost may become a deciding factor in selection. Using a thin layer of the expensive bearing material in bimetal or trimetal structures can reduce the total cost of the bearing.

The above discussion shows that the main requirements for a bearing material are:

1. Compressive strength at the operating temperature. This is a lower-limit property, which means that for a candidate material to be considered, its strength should exceed a given minimum value. This value is determined from the bearing design. In the present case study, the minimum strength of the bearing material at the operating temperature of 120°C (248°F) is:

$$S_m = S_c \star n$$

where S_m = minimum compressive strength of the bearing material
 n = factor of safety, which can be taken as 2 in the present case.

From the above design calculations:

$S_m = 1.86 \, \text{MPa}.$

As information on the strength of bearing materials at 120°C is not readily available, comparison between the different materials will be based on room temperature properties. For the present case study, the minimum allowable room-temperature compressive strength will be taken as 20 MPa. Most available metallic bearing alloys can meet this requirement.

2. Fatigue strength at the operating temperature. As in the case of the compressive strength, this is a lower-limit property. As fatigue is not expected to be the main selection criterion in the present case study, no special calculations are needed. It will be assumed that materials that satisfy the lower limit of the compressive strength will also satisfy the lower limit of the fatigue strength. Using a similar reasoning as in the case of compressive strength, the minimum allowable fatigue strength will be taken as 20 MPa.

3. Hardness is an upper-limit property which means that for a candidate material to be considered, its hardness should be below a given maximum value. This maximum value depends on the hardness of the journal material. For the present case study it will be

assumed that the maximum allowable bearing material hardness is 100 BHN. This will allow the use of most well-known bearing materials except the hardest copper-base alloys.

4. Young's modulus is also an upper-limit property. In this case, the maximum allowable Young's modulus is will be taken as 100 GPa, which will allow most well-known bearing materials to be considered.

5. Wear resistance is a lower-limit property which is system dependent. It depends on the journal material, lubricant, surface roughness of the journal and cleanliness of the service environment. This property is usually given as excellent (5), very good (4), good (3), fair (2) and poor (1). In the present case study materials with a rating of poor will not be considered.

6. Corrosion resistance is a lower limit property and is usually described by rating system similar to that used for wear resistance. In the present case study, materials with a corrosion resistance rating of poor will not be considered.

7. Thermal conductivity is a lower limit property and, in view of the high rotational speeds of the shaft, a relatively high minimum conductivity of 20 W/m K will be specified. This means that all nonmetallic bearing materials are excluded.

8. Cost of the bearing material, backing material and fabrication should be considered. In the present case study, a single value for the cost of the material on the job, which includes all the above factors, will be given. For the present case study, the maximum allowable cost will be taken as that of tin-base ASTM B23 grade 5.

Table 23.1 Composition of some bearing alloys (%)

Alloy grade	Sn	Sb	Pb	Cu	Fe	Zn	Al	Others
White metals ASTM B23 (tin-base)								
1	91	4.5	0.35	4.5	0.08	0.005	0.005	0.08 Bi, 0.1 As
2	89	7.5	0.35	3.5	0.08	0.005	0.005	0.08 Bi, 0.1 As
3	84	8	0.35	8	0.08	0.005	0.005	0.08 Bi, 0.1 As
4	75	12	10	3	0.08	0.005	0.005	0.15 As
5	65	15	18	2	0.08	0.005	0.005	0.15 As
White metals ASTM 23 (lead-base)								
6	20	15	63.5	1.5	0.08	–	–	0.15 As
7	10	15	75	0.5	0.1	–	–	0.6 As
8	5	15	80	0.5	–	–	–	0.2 As
10	2	15	83	0.5	–	–	–	0.2 As
11	–	15	Rem	0.5	–	–	–	0.25 As
15	1	15	Rem	0.5	–	–	–	1.4 As
Copper-base alloys SAE (copper-lead)								
48	0.25	–	28	Rem	0.35	0.1	–	1.5 Ag, 0.025 P
49	0.5	–	24	Rem	0.35	–	–	–
480	0.5	–	35	Rem	0.35	–	–	15 Ag
Copper-base alloys ASTM B22 (bronze)								
A	19	–	0.25	Rem	0.25	0.25	–	1 P
B	16	–	0.25	Rem	0.25	0.25	–	1 P
C	10	–	10	Rem	0.15	0.75	–	0.1 P, 1 Ni
Aluminum-base alloys								
770	6	–	–	1	0.7	–	Rem	1 Ni, 0.7 Si
780	6	–	–	1	0.7	–	Rem	0.5 Ni, 1.5 Si
MB7	7	–	–	1	0.6	–	Rem	1.7 Ni, 0.6 Si

23.4 CLASSIFICATION OF BEARING MATERIALS

Many alloy systems have been specially developed to accommodate the conflicting requirements that have to be satisfied by bearing materials. They are used in relatively small quantities and are produced by a relatively small number of manufacturers. Although the composition and processing methods of most commercial systems are of proprietary nature, widely used bearing materials can be classified into:

1. Whitemetals (babbitt alloys). These are either tin-base or lead-base alloys with additions of antimony and copper. Iron, aluminum, zinc and arsenic are also usually present in small amounts, as shown in Table 23.1. The relevant properties of a selected number of these alloys are given in Table 23.2.
2. Copper-base bearing alloys offer a wider range of strengths and hardness than white metals. Lead and tin are the main alloying elements, but silver, iron, zinc, phosphorus and nickel are sometimes found in small quantities. Tables 23.1 and 23.2 give the composition and properties of a selected number of copper-base bearing alloys.
3. Aluminum-base bearing alloys are suitable for high-duty bearings in view of their high strengths and thermal conductivities. They can be used in single metal, bimetal or trimetal systems. Tables 23.1 and 23.2 give the composition and properties of selected aluminum-base bearing alloys.

Table 23.2 Properties of some bearing alloys

Material grade	Yield strength (MPa)	Fatigue strength (MPa)	Hardness BHN	Corrosion resistance	Wear resistance	Thermal conduction (W/m k)	Young's modulus (GPa)	Relative cost
White metals ASTM B23 (tin-base)								
1	30	27	17	5	2	50.2	51	7.3
2	42.7	34	25	5	2	50.2	53	7.3
3	46.2	37	37	5	2	50.2	53	7.3
4	38.9	31	25	5	2	50.2	53	7.3
5	35.4	28	23	5	2	50.2	53	7.5
White metals ASTM 23 (lead-base)								
6	26.6	22	21	4	3	23.8	29.4	1.3
7	24.9	28	23	4	3	23.8	29.4	1.2
8	23.8	27	20	4	3	23.9	29.4	1.1
10	23.8	27	18	4	3	23.9	29.4	1
11	21.4	22	15	4	3	23.9	29.4	1
15	28.0	30	21	4	3	23.9	29.4	1
Copper-base alloys SAE (copper-lead)								
48	40	45	28	3	5	41.8	75	1.5
49	45	50	35	3	5	41.8	75	1.5
480	38	42	26	3	5	41.8	75	1.5
Copper-base alloys ASTM B22 (bronze)								
A	168	120	100	2	5	41.8	95	1.8
B	126	100	100	2	5	41.8	95	1.8
C	119	91	65	2	3	42	77	1.6
Aluminum-base alloys								
770	173	150	70	3	2	167	73	1.5
780	158	135	68	3	2	167	73	1.5
MB7	193	170	73	3	2	167	74	1.5

4. Nonmetallic bearing alloys are mostly based on polymers or polymer-matrix composites. They are widely used under conditions of light loading. The major disadvantage of this group of bearing materials is their low thermal conductivities. In view of the high speed of rotation encountered in the present case study, this group will not be considered further.

23.5 SELECTION OF THE OPTIMUM BEARING ALLOY

Based on the above analysis and design considerations, the weighting factors were estimated using the digital logic approach described in Section 17.7. Table 23.3 gives the different weighting factors for the present case study.

Table 23.3 Weighting factors of selection criteria for bearing material of centrifugal pump

Property	Weighting factor
Yield strength	0.2
Fatigue strength	0.14
Hardness	0.08
Corrosion resistance	0.11
Wear resistance	0.11
Thermal conductivity	0.2
Young's modulus	0.08
Cost	0.08
Total	1.0

The table shows that the yield strength is considered as one of the most important requirements. Ensuring that extensive yielding will not take place in the bearing material will ensure the uniformity of the lubricant film and would avoid vibrations at the high operating speeds of the pump. The thermal conductivity is considered equally important to the yield strength to ensure adequate conduction of heat away from the bearing-journal interface. With the relatively high speeds in the present case, sharp temperature rise as a result of temporary failure of the lubricant film could be serious. As no mention of excessive load fluctuation was made, it was assumed that fatigue is not expected to represent a serious problem and was, therefore, given a lower weighting factor than yield strength. Corrosion and weight resistances were given moderate weighting factors as the danger of contamination and foreign particles was not emphasized in the service conditions. Hardness was given one of the lowest weights as it is expected that the rotor shaft will be adequately hardened. Young's modulus is treated similarly as misalignment and deflection of the rotor shaft is not expected to be excessive. As the cost of the bearing material is expected to represent a small part of the total cost of the centrifugal pump, this factor was given a low weighting factor.

The candidate bearing materials of Tables 23.1 and 23.2 were evaluated using the limits on properties method described in Section 20.8. The lower limits and upper limits for the different properties were discussed in Section 25.3 and can be summarized as follows:

Lower limit of yield strength = 20 MPa
Lower limit of fatigue strength = 20 MPa

Table 23.4 Merit parameters and suitability of bearing materials

Material	Merit parameter (m)	Suitability
White metals ASTM B23 (tin-base)		
1	0.59	8
2	0.54	5
3	0.54	5
4	0.56	6
5	0.58	7
White metals ASTM 23 (lead-base)		
6	0.63	11
7	0.61	9
8	0.62	10
10	0.61	9
11	0.66	12
15	0.58	6
Copper-base alloys SAE (copper-lead)		
48	0.47	3
49	0.47	3
480	0.49	4
Copper-base alloys ASTM B22 (bronze)		
A	0.47	3
B	0.49	4
C	0.49	4
Aluminum-base alloys		
770	0.37	1
780	0.38	2
MB7	0.37	1

Lower limit on thermal conductivity	= 20 W/m K
Lower limit on corrosion resistance	= 2
Lower limit on wear resistance	= 2
Upper limit on hardness	= 100 BHN
Upper limit on Young's modulus	= 100 GPa
Upper limit on relative cost	= 7.5

The above lower and upper limits were used to calculate the merit parameters, m, of the different materials using (20.12). The results of evaluation are given in Table 23.4.

The results in Table 23.4 show that aluminum-base alloys are most suitable for the present application and were given suitability ratings of 1 and 2. The high conductivity, high compressive and fatigue strengths and moderate cost are their main attractions. Copper-base alloys would also be adequate and were given suitability ratings of 3 and 4.

BIBLIOGRAPHY AND FURTHER READING

Black, H. and Adams, O. E., *Machine Design*, 3rd ed., McGraw-Hill, London, 1983.
Boyer, H. E. and Gall, T. L., *Metals Handbook Desk Edition*, ASM, Ohio, 1985.
De Gee, A. W., 'Selection of materials for lubricated journal bearings', *Wear*, vol. 36, 1976, pp. 33–61.

Farag, M. M., *Materials and Process Selection in Engineering*, Applied Science Pub., London, 1979.

Forrester, P. G., 'Selection of plain bearing materials', in *Engineering Materials*, ed. H. J. Sharp, Heywood, London, 1964, pp. 255–70.

Jain, R. K., *Machine Design*, 3rd ed., Khanna, Delhi, 1983.

Shigley, J. E. and Mitchell, L. D., *Mechanical Engineering Design*, 4th ed., McGraw-Hill, London, 1983.

Chapter 24

Analysis of the Requirements and Selection of Materials for Tennis Rackets

24.1 INTRODUCTION

Many leading sports and recreational industries are now using sophisticated materials and high-technology production methods to manufacture their products. In addition, biomechanics is also being used to gain better understanding of the human body in order to enhance player comfort and to optimize equipment performance. As a result, the shape and the materials used in making many sports equipment and leisure products have undergone considerable changes. For example, tennis rackets are now available in many shapes and sizes, as shown in Fig. 24.1, even though they all comply with the ITF rules which limit the total length to a maximum of 32 in (81.3 cm) and the strung surface, called the head, to maximum dimensions of 15.5 in (39.4 cm) in length and 11.5 in (29.2 cm) in width. The rackets are also available in many materials including laminated wood, steel, aluminum alloys and fiber-reinforced composites. Although there are no limitations on the racket weight, it usually ranges between 13 and 15 ounce (368.5 and 425 g).

In this case study, the functional requirements of the tennis racket will be analyzed in order to allow the selection of the optimum racket material for a given player level.

24.2 ANALYSIS OF THE FUNCTIONAL REQUIREMENTS OF THE TENNIS RACKET

From an engineering point of view, a tennis racket can be considered as an implement for transmitting power from the arm of the player to the ball. This should be done as efficiently as possible in order to allow the player to deliver the fastest balls with the least effort. Tennis players usually call this characteristic the power of the racket. In addition to power, players evaluate rackets in terms of playability, which is a subjective evaluation of the overall performance of the racket. For a more objective evaluation, playability may be considered as a function of vibrations and control.

Vibrations take place in the strings as a result of hitting the ball and are then transmitted to the player's arm through the racket frame. If the racket material does not sufficiently dampen the vibrations, the player may develop tennis elbow.

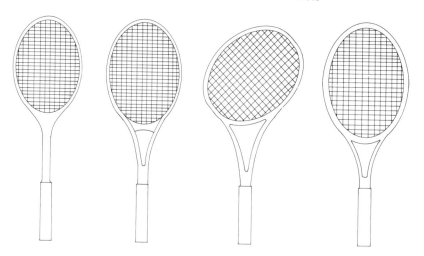

Figure 24.1 Examples of different tennis racket shapes.

Control is the ability to give the ball the desired speed and spin and to place it in the desired area of the court. Control can be considered a function of: (a) weight; (b) balance; (c) stability; and (d) area of the sweet spot. These parameters are mainly affected by the material, shape and design of the racket. The weight of the racket is mainly a function of the cross-sectional area of the racket, the shape and the density of the material. Using materials with high specific strength (strength/density) and high specific stiffness (tensile modulus/density) will allow the designer to use larger head or similar desirable design features without increasing the total weight of the racket. Vaccari (see bibliography) reports that most good players are sensitive to weight differentials of 10–15 g and pros can detect 5 g. Balance is a function of the position of the center of gravity of the racket in relation to the player's hand. According to Vaccari, a center of gravity within 20 mm ($\frac{3}{4}$ in) of the optimum is sometimes acceptable; pros can sense if it is 3 mm ($\frac{1}{8}$ in) off. Stability can be defined as the ability of the racket to resist twisting due to off-center hits. It depends primarily on the weight distribution in the racket head. In some cases, balancing weights are added to the sides of the racket head to improve stability. The sweet spot is defined as the area of maximum ball rebound velocity. The size of this area depends on the size and shape of the racket head.

In addition to the shape and the frame material, the performance of the racket is influenced by the material of the strings and the tension in the strings. The present case study will only consider the effect of the racket design and material of the frame on the performance of the racket.

Other factors that affect the selection of tennis rackets include durability and cost. Durability is related to the service life of the racket before it breaks or warps. The cost is a function of the cost of materials and manufacturing processes involved in making the racket.

24.3 DESIGN CONSIDERATIONS

From the stress analysis point of view, the tennis racket can be modeled as a cantilever with the handle as fixed end, as shown in Fig. 24.2. The stiffness of the racket, i.e. deflection as a

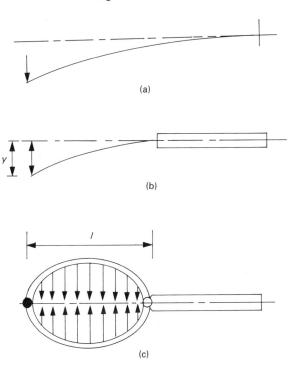

Figure 24.2 Simple modeling of tennis racket as a cantilever. (a) Deflection of a cantilever beam of uniform cross-section under a concentrated load acting on its end. (b) Deflection of tennis racket as a result of hitting the ball with its tip. (c) Forces acting on the racket head as a result of tension in the strings.

result of hitting the ball, will determine the power of the racket. A stiffer racket will absorb less energy in deflection and will deliver more power to the ball. The maximum deflection will take place when the ball is hit with the outermost tip of the head. As the cross-sectional area and dimensions of the racket head are normally much smaller than those of the handle, it can be assumed that most of the deflection will take place in the head, as shown in Fig. 24.2. The maximum deflection, y, is given by:

$$y = \frac{Fl^3}{3EI} \qquad (24.1)$$

where F = force acting on one side of the racket head as a result of hitting the ball, assumed constant for all racket designs and materials.

l = length of the racket head

E = Young's modulus of the racket material

I = moment of inertia of the cross-section of the racket head in the direction of the applied force.

Equation 24.1 shows that rackets with larger head lengths will suffer larger deflection, i.e. will be less powerful, unless materials with higher E are used in their manufacture. Figure 24.3 shows that the cross-sectional area of the racket head is complex, especially

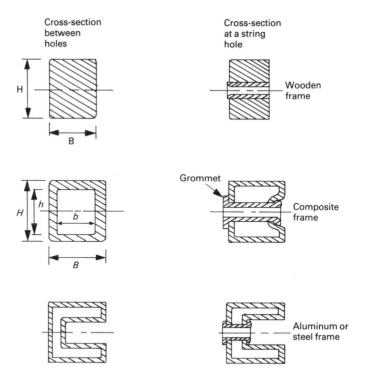

Figure 24.3 Examples of cross-sectional shapes of racket head.

in the case of hollow sections. For simplicity of analysis, it will be considered as a rectangle with outer dimensions of H and B for all materials. With the notations of Fig. 24.3,

$$I = \frac{BH^3}{12} \text{ for a solid rectangular cross-section}$$

$$I = \frac{BH^3 - bh^3}{12} \text{ for a hollow rectangular cross-section}$$

If the outer dimensions of the head cross-section, H and B, are assumed the same in all cases, it becomes clear that racket heads with hollow cross-sections will be less powerful unless materials with higher E are used in their manufacture.

Another parameter which affects the power of the racket is the tension in the strings. With higher tension, the strings will absorb less power in deflection and deliver more power to the ball. The limiting value of the tension in the strings is decided by their own strength and by the strength of the material of the racket head. To simplify the analysis, the racket head is considered to be made of two simply supported beams which are subjected to uniformly distributed static loads, as shown in Fig. 24.2. In this case, the maximum static stress, s_s, will take place in the middle of the beam and can be calculated from the relationship:

$$s_s = \frac{M}{Z} \tag{24.2}$$

where M = maximum bending moment = $\dfrac{Ll^2}{8}$

L = the uniformly distributed load per unit length
l = length of the head
Z = section modulus = $2I/H$

Hitting the ball with the strings will generate a dynamic stress, s_d, in the head. The value of s_d can be estimated approximately using (24.2), with M replaced by the dynamic force that results from hitting the ball. In this case, the total stress, s_t, is given by the summation of s_s and s_d. Failure of the racket head can be avoided if the yield strength of the racket material is higher than s_t by a suitable factor of safety.

The above analysis shows that higher power can be achieved if the racket is made of materials with higher Young's modulus and yield strength. Using such materials will also allow larger heads to be used for a given racket power. Larger heads mean larger sweet spot and better control.

As shown in Section 24.2, balance is a function of the position of the center of gravity. Changing the size of the racket head is expected to change the position of the center of gravity. This means that the density of the material has to be taken into account when designing the shape of the racket. The density also affects the balance of the racket. Using lighter materials will allow the use of balancing weights at the appropriate points of the head without increasing the total weight of the racket.

Combining the strength and weight requirements, it can be concluded that materials with higher specific stiffness and higher specific strength will allow the design of rackets with higher power and better control.

24.4   ANALYSIS OF THE RACKET MATERIAL REQUIREMENTS

The above discussions show that the racket material has to satisfy several requirements which include:

1. Specific strength. Materials with higher specific strength will allow the safe use of higher tension in the strings and larger head lengths. This means more power and control in the racket.
2. Specific stiffness. Materials with higher stiffness will allow the designer to combine better control and higher power by using larger heads without sacrificing stiffness.
3. Toughness. Rackets made of tough materials are less likely to fracture as a result of impact loading. Such loads can result from striking fast balls with the racket frame, rather than the strings, or accidentally hitting the ground, net posts and similar objects.
4. Durability. Materials that will not warp or creep as a result of temperature or humidity changes during storage or play will have better durability.
5. Vibration damping. Materials that absorb vibrations generated in the strings are desirable, as they will allow long sessions of play without causing the player a tennis elbow.
6. Processability, which is a measure of the ability of the material to be formed in the desired shape, is an important requirement in the present case. This is because the shape of the racket head should closely follow the required form to ensure reproducibil-

Table 24.1 Weighting factors of material requirements for a tennis racket

Property	Weighting factor	
	Beginner	Professional
Specific strength	0.15	0.3
Specific stiffness	0.15	0.3
Toughness and durability	0.25	0.05
Vibration damping	0.05	0.3
Total cost	0.4	0.05

ity of the racket characteristics and quality. In addition, the material should be capable of being formed into the complex cross-sectional shape of the racket head.

7. Cost is an important consideration in selecting a tennis racket, especially when catering for beginners and amateur players. In this case the cost of processing represents a large proportion of the total cost of the racket.

For simplicity of analysis, the toughness and durability will be combined as one property, designated as toughness. Processability will also be considered as a cost item and added to the total cost. A survey of the tennis racket prices shows that there is a wide variation between the different name-brands, sizes and materials. Price variations exist even for a given racket material and size. The average prices used in this case study are only given as guidelines.

The relative importance of the above properties depends on the level and type of intended user. Whereas the cost, toughness and control are important for a beginner, power and vibration damping are a necessity for a professional player. The relative importance of the racket material requirements is shown in Table 24.1 for the cases of rackets intended for beginners and professional players. The digital logic method, which was described in Section 17.7, was used to estimate the relative importance of each property, i.e. the weighting factors.

24.5 CLASSIFICATION OF RACKET MATERIALS

As discussed earlier, tennis rackets can be made of several widely different materials including various types of wood, aluminum alloys, steels and composite materials. The following discussion will give a brief outline of the characteristics and methods of fabrication which may be used in each case.

Wood

Laminated wood is the traditional material for making tennis rackets. Ash, beech, hickory, sycamore, mahogany and West African obeche are among the most commonly used types. Several types of wood may be combined according to the particular quality of each wood and the requirements of the racket. Table 24.2 gives the relevant properties of two representative types of wood. The major advantages of wood are its excellent damping capacity, high toughness and relatively low cost. Its major drawback is low specific stiffness.

Wood is normally used as laminations either of different types of wood or wood plus

Table 24.2 Properties of some candidate materials for tennis rackets

Material	Specific strength (MPa)[a]	Specific stiffness (MPa)[a]	Toughness[c]	Vibration damping[c]	Relative cost[b]
WOOD					
Ash	107.4	20.3	5	5	1
Hickory	104.5	20.9	5	5	1
ALUMINUM ALLOYS					
AA 2014 T6	15	25	5	2	1.5
AA 5052 O	34	26.1	5	2	1.8
STEELS					
AISI 1340	72.7	26.5	5	1	1.5
AISI 8650	89.5	26.5	5	1	1.7
COMPOSITE MATERIALS					
Epoxy + 35% glass fabric + 35% glass fibers	666.7	19	3	4	3
Epoxy + 33% carbon fabric + 30% carbon fibers	1242.2	62.1	2	3	5
Epoxy + 32% aramid fabric + 30% aramid fibers	1186.8	39.9	3	4	4
Epoxy + 62% carbon + aramid	1214.5	51	3	4	4.5
Epoxy + 63% carbon + aramid + glass	1032	40.3	3	4	4

[a] Specific strength and specific stiffness are calculated by dividing the yield strength elastic modulus by the specific gravity.
[b] Relative cost is based on wood as unity and includes the material and processing costs. As the processing cost is a considerable proportion of the total cost, the differences in the total cost are not as great as the differences between the raw materials costs.
[c] Both toughness and vibration damping are given as: excellent (5), very good (4), good (3), fair (2), poor (1).

composites. Wooden rackets are usually made by cutting relatively small pieces of wood and assembling them to give the required racket shape. Controlling the directions of fibers in the different pieces ensures no warping and gives the required properties. A transparent protective coating is normally applied to the wooden frame to improve its durability.

Aluminum alloys

Aluminum is a well established racket material because of its excellent toughness, good specific stiffness and moderate cost. Its major drawback is lack of vibration damping which could cause tennis elbow with long sessions of playing. Table 24.2 gives the properties of two representative aluminum alloys.

Aluminum racket frames are fabricated by bending extruded hollow sections of suitable thickness and shape (see Fig. 24.3). Anodizing is used for protection against corrosion and for coloring.

Steels

High-strength steels have better specific strength and stiffness than aluminum alloys. However, they are not as widely used as aluminum alloys as racket materials. This is because of inferior corrosion resistance, which necessitates the use of expensive finishes. Lack of vibration damping is also a major drawback. Table 24.2 gives the properties of two representative steels.

Steel racket frames are fabricated by bending suitable hollow sections, as in the case of aluminum alloys. Effective protection against corrosion is essential in this case.

Composite materials

Composite materials are relative newcomers to the field of sports equipment. However, they are rapidly gaining popularity as tennis racket materials. The major advantages of composite materials are their high specific strength and stiffness and reasonable vibration damping. Their composition can be varied to achieve a wide range of properties to suit a particular level and type of player. The major drawback of composite materials is their relatively high cost. Table 24.2 gives the properties of some commonly used fiber-reinforced composite materials.

    Composite frames are manufactured by winding prepregs of the required fibers round a mandril made of a fusible alloy, like indalloy. The orientation and material of the different plies is controlled to achieve the required racket characteristics. The frame is then placed in a mold and is then hot pressed to cure the resin matrix. During curing the core material melts and is poured out to give the required hollow section. In most cases the hollow section is filled with foam plastic like polyurethane. A polyurethane coating is normally given for protection and to give the required coloring.

24.6   EVALUATION OF RACKET MATERIALS

The different materials discussed in the earlier sections can be evaluated using the weighted properties method discussed in Section 20.6. The properties given in Table 24.2 are first scaled to get the dimensionless values, which are then multiplied by the weighting factors given in Table 24.1. The weighted properties are then added to give the material performance index (γ).

    Table 24.3 gives the calculated values of (γ) and the ranking of the different materials for the cases of a beginner and professional player. As shown in the table, wood has the highest performance index and occupies the top two ranks in the case of beginners. The major factors that contributed to the high performance index are the cost and toughness.

Table 24.3 Performance index and ranking of candidate materials for tennis rackets for beginners and professional players

Material	Beginners		Professional players	
	(γ)	Rank	(γ)	Rank
WOOD				
Ash	76.15	2	52.4	7
Hickory	76.36	1	52.6	6
ALUMINUM ALLOYS				
AA 2014 T6	59.7	3	32.8	9
AA 5052 O	56.1	6	33.2	8
STEELS				
AISI 1340	59.6	4	28.8	11
AISI 8650	57.1	5	29	10
COMPOSITE MATERIALS				
Epoxy + 35% glass fabric + 35% glass fibers	44.8	10	53.9	5
Epoxy + 33% carbon fabric + 30% carbon fibers	51	9	81	2
Epoxy + 32% aramid fabric + 30% aramid fibers	53	7	76.3	3
Epoxy + 62% carbon + aramid	40.8	11	82	1
Epoxy + 63% carbon + aramid + glass	51.2	8	72.7	4

Aluminum 2014 T6 alloy occupies the third rank followed by steel AISI 1340. Composite materials occupy lower ranks in this case in view of their higher cost. In the case of professional players, the results show that the carbon-Aramid hybrid has the highest performance index and occupies top rank. Close behind is carbon composite. The high specific strength and stiffness combined with good vibration damping were their main assets. The metal-frame rackets occupy the lowest ranks in view of their poor vibration damping and relatively low specific strength and stiffness.

BIBLIOGRAPHY AND FURTHER READING

Farag, M. M., *Materials and Process Selection in Engineering*, Applied Science Pub., London, 1979.

Farag, S., Sidky, B., Arafa, K., Nosseir, K., Moghrabi, M. and Idris, S., *Tennis racket project*, American University in Cairo, 1986.

Jones, C., *How to Play Tennis*, Hamlyn, London, 1981.

Lenel, U. R., 'Materials selection in practice', *The Institute of Metals Handbook*, London, 1986, pp. 165–78.

Lockwood, P. A., 'Composites for industry', *ASTM Standardization News*, Dec. 1983, pp. 28–31.

Vaccari, J. A., 'Scoring with materials innovations', *Design Engineering*, July, 1980, pp. 31–8.

Chapter 25

Design and Selection of Materials for a Surgical Implant

25.1 INTRODUCTION

Surgical implant materials are used in repairing many parts of the human body. The number of materials in current use as implant materials is large and includes metallic, polymeric, ceramic and composite materials. Both hard tissues, like bones, and soft tissues, like skin, can be restored or replaced with implants of similar mechanical properties, texture or color. For example, rigid metallic, ceramic and composite materials are used for fixing or replacing bones and joints, foams and gels are used for soft tissue supplementation, and elastic materials for replacement of skin and blood vessels. Although the mechanical requirements of implant materials are relatively simple, the biocompatibility requirements are stringent and more difficult to meet. Biocompatibility means that the material and its possible degradation products must be tolerated and cause no tissue disfunction at any time.

This case study will discuss the design and selection of materials for a hip joint prosthesis. Figure 25.1 shows the components used for a complete hip joint. In this case, the femoral head is replaced by a rigid pin which is installed in the shaft of the femur while the pelvic socket (acetabulum) is replaced by a rigid or soft cup which is fixed to the ilium. Both the pin and cup are normally fixed to the surrounding bone with an adhesive.

25.2 DESIGN CONSIDERATIONS

As the prosthesis is intended to replace the bone structure of the hip joint, it is important to study the bone structure and properties. Bone is a living tissue composed of inorganic and organic materials in dynamic equilibrium with the body fluids. Although the proportions vary from one part of the skeleton to another, water-free bone contains about two-thirds inorganic and one-third organic materials. The inorganic phase, primarily hydroxyapatite crystals, is hard and brittle and represents the main load-bearing component of the bone structure. The organic phase, primarily collagen fibers, is gelatine-like protein and its presence makes the bone tough, in addition to its biological functions. These phases are generally arranged in a complex structure to give maximum strength in the required direction. Figure 25.2 shows how the load-bearing phase is arranged in the direction of maximum stress in the head of the femur bone.

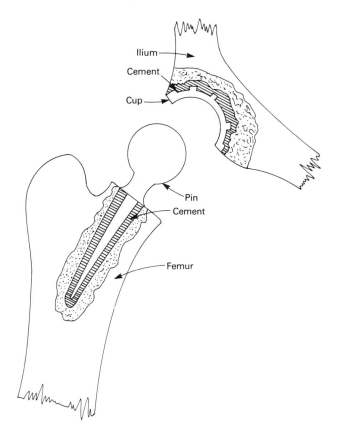

Figure 25.1 Components of a complete hip joint prosthesis.

The compressive strength of compact bone is about 140 MPa and the elastic modulus is about 14 GPa in the longitudinal direction and about one-third of that in the radial direction. These properties are modest in comparison with those of most engineering metallic and composite materials. However, live healthy bone is self-healing and has a great resistance to fatigue loading. The implant material, on the other hand, does not have this ability to repair itself, and usually has a finite fatigue life. For this reason, the implant material must be stronger than bone, especially under fatigue loading.

In the case of the hip joint, carrying the repeated loading caused by walking and similar activities is an important part of its function. The loading frequency ranges from 1 to 2.5 million cycles per year, depending on the activity and movement of the individual. Stress analysis of the forces acting on the hip joint shows that the fatigue loading is usually equal to 2.5–3 times the body weight and can possibly be higher depending on the posture of the individual. In addition to the repeated loading, the joint is subjected to a static loading as a result of muscle action which keeps the parts of the joint together. This static load is normally much smaller than the repeated loading. As the hip joint prosthesis is generally intended to be permanent, it is expected to resist fatigue fracture for years or decades.

In designing the hip joint prosthesis, it should be borne in mind that, regardless of the strength of the implant material properties, the dimensions of the prosthesis must match

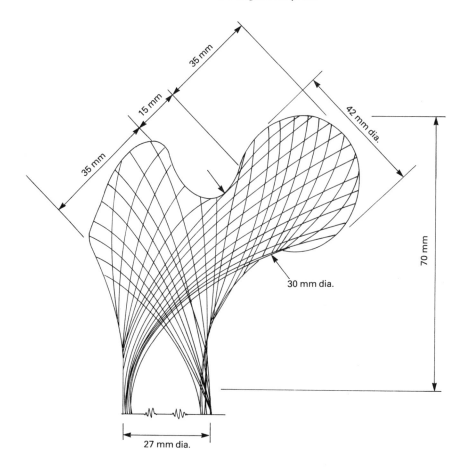

Figure 25.2 Main average dimensions and distribution of lead-bearing phases in the head of the femur bone.

the dimensions of the original bone. The design calculations in the present case study will be based on the dimensions given in Fig. 25.2 which represent approximate average values for an adult. Assuming that the weight of the person is 75 kg (165 lb), and taking the alternating load as three times the weight it can be shown that the hip prosthesis will be subjected to an alternating load of 2205 N. The maximum stress occurs in the prosthesis neck, which is about 30 mm diameter according to Fig. 25.2. From these values the alternating stress is estimated as 3.1 MPa. Assuming that the static load due to muscle contraction is about 300 N, the static stress at the neck of the prosthesis is estimated as 0.42 MPa. Using the modified Goodman relationship given by (11.23):

$$\frac{n_m K_t S_m}{\text{UTS}} + \frac{n_a K_f S_a}{S_e} = 1$$

where n_m = factor of safety for static load, taken as 2 in this case.
n_a = factor of safety for alternating loading, taken as 3 in this case.
K_t = stress-concentration factor for static load, taken as 2.2.

K_f = stress-concentration factor for alternating stress, taken as 3.5.
S_m = static stress = 0.42 MPa, as calculated above.
S_a = alternating stress = 3.1 MPa, as calculated above.
UTS = ultimate tensile strength of the prosthesis material
S_e = endurance limit of the prosthesis material.

Taking the endurance ratio of the prosthesis material as 0.35. The values of S_e = 0.35 UTS.

Substituting the above values in the above Goodman relationship, UTS is found to be about 95 MPa. From the assumed value of endurance ratio of 0.35, S_e is estimated as 33.25 MPa. These values represent the minimum strengths that an implant material must have in order to be considered for making the pin of the hip joint prosthesis. Using materials with higher strengths will not make it possible to reduce the cross-sectional area of the prosthesis, but will have the advantage of being less likely to fail in service.

Wear of the prosthesis parts (cup and pin) should also be considered when designing a hip joint prosthesis. Wear debris are often found in tissue surrounding joints and could cause adverse effects due to sensitivity of the patient to the material. In addition, material wear in the total hip prosthesis, especially the enlargement of the cup (acetabular concavity) causes poor articulation of the joint. The pressure between the mating surfaces of the prosthesis can be estimated by dividing the maximum force by the projected area of the cup. In the present case the maximum force is given by 2205 + 300 = 2505 N and the projected area of the cup is 1385 mm². This gives an average pressure of 1.8 MPa. If both the pin and cup are made of metallic materials, the high friction coefficient could lead to mechanical difficulties. For this reason, the cup is usually made of a low-friction plastic, such as high-density polyethylene and PTFE. The compressive strengths of most polymeric materials are higher than the calculated contact pressure, which means that selecting the material of the cup is mostly based on the basis of their biocompatibility.

25.3   ANALYSIS OF IMPLANT MATERIAL REQUIREMENTS

Earlier discussion has shown that the total hip joint prosthesis consists of a pin, which replaces the head of the femur bone, and a cup, which replaces the acetabular concavity. The pin and cup are fixed to the surrounding bone structure by an adhesive cement. As would be expected, the requirements for each of these components are different, as they perform different functions. In the present case study, only the material requirements for the pin will be discussed. Similar procedures can be applied to the cup and the cement.

In general, biological requirements represent the major constraints in selecting implant materials. The action of the implant material on the body tissues can range from toxicity, in which case the implant material is totally rejected, to inertness. Even with inert materials, their presence impedes the normal healing sequence at the implant site and leads to fibrocartilagenous membrane of low cellularity which isolates the implant from normal tissue. This membrane is the body's response to the stimulus of an inert foreign material which is impervious to body fluids. The thickness of the membrane is proportional to the degree of implant material dissolution. Toxicity is usually signalled by a large population of inflammatory cells. This requirement is usually called tissue tolerance and is difficult to quantify. For the sake of comparison, materials are given a rating of 10 for the best material and 1 for the worst. Only materials with tissue tolerance of 7 or more will be considered in the present case study. The tissue tolerance requirements apply equally well to the pin, cup and cement materials.

Another important implant material requirement is corrosion resistance. This is because body fluids are aqueous salt solutions with concentrations roughly comparable to seawater and with a pH value of 7.4. This environment is hostile and could cause corrosion of many metallic materials. Such corrosion should be avoided as it may induce deleterious effects in the surrounding tissues or even distant organs. The combination of corrosive action of the body fluids and fatigue loading could result in stress corrosion cracking, which imposes strict limitations on the implant material surface finish and structural homogeneity. As in the case of tissue tolerance, corrosion resistance is rated according to a scale of 10 to 1. Only materials with a corrosion resistance rating of 7 or better are considered in this case study. The corrosion resistance requirements apply equally well to the pin, cup and cement materials.

The design analysis of Section 25.2 shows that the strength requirements of implant materials are relatively easy to meet. Other material requirements include toughness, wear resistance, elastic compatibility and specific gravity. Brittle materials are undesirable in hip joint prosthesis in view of the possible shock loading which may result in service. High wear resistance is necessary to avoid the accumulation of wear debris in the surrounding tissue or other organs of the body. Both toughness and wear resistance will be rated on a scale of 10 to 1, as in the case of tissue tolerance and corrosion resistance. Only materials with a wear resistance equal to or better than 7 and toughness equal to or better than 2 are considered in the present application.

Elastic compatibility of an implant material with the surrounding bone structure is also an important requirement. This is because large mismatches can lead to deterioration of the interface between the two materials. Unfortunately, the elastic moduli of the currently available materials for bone replacement are much higher than that of bone. Although this problem can be partially overcome by selecting the appropriate cementing agents, it is preferable for the implant material to be elastically compatible with the bone. The elastic modulus of bone (14 GPa) will be taken as a target value when candidate implant materials are evaluated in Section 25.4.

Similarity between the specific gravity of the implant material and bone is also desirable in order to keep the weight of the implant material as close as possible to that of the original bone. The specific gravity of compact bone is about 2.1, which is less than that of the metallic implant materials in current use. The specific gravity of bone will be taken as a target value when candidate implant materials are evaluated in Section 25.4.

Reasonable cost is another requirement for the hip joint prosthesis. The total cost includes the cost of the stock material plus the cost of processing and finishing. As the volume of the prosthesis is independent of the material, it is more appropriate to compare materials on the basis of the cost per unit volume rather than the cost per unit mass. In view of the small numbers produced, mass production method cannot be used in making the hip joint prosthesis. This means that the costs of processing and finishing are expected to represent a large proportion of the total cost.

Based on the above discussion, the material requirements for the pin of the hip joint prosthesis can be summarized as shown in Table 25.1. Higher values of the first six properties are more desirable and should comply to the lower limits outlined in the above discussion. According to Section 20.8, these properties are lower-limit properties. Elastic modulus and specific gravity are target-value properties, as materials should match the bone as closely as possible. The cost is an upper limit property.

Table 25.1 Main requirements and weighting factors for pin of hip joint prosthesis

Property	Weighting factor
1. Tissue tolerance (lower limit property)	0.2
2. Corrosion resistance (lower limit property)	0.2
3. Tensile strength (lower limit property)	0.08
4. Fatigue strength (lower limit property)	0.12
5. Toughness (lower limit property)	0.08
6. Wear resistance (lower limit property)	0.08
7. Elastic modulus (target value property)	0.08
8. Specific gravity (target value property)	0.08
9. Cost (upper limit property)	0.08

The relative importance of the different material requirements are given in Table 25.1. The digital logic method, which was described in Sections 17.7, can be used to assist in allocating the weighting factors.

25.4   CLASSIFICATION OF MATERIALS FOR THE PROSTHESIS PIN

A survey of the literature shows that possible materials for the pin of the hip joint prosthesis include stainless steels, cobalt-chromium-base alloys, titanium alloys, tantalum, and fiber-reinforced plastics. The latter materials have been under study for several years and are now being considered for approval by the FDA for use in the USA. Although several ceramic materials have been developed for surgical implants, their use in hip joint prosthesis is still in the experimental stage. These materials should be seriously considered when their performance is better characterized.

Table 25.2 Properties of selected surgical implant materials

Material	Tissue tolerance	Corrosion resistance	Tensile strength (MPa)	Fatigue strength (MPa)	Elastic modulus (GPa)	Toughness	Wear resistance	Specific gravity	Relative total cost (3)
Stainless steels									
316 (annealed)	10	7	517	350	200	8	8	8.0	1.0
317 (annealed)	9	7	630	415	200	10	8.5	8.0	1.1
321 (annealed)	9	7	610	410	200	10	8	7.9	1.1
347 (annealed)	9	7	650	430	200	10	8.4	8.0	1.2
Co-Cr alloys									
Cast alloy (1)	10	9	655	425	238	2	10	8.3	3.7
Wrought alloy (2)	10	9	896	600	242	10	10	9.13	4.0
Titanium alloys									
Unalloyed Ti	8	10	550	315	110	7	8	4.5	1.7
Ti-6Al-4V	8	10	985	490	124	7	8.3	4.43	1.9
Composites (fabric reinforced)									
Epoxy-70% glass	7	7	680	200	22	3	7	2.1	3
Epoxy-63% carbon	7	7	560	170	56	3	7.5	1.61	10
Epoxy-62% aramid	7	7	430	130	29	5	7.5	1.38	5

1. Composition: 27-30 Cr, 2.5 Ni, 5-7 Mo, 0.75 Fe, 0.36 C, 1 max Mn and Si, Rem. Co.
2. Composition: 20 Cr, 10 Ni, 15 W, 0.13 Mo, 3 max Fe, 0.1 C, 2 max Mn, 0.48 Si, Rem. Co.
3. Total cost includes cost of material per unit volume, processing cost and finishing cost.

When the pin is made of a metallic material, it can be manufactured either by casting or by forging. In both cases, the part is finished to the required final dimensions and given a mirror finish. This means that both cast and wrought alloys can be considered for the pin of the hip joint prosthesis. In the case of fiber-reinforced plastics, a processing method should be devised to orient the fibers in the direction of maximum loading. The matrix material can then be cast to infiltrate the fibers.

Table 25.2 gives the properties of representative examples of possible candidates for the pin of the hip joint prosthesis. The composition and characteristics of these materials are discussed in detail in Part I of this book.

25.5   EVALUATION OF CANDIDATE MATERIALS

The limits on properties method, given in Section 20.8, was used to evaluate the candidate materials listed in Table 25.2. Following the notations of (20.12), the lower limit, Y_i, upper limit, Y_j, and target values, Y_k, used in the calculations were as follows:

1. Tissue tolerance, lower limit, $Y_i = 7$.
2. Corrosion resistance, lower limit, $Y_i = 7$.
3. Tensile strength, lower limit, $Y_i = 95$ MPa.
4. Fatigue strength, lower limit, $Y_i = 33.25$ MPa.
5. Toughness, lower limit, $Y_i = 2$.
6. Wear resistance, lower limit, $Y_i = 7$.
7. Elastic modulus, target value, $Y_k = 14$ GPa.
8. Specific gravity, target value, $Y_k = 2.1$.
9. Relative total cost, upper limit, $Y_j = 10$.

Equation 20.12 and the above limits were used to calculate the merit parameter, m, for the candidate materials and the results are shown in Table 25.3. The table shows that the Ti-6 Al-4 V alloy ranks as number one followed by Co-Cr alloy. Stainless steels, although commonly used as metal plates for repairing fractures, ranked 4, 5, 6 and 7 for the present application. If less weight is given to corrosion resistance and more weight is given to the cost of the prosthesis, as in the case of temporary implants, stainless steels would occupy top ranks.

Table 25.3 Merit parameter (m) and ranking of candidate materials for the pin of a hip joint prosthesis

Material	Merit parameter (m)	Rank
Ti-6 Al-4 V	0.554	1
Co-Cr wrought alloy	0.555	2
Unalloyed titanium	0.563	3
316 stainless steel	0.593	4
347 stainless steel	0.597	5
317 stainless steel	0.607	6
321 stainless steel	0.608	7
Epoxy-70% glass fabric	0.615	8
Co-Cr cast alloy	0.622	9
Epoxy-62% aramid fabric	0.649	10
Epoxy-63% carbon fabric	0.720	11

BIBLIOGRAPHY AND FURTHER READING

Bement, A. L., ed., *Biomaterials*, Univ. of Washington Press, Seattle, 1971.

Farag, M. M., *Materials and Process Selection in Engineering*, Applied Sci. Pub. London, 1979.

Graham, J. W., 'Biomedical materials emerge through teamwork', *Advanced Materials and Processes*, Jan. 1988, p. 41.

Homsy, C. and Armeniades, C. D., eds, *Biomaterials for Skeletal and Cardiovascular Applications*, Interscience Pub., New York, 1972.

ISO5839, 'Orthopaedic joint prostheses', *ISO Bulletin*, Sept., 1985, p. 5.

Stanley, H. E., ed., *Biomedical Physics and Materials Science*, The MIT Press, Cambridge, Mass., 1972.

'Surgical implants', *ISO Bulletin*, January 1986, p. 6.

PART V

APPENDICES

Appendix A

Properties and Composition of Selected Engineering Materials

Table A.1 Mechanical properties of some carbon steels

Specification and grade		Tensile strength		Yield strength		Elongation (%)
		(MPa)	(ksi)	(MPa)	(ksi)	
HOT ROLLED SHEET AND STRIP STRUCTURAL QUALITY LOW-CARBON STEEL						
ASTM No. A570	A	310	45	170	25	23–27
	B	340	49	205	30	21–25
	C	360	52	230	33	18–23
	D	380	55	275	40	15–21
	E	400	58	290	42	13–19
SPECIAL QUALITY HOT ROLLED STEEL BARS						
ASTM No. A675	45	310–380	45–55	155	22.5	33
	50	345–415	50–60	170	25	30
	60	415–495	60–72	205	30	22
	70	485–585	70–85	240	35	18
	80	550 min	80 min	275	40	17
STEEL CASTINGS						
ASTM A27-77	60-30	415	60	205	30	24
	70-36	485	70	250	37	22
A148-73	80-40	552	80	276	40	18
	90-60	621	90	414	60	20
	120-95	827	120	655	95	14
	175-145	1207	175	1000	145	6

Table A.2 Mechanical properties of selected HSLA steels

ASTM No. and type	UNS designation	Tensile strength		Yield strength		Elongation (%)
		(MPa)	(ksi)	(MPa)	(ksi)	
A242 Type 1	K11510	435–480	63–70	290–345	42–50	21
A572 Grade 50	–	450	65	345	50	21
A607 Grade 60	–	520	75	415	60	16–18
Grade 70	–	590	85	485	70	14
A618 Grade 1	K02601	483	70	345	50	22
A717 Grade 60	–	485	70	415	60	20–22
Grade 70	–	550	80	485	70	18–20
Grade 80	–	620	90	550	80	16–18

Table A.3 Mechanical properties of some ultrahigh-strength steels

Designation or grade	Tempering temp.		Tensile strength		Yield strength		Elongation (%)
	(°C)	(°F)	(MPa)	(ksi)	(MPa)	(ksi)	
MEDIUM-CARBON LOW-ALLOY WATER QUENCHED AND TEMPERED							
4130	205	400	1765	256	1520	220	10
	315	600	1570	228	1340	195	13
	425	800	1380	200	1170	170	16.5
4340	205	400	1980	287	1860	270	11
	315	600	1760	255	1620	235	12
	425	800	1500	217	1365	198	14
MEDIUM-ALLOY AIR-HARDENING STEEL							
H13	527	980	1960	284	1570	228	13
	593	1100	1580	229	1365	198	14.4
18Ni MARAGING STEELS [solution treat 1 h at 820°C (1500°F), then age 3 h at 480°C (900°F)]							
18Ni(200)			1500	218	1400	203	10
18Ni(250)			1800	260	1700	247	8
18Ni(300)			2050	297	2000	290	7

Table A.4 Composition and properties of selected stainless steels

AISI	Norminal composition (%)						Tensile strength (MPa)	(ksi)	Yield strength (MPa)	(ksi)	Elongation (%)	Hardness (BHN)
	C	Cr	Ni	Mn	Si	Other[a]						
Austenitic												
201	0.15	17	4.5	6.50	1.0		805	117	385	56	55	185
301	0.15	17	7	2.0	1.0		770	112	280	41	60	162
302	0.15	18	9	2.0	1.0		630	91	280	41	50	162
304	0.08	19	9.5	2.0	1.0		588	85	294	43	55	150
316	0.08	17	12	2.0	1.0	2.5 Mo	588	85	294	43	50	145
330	0.08	18.5	35.5	2.0	1.0		630	91	266	39	45	150
Ferritic												
405	0.08	13	–	1.0	1.0	0.2 Al	490	71	280	41	30	150
430	0.12	17	–	1.0	1.0		525	76	315	46	30	155
442	0.2	21.5	–	1.0	1.0		560	81	315	46	20	185
Martensitic												
403	0.15	12.5	–	1.0	0.5		525	76	280	41	35	153
416	0.15	13	–	1.25	1.0	0.6 Mo	525	76	280	41	30	153
431	0.2	16	2.0	1.0	1.0		875	127	665	97	20	260
502	0.1	5	–	1.0	1.0	0.55 Mo	455	66	175	25	30	150

[a] Most steels contain 0.035 S and 0.04–0.06 P except for type 416 which contains 0.15 S.

Table A.5 Composition and mechanical properties of some tool steels

Grade	Composition %						Hardness (RC)		Toughness	
	C	Cr	V	W	Mo	Other	R.T.	560°C (1040°F)	(J)	(ft lb)
Water-hardening										
W1	0.6–1.4						63	10	68	50.2
W2	0.6–1.4		0.25				63	10	68	50.2
Shock-resisting										
S1	0.50	1.5		2.5			60	20	95	70.1
S2	0.50				0.50	1.0 Si	63	20	95	70.1
Oil-hardening										
O1	0.9	0.5		0.5		1.0 Mn	63	20	54	39.9
Air-hardening										
A2	1.0	5.0			1.0		63	30	48	35.4
Tungsten high-speed										
T1	0.70	4.0	1.0	18.0			66	52	61	45
T2	0.85	4.0	2.0	18.0			65	52	61	45
Molybdenum high-speed										
M2	0.85	4.0	2.0	6.25	5.00		65	52	68	50.2
M3	1.00	4.0	2.4	6.00	5.00		67	52	48	35.4
M4	1.3	4.0	4.0	5.50	4.50		67	52	48	35.4
M10	0.85	4.0	2.0		8.00		65	52	68	50.2

Table A.6 Composition and properties of selected cast irons

Specification	Class or grade	UNS No.	Composition (CE = %C + 0.3 (%Si + P))	Tensile strength (MPa)	(ksi)	Yield strength (MPa)	(ksi)	Average BHN	Elongation (%)
Gray iron ASTM A48-74	20		CE = 4.34	140	20.3	140	20.3	170	—
	25		CE = 4.08	175	25.38	175	25.38	190	—
	35		CE = 3.77	245	35.53	245	35.53	220	—
	40		CE = 3.65	280	40.6	280	40.6	225	—
	50		CE = 3.45	350	50.76	350	50.76	250	—
	60		CE = 3.37	420	60.9	420	60.91	270	—
Nodular (ductile) iron ASTM A536	60-40-18	F32800	Chemical composition is subordinate to	420	60.9	280	40.6	170	18
	65-45-12	F33100	mechanical properties. However, the content	455	65.98	315	45.68		12
	80-55-06	F33800	of any chemical element may be specified by	560	81.2	385	55.8	215	6
	100-70-03	F34800	mutual agreement	700	101.5	490	71.06		3
	120-90-02	F36200		840	121.8	630	91.36	270	2
ASTM A395	60-40-18	F32800	CE = 3.77	420	60.9	280	40.6	165	18
ASTM A476	80-60-03	F34100	CE = 3.8-4.5	560	81.21	420	60.9	201 min	3
Malleable iron ASTM	32510		TC, 2.5%; Si, 1.3%; S, 0.11%, P, 0.18% max.	350	50.76	230	33.35	110–145	10
A47 (ferritic)	35018		TC, 2.3%; Si, 1.2%; S, 0.11%; P, 0.18% max.	370	53.66	245	35.5		18
ASTM A220 (pearlitic)	40010		TC, 2.3%; Si, 1.3%; S, 0.11%; P, 0.18% max.	420	60.9	280	40.6	180–240	10

Table A.7 Composition and properties of selected aluminum alloys

Alloy	Temper	Nominal composition (%)	Tensile strength (MPa)	(ksi)	Yield strength (MPa)	(ksi)	Elongation (%)	Hardness (BHN)
Wrought alloys								
1060	0	99.6 + Al	70	10	28	4	43	19
	H18		133	19	126	18	6	35
2014	0	4.4 Cu, 0.8 Si, 0.8 Mn, 0.4 Mg	189	27	98	14	18	45
	T6		490	71	420	61	13	135
3003	0	1.2 Mn	112	16	42	6	40	28
	H18		203	29	189	27	10	55
4032	T6	12.5 Si, 1.0 Mg, 0.9 Cu, 0.9 Ni	385	56	322	47	9	120
5052	0	2.5 Mg, 0.25 Cr	196	28	91	13	30	47
	H38		294	43	259	38	8	77
6061	0	1.0 Mg, 0.6 Si, 0.25 Cu, 0.25 Cr	126	18	56	8	30	30
	T6		315	46	280	41	17	95
7075	0	5.5 Zn, 2.5 Mg, 1.5 Cu, 0.3 Cr	231	34	105	15	16	60
	T6		581	84	511	74	11	150
Casting alloys								
208.0	Sand-cast	4 Cu, 3 Si	147	21	98	14	2.5	55
356.0	T51	7 Si, 0.3 Mg	175	25	140	20	2	60
	T6		231	34	168	24	3.5	70
B443.0	Sand-cast	5 Si	133	19	56	8	8	40
	Die-cast		231	34	112	16	9	50
520.0	T4	10 Mg	336	49	182	26	16	75
850.0	T5	6.5 Sn, 1 Cu, 1 Ni	161	23	77	11	10	45

Table A.8 Composition and properties of selected magnesium alloys

Alloy	Temper	Nominal composition (%)	Tensile strength (MPa)	(ksi)	Yield strength (MPa)	(ksi)	Elongation (%)	Hardness (BHN)
Wrought alloys								
AZ31B	0	3.0 Al, 0.2 Mn, 1.0 Zn	224	33	115	17	11	56
	H24		255	37	165	24	7	73
AZ61A	F	6.5 Al, 0.15 Mn, 1.0 Zn	266	39	140	20	8	55
HK31A	H24	0.7 Zr, 3.2 Th	235	34	175	25	4	57
HM21A	T8	0.8 Mn, 2.0 Th	224	33	140	20	6	55
ZK40A	T5	4.0 Zn, 0.45 Zr	280	41	250	36	4	60
Casting alloys								
AM100A	F	10.0 Al, 0.1 Mn	140	20	70	10	6	53
	T4		238	35	70	10	6	52
	T6		238	35	105	15	2	52
AZ63A	F	6.0 Al, 0.15 Mn, 3.0 Zn	182	26	77	11	4	50
	T6		238	35	112	16	3	73
EZ33A	T5	2.6 Zn, 0.7 Zr, 3.2 Re	140	20	98	14	2	50
HK31A	T6	0.7 Zr, 3.2 Th	189	27	91	13	4	55
HZ32A	T5	2.1 Zn, 0.7 Zr, 3.2 Th	189	27	91	13	4	55
ZK61A	T6	6.0 Zn, 0.8 Zr	280	41	182	26	5	70

Table A.9 Properties and applications of selected wrought titanium alloys

Alloy and composition	Tensile strength (MPa)	(ksi)	Yield strength (MPa)	(ksi)	Elongation (%)	Hardness (RC)	Application
Unalloyed							
ASTM Grade 1	240	35	170	25	–	–	Excellent corrosion resistance
ASTM Grade 4	550	80	480	70	–	–	
Alpha alloys							
5 Al, 2.5 Sn	875	127	819	119	16	36	Weldable, aircraft engine
8 Al, 1 Mo, 1 V	1029	149	945	137	16	–	compressor blades and ducts, steam turbine blades
Alpha + beta alloys							
3 Al, 2.5 V	700	101	595	86	20	–	Aircraft hydraulic tubes
6 Al, 4 V	1008	146	939	136	14	36	Rocket motor cases, blades and discs for turbines
7 Al, 4 Mo	1120	162	1050	152	16	38	Airframes and jet engine parts
6 Al, 2 Sn, 4 Zr, 6 Mo	1288	187	1190	173	10	42	Components for advanced jet engines
10 V, 2 Fe, 3 Al	1295	188	1218	177	10	–	Airframe structures requiring toughness and strength
Beta alloys							
13 V, 11 Cr, 3 Al	1239	180	1190	173	8	–	High strength fasteners
8 Mo, 8 V, 2 Fe, 3 Al	1330	193	1260	183	8	40	Aerospace components
3 Al, 8 V, 6 Cr, 4 Mo, 4 Zr	1470	213	1400	203	7	42	High strength fasteners
11.5 Mo, 6 Zr, 4.5 Sn	1407	204	1337	194	11	–	High strength sheets for aircraft

Table A.10 Composition and properties of selected wrought copper alloys

Alloy	Nominal composition	Treatment	Tensile strength (MN/m²)	(psi)	Yield strength (MN/m²)	(psi)	Elongation (%)	Rockwell hardness
Pure copper								
C10200	99.95 Cu	–	221–455	33–66	69–365	10–53	55–4	–
Dilute copper alloys								
Beryllium copper	97.9 Cu, 1.9 Be 0.2 Ni or Co	Annealed	490	71	–	–	35	RB 60
		HT(hardened)	1400	203	1050	152	2	RC 42
Brass								
Gilding, 95%	95 Cu, 5 Zn	Annealed	245	36	77	11	45	RF 52
		hard	392	57	350	51	5	RB 64
Red brass, 85%	85 Cu, 15 Zn	Annealed	280	41	91	13	47	RF 64
		hard	434	63	406	59	5	RB 73
Cartridge brass, 70%	70 Cu, 30 Zn	Annealed	357	52	133	19	55	RF 72
		hard	532	77	441	64	8	RB 82
Muntz metal	60 Cu, 40 Zn	Annealed	378	55	119	17	45	RF 80
		half-hard	490	71	350	51	15	RB 75
High lead brass	65 Cu, 33 Zn, 2 Pb	Annealed	350	51	119	17	52	RF 68
		hard	318	46	420	61	7	RB 80
Bronze								
Phosphor bronze, 5%	95 Cu, 5 Sn	Annealed	350	51	175	25	55	RB 40
		hard	588	85	581	84	9	RB 90
Phosphor bronze, 10%	90 Cu, 10 Sn	Annealed	483	70	250	36	63	RB 62
		hard	707	103	658	95	16	RB 96
Aluminum bronze	95 Cu, 5 Al	Annealed	420	61	175	25	66	RB 49
		cold-rolled	700	102	441	64	8	RB 94
Aluminum bronze (2)	81.5 Cu, 9.5 Al, 5 Ni, 2.5 Fe, 1 Mn	Soft	630	91	–	–	12	–
		hard	735	107	420	61	12	RB 105
High silicon bronze	96 Cu, 3 Si	Annealed	441	64	210	31	55	RB 66
		hard	658	95	406	59	8	RB 93
Copper nickel								
Cupro-nickel, 30%	70 Cu, 30 Ni	Annealed	385	56	126	18	36	RB 40
		cold-rolled	588	85	553	80	3	RB 86
Nickel silver								
Nickel silver (German silver)	65 Cu, 23 Zn, 12 Ni	Annealed	427	62	196	28	35	RB 55
		hard	595	86	525	76	4	RB 89

Table A.11 Composition and properties of some zinc die-casting alloys

	Zamak 3 ASTM AG40A(XXIII)	Zamak 5 ASTM AG41A(XXV)
COMPOSITION (%)		
Copper	0.25	0.75–1.25
Aluminum	3.50–4.30	3.50–4.30
Magnesium	0.02–0.05	0.03–0.08
Iron (max.)	0.10	0.10
Zinc	Rem.	Rem.
PROPERTIES		
Tensile strength (MPa) (ksi)	287 (42)	329 (48)
Elongation (%)	10.00	7.00
Charpy impact, joule (ft lb)	58.3 (43)	66(49)
Hardness, BHN	82	91
Specific gravity	6.6	6.7

Table A.12 Chemical composition of selected Ni-base and Co-base alloys

Material	Nominal composition, weight %										Other
	C	Mn	Si	Cr	Ni	Co	W	Nb	Zr	Fe	
Fe-Ni-BASE ALLOYS											
Incoloy 80	0.05	0.8	0.5	21	32.5					45.7	0.38 Al, 0.38 Ti
Inconel 718	0.08			19	52.5			5.1		18.5	3 Mo, 0.9 Ti, 0.5 Al, 0.15 Cu
NICKEL-BASE ALLOYS											
DS-Nickel					bal						2.1 ThO2
Hastelloy X	0.1	1.0	1.0	21.8	bal	1.5	0.6			18.5	
Inconel 600	0.08	0.5	0.2	15.5	76					8.0	
Inconel 617	0.1	0.5	0.5	22.0	52	12.5				1.5	9 Mo, 0.3 Ti, 1.2 Al, 0.2 Cu
MAR-M200, c	0.15			9.0	bal	10	12.5	1.8	0.05		2 Ti, 5 Al, 0.015 B
TRW VI A, c	0.13			6.0	bal	7.5	5.8	0.5	0.13		2 Mo, 1 Ti, 0.02 B, 5.4 Al, 9 Ta, 0.5 Re, 0.4 Hf
COBALT-BASE ALLOYS											
AiResist 13, c	0.45	0.5		21	1	bal	11	2		2.5	3.5 Al, 0.1 Y
X-40, c	0.5	0.5	0.5	25	10	bal	7.5			1.5	
MAR-M302, c	0.85	0.1	0.2	21.5		bal	10		0.15		9 Ta, 0.005 B
MAR-M918	0.05	0.2	0.2	20	20	bal			0.1	0.5	7.5 Ta

c = cast alloy.

Table A.13 Rupture strength of selected nickel-base and cobalt-base alloys

Material	Rupture strength MPa (ksi)					
	650°C (1200°F)		815°C (1500°F)		1093°C (2000°F)	
	100 h	1000 h	100 h	1000 h	100 h	1000 h
NICKEL-BASE ALLOYS						
DS-Nickel	162 (23.5)	155 (22.5)	131 (19)	120.6 (17.5)	61.3 (8.9)	51 (7.4)
Hastelloy X	330 (48)	234 (34)	96 (14)	69 (10)	8.2 (1.2)	4.1 (0.6)
Inconel 600			55.2 (8)	38.6 (5.6)	9.6 (1.4)	6.2 (0.9)
Inconel 617	414 (60)	324 (47)	145 (21)	96 (14)	18.6 (2.7)	10.3 (1.5)
MAR-M200, c			524 (76)	414 (60)	75.8 (11)	44.8 (6.5)
TRW VI A, c	1000 (145)	896 (130)	552 (80)	420.6 (61)	82.7 (12)	
COBALT-BASE ALLOYS						
AiResist 13, c			172.4 (25)	117.2 (17)	30.3 (4.4)	
X-40, c	390 (57)	339 (49)	179.3 (26)	137.9 (20)	27.6 (4)	
MAR-M302, c			276 (40)	206.9 (30)	41.4 (6)	27.6 (4)
MAR-M918	462 (67)		206.9 (30)	137.9 (20)	17.2 (2.5)	

c = cast alloy.

Table A.14 Composition and properties of some refractory metals and alloys

Alloy	Nominal additions (%)	Test temp.		UTS at test temp.		10 h rupture stress	
		(°C)	(°F)	(MPa)	(ksi)	(MPa)	(ksi)
NIOBIUM AND ALLOYS							
Unalloyed Nb		1366	2491	70	10	38	5.5
SCB291	10 Ta, 10 W	1366	2491	224	33	63	9.1
C129Y	10 W, 10 Hf, 0.1 Y	1590	2894	182	26	105	15.2
FS 85	28 Ta, 11 W, 0.8 Zr	1590	2894	161	23	84	12.2
MOLYBDENUM AND ALLOYS							
Unalloyed Mo		1366	2491	182	26	102	14.8
TZM	0.5 Ti, 0.08 Zr, 0.015 C	1590	2894	371	54	154	22.3
WZM	25 W, 0.1 Zr, 0.03 C	1590	2894	504	73	105	15.2
TANTALUM AND ALLOYS							
Unalloyed		1590	2894	60	8.7	17.5	2.5
Ta-10 W	10 W	1590	2894	350	50.8	140	20.3
T-222	9.6 W, 2.4 Hf, 0.01 C	1590	2894	280	40.6	266	38.6
TUNGSTEN AND ALLOYS							
Unalloyed		1922	3492	175	25.4	48	7.0
W-2 ThO2	2 ThO2	1922	3492	210	30.5	126	18.3
W-15 Mo	15 Mo	1922	3492	252	36.6	84	12.2
GE 218	Doped	1922	3492	406	58.9	–	–

Table A.15 Some mechanical properties of selected plastics

Material	Tensile strength		Tensile modulus		Elongation (%)	Hardness	Izod impact	
	(MN/m²)	(ksi)	(MN/m²)	(ksi)			(J)	(ft lb)
			1-thermoplastics					
Polyethylene								
Low density	7–21	1.0–3.0	140	20.0	50–800	Rr 10	27	19.93
Medium density	14–21	2.0–3.0	280	40.0	50–800	Rr 14	2.7–22	1.99–16.24
High density	21–35	3.0–5.0	700–1400	100–200	1030	Rr 65	1.36–6.8	1–5.02
Polypropylene								
Homopolymers	35	5.0	1200	171.4	150	Rr 90	0.54–2.0	0.4–1.48
High impact copolymers	27	3.8–6	–	–	400	Rr 65	5.4–8.1	4–5.98
Talc filled, 40%	35	5.0	3600	514.2	5.0	Rr 95	0.54	0.4
Polystyrene								
General purpose	35–56	5.0–8.0	3300	471.4	1.5–4.0	Rm 74	0.27–0.54	0.2–0.4
High impact	23–32	3.3–4.6	2200	314.2	25–60	Rm 60	1.36–3.8	1–2.8
30% glass fiber	77–100	11.0–14.3	8800	1257.1	–	Rm 90	3.8	2.8
Polyvinylchloride								
General purpose	7–28	1.0–4.0	20	2.8	400	Sa 75	–	–
Flexible PVC	14–21	2.0–3.0	15	2.1	250	Sa 85	–	–
Rigid PVC	35–63	5.0–9.0	2000–4200	285.7–600	100	Rr 115	1.4–2.7	1.03–1.99
Acrylic (MMA)								
Cast, general purpose	42–84	6.0–12.0	2800	400	–	Rm 91	0.68	0.5
Molding grade	63–77	9.0–11.0	2800	400	–	Rm 95	0.5	0.37
High impact	42–63	6.0–9.0	2100	300	25–40	Rm 45	1.4–5.4	1.03–3.99
Nylons								
Nylon 6/6	84	12	3300	471.4	60–300	Rr 118	1.36–2.7	1–1.99
Nylon 6/12	62	8.8	21000	3000	150–340	Rr 114	1.36–2.7	1–1.99
Acetal								
Homopolymer	70	10.0	3700	528.6	25–75	Rm 94	1.9–3.1	1.4–2.29
Homopolymer 22% TFE fiber	53	7.5	2800	400.0	12–21	Rm 78	0.95–2.3	0.7–1.7
Copolymer	62	8.8	2900	414.2	60–75	Rm 80	1.63–3.5	1.2–2.58

Material								
Polycarbonate								
Unfilled	63	9.0	2400	342.8	110	Rm 70	16–22	11.81–16.24
ABS								
Medium impact	46	6.6	2500	357.1	6–14	Rr 111	5.4	3.99
High impact	42	6.0	2300	328.5	10–35	Rr 103	8.8	6.49
Very high impact	33.6	4.8	1750	250.0	15–50	Rr 88	10.8	7.97
Heat resistant	50.4	7.2	2450	350.0	5–20	Rr 111	3.12	2.3
Fluoroplastics								
PCTFE	28–42	4.0–6.0	1400	200.0	160	Sd 76	4	2.95
PTFE	14–49	2.0–7.0	700	100.0	100–450	Sd 58	6.1	4.5
High-temperature plastics								
Polymide unfilled	91	13.0	3150	450.0	7–9	Rm 97	1.36	1
Polymide 40% graphite	53	7.6	5300	757	2–3	Rm 73	–	–
Polysulfone	71.4	10.2	2500	357	50–100	Rr 120	1.63	1.2
II-Thermosetting plastics								
Phenolic								
General purpose	35–63	5.0–9.0	800	114.2	–	Re 95	0.44	0.32
Shock and heat	28–63	4.0–9.0	14000	2000	–	Re 85	2.17	1.6
Heat (mineral)	35–49	5.0–7.0	11000	1571.4	–	Re 85	0.54	0.4
Electrical (mineral)	28–56	4.0–8.0	6000	857.1	–	Re 87	0.41	0.59
Epoxy								
Cast rigid	63–105	9.0–15.0	3200	457.1	–	Rm 106	0.5	0.37
Molded	56–140	8.0–20.0	14000	2000	–	B78	1.3–5.5	0.96–4.06
High strength laminate	350–490	50.0–70.0	32000	4571.4	–	B71	1.36–41	1–30.26
Polyester								
Unfilled	56	8.0	2400	342.8	200–300	Rm 117	1.63	1.2
30% glass fiber	123	17.6	7700	1100	3	Rm 90	2.3	1.7
Alkyd								
Granular (mineral)	42–63	6.0–9.0	16000	2285.7	–	Re 85	0.42	0.31
Silicone								
30% glass fiber	42	6	17500	2500	–	Rm 90	4–20.3	2.95–14.98
Silica reinforced	28	4	11000	1571.4	–	Rm 82	0.41	0.3

Table A.16 Some physical properties and uses of selected plastics

Material	Expansion coefficient		Heat deflection temperature		Specific gravity	Relative cost
	$(\star 10^5$ m/m/°C)	$(\star 10^5$ in/in/°F)	(°C)	(°F)		
II-Thermoplastics						
Polyethylene						
Low density	19.8	11	36	96.8	0.92	1.0
Medium density	16.2	9	44.4	111.92	0.93	1.0
High density	13.5	7.5	49	120.2	0.96	1.0
Polypropylene						
Homopolymers	8.6	4.78	57	134.6	0.91	0.8
High-impact copolymers	–	–	50	122	0.9	1.1
Talc filled, 40%	–	–	81	177.8	1.23	1.0
Polystyrene						
General purpose	7.4	4.11	67–100	152.6–212	1.05	1.0
High impact	7.0	3.89	67–97	152.6–206.6	1.04	1.0
30% glass fiber	3.2	1.78	97	206.6	1.29	1.75
Polycarbonate						
Unfilled	6.75	3.75	132	269.6	1.2	3.6
ABS						
Medium impact	8.46	4.7	94	201.2	1.05	1.4
High impact	9.5	5.28	99	210.2	1.04	1.6
Very high impact	11.0	6.11	96	204.8	1.02	1.7
Heat resistant	6.7	3.72	114	237.2	1.05	1.8
Fluoroplastics						
PCTFE	4.5	2.5	–	–	2.1	20
PTFE	9.9	5.5	–	–	2.16	10
High-temperature plastics						
Polymide unfilled	5.0	2.78	360	680	1.43	–
Polymide 40% graphite	3.2	1.78	360	680	1.65	–
Polysulphone	5.6	3.11	174	345.2	1.24	6.3
Polyvinyl chloride						
General purpose	14.4	8	–	–	1.40	1.6
Flexible PVC	14.4	8	–	–	1.35	1.6
Rigid PVC	9–11	5–6.11	72	161.6	1.40	1.3
Acrylic (MMA)						
Cast, general purpose	5.4–7.2	3–4	99	210.2	1.19	1.4
Molding grade	5.4–7.2	3–4	88	190.4	1.18	1.7
High impact	7.2–11	4–6.11	77	170.6	1.11	2.2
Nylons						
Nylon 6/6	8.1	4.5	104	219.2	1.14	3.2
Nylon 6/12	9	5	82	179.6	1.07	5.8
Acetal						
Homopolymer	9–14.4	5–8	124	255.2	1.42	2.5
Homopolymer 22% TFE fiber	9–14.4	5–8	100	212	1.52	15
Copolymer	8.46	4.7	110	230	1.41	1.9
II-Thermosetting plastics						
Phenolic						
General purpose	3.8	2.11	174	345.2	1.38	–
Shock and heat	2.3	1.28	154	309.2	1.83	–
Heat (mineral)	1.8	1	177	350.6	1.53	–
Electrical (mineral)	3.96	2.2	154–204	309.2–399.2	1.52–1.67	–
Epoxy						
Cast rigid	5.6	3.11	166	330.8	1.20	1.3
Molded	3.6	2	191	375.8	1.91	1.3
High-strength laminate	3.6	2	–	–	1.84	1.3

Table A.16 – *contd.*

Material	Expansion coefficient		Heat deflection temperature		Specific gravity	Relative cost
	$(\star 10^5$ m/m/°C)	$(\star 10^5$ in/in/°F)	(°C)	(°F)		
Polyester						
Unfilled	11.3	6.28	54	129.2	1.31	2.7
30% glass fiber	2.3–9.7	1.28–5.39	213	415.4	1.54	2.9
Alkyd						
Granular (mineral)	2.9	1.61	146–191	294.8–375.8	2.2	1.4
Silicone						
30% glass fiber	5.76	3.2	482	899.6	1.88	–
Silica reinforced	4–8	2.22–4.44	482	899.6	1.93	–

Table A.17 Some properties and applications of selected rubbers

Material	Tensile strength		Elongation (%)	Hardness (Shore A)	Specific gravity
	(MN/m²)	(ksi)			
Natural rubber	28	40	700	30–90	0.92
Styrene butadiene	24.5	35	600	40–90	0.94
Neoprene	28	40	600	30–90	1.24
Butyl	21	30	800	40–80	0.92
Silicone	8.4	12	700	30–85	0.98
Fluorocarbon	17.5	25	300	60–90	1.85
Hypalon	21	30	500	50–90	1.18

Table A.18 Properties and preparation of some common adhesives

Adhesive	Curing temp.		Service temp		Lap-shear strength[a]		Peel strength at R.T.	
	(°C)	(°F)	(°C)	(°F)	(MPa at °C)	(ksi at °F)	(N/cm)	(lb/in)
Acrylics	R.T.	R.T.	up to 120	up to 250	17.2–37.9	2.5–5.5	17.5–105	10–60
Anaerobics	R.T.	R.T.	up to 166	up to 330	15.2–27.6	2.2–4.0	17.5	10
Butyral-phenolic	135–177	275–350	−51 to 79	−60 to 175	17.2	2.5	17.5	10
					6.9(80)	1.0(175)		
Cyanoacrylates	R.T.	R.T.	up to 166	up to 330	15.2–27.6	2.2–4.0	17.5	10
Epoxy (R.T. cure)	R.T.	R.T.	−51 to 82	−60 to 180	17.2	2.5	7	4
					10.3(80)	1.5(175)		
Epoxy (H.T. cure)	90–175	195–350	−51 to 175	−60 to 350	17.2	2.5	8.8	5
					10.3(175)	1.5(350)		
Epoxy-nylon	120–175	250–350	−250 to 82	−420 to 180	41	6.0	123	70
					13.8(80)	2.0(180)		
Epoxy-phenolic	120–175	250–350	−250 to 260	−420 to 500	21–28	3.0–4.0	8.8	5
					6.9(80)	1.0(175)		
Polyurethanes	149	300	up to 66	up to 150	24 `	3.5	123	70
Silicones	149	300	up to 260	up to 500	0.3	0.04	43.8	25
Hot melts (general)	–	–	up to 120	up to 250	1.4–4.8	0.2–0.7	35	20
(polyimides)	260–370	500–700	up to 315	up to 600	13.8	2.0	17.5	10

[a]R.T. if temperature is not specified.

Table A.19 Some properties of oxide and carbide ceramics

Material	Specific gravity	Thermal expansion coefficient		Tensile strength		Compressive strength		Young's modulus	
		$(10^{-6}/°C)$	$(10^{-6}/°F)$	(MPa)	(ksi)	(MPa)	(ksi)	(GPa)	$(lb/in^2 \times 10^6)$
I-Oxides									
Alumina (85%)	3.39	–	–	155	22.5	1930	280	221	32
(90%)	3.60	–	–	221	32	2482	360	276	40
(99.5%)	3.89	8.8	4.9	262	38	2620	380	372	54
Beryllia (98%)	2.90	9.0	5.0	85	12.3	1980	287	230	33
(99.5%)	2.88	9.0	5.0	98	14.2	2100	304	245	35.5
Magnesia	3.60	13.5	7.5	140	20.3	840	122	280	40.6
Zirconia	5.8	9.8	5.4	147	21.3	2100	304	210	30.4
Thoria	9.8	9.2	5.1	52	7.5	1400	203	140	20.3
II-Carbides									
SiC (Silicate bond)	2.57	4.68	2.6	–	–	105	15.2	21	3.0
SiC (SiN bond)	2.62	4.68	2.6	–	–	150	21.7	44	6.4
SiC (Self bonded)	3.10	3.78	2.1	–	–	1400	203	175	25.4
Boron carbide	2.51	4.5	2.5	–	–	2900	420	308	44.6
Tungsten carbide	16.0	5.95	3.3	–	–	–	–	–	–

Table A.20 Properties of selected whitewares

Property	Cordierite	Zircon	Steatite
PHYSICAL PROPERTIES			
Melting point °C (°F)	1490 (2713)	1567 (2853)	1567 (2853)
Max. service temp. °C (°F)	1017 (1863)	1107 (2025)	1017 (1863)
Specific gravity	1.9	3.6	2.6
Thermal expansion coefficient $10^6/°C$ $(10^6/°F)$	2.7 (1.5)	3.8 (2.1)	8.0 (4.4)
MECHANICAL PROPERTIES			
Tensile strength (MPa)	28–56	35–77	35–70
(ksi)	(4.1–8.1)	(5.1–11.2)	(5.1–10.2)
Compressive strength (MPa)	350–670	420–700	450–630
(ksi)	(50.8–97.2)	(60.9–102)	(62.3–91.4)
Modulus of elasticity (GPa)	49	147	105
$(lb/in^2 \times 10^6)$	7.1	21.3	15.2
Hardness (Moh's scale)	7	8	7.5
ELECTRICAL PROPERTIES			
Volume resistance (ohm cm)	$>10^{14}$	$>10^{14}$	$>10^{14}$
Dielectric strength (V/mm)	5600–9200	2400–1200	1000
Dielectric constant (1 megacycle)	5.1	9.3	5.8
Power factor, % (1 megacycle)	0.45	0.27	0.17

Table A.21 Typical compositions and properties of commercial glasses

	Fused silica	96% silica (Vycor)	Soda-lime-silica	Boro-silicate (Pyrex)	Alumino-silicate	Lead-alkali
Chemical composition (%)						
SiO_2	99.5+	96	72.6	80.2	57	56.5
Al_2O_3			1.6	2.6	20.6	1.4
K_2O			0.9	0.3		8.25
Na_2O			13.1	4.5	1.0	4.25
CaO			3.7	0.1	5.4	
MgO			8.0		12.0	
B_2O_3		4		12.3	4	
PbO						29.6
Fe_2O_3			0.1			
Properties						
Young's modulus						
$\quad$ GPa (lb/in$^2 \times 10^6$)	73.5 (10.7)	67.2 (9.8)	70 (10.2)	66.5 (9.6)	89.6 (13)	63 (9.1)
Poisson's ratio	0.17	0.18	0.24	0.20	0.26	–
Specific gravity	2.20	2.18	2.46	2.23	2.53	3.04
Linear expansion coefficient						
$\quad \times 10^7/°C$ (/°F)	5.6 (3.1)	8 (4.4)	92 (51)	32.5 (18)	42 (23)	90 (50)
Maximum service						
$\quad$ temperature °C (°F)	1197 (2187)	1097 (2007)	460 (860)	490 (914)	637 (1179)	380 (716)
Volume resistivity						
$\quad$ (ohm cm) at 250°C (482°F)	1.6×10^{12}	5×10^9	2.5×10^6	1.3×10^8	3.2×10^{13}	8×10^8
Dielectric constant	3.8	3.8	7.2	4.6	6.3	6.7
Refractive index	1.458	–	1.510	1.474	1.634	1.583

Table A.22 Some properties of selected composite materials

Matrix material	Fiber material	ff (%)	Specific gravity	Tensile strength		Elastic modulus	
				(MPa)	(ksi)	(GPa)	(lb/in² × 10⁶)
Epoxy	S glass a, b	70	2.11	2100	304.4	62.3	9.0
	S glass e	70	2.11	680	98.6	22	3.2
	S glass a, b	14	1.38	500	72.5	–	–
	E glass a, b	73	2.17	1642	238	55.9	8.1
	E glass a, b	56	1.97	1028	149	42.8	6.2
	E glass a, c	56	1.97	34.5	5	10.4	1.5
	Carbon a, b	63	1.61	1725	250	158.7	23
	Carbon a, c	63	1.61	41.4	6	11.0	1.6
	Aramid a, b	62	1.38	1311	190	82.8	12
	Aramid a, c	62	1.38	39.3	5.7	5.5	0.8
Polyester	E glass e	65	1.8	340	49.3	19.6	2.8
	Glass d	40	1.55	140	20.3	8.9	1.3
Polystyrene	Glass d	30	1.28	97	14	8.2	1.2
Polycarbonate	Glass d	20	1.31	107	15.5	6.2	0.9
	Glass d	40	1.44	131	19	10.4	1.5
Nylon 66	Glass d	20	1.31	152	22	8.3	1.2
	Glass d	40	1.41	200	29	11.0	1.6
	Glass d	70	–	207	30	21.4	3.1
Aluminum	Carbon a, b	40		1242	180		
	SiO2 a, b	48		870	126		
	B a, b	10		297	43		
Nickel	B a, b	8		2650	384		
	W a, b	40		1100	159		
Copper	Carbon a, b	65		794	115		
	W a, b	77		1760	255		

a. Continuous fibers.
b. Fibers aligned in loading direction.
c. Fibers at 90° to loading direction.
d. Discontinuous fibers.
e. Fabric.

Table A.23 Some properties of selected fiber materials

Fiber material	Specific gravity	Tensile strength		Elastic modulus		Relative cost*
		(GPa)	(ksi)	(GPa)	(lb/in² × 10⁶)	
E glass	2.54	3.5	507	73.5	10.7	1
S glass	2.49	4.6	667	85.5	12.4	3–4
Carbon (HS)	1.9	2.5	263	240	34.8	20–40
Carbon (HM)	1.9	2.1	305	390	56.6	20–40
Aramid	1.5	2.8–3.4	406–493	66–130	9.6–18.8	4–12
Steel	7.8	4.2	609	207	30	
W	19.4	4.1	594	413	59.9	
Rene 41	8.26	2.0	290	168	24.3	
Mo	10.2	2.2	319	364	52.8	
Boron	2.6	3.5	507	420	60.9	
SiC	4	2.1	305	490	71	
Al2O3	3.15	2.1	305	175	25.4	
SiO2	2.19	6	870	73.5	10.7	

*Cost of E glass is taken as unity

Appendix B

Conversion of Units and Hardness Values

Table B.1 Conversions to SI units

Quantity	Multiply number of	by	to obtain number of
Length	inches	25.4	mm
	feet	0.3048	metres (m)
	yards	0.9144	m
Area	square inches	645.16	mm^2
	square feet	0.092903	m^2
	square yards	0.836130	m^2
Volume	cubic inch	16387.1	mm^3
	cubic feet	0.0283168	m^3
	cubic yard	0.764555	m^3
Mass	ounces	0.0283495	kilograms (kg)
	pounds (lb)	0.45359237	kg
	short tons	907.185	kg
	long tons	1016.05	kg
Density	lb/in^3	27679.9	kg/m^3
	lb/ft^3	16.0185	kg/m^3
Force	pounds force (lbf)	4.44822	newtons (N)
	tons force (long)	9964.02	N
	dynes	10^{-5}	N
	kgf	9.80665	N
Stress	lbf/in^2	6894.76	N/m^2
	tonf/in^2	15.4443×10^6	N/m^2
	kgf/cm^2	98.0665×10^3	N/m^2
Work	ft lbf	1.35582	joules (J)
	hp/h	2.68452×10^6	J
	BTU	1.05506×10^3	J
	kw/h	3.6×10^6	J
	kcal	4.1868×10^3	J
	kgf/m	9.80665	J
Power	ft/lbf s	1.35582 watts (W)	
	horsepower (hp)	745.7	W
	metric hp (CV)	735.499	W
	BTU/h	0.293071	W
Thermal conductivity	BTU/h ft °F	1.73073	W/m^3
	BTU in/h ft^2°F	0.144228	W/m/K
	kcal/mh °C	1.163	

Temperature °C = 5/9 (°F − 32).
°F = 9/5 °C + 32.

Table B.2 Hardness conversions.
Soft steel, gray and malleable cast iron, and most nonferrous metals

Rockwell scale			BHN 500 kg (10 mm ball)	VHN 10 kg and BHN 3000 kg	Tensile strength MN/m^2
B	A	30T			
100	61.5	82.0	201	240	800
98	60.0	81.0	189	228	752
96	59	80	179	216	710
94	57.5	78.5	171	205	676
92	56.5	77.5	163	195	641
90	55.5	76.0	157	185	614
88	54	75	151	176	586
86	53	74	145	169	559
84	52	73	140	162	538
82	50.5	71.5	135	156	517
80	49.5	70	130	150	497
78	48.5	69	126	144	476
74	46	66	118	135	448
70	44	63.5	110	125	420
66	42	60.5	104	117	392
62	40.5	58	98	110	–
58	38.5	55	92	104	–
54	37	52.5	87	98	–
50	35	49.5	83	93	–
46	33.5	47	79.5	88	–
42	31.5	44	76	86	–
38	30	41.5	73	–	–
34	28	38.5	70	–	–
30	26.5	36	67	–	–
20	22	29	61.5	–	–
10	–	22	57	–	–

Table B.3 Hardness conversions
Hardened steel and hard alloys

C	Rockwell scale A	30T	VHN 10 kg	BHN 3000 kg	Tensile strength MN/m^2
80	92	92	1865	–	–
75	89.5	89	1478	–	–
70	86.5	86	1076	–	–
65	84	82	820	–	–
64	83.5	81	789	–	–
62	82.5	79	739	–	–
60	81	77.5	695	614	2310
58	80	75.5	655	587	2205
56	79	74	617	560	2065
54	78	72	580	534	2006
52	77	70.5	545	509	1889
50	76	68.5	513	484	1758
48	74.5	66.5	485	460	1634
46	73.5	65	458	437	1524
44	72.5	63	435	415	1427
42	71.5	61.5	413	393	1338
40	70.5	59.5	393	372	1255
38	69.5	57.5	373	352	1179
36	68.5	56	353	332	1117
34	67.5	54	334	313	1054
32	66.5	52	317	297	992.9
30	65.5	50.5	301	283	937.7
28	64.5	48.5	285	270	889.4
26	63.5	47	271	260	848
24	62.5	45	257	250	807
22	61.5	43	246	240	772
20	60.5	41.5	236	230	745

Appendix C

Principles of Engineering Statistics

The use of statistics in design provides a means of taking into account the variability and uncertainty in the values of loads acting on the component, its shape and dimensions and the properties of the material used in making it. Although these uncertainties can be accounted for by using an appropriate factor of safety, as discussed in Chapter 9, statistics are an essential tool in estimating reliability and life of a component, an assembly of components or a whole system, as discussed in Chapter 13. Statistical principles are also used extensively in quality control and manufacturing. The importance of statistics is recognized in many engineering curricula and many engineering statistics textbooks are now available. This appendix is only intended as a review of the statistical principles that are used in other sections of the book. Emphasis will be placed on the applications rather than the underlying mathematical concepts. For more in-depth treatment of this subject, reference should be made to the specialized books in the bibliography.

FREQUENCY DISTRIBUTION

Consider a group of 80 hardness readings taken in different locations of a steel casting, Table C.1.

As would be expected, the readings show differences in hardness values due to variations in cooling rate, segregation of elements and errors in testing. With such a large number of measurements, a method is needed to characterize the data. A common method is to arrange the measurements according to magnitude and then group them into equal-valued class or step intervals. The number of measurements in each interval, called the class frequency, f, is determined. The relative frequency rf, which is the number of measurements in each interval divided by the total number of measurements, can be used to determine the fraction of measurements below or above a specified limit. In Table C.2, the hardness results have been arranged in 10 class intervals. Out of the 80 BHN measurements, four values fell between 130 and 132, six fell between 133 and 135, etc.

The above results can be represented graphically by plotting the frequency f against the class intervals, as shown in Fig. C.1. This type of bar diagram is known as a frequency histogram. As the number of observations increases, the size of the class interval can be reduced until the limiting frequency-distribution curve is obtained, Fig. C.2. When the relative frequency rf is used instead of the frequency f, the area under the frequency-distribution curve is equal to unity, Fig. C.3. The probability that a random BHN measurement will be between the values x_1 and x_2 is given by the area under the relative

Table C.1 BHN measurements on a steel casting

Test	BHN	Test	BHN	Test	BHN	Test	BHN
1	150	21	146	41	130	61	143
2	145	22	133	42	146	62	154
3	140	23	148	43	148	63	141
4	131	24	146	44	138	64	159
5	147	25	139	45	150	65	142
6	148	26	140	46	152	66	143
7	135	27	148	47	141	67	150
8	140	28	146	48	144	68	151
9	152	29	133	49	147	69	144
10	145	30	136	50	149	70	144
11	146	31	134	51	151	71	148
12	146	32	139	52	148	72	152
13	154	33	142	53	147	73	157
14	152	34	144	54	147	74	152
15	149	35	136	55	145	75	150
16	145	36	132	56	142	76	146
17	142	37	134	57	149	77	146
18	144	38	132	58	145	78	150
19	144	39	133	59	140	79	151
20	145	40	136	60	141	80	140

Table C.2 Statistical analysis of BHN measurements

(1) BHN class interval	(2) Class midpoint (x)	(3) Frequency (f)	(4) Relative frequency (rf)	(5) Cumulative frequency (cf)	(6) Cumulative relative frequency	(7) $(x \star rf)$
130–132	131	4	0.050	4	0.050	6.55
133–135	134	6	0.075	10	0.125	10.05
136–138	137	4	0.050	14	0.175	6.85
139–141	140	10	0.125	24	0.300	17.50
142–144	143	12	0.150	36	0.450	21.45
145–147	146	18	0.225	54	0.675	32.85
148–150	149	14	0.175	68	0.850	26.075
151–153	152	8	0.100	76	0.950	15.20
154–156	155	2	0.025	78	0.975	3.875
157–159	158	2	0.025	80	1.000	3.950
Total		80	1.000			144.35

frequency-distribution curve bounded by those limits. The probability that a random observation will be less than x_1 is given by the area under the curve to left of x_1 and the probability that a random observation will be greater than x_2 is given by the area to the right of x_2. Calculation of the areas under the curve is made easier by fitting the data to a standard statistical distribution such as normal or log-normal distribution, as will be discussed later.

Another way to present the hardness values in our example is to plot the cumulative relative frequency against BHN, as shown in Fig. C.4. A cumulative relative frequency value corresponding to a given BHN represents the probability that the hardness will be less or equal to that value.

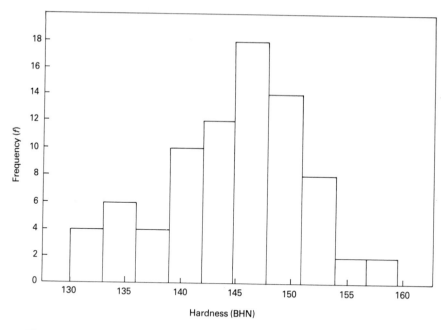

Figure C.1 Frequency histogram of the hardness data in Tables C.1 and C.2.

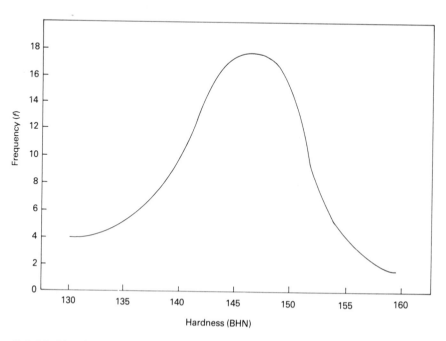

Figure C.2 Limiting frequency distribution curve of the hardness data in Tables C.1 and C.2.

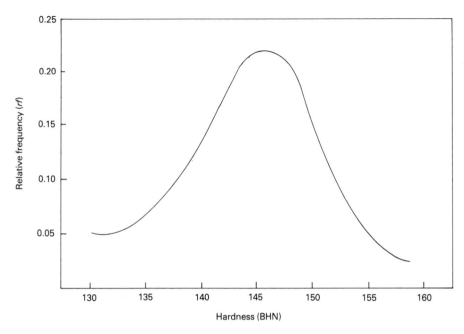

Figure C.3 Relative frequency distribution curve of the hardness data in Tables C.1 and C.2.

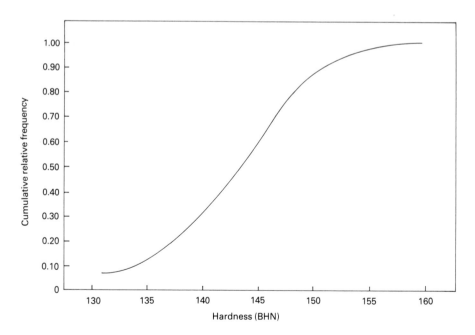

Figure C.4 Cumulative relative frequency distribution curve of the hardness data in Tables C.1 and C.2.

CENTRAL TENDENCY AND DISPERSION

The frequency distribution in Fig. C.2 can be characterized by its central location and the dispersion of the results away from the central region. The central location (tendency) is commonly described by the arithmetic mean, or average (μ).

$$\mu = \frac{x_1 + x_2 + x_3 + \ldots + x_i}{N} = \frac{1}{N} \sum_{i=1}^{N} x_i \qquad \text{(C 1)}$$

N = total number of measurements in the entire group or population.

$x_1, x_2, x_3 \ldots x_i$ are 1st, 2nd, 3rd . . . ith measurements.

For the BHN results in Table C.1, $x_1 = 150$, $x_2 = 145$, . . .

$$x_{80} = x_N = 140, \text{ and } N = 80.$$

According to (C 1), the arithmetic mean for the BHN measurements is 144.25.

With only a slight loss of accuracy, μ can be calculated from the grouped data of Table C.1 by using the relationship:

$$\mu = \sum_{i=1}^{k} [(\text{relative frequency}) \star (\text{midpoint BHN})] \qquad \text{(C 2)}$$

where k is the number of class mid-points.

According to (C 2) the approximate arithmetic mean is 144.35, as shown in Table C.2.

As the arithmetic mean is affected by every item, it can be distorted by unusually large or small values at the extremes.

When the number of elements in a population is very large, or when measurements involve destructive tests, it becomes impractical to measure the characteristics of each member of the population. In such cases a sample, which is a small part of the population, is selected. If n is the number of elements in the sample, the arithmetic mean of the sample ($\bar{x}$) is given as:

$$\bar{x} = \frac{x_1 + x_2 + \ldots + x_n}{n} = \frac{1}{n} \sum_{i=1}^{n} x_i \qquad \text{(C 3)}$$

For the BHN measurements of Table C.1, a sample of eight values is taken from readings number 5, 15, 25, 35, 45, 55, 65, and 75. The average of this sample is 144.75, which is about 0.35 percent different from the arithmetic mean calculated according to (C 1) from the total population.

Two other measures of the central tendency are the mode and the median. The mode applies to grouped data and is the value that occurs most frequently and may also be taken as midpoint of the class interval with the highest frequency. In the BHN measurements, the class interval with the highest frequency is 145–147 with a midpoint of 146. The value which occurs most frequently is also 146. The median is the value of the middle item in an array. In a distribution where the central values are closely grouped, the median is typical of the data since it is not affected by unusual terminal values. In the BHN measurements the median is 145. The median falls between the arithmetic mean and mode. For a symmetrical unimodal distribution curve the arithmetic mean, the median and mode all have the same value.

A measure of dispersion is the variance, which can be taken as the average of the squares of the deviations of the numbers from their arithmetic mean:

$$\sigma^2 = \frac{1}{N} \sum_{i=1}^{N} (x_i - \mu)^2 \tag{C 4}$$

where σ^2 = variance

x_i = value of the ith measurement

μ = arithmetic mean of the measurements

N = number of measurements

When the mean of a sample rather than the mean of the entire population is employed, it is more common to divide by the number of degrees of freedom, $n - 1$, instead of the number of items, n. This is known as Bessel's correction, and the estimate of the variance becomes:

$$S^2 = \frac{1}{n-1} \sum_{i=1}^{n} (x_i - \bar{x})^2 \tag{C 5}$$

where $\bar{x}$ = mean value of the sample

n = number of measurements in the sample.

In dealing with the dispersion of data it is usual to work with the standard deviation, which is defined as the positive square root of the variance. According to (C 4) the standard deviation of the population, σ, is given by:

$$\sigma = \left[\frac{1}{N} \sum_{i=1}^{N} (x_i - \mu)^2 \right]^{1/2} \tag{C 6}$$

And according to (C 5) the standard deviation of the sample is given by:

$$S = \left[\frac{1}{n-1} \sum_{i=1}^{n} (x_i - \bar{x})^2 \right]^{1/2} \tag{C 7}$$

Since comparisons of dispersions are difficult when expressed in absolute measurements, expressing the variability as a ratio would be more appropriate. In this case the coefficient of variation, v, is used. For the case of a sample of the population:

$$v = \frac{S}{\bar{x}} \tag{C 8}$$

COMBINING POPULATIONS

Cases often arise in design where two random variables are combined in some specified manner. An example is the assembly of a population of shafts, whose diameters have a mean μ_1 and a standard deviation σ_1, into a population of sleeves, whose diameters have a mean of μ_2 and a standard deviation of σ_2. The probability of being able to fit a randomly selected shaft into a randomly selected sleeve can be represented by a distribution whose mean is μ and standard deviation is σ, where:

$$\mu = \mu_1 - \mu_2 \tag{C 9}$$

$$\sigma = \sqrt{(\sigma_1^2 + \sigma_2^2)} \tag{C 10}$$

Generally, the means of two or more populations can either be added to subtracted to obtain the mean of the resulting distribution and the standard deviations can be combined according to Pythagorean theorem. In the case of multiplication of two populations:

$$\mu = \mu_1 \star \mu_2 \tag{C 11}$$

$$\sigma = (\mu_1^2 \sigma_2^2 + \mu_2^2 \sigma_1^2)^{1/2} \tag{C 12}$$

In the case of the division of two populations:

$$\mu = \frac{\mu_1}{\mu_2} \tag{C 13}$$

$$\sigma = \frac{(\mu_1^2 \sigma_2^2 + \mu_2^2 \sigma_1^2)^{1/2}}{\mu_2^2} \tag{C 14}$$

THE NORMAL DISTRIBUTION

One of the important distributions which are used in the study of statistics is the normal, or Gaussian, distribution. This is because many physical measurements follow the symmetrical bell-shaped curve of the normal frequency distribution. The yield strength, tensile strength and reduction in area from tension tests are known to follow the normal distribution curve. The normal distribution curve can be described by the equation:

$$f(x) = \frac{1}{\sigma\sqrt{2\pi}} \exp\left[-\frac{1}{2}\left(\frac{x-\mu}{\sigma}\right)^2 \right] \tag{C 15}$$

where $f(x)$ is the frequency function. The total area under this curve, from $x = -\infty$ to $x = +\infty$, is one square unit. Therefore the area between any two limits represents the proportion of cases which lie between them. If the standard deviation σ is small, the curve is tall and thin, if it is large the curve is short and broad as shown in Fig. C.5. A relationship exists between the standard deviation and the area under the normal curve as shown in Fig. C.6. It is seen that 68.26 percent of the population values fall between the limits of $\mu + 1\sigma$ and $\mu - 1\sigma$; 95.46 percent of the values fall between the limits of $\mu + 2\sigma$ and $\mu - 2\sigma$ and 99.73 percent of the values fall between the limits of $\mu + 3\sigma$ and $\mu - 3\sigma$. These percentages hold true regardless of the shape of the normal distribution curve.

In order to standardize the different normal distributions, the curve is frequently expressed in terms of the standard normal deviate z:

$$z = \frac{x-\mu}{\sigma} \tag{C 16}$$

where z is normally distributed with a mean of zero.

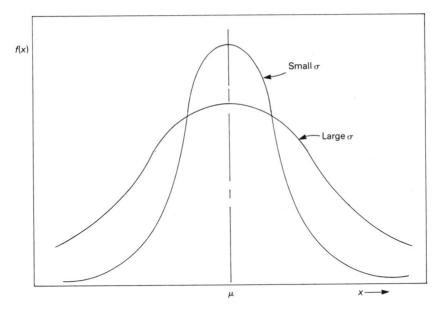

Figure C.5 Effect of the value of the standard deviation on the shape of the normal distribution curve.

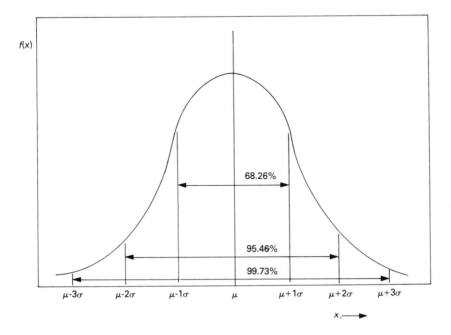

Figure C.6 Relationship between the standard deviation and the area under the normal distribution curve.

Appendix C

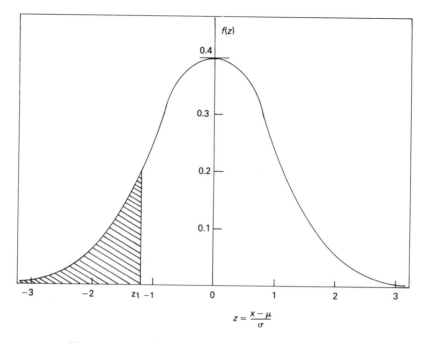

$$z = \frac{x - \mu}{\sigma}$$

Figure C.7 Standardized normal frequency distribution.

Equation C 15 of the standard normal distribution curve can be written in terms of z as:

$$f(z) = \frac{1}{\sqrt{2\pi}} \exp\left(-\frac{z^2}{2}\right) \qquad \text{(C 17)}$$

This distribution is shown in Fig. C.7 and has a mean of 0 and a standard deviation of 1. The total area under the curve is unity. An area, a, under the curve between z_1 and $-\infty$ represents the cumulative relative frequency of z_1 and can be found from statistical tables, e.g. Montgomery, D. C.: 'Statistical quality control', John Wiley and Sons, New York, 1985, p. 501. Table C.3 gives a sample of the areas under the curve for selected values of z.

Table C.3 Areas under the standardized normal distribution curve

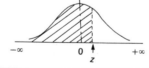

z	0.0	0.1	0.2		0.4	0.5	0.6	0.7	0.8	0.9
-3	0.0014	0.0010	0.0007	0.0005	0.0003	0.0002				
-2	0.0228	0.0179	0.0139	0.0107	0.0082	0.0062	0.0047	0.0035	0.0026	0.0019
-1	0.1587	0.1357	0.1151	0.0968	0.0808	0.0668	0.0548	0.0446	0.0359	0.0287
-0	0.5000	0.4602	0.4207	0.3821	0.3446	0.3085	0.2743	0.2420	0.2119	0.1841
$+0$	0.5000	0.5398	0.5793	0.6179	0.6554	0.6915	0.7257	0.7580	0.7881	0.8159
$+1$	0.8413	0.8643	0.8849	0.9032	0.9192	0.9332	0.9452	0.9554	0.9641	0.9713
$+2$	0.9773	0.9821	0.9861	0.9893	0.9918	0.9938	0.9953	0.9965	0.9974	0.9981
$+3$	0.9987	0.9990	0.9993	0.9995	0.9997	0.9998				

In the case of the BHN results of Table C.1, knowing that the mean is 144.25 and the standard deviation is 6.306, the number of hardness values below a certain value can be determined from (C 16) and Table C.3. For example the number of hardness values below 141 is:

$$z_{141} = \frac{141 - 144.25}{6.306} = -0.515$$

From statistical tables, the area under the curve is 0.3015, which is close to the value of 0.300 given in Table C.2 for the cumulative relative frequency at BHN 141.

WEIBULL DISTRIBUTION

In spite of its wide use, the normal distribution has the limitations of being symmetrical and unbounded with its ends extending from $-\infty$ to $+\infty$. Many random variables in engineering follow a nonsymmetrical bounded distribution and cannot, therefore, be described by the normal distribution. An example is the distribution that describes the performance of a well-designed component in service where the life is always positive and the components that last a long time are more than those that fail prematurely. The Weibull distribution is often used to describe such cases and is described by:

$$f(x) = \left(\frac{m}{\theta - x_o}\right)\left(\frac{x - x_o}{\theta - x_o}\right)^{m-1} \exp -\left[\frac{(x - x_o)}{(\theta - x_o)}\right]^m \text{for } x > 0 \qquad \text{(C 18)}$$

where $f(x)$ = frequency distribution of the random variable x which is always positive and has a lower bound value x_o.

  m = shape parameter or Weibull slope. The greater the slope, m, the smaller the scatter in the random variable x.

  θ = scale parameter or characteristic value is related to the median and thus is an indicator of central tendency.

Figure C.8 gives the Weibull distribution for $\theta = 1$, $x_o = 0$, and different values of m. The figure illustrates the flexibility of the Weibull distribution which assumes a wide variety of shapes with the change of m. When $m = 1$ the Weibull distribution reduces to the exponential distribution and when $m = 3.44$ the Weibull distribution is a good approximation to the normal distribution.

When x_o is 0, the cumulative frequency distribution is given by:

$$F(x) = 1 - \exp\left(\frac{x}{\theta}\right)^m \qquad \text{(C 19)}$$

This equation can be rewritten as:

$$\ln\left(\ln\frac{1}{1 - F(x)}\right) = m \ln x - m \ln \theta \qquad \text{(C 20)}$$

which is a straight line of the form $Y = mx + C$.

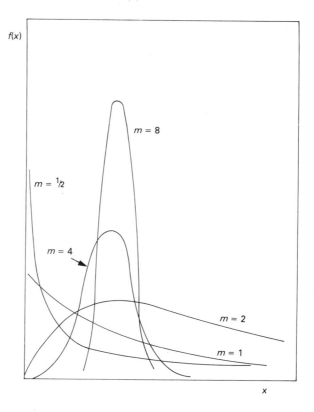

Figure C.8 The Weibull distribution for different values of m.

Special Weibull probability paper is available to assist in the analysis according to C 20. A set of data which follows the Weibull distribution will plot as a linear curve on this paper. The slope of the line is m and the intercept on the Y axis is ($m \ln \theta$). If the plot is nonlinear it could be that the data cannot be represented by the Weibull distribution, or it may be that x_o is greater than 0. In the latter case x_o can be estimated by trial and error according to the procedure outlined by Carter, A. D. S.: 'Mechanical reliability', 2nd Ed, Macmillan Press Ltd., London, 1986.

EXPONENTIAL DISTRIBUTION

As shown earlier, the exponential distribution can be considered as a special case of the Weibull distribution when $m = 1$ and $x_o = 0$. When the exponential distribution is used to describe the failure of equipment in service x is taken as the life t and the distribution function can be written as:

$$f(t) = \lambda \exp(-\lambda t) \tag{C 21}$$

The constant λ has the dimensions of $(\text{time})^{-1}$ and $1/\lambda$ is the mean life. The exponential distribution is frequently used to model the probability of failure of complex maintained equipment as discussed in Chapter 13.

BIBLIOGRAPHY

Blank, L., *Statistical Procedures for Engineering, Management, and Science*, McGraw-Hill, New York, 1980.

Box, G. P., Hunter, W. G. and Hunter, J. S., *Statistics for Engineers*, John Wiley & Sons, New York, 1978.

Miller, I. and Freund, J. E., *Probability and Statistics for Engineers*, 2nd ed., Prentice Hall, N.J., 1977.

Montgomery, D. C., *Statistical Quality Control*, John Wiley & Sons, New York, 1985.

Index

ABS plastics, 76, 80, 84
acetal, 85
acrylic, 80, 83, 84
adaptive control (AC), 342
adhesives, 88
    joint design, 283
    materials, 88
    processes, 64
    properties, 509
allowances on dimensions, 41
alloy steels see steels
alumina, 106
    properties, 510
aluminum and alloys, 29
    alloy designation, 15
    alloy properties, 30, 501
    corrosion resistance, 189
    cost relative to other materials, 408, 409
    temper designation, 19
    use relative to other materials, 13
anisotropy, 173
annealing, 20
ausforming, 20, 286
austempering, 20
austinite, 19
automated assembly, 292, 355
automated drafting, 321
automated factory, 355

bearings, 467
    design, 468
    materials, 473
    selection, 474
bendability, 278
bending processes, 56
benefit-cost analysis, 373
beryllia, 106
    properties, 510
biocompatibility, 486
bone, 487
brasses, 32, 188, 503
brazing, 64, 283
break-even analysis, 362
British Standards numbering system for steels, 15
brittle fracture, 149

bronze, 32, 188, 503
buckling, 178, 249
bulk molding compound (BMC), 132

CAD see computer aided design
CAM see computer aided manufacture
carbides, 107
    properties, 510
carbon fiber reinforced plastics (CFRP), 126
carbon fibers, 126
    properties, 512
carbon steels, 21
    composition range, 22
    properties, 497
cargo trailer, 457
    design, 457
    materials, 459
    selection, 463
cast irons, 27
    compacted graphite (CG), 28
    gray, 27
    nodular, 28
    properties, 500
    wear resistance, 196
casting, 47
    centrifugal, 49
    ceramics, 110
    defects, 51, 270
    die casting, 49
    investment casting, 49
    plastics, 96
    quality, 49
    sand casting, 47
    shell casting, 47
cellulosics, 86
ceramics, 103
    casting, 110
    characteristics, 103
    chemical properties, 106
    classification, 103
    clay products, 108
    corrosion resistance, 190
    creep resistance, 194
    design considerations, 114
    extrusion, 112

firing, 113
glass, 108
jiggering, 112
mechanical properties, 103, 510, 511
molding, 111
physical properties, 105
processing, 110
refractories, 106
thermal shock resistance, 105, 194
wear resistance, 197
whitewares, 107–8, 510
cermets, 52
CFRP *see* carbon fiber reinforced plastics
chevron pattern, 151
clay products, 108
cobalt and alloys, 33, 193, 504
properties, 505
wear resistance, 196, 201
codes of practice, 211
cold working, 53
composites, 117–40
design considerations, 134
dispersion strengthened, 118
elastic modulus, 121
fiber strengthened, 121
hybrid, 126, 135
laminates, 127
manufacturing of components, 130
particulate strengthened, 120
rule of mixtures, 118, 122
tensile strength, 122, 512
compression molding, 90
computer aided design (CAD), 316–28
applications in industry, 326
automated drafting, 321
benefits, 327
computer aided manufacture (CAM), 341–57
applications, 341
automated factory, 355
flexible manufacturing systems, 353
group technology, 346
numerical control, 341
robots, 351
computer aided materials selection, 433
computer aided process planning (CAPP), 343
computer aided quality control, 345
computer numerical control (CNC), 342
condition monitoring, 312
copper and alloys, 32
corrosion resistance, 188
properties, 30, 503
wear resistance, 198
corrosion, 160
atmospheric, 160
crevice, 162
fatigue, 163
galvanic, 162
intergranular, 160
protection by surface treatment, 69
resistance, 185
stress, 163
cost
analysis, 377

direct, 389
estimation in manufacturing industries, 384
estimation in process industries, 382
fixed, 361
indices, 383
indirect, 389
labor, 389
manufacturing, 361
materials, 388, 402
overhead, 362, 389
standard, 394
variable, 361
variance, 394
cost-effectiveness analysis, 375
cost per unit property, 422
creep, 165, 191, 263
critical path method (CPM), 225

data banks, 435
decision making, 221–39
matrix, 222
trees, 224
deep drawing, 56
depreciation, 370
derating factor, 216
design, 205–20
codes, 211
effect of manufacturing processes, 269
effect of material properties, 240
phases, 208
probabilistic, 213
reliability, 305
review, 6, 309, 325
design considerations
adhesive bonding, 283
automated assembly, 292
brazed joints, 283
cast products, 269
ceramic products, 114
composite materials, 134
forged components, 274
heat treatment, 285
machined parts, 287
metallic materials, 35
molded plastics, 271
plastics, 100
powder metallurgy parts, 275
sheet metal parts, 278
welded joints, 279
designing against corrosion, 289
designing against fatigue, 255
designing for static strength, 246
designing under high temperature conditions, 263
designing with high-strength low-toughness
    materials, 250
die-casting, 49
diffusion bonding, 59
digital logic method, 377
dimensional accuracy, 40
direct costs, 389
direct numerical control (DNC), 342
dispersion strengthening, 18, 118
ductile fracture, 149

economic analysis, 361–81
  annual worth, 368
  capitalized worth, 369
  payback period, 369
  present worth, 368
economics of manufacturing, 382–400
economics of materials, 401–13
economics of metal cutting, 391
elastic modulus
  aluminum, 16
  composites, 121, 176, 512
  magnesium, 31, 176
  plastics, 81, 506
  steels, 16, 176
  titanium, 176
elastomers, 87
  properties, 508
elongation percent
  values for different materials, 497–512
enamels, 199
endurance limit, 156, 256
endurance ratio, 156
engineering design *see* design
epoxy, 83, 87, 190
Euler formula, 178, 250
extrusion
  ceramics, 112
  metals, 54
  plastics, 93

factor of safety, 214
fail-safe design, 219, 251, 262
failure, 141
  analysis, 142, 167
  classification, 141
  corrosion, 160
  corrosion fatigue, 163
  creep, 165
  fatigue, 153
  high temperature, 164
  mechanical, 142
  stress corrosion, 163
  thermal fatigue, 167
  wear, 159
fastening processes, 60
fatigue, 153
  crack initiation, 156
  crack propagation, 158
  endurance ratio, 156
  limit, 156
  resistance, 179
  S–N curves, 156
  strength, 155
  surface finish, 255, 257
  thermal, 167
fault tree analysis, 303
feasibility studies, 4
fibers, 125
  aramid *see* Kevlar
  carbon, 126, 512
  glass, 125, 512
  graphite *see* carbon
  Kevlar, 126, 512
filament winding, 132

finite element analysis, 322
fits and tolerances, 41
flexible manufacturing systems (FMS), 353
forging, 53
fracture
  brittle, 148
  ductile, 148
  fatigue, 153
  mechanics, 143
fracture toughness, 143
free machining steels, 67

galvanic series, 161
Gaussian distribution, 522
geometric modeling, 319
geometric programming, 237
glass fibers, 125
  properties, 512
glass reinforced plastics (GRP), 117
glasses, 108
  corrosion resistance, 190
  forming, 113
  properties, 511
graphite fibers, 126
group technology, 346

hard facing alloys, 201
hardenability, 19, 285
hardness
  conversion of units, 514
  relation to tensile properties, 514
hazard analysis, 302
heat affected zone (HAZ), 63
heat deflection temperature, 81
heat treatment, 19, 285
heavy nonferrous metals *see* specific alloys
high speed steels, 27
  properties, 499
high-strength low-alloy steels (HSLA), 25
  properties, 497
hot working, 53
human reliability, 303
hybrid composites, 126, 135

inconel alloys, 33, 188, 504
indirect costs, 389
injection molding, 92

jiggering, 112
jigs and fixtures
  economic justification, 390
  welding, 63
joining processes, 59, 279

Kevlar fibers, 126
  properties, 512

labor costs, 389
laminated composites, 127
Lang factor, 383
Larson-Miller parameter, 266
lead and alloys, 34
  corrosion resistance, 189
learning curve, 395

liability, 312
life cycle costing (LCC), 399
life cycle of a product, 8
light alloys, 29
limits on properties method, 430
linear programming, 234
low melting alloys, 34

machinability, 67
machining processes, 65
    economics, 391
    nontraditional methods, 68
    tolerances, 66
magnesium and alloys, 30
maintainability, 312
maintenance and condition monitoring, 311
Manson-Haferd parameter, 266
manufacturing
    data base, 325
    economics, 382
    expenses, 389
    systems, 331
    time, 385
maraging steel, 26
    properties, 498
martempering, 20, 286
martensite, 19, 20
materials
    comparison on cost basis, 407
    cost, 388, 402
    data bases, 435
    economics, 401
    performance requirements, 419
    prices, 404
    selection, 418
    sources of information, 437
    utilization, 410
materials requirement planning (MRP), 333
metallic materials, 13–37
    classification, 13
    see also specific metals and alloys
MICLASS coding system, 350
Miner's rule, 261, 267
minimum cost analysis, 377
modeling, 217
    analog, 218
    geometric, 319
    iconic, 218
    symbolic, 217
monel alloys, 33, 188

nickel and alloys, 33, 193, 201, 504
    corrosion resistance, 188
    properties, 505
nitrides, 107
normal distribution, 522
numerical control (NC), 341
nylon, 76, 80, 190, 506, 509

OPITZ coding system, 348
optimization techniques, 229
    differential calculus, 231
    geometric programming, 237
    linear programming, 234

search methods, 231
overhead costs, 362, 389
oxidation, 164

Palmgren–Miner rule, 261, 267
performance requirements, 419
PERT see program evaluation and review technique
planning and scheduling models, 225
planning for manufacture, 332
plastics, 74–102
    additives, 76
    blow molding, 95
    casting, 96
    characteristics, 77
    chemical properties, 83
    classification, 74
    compression molding, 90
    corrosion resistance, 189
    cost relative to other materials, 408, 409, 508
    deflection temperature, 191
    design considerations, 100
    extrusion, 93
    fastening, 97
    finishing, 98
    high temperature plastics, 192
    injection molding, 92
    joining, 97
    mechanical behavior, 77, 506
    optical properties, 75, 83
    physical properties, 81, 508
    structure, 75
    thermoforming, 94
    thermoplastics, 83
    thermosetting plastics, 86
    transfer molding, 92
    wear resistance, 196
phenolics, 86
polyamides see nylon
polycarbonates, 76, 80, 83, 85, 190, 507, 508
polyester, 86, 87, 507, 509
polyethylene, 80, 83, 506, 508
polymers see plastics, elastomers or adhesives
polymethylmethacrylate (PMMA), 76
polyphenylene sulfide, 190
polypropylene, 80, 83, 84, 506, 508
polystyrene, 80, 83, 84, 506, 508
polysulfone, 190, 508
polytetrafluorethylene (PTFE), 80, 85, 507, 508
polyurethane, 81, 83, 86, 87
polyvinylchloride (PVC), 76, 83, 84, 506, 508
porcelain, 107
    enamels, 199
    properties, 510
porosity, 52
powder metallurgy, 52
precipitation hardening, 18
probabilistic design, 213
process planning, 333
process sheets, 333
processability requirements, 420
profit, 398
product liability, 312
product life cycle, 8

production
  control, 335
  modeling, 336
  planning, 332
  systems, 329
productivity, 337
profit, 398
program evaluation and review technique (PERT), 225
proof stress, 172
protective coatings, 198
pultrusion, 133

refractory metals and alloys, 34, 193
  properties, 505
reliability, 294–315
  assessment, 296
  cost, 308
  hazard analysis, 302
  human, 303
  role of design, 305
  role of manufacture, 309
  role of materials, 309
  role of user, 311
rigidity modulus, 175, 247
riveting, 61
robots, 351
rolling, 53
rubber see elastomers
rubber pad forming, 58

safe-life, 219, 261
sand casting, 47
sandwich materials, 128
selection of materials, 417–44
  computer aided methods, 433
  cost per unit property method, 422
  incremental return method, 429
  limits on properties method, 430
  weighted properties method, 424
selling price, 396
sensitivity analysis, 236
service life, 219, 301
shearing operations, 56
sheet metal forming, 55
SiC, 107, 510
silicones, 87
simulation, 218
sintered aluminum powder (SAP), 119
slip casting, 110
soldering, 64
sources of information on material properties, 437
specific gravity
  aluminum, 29
  brass, 35
  magnesium, 30
  plastics, 81
  steels, 35
  titanium, 31
  zinc alloys, 504
specific stiffness, 178
specific strength, 174
specifications, 212

spinning operations, 58
stainless steels, 26, 186, 193
  cost relative to other materials, 408, 409
  properties, 498
  sensitization, 27
  wear resistance, 195
standard cost, 394
standard deviation, 521
standard time, 387, 389
standards, 212
statistical factor of safety, 216
statistical variations of material properties, 214, 240
statistics, 516–27
  central tendency, 520
  combining populations, 521
  dispersion, 520
  exponential distribution, 526
  frequency distribution, 516
  normal distribution, 522
  standard deviation, 521
  Weibull distribution, 525
steels
  alloy steels, 24
  carbon steels, 21, 497
  cost relative to other materials, 408, 409
  designation system, 14
  effect of alloying elements, 25
  EX steels, 25
  free machining, 24
  heat treatment, 19, 286
  high-strength low-alloy (HSLA), 25, 497
  maraging steels, 26, 498
  stainless, 26, 193, 498
  tool steels, 27, 499
  ultrahigh strength, 26, 498
  use relative to other materials, 13
  wear resistance, 195
stiffness, 174
stitching, 61
strengthening of metallic materials, 16
  cold working, 19
  grain size control, 19
  precipitation hardening, 18
  solid solution, 17
stress
  concentration, 242
  corrosion, 163
  intensity factor, 143
superalloys, 33, 193, 504
  properties, 505
superplastic forming, 58
supply and demand, 396
surface coating, 69, 198
surface hardening, 71
surface roughness, 45
surface treatment, 69, 199
  corrosion protection, 69
  wear resistance, 71, 201
surgical implants, 486
  design considerations, 486
  materials, 491
  selection, 492

tax considerations, 370

Taylor tool life, 345, 392
tennis racket
    design considerations, 478
    materials, 482
    selection, 484
tensile strength
    values for specific materials, 497–512
thermal fatigue, 167
thermal shock, 105, 194
thermoforming, 94
thermoplastics, 83
thermosetting plastics, 86
thoria, 107
    properties, 510
time value of money, 365
tin and alloys, 34, 188
titanium and alloys, 31, 192
    corrosion resistance, 189
    properties, 30, 502
tolerances, 41
    welded joints, 63
tool life, 345, 392
tool steels, 27
    properties, 499
toughness, 181
turnbuckle, 445
    design, 446
    materials, 453

ultimate tensile strength *see* tensile strength
Unified Numbering System for steels, 15

value analysis, 380, 407
vitreous enamels, 199

wear
    failure, 159
    resistance, 194
Weibull distribution, 525
weighted properties method, 424
weighting factor, 424, 431
weldability, 63
welding processes, 61
    joint design, 279
whitewares, 107–8
    properties, 510
wire drawing, 55

yield strength
    values for specific materials, 497–512
Young's modulus *see* elastic modulus

zinc and alloys, 32
    properties, 504
zirconia, 106
    properties, 510

Russell C. Thomas

Adventure Unlimited